Uwe T. Bornscheuer and
R. Joseph Kazlauskas

Hydrolases in Organic Synthesis

Related Titles

Breslow, R. (ed.)
Artificial Enzymes
2005
ISBN 3-527-31165-3

Cao, L.
Carrier-bound Immobilized Enzymes
Principles, Application and Design
2005
ISBN 3-527-31232-3

Bommarius, A. S., Riebel, B. R.
Biocatalysis
Fundamentals and Applications
2004
ISBN 3-527-30344-8

Brakmann, S., Schwienhorst, A. (eds.)
Evolutionary Methods in Biotechnology
Clever Tricks for Directed Evolution
2004
ISBN 3-527-30799-0

Fuhrhop, J.-H., Li, G.
Organic Synthesis
Concepts and Methods
2003
ISBN 3-527-30272-7

Schmalz, H.-G., Wirth, T. (eds.)
Organic Synthesis Highlights V
2003
ISBN 3-527-30611-0

Uwe T. Bornscheuer and Romas J. Kazlauskas

Hydrolases in Organic Synthesis

Regio- and Stereoselective Biotransformations

2nd edition

WILEY-VCH

WILEY-VCH Verlag GmbH & Co. KGaA

The Authors

Prof. Dr. Uwe T. Bornscheuer
Institute for Chemistry and Biochemistry
Ernst-Moritz-Arndt-University
Soldmannstrasse 16
17489 Greifswald
Germany

Prof. Dr. Romas J. Kazlauskas
University of Minnesota
Dept. of Biochemistry,
Molecular Biology & Biophysics
174 Gortner Lab., 1479 Gortner Avenue
St. Paul, MN 55108
USA

Cover :
Xray structure showing a phosphonate transition-state analog bound in the active site of Pseudomonas cepacia lipase (PDB code: 1ys2)

All books published by Wiley-VCH are carefully produced. Nevertheless, authors, editors, and publisher do not warrant the information contained in these books, including this book, to be free of errors. Readers are advised to keep in mind that statements, data, illustrations, procedural details or other items may inadvertently be inaccurate.

Library of Congress Card No.:
applied for

British Library Cataloguing-in-Publication Data
A catalogue record for this book is available from the British Library.

Bibliographic information published by Die Deutsche Bibliothek
Die Deutsche Bibliothek lists this publication in the Deutsche Nationalbibliografie; detailed bibliographic data is available in the Internet at <http://dnb.ddb.de>.

© 2006 WILEY-VCH Verlag GmbH & Co. KGaA, Weinheim

All rights reserved (including those of translation into other languages). No part of this book may be reproduced in any form – by photoprinting, microfilm, or any other means – nor transmitted or translated into a machine language without written permission from the publishers. Registered names, trademarks, etc. used in this book, even when not specifically marked as such, are not to be considered unprotected by law.

Printed in the Federal Republic of Germany
Printed on acid-free paper

Coverdesign: Schulz Grafikdesign, Fußgönheim
Printing: Strauss GmbH, Mörlenbach
Bookbinding: J.Schäffer GmbH, Grünstadt

ISBN-13: 978-3-527-31029-6
ISBN-10: 3-527-31029-0

To

Aleli and Tanja

and to Jonas and Stasė

and also to

Lottie, Anna, Jonas and Annika

Preface for the 2nd edition

Over the decade, hydrolases have become key parts of the growing area of industrial biotechnology (Schmid et al., 2001; Shoemaker et al., 2003). We hope that the first edition of *Hydrolases in Organic Syntheses* contributed to this success. It was a key reference in our laboratories and an excellent starting point for new researchers in the field. For this success to continue, we felt that we had to keep it current with new developments in the area – directed evolution and protein design, dynamic kinetic resolutions, and the use of structures to explain selectivity.

Planning the 2nd edition presented a challenge – more than 900 publications in the area since 1999. We chose to focus on new areas, new insights, and new frontiers and not to focus on comprehensive lists of substrates. For this information, databases are more efficient than a book. We've added a new chapter on protein sources and optimization of biocatalyst performance for organic synthesis, an expanded chapter about directed evolution and a new chapter about catalytic promiscuity. In addition, we expanded sections dynamic kinetic resolution and immobilization and added new classes of hydrolases, such as haloalkane dehalogenases and organophosphorus hydrolases. In turn, we shortened some sections including a major shortening of the sections on lipid modification, which is of limited interest to organic chemists. These changes required a rearrangement of the book chapters. In addition, many sections contain minor updates.

We sincerely hope that these changes will extend and enhance the usefulness of this book.

Minneapolis/Greifswald, August 2005   Romas J. Kazlauskas, Uwe T. Bornscheuer

Preface for the 1st edition

Each traveller to a city seeks something different. One wants to see that special painting in the museum, another wants to drink the local beer, a third wants to meet a soulmate.

Each organic chemist also seeks something different from the field of biocatalysis. One wants high enantioselectivity, another wants reaction under mild conditions, a third wants to scale up to an industrial scale. We hope this book can be a guide to organic chemists exploring the field of biocatalysis. Enzyme-catalyzed reactions, especially hydrolase-catalyzed reactions, have already solved hundreds of synthetic problems usually because of their high stereoselectivity.

The organization is aimed at the chemist – by reaction type and by different functional groups. This information should help organic chemists identify the best hydrolase for their synthetic problem. In addition, we suggest how to choose an appropriate solvent, acyl donor, immobilization technique and other practical details. We hope that learning how others solved synthetic problems will generate ideas that solve the next generation of problems.

Although this book has more than 1 700 references, we might have missed important hydrolase-catalyzed reactions. The choices on what to include usually reflect our own research interests, but were sometimes arbitrary or even inadvertent. We will post corrections and additions on a web site: *http://pasteur.chem.mcgill.ca/hydrolases.html*.

Montreal/Stuttgart, February 1999   Romas J. Kazlauskas, Uwe T. Bornscheuer

Acknowledgments

A number of people helped to make the 2nd edition possible. UTB especially thanks Prof. Karl-Erich Jäger for figures about lipase evolution. He is grateful to Birgit Heinze, Anett Kirschner, Mark Bertram, Patrick Möbius and Dr. Aurelio Hidalgo for proofreading the manuscript. RJK thanks Kurt Faber for helpful suggestions. We thank the staff at Wiley-VCH, especially Dr.-Ing. Waltraud Wüst.

Contents

1	**Introduction**	1
2	**Designing Enantioselective Reactions**	5
2.1	Kinetic Resolutions	5
2.1.1	Recycling and Sequential Kinetic Resolutions	6
2.1.2	Dynamic Kinetic Resolutions	9
2.1.2.1	Introduction	9
2.1.2.2	Racemization by Protonation/Deprotonation	10
2.1.2.3	Racemization by Addition/Elimination	14
2.1.2.4	Racemization by Nucleophilic Substitution	17
2.1.2.5	Racemization by Oxidation/Reduction	17
2.1.2.6	Related Strategies	22
2.2	Asymmetric Syntheses	23
3	**Choosing Reaction Media: Water and Organic Solvents**	25
3.1	Hydrolysis in Water	25
3.2	Transesterifications and Condensations in Organic Solvents	25
3.2.1	Increasing the Catalytic Activity in Organic Solvents	27
3.2.1.1	Choosing the Best Organic Solvent for High Activity	27
3.2.2	Increasing the Enantioselectivity in Organic Solvents	28
3.2.3	Water Content and Water Activity	30
3.3	Other Reaction Media	32
3.3.1	Ionic Liquids	32
3.3.2	Reverse Micelles	34
3.3.3	Supercritical Fluids	36
3.4	Immobilization	39
3.4.1	Introduction	39
3.4.1.1	Increasing the Surface Area to Increase Catalytic Activity	39
4	**Protein Sources and Optimization of Biocatalyst Performance**	43
4.1	Accessing Biodiversity	43
4.2	Creating Improved Biocatalysts	44
4.2.1	Directed Evolution	45
4.2.1.1	Methods to Create Mutant Libraries	45
4.2.1.2	Assay Systems	47
4.2.1.3	Selected Examples	50
4.2.2	Focused Directed Evolution	52
4.3	Catalytic Promiscuity in Hydrolases	54
4.3.1	Reactions Involving Functional Group Analogs	54
4.3.1.1	Perhydrolases	55

4.3.2	Aldol and Michael additions Catalyzed by Hydrolases	56
4.3.2.1	Aldol Additions 56	
4.3.2.2	Michael-Type Additions 57	
4.3.3	Modifications to Introduce New Reactivity in Hydrolases	58
4.3.3.1	Enantioselective Reduction of Hydroperoxides with Selenosubtilisin	58
4.3.3.2	Vanadate-Modified Phosphatases as Peroxidases 59	
5	**Lipases and Esterases 61**	
5.1	Availability, Structures and Properties 61	
5.1.1	Lipases 62	
5.1.1.1	Classification of Lipases 65	
5.1.1.2	General Features of PPL, PCL, CRL, CAL-B, and RML 66	
5.1.2	Esterases 68	
5.1.3	Lipases and Esterases are α/β Hydrolases 68	
5.1.4	Lid or Flap in Interfacial Activation of Lipases 70	
5.1.5	Substrate Binding Site in Lipases and Esterases 72	
5.1.6	Designing Reactions with Lipases and Esterases 74	
5.1.6.1	Acyl Donor for Acylation Reactions 74	
5.1.7	Assays for Lipases and Esterases 77	
5.1.7.1	Requirements for a Suitable Assay 77	
5.1.7.2	How to Distinguish Between Lipase, Esterase, and Protease 83	
5.2	Survey of Enantioselective Lipase-Catalyzed Reactions 84	
5.2.1	Alcohols 84	
5.2.1.1	Secondary Alcohols 84	
5.2.1.2	Primary Alcohols 106	
5.2.1.3	Other Alcohols, Amines, and Alcohol Analogs 113	
5.2.2	Carboxylic Acids 124	
5.2.2.1	General Considerations 124	
5.2.2.2	Carboxylic Acids with a Stereocenter at the α-Position 124	
5.2.2.3	Carboxylic Acids with a Stereocenter at the β-Position 128	
5.2.2.4	Other Carboxylic Acids 130	
5.2.2.5	Double Enantioselection 131	
5.2.2.6	Anhydrides 132	
5.2.3	Lactones 133	
5.2.4	Commercial Enantioselective Reactions 138	
5.2.4.1	Enantiomerically-Pure Chemical Intermediates 138	
5.2.4.2	Enantiomerically-Pure Pharmaceutical Intermediates 138	
5.3	Chemo- and Regioselective Lipase-Catalyzed Reactions 141	
5.3.1	Protection and Deprotection Reactions 141	
5.3.1.1	Hydroxyl Groups 141	
5.3.1.2	Amino Groups 159	
5.3.1.3	Carboxyl Groups 160	
5.3.2	Lipid Modifications 163	
5.3.3	Oligomerization and Polymerizations 163	
5.3.4	Other Lipase-Catalyzed Reactions 165	
5.4	Reactions Catalyzed by Esterases 166	
5.4.1	Pig Liver Esterase 166	

5.4.1.1	Biochemical Properties	166
5.4.1.2	Recombinant PLE	167
5.4.1.3	Overview of PLE Substrate Specificity and Models	169
5.4.1.4	Asymmetrization of Carboxylic Acids with a Stereocenter at the α-Position	171
5.4.1.5	Asymmetrization of Carboxylic Acids with Other Stereocenters	172
5.4.1.6	Asymmetrization of Primary and Secondary *meso*-Diols	172
5.4.1.7	Kinetic Resolution of Alcohols or Lactones	173
5.4.1.8	Kinetic Resolution of Carboxylic Acids	174
5.4.1.9	Reactions Involving Miscellaneous Substrates	174
5.4.2	Acetylcholine Esterase	175
5.4.2.1	Biochemical Properties and Applications in Organic Syntheses	175
5.4.3	Other Mammalian Esterases	176
5.4.4	Microbial Esterases	177
5.4.4.1	Carboxylesterase NP	181
5.4.4.2	Other Microbial Esterases	183

6 Proteases and Amidases 185

6.1	Occurrence and Availability of Proteases and Amidases	185
6.2	General Features of Subtilisin, Chymotrypsin, and Other Proteases and Amidases	186
6.2.1	Substrate Binding Nomenclature in Proteases and Amidases	186
6.2.2	Synthesis of Amide Bonds Using Proteases and Amidases	187
6.2.3	Subtilisin and Related Proteases	189
6.2.4	Chymotrypsin	190
6.2.5	Thermolysin	190
6.2.6	Penicillin G Acylase	191
6.2.7	Amino Acid Acylases	191
6.2.8	Protease Assays	191
6.3	Structures of Proteases and Amidases	193
6.3.1	Serine Proteases – Subtilisin and Chymotrypsin	193
6.4	Survey of Enantioselective Protease- and Amidase-Catalyzed Reactions	195
6.4.1	Alcohols and Amines	195
6.4.1.1	Secondary Alcohols and Primary Amines	195
6.4.1.2	Secondary amines	198
6.4.1.3	α-Amino Acids via Reactions at the Amino Group	199
6.4.2	Carboxylic Acids	200
6.4.2.1	α-Amino Acids via Reactions at the Carboxyl Group	200
6.4.2.2	Other Carboxylic Acids	207
6.4.2.3	Commercial Enantioselective Reactions	209

7 Phospholipases 211

7.1	Phospholipase A_1	211
7.2	Phospholipase A_2	211
7.3	Phospholipase C	212
7.4	Phospholipase D	212

8	**Epoxide Hydrolases**	**215**
8.1	Introduction	215
8.2	Mammalian Epoxide Hydrolases	218
8.3	Microbial Epoxide Hydrolases	220
8.3.1	Bacterial Epoxide Hydrolases	220
8.3.2	Fungal and Yeast Epoxide Hydrolases	223
9	**Hydrolysis of Nitriles**	**227**
9.1	Introduction	227
9.2	Mild Conditions	233
9.3	Regioselective Reactions of Dinitriles	236
9.4	Enantioselective Reactions	237
10	**Other Hydrolases**	**241**
10.1	Glycosidases	241
10.2	Haloalcohol Dehalogenases	245
10.3	Phosphotriesterases	246
10.3.1	Organophosphorus Hydrolases	246
10.3.2	Diisopropyl Fluorophosphatases	248

Abbreviations 249

References 251

Index 347

1 Introduction

Hydrolases are the group of enzymes that catalyze bond cleavage by reaction with water. The natural function of most hydrolases is digestive – to break down nutrients into smaller units for digestion. For example, proteases hydrolyze proteins to smaller peptides and then to amino acids and lipases hydrolyze lipids (triglycerides) to glycerol and fatty acids (Fig. 1). Because of the need to break down a wide range of nutrients, hydrolases usually have a broad substrate specificity.

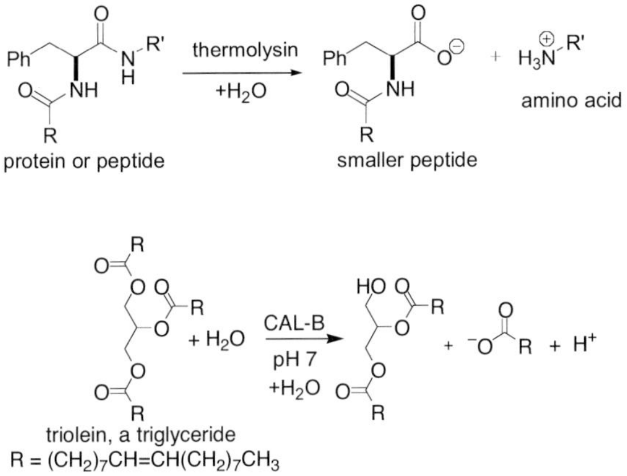

Fig. 1. The natural role of most hydrolases is digestive – to break down nutrients into smaller units. Thermolysin, a protease secreted by thermophilic bacteria, catalyzes the hydrolysis of proteins to peptides and then further to amino acids. Lipase B from *Candida antarctica* (CAL-B) catalyzes the stepwise hydrolysis of triglycerides (e.g., triolein) to fatty acids and glycerol. The reaction shows only the first step from a triglyceride to a diglyceride.

Several characteristics make hydrolases useful to the organic chemist. First, because of their broad substrate specificity, hydrolases often accept as substrates various synthetic intermediates. Second, hydrolases often show high stereoselectivity, even toward unnatural substrates. Third, besides hydrolysis, hydrolases also catalyze several related reactions – condensations (reversal of hydrolysis) and alcoholysis (a cleavage using an alcohol in place of water). Two examples from industry are shown in Fig. 2. Thermolysin catalyzes the condensation of two amino acid derivatives to make an aspartame derivative (Isowa et al., 1979). The reaction proceeds in the condensation direction because the product precipitates from solution. The high enantioselectivity permits using racemic starting materials and the high regioselectivity of thermolysin eliminates the need to protect the β-carboxyl group of the aspartic acid derivative. The second example is an alcoholysis reaction (Morgan et al., 1997a). The ester, vinyl acetate, is cleaved not by water, but by the substrate alcohol. The liberated vinyl alcohol (not shown) tautomerizes to acetaldehyde. These alcoholysis reactions are also called transesterification reactions.

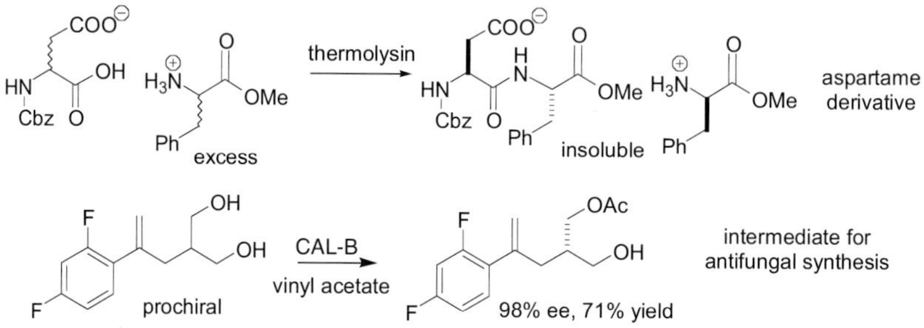

Fig. 2. Several unnatural, synthetically useful reactions catalyzed by hydrolases. Thermolysin catalyzes the regio- and enantioselective coupling of N-benzyloxycarbonyl-L-aspartate with L-phenylalanine methyl ester. Precipitation of the product drives this reaction in the condensation direction instead of the normal hydrolysis direction. This condensation is a key step in the manufacture of aspartame, a low-calorie sweetener. Because of the high enantioselectivity of thermolysin, racemic substrates may be used. Because of the high regioselectivity of thermolysin for the α-carboxyl group, the β-carboxyl group in the aspartic acid derivative needs no protection. Lipase B from *Candida antarctica* (CAL-B) catalyzes the enantioselective acetylation of a prochiral diol yielding an intermediate for the synthesis of antifungal agents. This example is an alcoholysis where the ester, vinyl acetate, is cleaved not by water, but by the substrate alcohol. This reaction is run in an organic solvent to avoid the competing hydrolysis.

Several other features make hydrolases convenient to use as synthetic reagents. Many hydrolases (approximately several hundred) are commercially available. They do not require cofactors and they tolerate the addition of water-miscible solvents (e.g., DMSO, DMF). Lipases, esterases and some proteases are also stable and active in neat organic solvents.

Enzymes are often classified according to the reaction catalyzed using an Enzyme Commission (EC) number. According to this classification, hydrolases form group 3 and are further classified according to the type of bond hydrolyzed. For example enzymes in the group 3.1 hydrolyze ester bonds (Tab. 1). Further classification into subcategories yields a four digit EC number. For example, lipases have the number EC 3.1.1.3. Classification of the more useful enzymes for organic synthesis is given in Tab. 1. A convenient web site to look up numbers and classification is at *http://www.expasy.ch/enzyme*. One disadvantage of this classification is that all enzymes catalyzing the same reaction have the same number, even though they may have very different structures, properties and other characteristics. For example, all lipases have the same number even though there are more than one hundred different lipases.

Tab. 1. Selected Hydrolases Useful in Organic Synthesis.

EC Number	Type of bond hydrolyzed	Examples
3.1	**Ester**	
3.1.1	in carboxylic acid esters	triacylglycerol lipase, acetylcholine esterase, phospholipase A_1, phospholipase A_2, gluconolactonase, lipoprotein lipase
3.1.3–4	in phosphoric acid mono- or diesters	phospholipase C, phospholipase D
3.2	**Glycosidic**	
3.2.1	in *O*-glycosides	α-amylase, oligo-1,6-glucosidase, lysozyme, neuraminidase, α-glucosidase, β-galactosidase, α-mannosidase, N-acetyl-β-glucosaminidase, sucrose α-glucosidase, nucleosidases
3.3	**Ether**	
3.3.2	in epoxides	epoxide hydrolase
3.4	**Peptide**	
3.4.11	aminopeptidase	leucine aminopeptidase
3.4.16, 21	serine proteinase	subtilisin, chymotrypsin, thermitase
3.4.18, 22	cysteine proteinase	papain
3.4.17, 24	metalloproteinase	thermolysin
3.5	**Other amides**	
3.5.1	in linear amides	penicillin amidase (penicillin G acylase)
3.5.2	in cyclic amides	hydantoinase
3.5.5	in nitriles	nitrilase[a]
3.8	**Halide bonds**	
3.8.1	carbon-halide bonds	haloalkane dehalogenase

[a]Nitrile hydratase (EC 4.2.1.84), which catalyzes addition of water to a nitrile yielding an amide, is not a hydrolase, but a lyase.

This book describes the application of lipases and proteases in organic syntheses, but also surveys esterases, epoxide hydrolases, nitrile hydrolyzing enzymes and glycosidases. The emphasis is on examples that are synthetically useful, especially those that exploit the regio- and stereoselectivity of hydrolases.

2 Designing Enantioselective Reactions

2.1 Kinetic Resolutions

In a kinetic resolution, the enantiomeric purity of the product and starting material varies as the reaction proceeds (reviewed by Kagan and Fiaud, 1988). Thus, comparing enantiomeric purities for two kinetic resolutions is meaningful only at the same extent of conversion. To more conveniently compare kinetic resolutions, Charles Sih's group developed equations to calculate their inherent enantioselectivity (Chen et al., 1982; 1987; reviewed by Sih and Wu, 1989). This enantioselectivity, called the enantiomeric ratio, E, measures the ability of the enzyme to distinguish between enantiomers. A non-selective reaction has an E of 1, while resolutions with E's above 20 are useful for synthesis. To calculate E, one measures two of the three variables: enantiomeric purity of the starting material (ee_s), enantiomeric purity of the product (ee_p), and extent of conversion (c) and uses one of the three equations below (Eq. 1). Often enantiomeric purities are more accurately measured than conversion; in these cases, the third equation is more accurate.

$$E = \frac{\ln[1 - c(1 + ee_p)]}{\ln[1 - c(1 - ee_p)]}; \quad E = \frac{\ln[(1 - c)(1 - ee_s)]}{\ln[(1 - c)(1 + ee_s)]}; \quad E = \frac{\ln\left[\frac{1 - ee_s}{1 + (ee_s/ee_p)}\right]}{\ln\left[\frac{1 + ee_s}{1 + (ee_s/ee_p)}\right]} \quad (1)$$

High E values (≥ 100) are less accurately measured than low or moderate E values because the enantiomeric ratio is a logarithmic function of the enantiomeric purity. At $E \geq 100$, small changes in the measured enantiomeric purities give large changes in the enantiomeric ratio. Thus, the survey below avoids reporting E values above 100. In practice, we found that even E values near 50 were sometimes difficult to measure more precisely than ± 10. A simple program to calculate enantiomeric ratio using the above equations is freely available at *http://www.orgc.tugraz.at/* (Kroutil et al., 1997a). In spite of the fact that these equations include assumptions such as an irreversible reaction, one substrate and product, and no product inhibition, they are reliable in the vast majority of cases, especially for screening studies. Recently, faster spectrophotometric methods for measuring the enantiomeric ratio (Janes and Kazlauskas, 1997; Janes et al., 1998) using samples of pure enantiomers were developed. One method, called Quick E, is restricted to *p*-nitrophenyl derivatives of chiral carboxylic acids, but a more recent method can be used for any ester and identification of active and enantioselective hydrolases is based on a pH change using *p*-nitrophenol as pH indicator.

For careful optimization of reactions, three situations require a more careful approach. First, when the biocatalyst is a mixture of enzymes, for example, isozymes, which all act on the substrate, then the calculated E value reflects a weighted average of all the enzymes (Chen et al., 1982). When these enzymes differ significantly in their affinity for the substrate, then different enzymes will dominate the activity at different substrate concentrations. Thus, the apparent enantioselectivity may vary as the reaction depletes the substrate or when the reaction is carried out with different initial substrate concentrations. When enzymes differ in their stability, apparent enantioselectivities for long vs. short reaction times may differ. To measure the true E-value, one must purify the enzymes and measure E separately.

Second, when product inhibits the reaction the apparent enantioselectivity can change (Rakels et al., 1994a; van Tol et al., 1995a; b). For example, addition of 4 v/v% ethanol to a carboxylesterase NP-catalyzed hydrolysis of ethyl 2-chloropropionate increased the enantioselectivity from 4.7 to 5.4 (see also Sect. 5.4.4.1). Rakels et al. (1994a) attributed this change not to changes in the inherent selectivity of the enzyme, but to selective inhibition of one of the enantiomers by ethanol. In another example (van Tol et al., 1995a; b), could not recover enantiomerically pure starting material in the PPL-catalyzed hydrolysis of glycidol butyrate even at high conversion. The enantiomeric purity of the remaining glycidol butyrate reached 95% ee at 70% conversion, but did not increase further even at 90% conversion. In other words, the apparent enantioselectivity dropped from 20 at 31% conversion to 2.7 at 95% conversion. van Tol et al. (1995a; b) attributed this plateau to product inhibition promoting the reverse reaction for the product enantiomer. To include product inhibition in the quantitative analysis, reseachers used more complex equations which take into account the mechanism of lipase-catalyzed reactions (ping-pong bi-bi). Until now few researchers included product inhibition in their analysis, but a readily available computer program (Anthonsen et al., 1995) simplifies this task.

Third, when the reaction is reversible, such as transesterification, one must include the equilibrium constant for the reaction (Chen et al., 1987). One can first measure the equilibrium constant in a separate experiment and then determine E from measurements of ee_s and ee_p. Anthonsen et al. (1995) developed a simpler approach where they determine both K and E by fitting a series of ee_s and ee_p measurements.

2.1.1 Recycling and Sequential Kinetic Resolutions

To enhance the enantiomeric purity, the enriched material can be isolated and resolved again. This double resolution is called recycling. Chen et al. (1982) derived an equation to predict the optimum degree of conversion in recycling reactions and many researchers have used this strategy (for an example see: Johnson et al., 1995). Brown et al. (1993) and Kanerva and Vänttinen (1997) reported several examples. A computer program for calculations is available at *http://www.orgc.tu-graz.ac.at* (Kroutil et al., 1997b). Guo (1993) reported plots to predict the maximum chemical yield in various situations. To minimize the work in recycling reactions, several groups used *in situ* recycling where the two resolutions are carried out stepwise, but without isolation of the intermediate products (Chen and Liu, 1991; Majeric and Sunjic, 1996; Sugai et al., 1996). Some authors called these reactions sequential kinetic resolutions, but we favor *in situ* recycling and reserve the term sequential kinetic resolution only for those reactions where both steps occur at the same time, such as the acylation of diols.

Like recycling reactions, sequential kinetic resolutions enhance the enantiomeric purity of the products (Caron and Kazlauskas, 1991; Guo et al., 1990; Kazlauskas, 1989). For example, hydrolysis of *trans*-1,2-diacetoxycyclohexane proceeds stepwise – first hydrolysis to the monoacetate, then to the diol (Fig. 3) (Caron and Kazlauskas, 1991). Both reactions favor the same enantiomer, thus, the two resolutions reinforce each other. Maximum reinforcement occurs when both reactions occur at comparable rates with an overall enantioselectivity of approximately $(E_1 \times E_2)/2$ (Caron and Kazlauskas, 1991). In addition, sequential kinetic resolutions yield both the starting material and product in high enantiomeric purity at the same extent of conversion because the 'mistakes' remain in the intermediate product (monoacetate in the example in Fig. 3). In contrast, single step kinetic resolutions yield high enantiomeric purity for the product at < 50% conversion, but high enantiomeric purity for the starting material requires > 50% conversion.

Fig. 3. Sequential kinetic resolution enhances the enantiomeric purity of the product through two enantioselective steps.

C_2-symmetric diols are especially well suited to sequential kinetic resolution because both steps are likely to have the same enantiopreference (Fig. 4).

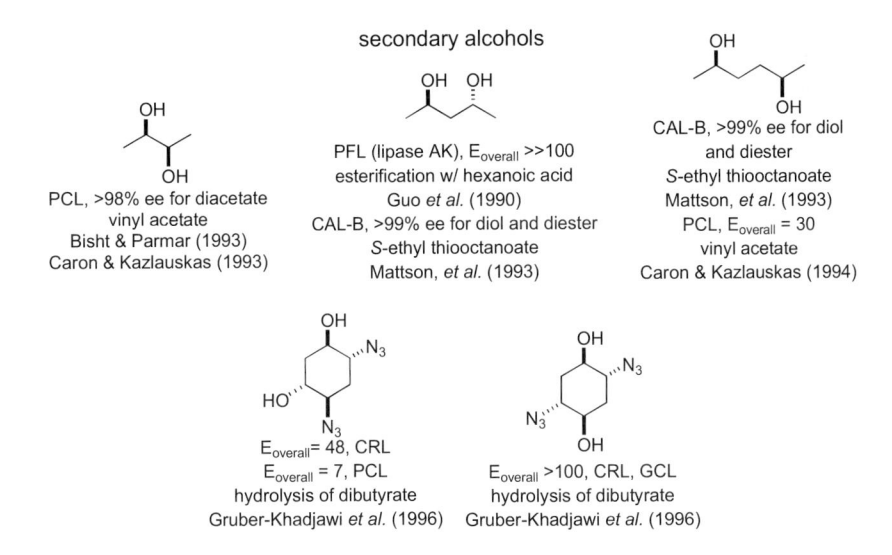

Fig. 4. Examples of C_2-symmetric diols resolved by sequential kinetic resolution include secondary and primary alcohols as well as diols with axial chirality.

2.1 Kinetic Resolutions

primary alcohols   axially chiral diol

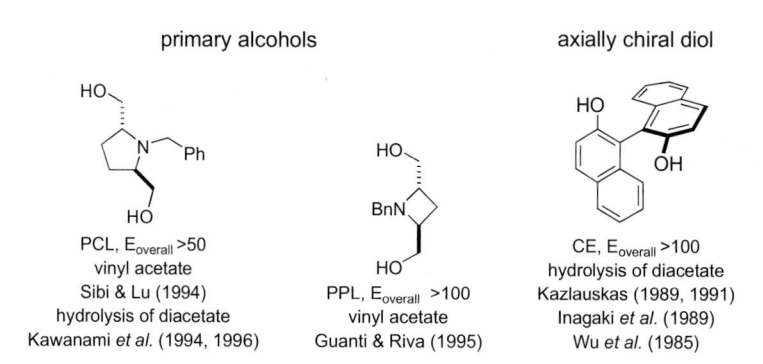

PCL, $E_{overall}$ >50
vinyl acetate
Sibi & Lu (1994)
hydrolysis of diacetate
Kawanami et al. (1994, 1996)

PPL, $E_{overall}$ >100
vinyl acetate
Guanti & Riva (1995)

CE, $E_{overall}$ >100
hydrolysis of diacetate
Kazlauskas (1989, 1991)
Inagaki et al. (1989)
Wu et al. (1985)

Fig. 4. Examples of C_2-symmetric diols resolved by sequential kinetic resolution include secondary and primary alcohols as well as diols with axial chirality (continued).

Unsymmetrical diols can also undergo a sequential kinetic resolution (Fig. 5).

PCL, E high
vinyl acetate
R = Ph, n-Pr to n-Hx
Kim et al. (1995b)

PCL, E high
vinyl acetate
R = OTr, CH_2OTr
Kim et al. (1995b)

Fig. 5. Sequential kinetic resolution of non-C_2 symmetric diols.

Only one dicarboxylic acids was resolved by lipase-catalyzed sequential kinetic resolution and this was a special case. Node et al. (1995) hydrolyzed a racemic C_2-symmetric tetraester. The non-conjugated ester groups reacted selectively followed by spontaneous decarboxylation. Interestingly, CRL and RJL favored opposite enantiomers. Although Node et al. (1995) suggested possible racemization of the starting tetraester, which would allow a dynamic kinetic resolution (Sect. 2.1.2), they did not report yields over 50%. The lack of carboxylic acid examples may be due to more efficient resolution of alcohols by lipases, or to the slow hydrolysis of monoesters containing a charged carboxylate group by lipases (Fig. 6).

hydrolysis
decarboxylation

CRL, 32% yield, 100% ee
RJL, 20% yield, 90% ee
(opposite enantiomers)
Node et al. (1995)

Fig. 6. Sequential kinetic resolution of a chiral diacid.

For substrates with a single functional group, researchers demonstrated a sequential kinetic resolution by *in situ* hydrolysis of an ester and reesterification to a new ester (Macfarlane et al., 1990). However, reversibility of these reactions limited the enhancement of enantioselectivity (Straathof et al., 1995). In these cases, an *in situ* recycling reaction (see above) is probably a better way to enhance the enantiomeric purity.

Enantioselective reactions can also separate diastereomers. For example, Wallace et al. (1992) used the (*R*)-enantioselectivity of PCL to separate a mixture of *meso* and racemic diols. The (*R*,*R*)-diol reacted to the diacetate, the (*R*,*S*)-diol to the monoacetate, and the (*S*,*S*)-diol did not react (Fig. 7).

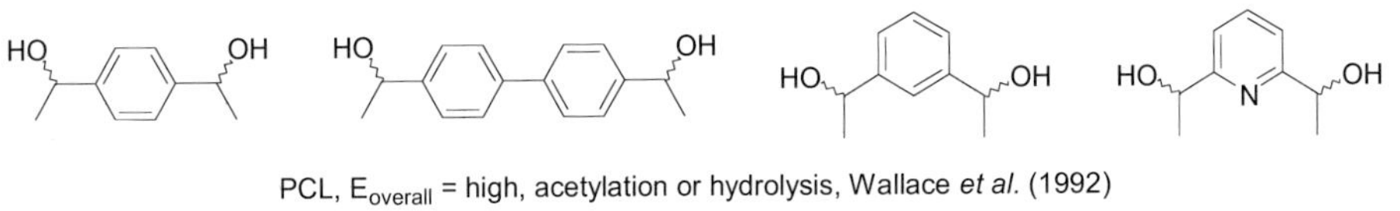

PCL, $E_{overall}$ = high, acetylation or hydrolysis, Wallace *et al.* (1992)

Fig. 7. Enantioselective reactions separated diastereomers as well as enantiomers.

2.1.2 Dynamic Kinetic Resolutions

2.1.2.1 Introduction

Kinetic resolution limits the yield of each enantiomer to 50%. However, if the substrate racemizes quickly in the reaction mixture, then the yield of product enantiomer can be 100% (Fig. 8). This resolution with *in situ* racemization is called dynamic kinetic resolution or asymmetric transformation of the second kind (for reviews see: Stecher and Faber, 1997; Ward, 1995; Faber, 2001; Pellissier, 2003; Schnell et al., 2003). The requirements for a dynamic kinetic resolution are: (1), the substrate must racemize at least as fast as the subsequent enzymatic reaction, (2), the product must not racemize, and (3), as in any asymmetric synthesis, the enzymic reaction must be highly stereoselective. The equations relating product enantiomeric purity and enantioselectivity are the same as those for asymmetric syntheses (Sect. 2.2): $ee_P = (E - 1)/(E + 1)$ and $E = (1 + ee_P)/(1 - ee_P)$, where E is the enantiomeric ratio and ee_P is the enantiomeric purity of the product.

The key step in dynamic kinetic resolutions is the *in situ* racemization. A number of reactions can racemize organic substrates (review: Ebbers et al., 1997), but most conditions are too harsh to allow a simultaneous enzyme-catalyzed reaction. In the past, the difficulty of racemizing normal alcohols and carboxylate esters restricted these dynamic kinetic resolutions to special cases such as the one in Fig. 8b. However, recent discoveries of organometallic catalysts that can racemize a wide range of secondary alcohols have extended the range of these reactions. The sections below group the dynamic kinetic resolution examples according to the racemization mechanism.

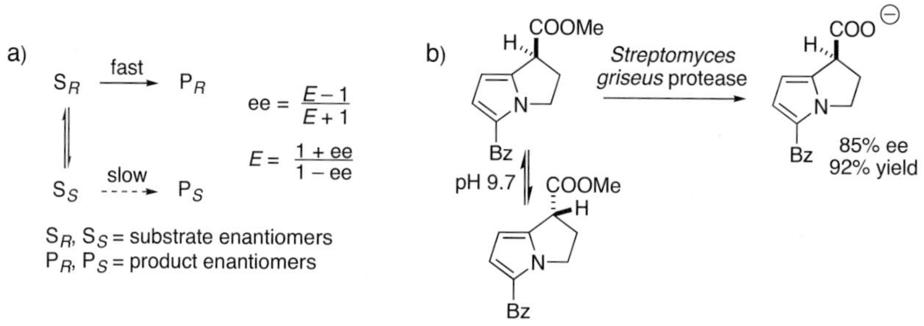

Fig. 8. Dynamic kinetic resolution. a) Dynamic kinetic resolution involves an *in situ* racemization of the substrate (S_R and S_S) combined with an enantioselective reaction of one substrate enantiomer (P_R favored in this scheme). b) An example of a dynamic kinetic resolution. The substrate ester contains a moderately acidic hydrogen α to the carbonyl and aromatic ring. Deprotonation to the achiral enolate and reprotonation racemized this ester. The protease catalyzed the enantioselective hydrolysis of one enantiomer. The product carboxylate did not racemize because the negative charge on the carboxylate made deprotonation less favorable (Fülling and Sih, 1987).

2.1.2.2 Racemization by Protonation/Deprotonation

The earliest dynamic kinetic resolutions involved 5-arylsubstituted hydantoins, which racemize spontaneously at pH > 8 via an enolate (Olivieri et al., 1981; Takahashi et al., 1979; Tsugawa et al., 1966). Hydantoinases catalyze the highly enantioselective hydrolysis of 5-monosubstituted hydantoins to *N*-carbamoyl-α-amino acids (Fig. 9).

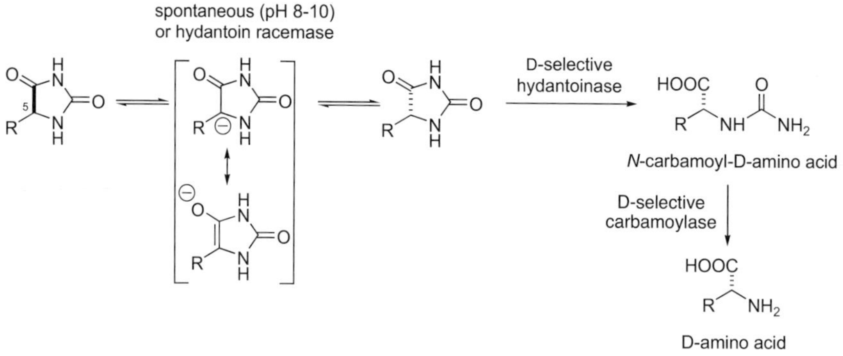

Fig. 9. Hydantoinases catalyze the hydrolysis of 5-monosubstituted hydantoins to *N*-carbamoyl α-amino acids. 5-Substituted hydantoins, especially 5-aryl hydantoins, racemize readily either enzymatically or chemically at pH 8–10 via the enolate. The figure shows a D-selective hydanoinase; L-selective hydanoinases are rare. The last step, removal of the *N*-carbamoyl group, may also contribute to the overall enantioselectivity.

The most important amino acids produced by this route are D-phenylglycine and D-4-hydroxyphenylglycine for production of the semisynthetic penicillins, ampicillin and amoxicillin, respectively (May et al., 2002). A similar process could also yield the L-series of α-amino acids, but the L-selective hydantoinases are rare. See also Sect. 4.2.1.3 for directed evolution and Sect. 6.4.2.1 for further information on hydantoinases.

Another dynamic kinetic resolution route to D-phenylglycine starts with (±)-phenylglycine methyl ester and yields the amide (Wegman et al., 1999). Phenylglycine methyl ester racemizes in the presence of aldehydes via a Schiff-base intermediate. Combining this racemization with a CAL-B-catalyzed ammonolysis at −20°C yielded D-phenylglycine amide in 85% ee and 85% yield.

The structurally similar 4-substituted-2-phenyloxazolin-5-ones also racemize readily, but finding enantioselective hydrolases has been more difficult (Eq. 2; Tab. 2). For R = Me, Bn, and several others, Bevinakatti et al. (1990; 1992) used a RML-catalyzed alcoholysis in organic solvents, but the enantioselectivity was only E = 3–5. Sih's group screened a dozen lipases for hydrolysis of the phenylalanine derivative (R = Bn) and found that PPL favored the natural (R)-enantiomer (E > 100), while ANL favored the unnatural (S)-enantiomer (E > 100) (Crich et al., 1993; Gu et al., 1992) (Tab. 2). However, these enzymes were less enantioselective toward other, similar derivatives. Several *Pseudomonas* lipases (PCL, Amano AK, Amano K-10) at 50°C in *t*-BuOMe catalyzed methanolysis of a variety of 4-substituted 2-phenyloxazolin-5-ones with enantioselectivities of 5–39, usually favoring the (S)-enantiomer.

$$\text{(2)}$$

Tab. 2. Lipase-Catalyzed Ring-Opening of 2-Phenyloxazolin-5-ones.

Lipase	Reaction	R =	E	References
RML	alcoholysis	Bn, Me, *n*-Pr, CH$_2$*i*-Pr	3–5 (S)	Bevinakatti et al. (1990; 1992)
PPL	hydrolysis	Bn	> 100 (S)	Gu et al. (1992)
ANL	hydrolysis	Bn	> 100 (R)	Gu et al. (1992)
ANL, PPL	hydrolysis	Ph, 4-OMePh, CH$_2$CH$_2$Ph, several CH$_2$Ar, CH$_2$*i*-Pr, CH$_2$CH$_2$SMe	2–12	Crich et al. (1993); Gu et al. (1992)
PXL[a]	alcoholysis	13 different examples	5–39	Crich et al. (1993)
RML	alcoholysis	*t*-Bu	> 100 (S)	Turner et al. (1995)
CAL-B	alcoholysis	Bn, CH$_2$*i*-Pr, *i*-Pr, indole-methylene	≥19	Brown et al. (2000)

[a] One of several *Pseudomonas* lipases: PCL, Amano AK, or Amano K-10. Most reactions favored the (S)-enantiomer, but in some cases the enantiopreference was either (R) or (S) depending on the amount of added water.

In several cases, the enantioselectivity reversed depending on whether the reaction mixture contained added water or not. The lipase usually hydrolyzed substrates with larger R groups (e.g., Ph, CH$_2$*i*-Pr) more selectively than small ones (e.g., Me). For preparative use, Crich et al. (1993) further resolved the enantiomerically-enriched methyl esters of *N*-benzoyl amino acids by protease-catalyzed cleavage of the ester. Turner et al. (1995) found that RML-catalyzed alcoholysis of the *t*-butyl derivative was highly enantioselective (99.5% ee, 94% yield), but only when the reaction mixture con-

2.1 Kinetic Resolutions

tained a catalytic amount of triethylamine. The authors suggested that the triethylamine may increase enantioselectivity by forming an ion pair with a side product formed by hydrolysis instead of alcoholysis – N-benzoyl amino acid (Brown et al., 2000). Without ion pair formation, this carboxylic acid seems to lower the enantioselectivity of the reaction.

Simple esters of chiral carboxylic acids also undergo base-catalyzed racemization if the acid contains electron-withdrawing substituents. Fülling and Sih (1987) reported the first enzyme-catalyzed example using a *Streptomyces griseus* protease, which was described in Fig. 8 above. Acyloin derivatives also racemize quickly in the presence of a catalytic amount of triethylamine (Taniguchi et al., 1997; Taniguchi and Ogasawara, 1997), while butenolides racemize readily at room temperature and pyrrolinones racemize at 69°C, Fig. 10 (Thuring et al., 1996a; van der Deen et al., 1996). In each case, lipases catalyzed selective acetylation of one enantiomer in excellent yield.

PCL, E>50, n = 1,2
vinyl acetate/NEt$_3$
Taniguchi & Ogasawara (1997)
Taniguchi et al. (1997)

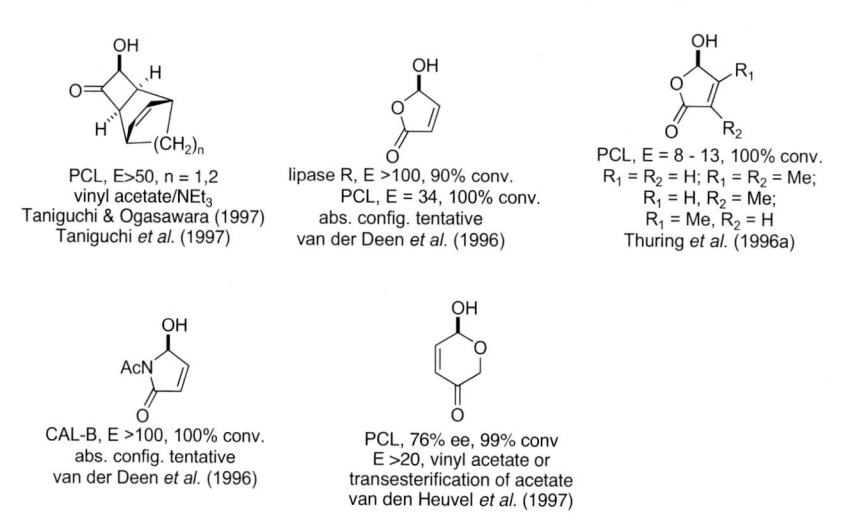

lipase R, E >100, 90% conv.
PCL, E = 34, 100% conv.
abs. config. tentative
van der Deen et al. (1996)

PCL, E = 8 - 13, 100% conv.
$R_1 = R_2 = H$; $R_1 = R_2 = Me$;
$R_1 = H$, $R_2 = Me$;
$R_1 = Me$, $R_2 = H$
Thuring et al. (1996a)

CAL-B, E >100, 100% conv.
abs. config. tentative
van der Deen et al. (1996)

PCL, 76% ee, 99% conv
E >20, vinyl acetate or
transesterification of acetate
van den Heuvel et al. (1997)

Fig. 10. Dynamic kinetic resolution of alcohols that racemize by deprotonation/protonation. The butenolides, acyloins, and pyrrolinones contain a carbonyl group or a vinylogous carbonyl group α to the stereocenter making deprotonations more facile. In each case, the product isolated was the corresponding acetate.

Tan et al. (1995) resolved 2-(phenylthio)propanoic acid by PCL-catalyzed hydrolysis of the thioester in the presence of trioctylamine (Eq. 3). Both the thioester and the trioctylamine promote racemization via an enolate mechanism. Similarly, Um and Drueckhammer (1998) resolved thioesters of 2-aryl and 2-aryloxypropanoic acids using subtilisin (E = 7-11) combined with *in situ* racemization promoted by trialkylamines. Chosing the type of thioester (e.g., 2,2,2-trifluoroethyl thiol ester) was the key to rapid racemization. Vörde et al. (1996) suggested that even simple esters may racemize in the presence of both CAL-B and α-phenylethylamine (Eq. 4). They did not detect racemization in the presence of only one of these.

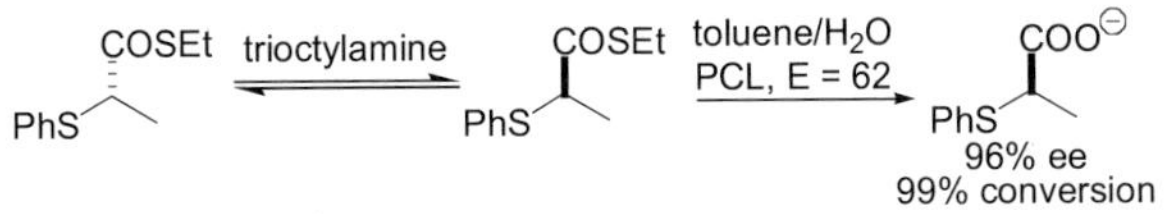

$$\underset{PhS}{\overset{COSEt}{\bigwedge}} \xrightleftharpoons[]{\text{trioctylamine}} \underset{PhS}{\overset{COSEt}{\bigwedge}} \xrightarrow[\text{PCL, E = 62}]{\text{toluene/H}_2\text{O}} \underset{\substack{PhS \\ 96\% \text{ ee} \\ 99\% \text{ conversion}}}{\overset{COO^\ominus}{\bigwedge}} \quad (3)$$

$$\underset{n\text{-C}_6\text{H}_{13}}{\overset{COOEt}{\bigwedge}} \xrightleftharpoons[]{} \underset{n\text{-C}_6\text{H}_{13}}{\overset{COOEt}{\bigwedge}} \xrightarrow[\text{70°C}]{\substack{H_2N-\overset{Ph}{\bigwedge} \\ \text{CAL-B}}} \underset{\substack{n\text{-C}_6\text{H}_{13} \\ 45\% \text{ de} \\ 99\% \text{ conversion}}}{\overset{CO-N-\overset{Ph}{\bigwedge}}{\bigwedge}} \quad (4)$$

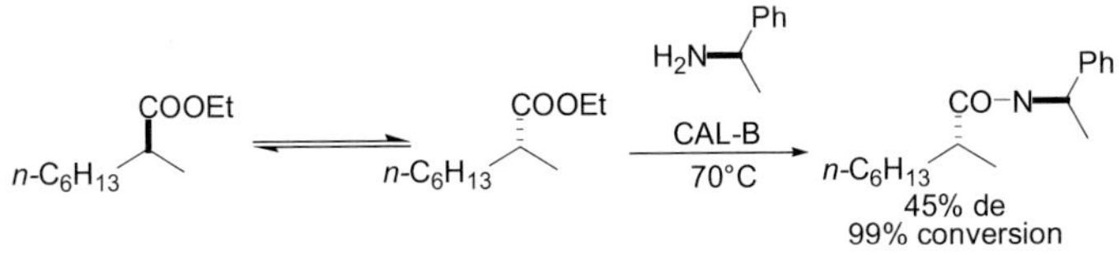

Racemases are an ideal way to convert kinetic resolutions to dynamic kinetic resolutions (review: Schnell et al., 2003). Research in this area accelerated recently and several examples have been reported. For example, an N-acetyl-α-amino acid racemase converts the acylase-catalyzed kinetic resolution of N-acetyl-α-amino acids (see Sect. 6.4.1.3) into a dynamic kinetic resolution (Tokuyama and Hatano, 1996; Tokuyama, 2001; Verseck et al., 2001; May et al., 2002). In the kinetic resolution approach, the unreacted D-N-acetyl-α-amino acid is separated, racemized chemically *ex-situ* and added to the next kinetic resolution. However, adding an N-acetyl-α-amino acid racemase eliminates the separation and racemization steps and allows use of a continuous bioreactor (Fig. 11).

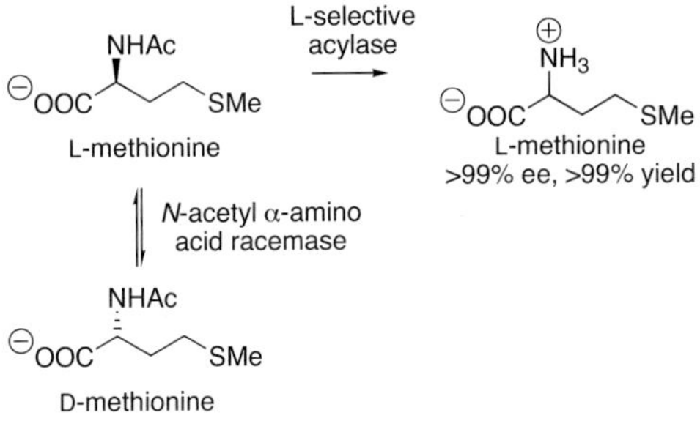

Fig. 11   Dynamic kinetic resolution of N-acetyl methionine using N-acetyl-α-amino acid racemase and an L-selective acylase (Tokuyama and Hatano, 1996). Replacing the L-selective acylase with a D-selective acylase yields D-methionine.

Interestingly, the N-acyl amino acid racemase from *Amycolaptosis* sp. discovered by Tokuyama and Hatano (1996) may be a case of mistaken identity due to catalytic promiscuity (see Sect. 1.1). This racemase is one-thousand times more efficient as a catalyst for dehydration to form *o*-succinylbenzoate suggesting that succinylbenzoate formation is its true role, Fig. 12 (Palmer et al., 1999; Ringia et al., 2004). By changing the N-acyl amino acid from N-acetyl methionine (the previous best substrate for racemase activity) to N-succinyl phenylglycine, which better resembles the succinylbenzoate precursor, the efficiency of the racemization reaction increased one-thousand fold making it similar to the succinylbenzoate reaction.

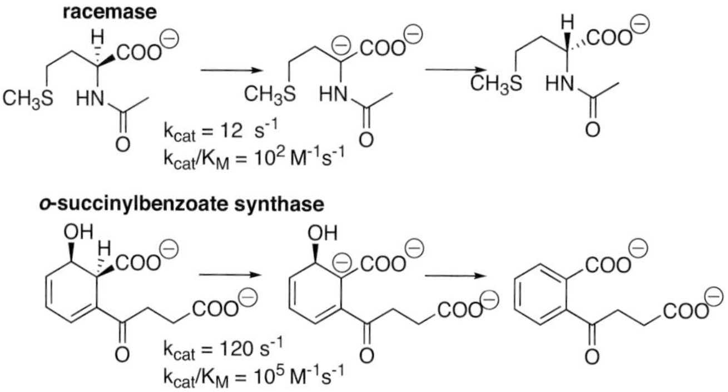

14   2.1 Kinetic Resolutions

Fig. 12. An enzyme discovered as an N-acyl amino acid racemase is one thousand fold more efficient in the dehydration to form o-succinylbenzoate. Both reaction mechanisms involve a similar anion intermediate.

Several groups discovered an α-amino-ε-caprolactam racemase that also racemizes α-amino acid amides (Asano and Yamaguchi, 2005; Boesten et al., 2003). These amides come from the classic Strecker synthesis of amino acids and are therefore key precursors of α-amino acids. Asano and Yamaguchi (2005) used this racemase in combination with a D-selective amidases in a dynamic kinetic resolution of the unnatural D-α-amino acids.

Both mandelate racemase (Felfer et al., 2005) and lactate racemase (Glueck et al., 2005) accept a range of α-hydroxy carboxylic acid substrate analogs. Their application in a dynamic kinetic resolution with a hydrolase would require their use in nonaqueous solvents. The substrate α-hydroxy carboxylic acid must be enantioselectively esterified at the carboxylate or acylated on the hydroxy group. Unfortunately, mandelate racemase is inactive in nonaqueous solvents, but lactate racemase has not yet been tested.

2.1.2.3 Racemization by Addition/Elimination

Reversible base-catalyzed addition of HCN to aldehydes formed racemic cyanohydrins (Inagaki et al., 1991; 1992). Enantioselective acetylation of the (S)-cyanohydrin catalyzed by PCL yielded the acetate in good to moderate yields and enantiomeric purity. In general, PCL showed higher enantioselectivity toward cyanohydrins derived from aromatic aldehydes than from aliphatic aldehydes.

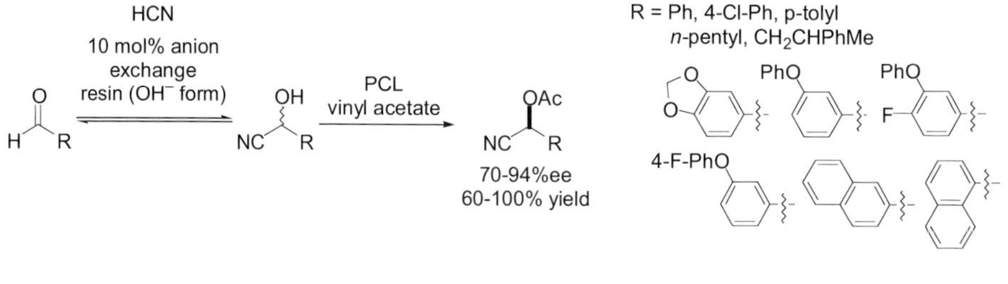

Fig. 13. Dynamic kinetic resolution of cyanohydrins via reversible addition of hydrogen cyanide to aldehydes.

CAL-B also shows high enantioselectivity toward cyanohydrins (Hanefeld et al., 2000). PCL did not catalyze acylation of the HCN donor, acetone cyanohydrin, a tertiary alcohol, presumably because it is too hindered (Fig. 13).

Veum et al. (2002), but dynamic kinetic resolutions with this lipase stopped at only 16% conversion. The culprit was water in the reaction mixture, which caused hydrolysis of the acyl donor (vinyl acetate) to form acetic acid. This acid neutralized the basic racemization catalyst and inactivated the enzyme. Adjusting the reaction conditions by using dry conditions, extra added base and a solid support that absorbs water strongly improved the reaction dramatically (Li et al., 2002; Veum et al., 2005). Under these optimized conditions, (S)-mandelonitrile acetate formed in 97% yield and 98% ee.

Similar reversible addition of thiols to aldehydes (catalyzed by silica gel) formed racemic hemithioacetals (Brand et al., 1995) (Fig. 14). Resolution using a PFL-catalyzed acetylation yielded precursors for nucleoside analogs.

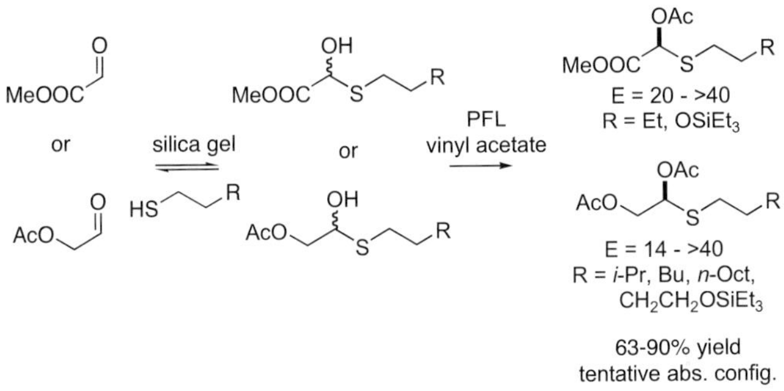

Fig. 14. Dynamic kinetic resolution of hemithioacetals via reversible addition of thiols to aldehydes.

Fig. 14 above included several examples of the related cyclic hemiacetals. These hemiacetals may racemize either via a protonation/deprotonation mechanism (most likely) or via an addition/elimination mechanism.

Another example of a dynamic kinetic resolution involving an addition/elimination mechanism is the reversible Michael addition to form the aryl isoxazoline combined with a PCL-catalyzed hydrolysis of the (R)-thioester, Fig. 15 (Pesti et al., 2001).

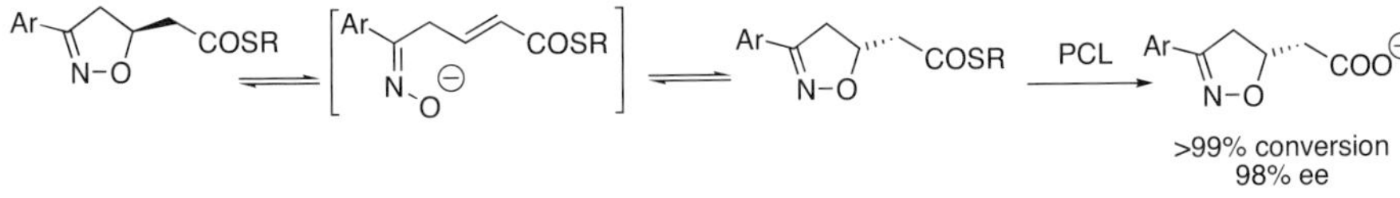

Fig. 15. Combination of a reversible Michael addition and lipase-catalyzed hydrolysis yields a DKR.

A major advance in dynamic kinetic resolutions was the discovery that lipases are compatible with some organometallic racemization catalysts. These catalysts can racemize a broad range of secondary alcohols so similar reaction conditions apply to a broad range of substrates. These organometallic racemization catalysts contain either palladium (for racemization of allylic acetates) or ruthenium (for racemization of secondary alcohols). The palladium catalysts follow an addition/elimination mechanism and

2.1 Kinetic Resolutions

are discussed below. The ruthenium racemization catalysts involve addition/elimination of hydrogen, which will be discussed in Sect. 2.1.2.5 below.

Palladium(0) or palladium(II) complexes add reversibly to allylic acetates to form a π-allyl palladium complex. This addition can racemize allylic acetates (Eq. 5). The details of the mechanism may differ depending of the substrate and reaction conditions (Granberg and Bäckvall, 1992) and rearrangement reactions are a possible side reaction. Enzyme-catalyzed hydrolysis or transesterification of allylic acetates yields allylic alcohols, which do not undergo this racemization.

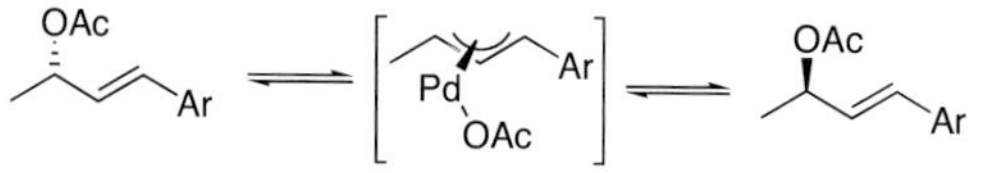

$$\text{Ar-CH=CH-CH(OAc)-} \rightleftharpoons [\text{Pd π-allyl OAc}] \rightleftharpoons \text{-CH(OAc)-CH=CH-Ar} \qquad (5)$$

Allen and Williams (1996) used the Pd(II)-catalyzed racemization in water for a dynamic kinetic resolution of a cyclic allylic acetate: 1-acetoxy-2-phenyl-2-cyclohexene (Fig. 16). Although slow racemization limited the rate of the reaction, both the yield and enantioselectivity were good.

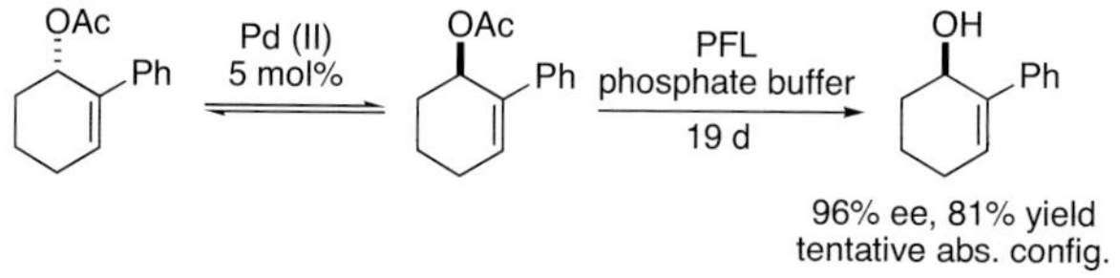

Cyclohexenyl-OAc-Ph → (Pd(II) 5 mol%) → Cyclohexenyl-OAc-Ph → (PFL, phosphate buffer, 19 d) → Cyclohexenyl-OH-Ph

96% ee, 81% yield
tentative abs. config.

Fig. 16. Dynamic kinetic resolution of an allylic acetate via palladium-catalyzed racemization followed by an enantioselective lipase-catalyzed hydrolysis of the acetate.

Similar dynamic kinetic resolution of another cyclic allylic acetate – 5-acetoxy-cyclohex-3-enecarboxylic acid methyl ester – gave the cyclohexenol in Fig. 17. Choi et al. (1999) extended these reactions to acyclic allylic acetates (Fig. 17) and decreased the reaction times from 11-19 d to 1.5-6 d by using a palladium(0) complex in organic solvent. To minimize formation of undesired rearrangement products, Choi and coworkers added the racemization catalyst after 45-50% conversion to minimize side reactions (elimination to form the diene or replacement of acetate by isopropanol). As an alternative, the corresponding acyclic allylic alcohols can also be racemized by oxidation-reduction using ruthenium catalysts, see Sect. 2.1.2.5 below. The final product is the acetate in one case, the alcohol in the other, but in both cases it is the same enantiomer.

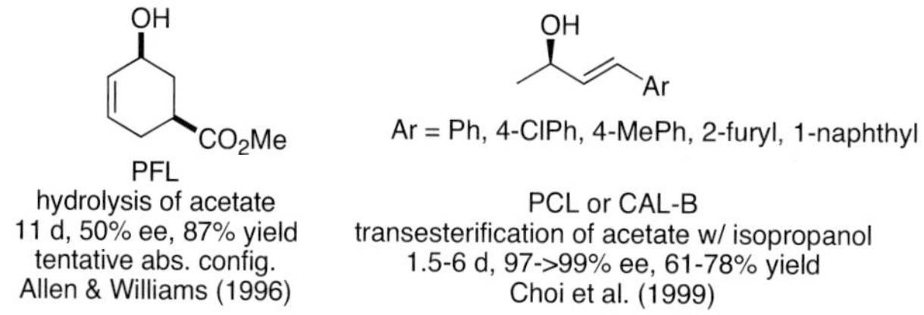

Cyclohexenol-CO₂Me
PFL
hydrolysis of acetate
11 d, 50% ee, 87% yield
tentative abs. config.
Allen & Williams (1996)

OH-CH-CH=CH-Ar
Ar = Ph, 4-ClPh, 4-MePh, 2-furyl, 1-naphthyl
PCL or CAL-B
transesterification of acetate w/ isopropanol
1.5-6 d, 97->99% ee, 61-78% yield
Choi et al. (1999)

Fig. 17. Dynamic kinetic resolution of allylic acetates yielded allylic alcohols. Palladium complexes catalyzed the racemization of the starting allylic acetates.

2.1.2.4 Racemization by Nucleophilic Substitution

An S_N2 displacement of halide by halide inverts the configuration at that center and can racemize chiral organic halides. Williams and coworkers used such a racemization in a dynamic kinetic resolution α-bromo and α-chlorophenylacetate esters with CRL (Jones and Williams, 1998; Haughton and Williams, 2001), Eq. 6. Surprisingly, the enantiopreference reversed for the bromide and chloride derivatives.

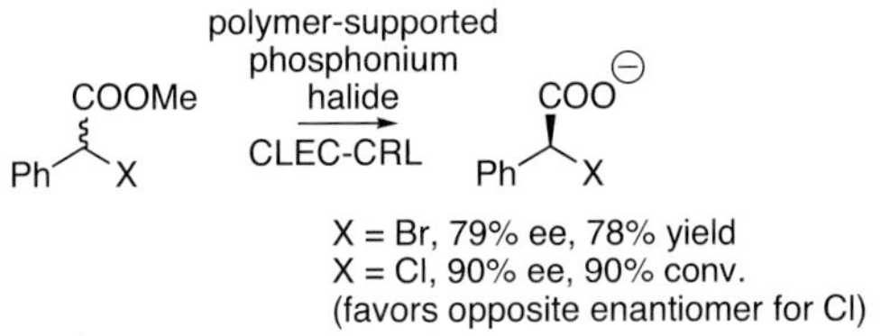

$$\underset{Ph}{COOMe}\underset{X}{\overset{\text{polymer-supported phosphonium halide}}{\xrightarrow{\text{CLEC-CRL}}}} \underset{Ph}{COO^{\ominus}}\underset{X}{} \qquad (6)$$

X = Br, 79% ee, 78% yield
X = Cl, 90% ee, 90% conv.
(favors opposite enantiomer for Cl)

Both chemical structure and the reaction conditions likely contributed to the different racemization rates for the starting ester (fast racemization) and product carboxylate (slow racemization). The authors suggest that S_N2 displacement is faster for the ester because the π* orbital of the ester carbonyl can accept electron density thereby stabilizing the transition state. One the other hand, the product carboxylate is electron rich carboxylate is less able to accept electron density. The use of the polymer-supported phosphonium halide as the racemization catalyst likely also enhanced the differences in racemization rates. The hydrophobic ester may partition near the polymer-supported racemization catalyst, while the carboxylate may favor the aqueous phase containing less halide ion.

2.1.2.5 Racemization by Oxidation/Reduction

Several ruthenium catalysts are compatible with hydrolases and catalyze the racemization of secondary alcohols. This racemization allows dynamic kinetic resolution of simple alcohols such as 1-phenylethanol, Fig. 18 (Dinh et al., 1996; Larsson et al., 1997).

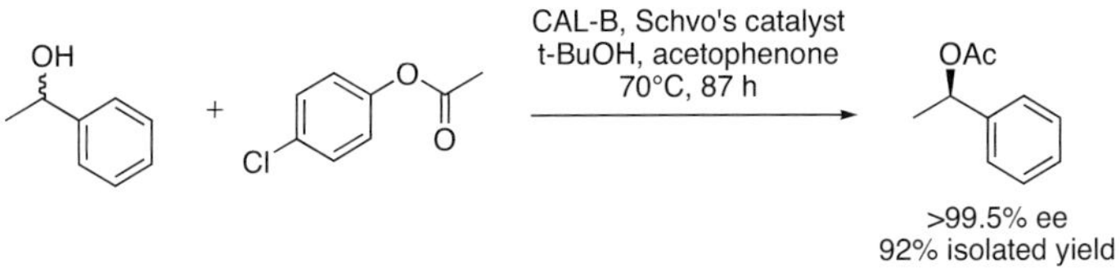

CAL-B, Schvo's catalyst
t-BuOH, acetophenone
70°C, 87 h

>99.5% ee
92% isolated yield

Fig. 18. Dynamic kinetic resolution of 1-phenylethanol with CAL-B yields the (R)-acetate.

These dynamic kinetic resolutions apply to a wide range of secondary alcohols including various 1-arylethanols, alkyl secondary alcohols, β-chloroalcohols (precursors of epoxides), β-azidoalcohols (precursors of aziridines and β-amino alcohols) and secondary alcohol diols, Tab. 3 (reviews: Huerta et al., 2001; Pamies and Bäckvall, 2003; 2004). All reactions use either PCL or CAL-B and thus the products are normally the acetates of the (R)-alcohol.

The racemization mechanism is an oxidation of the secondary alcohol to the ketone followed by reduction back to the alcohol, shown schematically in Fig. 19A. Shvo's catalyst, a ruthenium dimer shown in Fig. 19B, is a common first generation catalyst. The mechanism starts with a rate-limiting dissociation of the dimer to create the catalytically active species, Fig. 19C. This dissociation is require 70°C and 3-4 days for a typical dynamic kinetic resolution. This high temperature also prevents the use of subtilisin, which has an opposite enantiopreference for secondary alcohols and is needed to prepare the (S)-enantiomers of secondary alcohols using a dynamic kinetic resolution. Other first generation ruthenium catalyst require strong base in the reaction mixture, which can catalyze non-enantioselective chemical acylation and lower the enantiomeric purity of the product.

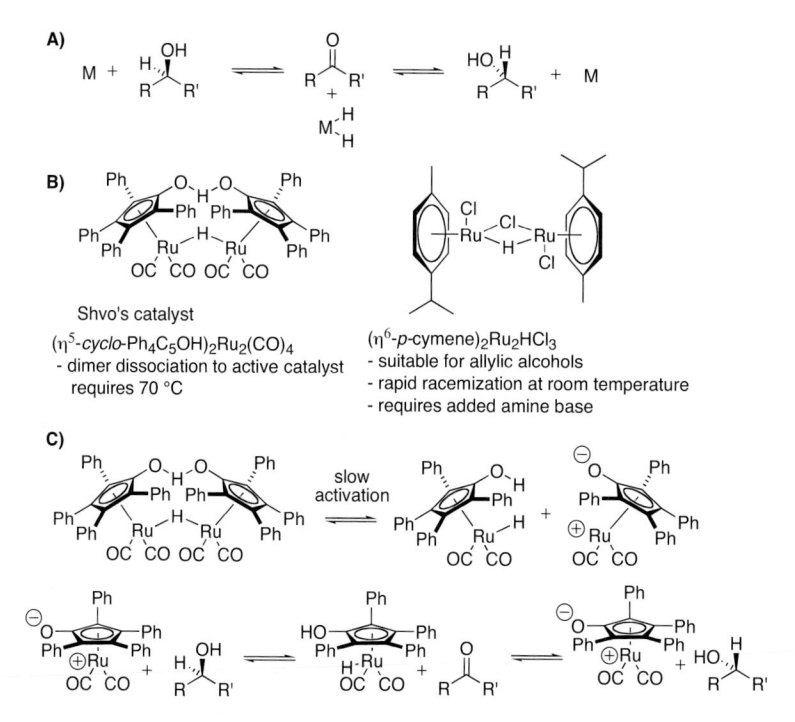

Fig. 19. Racemization of secondary alcohols by hydrogen transfer catalysts. A) A general mechanism showing a metal catalyst, M, oxidizes secondary alcohols to ketone B) examples of first generation ruthenium catalysts C) Mechanism of the racemization with Shvo's catalyst involves rate-limiting dissociation of the dimer to form the active catalyst and formation of free ketone.

The mechanism also involves a free ketone intermediate, which causes several disadvantages. First, oxidation of the secondary alcohol to the ketone is a possible side reaction. To minimize this side reaction, researchers add external ketone to shift the equilibrium. For simple secondary alcohols adding the corresponding ketone is a simple solution, but for more complex secondary alcohols, it can increase cost significantly. One alternative is not to use the secondary alcohol, but to start with the ketone (Jung et al., 2000).

Another consequence of ketone formation is the need for a special acyl donor. The normal acyl donors – vinyl acetate or isopropenyl acetate – form acetaldehyde or acetone after acetyl transfer. Reduction of these carbonyl compounds by the racemization catalyst yields an alcohol, which can also undergo acetylation and also promote oxidation of the substrate. A solution is 4-chlorophenyl acetate as the acetyl donor. This special acyl donor avoids formation of acetaldehyde, but removal of the byproduct 4-chlorophenol can be problematic.

Both the Kim and Park group (Choi et al., 2002; 2004) and the Bäckvall group (Martín-Matute et al., 2004; 2005) recently reported very similar second generation racemization catalysts that solve many of these limitations, Fig. 20. The catalysts differ by only a ring substituent – an isopropyl amino vs. a phenyl group.

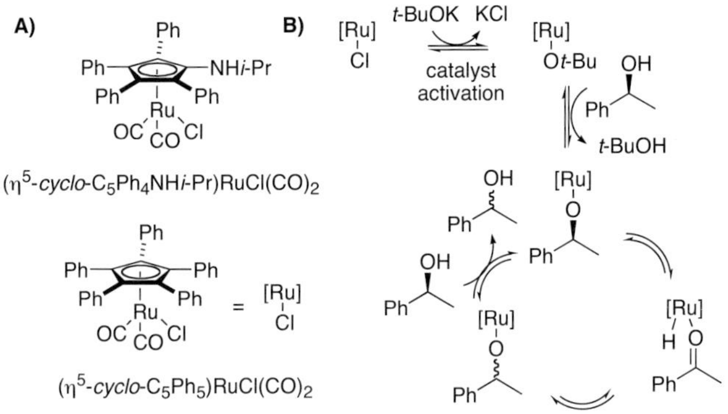

Fig. 20. Second-generation ruthenium hydrogen-transfer catalysts that racemize secondary alcohols by oxidation to the ketone followed by reduction. A) Second generation catalysts are monomeric so the racemization rate is not limited by dissociation of a dimer. B) The proposed mechanism of the second generation catalysts involves preactivation by potassium tert-butoxide and no dissociation of the intermediate ketone from the catalyst. This tight binding of the ketone to the catalyst allows the use of isopropenyl acetate as an acyl donor.

The second-generation catalysts are monomeric, thus eliminating the rate-limiting dissociation of the dimeric catalyts. The reactions are therefore faster and can be carried out at lower temperatures. Dynamic kinetic resolutions using subtilisin are now feasible thus permitting access to the (S)-secondary alcohols (Kim et al., 2003). The intermediate ketone does not dissociate, thus eliminating the side product of oxidation, the need to add ketone and the need for a special acyl donor.

Several issues still limit large scale applications of these dynamic kinetic resolutions involving ruthenium-catalyzed racemization. Although the racemization is fast, it slows considerably in the presence of enzyme, possibly due to coordination of the enzyme to the catalyst. The ruthenium complexes are sensitive to air, which makes handing more difficult. Finally, the subtilisin-catalyzed dynamic kinetic resolutions are still slower and less enantioselective that those with lipases.

A similar oxidation of primary amines to the imine followed by reduction should also racemize primary amines. Reetz and Schimossek (1996) demonstrated this concept in a palladium-catalyzed racemization of 1-phenylethylamine during a CAL-B catalyzed resolution. The reaction required a lot of catalyst (10 wt% Pd/C), ethyl amine as solvent and eight days reaction time and gave 75-77% yield with 99% ee. Unfortunately, no-one has extended this reaction to other amines.

Tab. 3. Examples of Dynamic Kinetic Resolution of Secondary Alcohols by Lipase-Catalyzed Acylation with in Situ Racemization using Ruthenium Hydrogen Transfer Catalysts.

Substrate Alcohol	Racemization Catalyst	Product Ester Isolated	Reaction Conditions	References
OH / Ph	$Rh_2(OAc)_4$, o-phenanthroline	OH / Ph; 98% ee, 60% conv.	PFL, acetophenone, vinyl acetate, 20°C, 3 d	Dinh et al. (1996)
OH / Ph, indanol-OH	Shvo's catalyst[a]	OH / Ph, indanyl-OH; >99.5% ee, 92% yield or 81% conv.	CAL-B, ketone, p-chlorophenyl acetate, toluene, 70°C, 3-4 d	Larsson et al. (1997)
OH / Ar, OH / Bn, indanyl-OH	(β5-indenyl)RuCl(PPh$_3$)$_2$[b]	OAc / Ar, OAc / Bn, indanyl-OAc; 82-99% ee, 60-98% yield	PCL, p-chlorophenyl acetate, NEt$_3$, O$_2$, 60°C, 43 h	Koh et al. (1999)
OH / Ar, OH / R, OH / R-or-Ar	Shvo's catalyst	OAc / Ar, OAc / R, OAc / R-or-Ar; 79->99% ee, 63-88% yield	CAL-B, p-chlorophenyl acetate, toluene, 70°C, 2-5 d	Persson et al. (1999a; b)
OH / =/R	(p-cymene)$_2$Ru$_2$HCl$_3$	OAc / =/R; 95->99% ee, 81-88% yield	PCL, p-chlorophenyl acetate, NEt$_3$, room temp., 2 d	Lee et al. (2000)
OAc / =/Ar, R, others	Shvo's catalyst	OAc / Ar, R, OAc / others; 47-99% ee, 74-91% yield	CAL-B, H$_2$ donor, toluene, 70°C, 2-5 d	Jung et al. (2000)
O / Ar, R, others	Shvo's catalyst	OAc / Ar, R, OAc / others; 47-99% ee, 71-89% yield	CAL-B, H$_2$ donor, ethyl acetate, 70°C, 2 d	Jung et al. (2000)

Tab. 3. Examples of Dynamic Kinetic Resolution of Secondary Alcohols by Lipase-Catalyzed Acylation (continued).

Substrate Alcohol	Racemization Catalyst	Product Ester Isolated	Reaction Conditions	References
MeO-C(O)-CH(OH)-Ar	Shvo's catalyst[a]	MeO-C(O)-CH(OAc)-Ar, 94–98% ee, 69–80% yield	PCL-C, p-chlorophenyl acetate, 60°C, 2-3 d	Huerta et al. (2000)
N₃-CH(OH)-CH₂-O-Ar	Shvo's catalyst	N₃-CH(OAc)-CH₂-O-Ar, 85–86% ee, 70–71% yield	CAL-B, p-chlorophenyl acetate, 60°C, 3 d	Pàmies and Bäckvall (2001)
N₃-CH(OH)-CH₂-Ar or Bn	Shvo's catalyst	N₃-CH(OAc)-CH₂-Ar or Bn, 96–>99% ee, 80–87% yield	CAL-B, p-chlorophenyl acetate, 80°C, 1-2 d	Pàmies and Bäckvall (2001)
EtO-C(O)-CH₂-C(O)-CH(OH)-R [R = Ph, 4-MeO-C₆H₄, Bn] Formed by an aldol addition in the same pot	Shvo's catalyst	EtO-C(O)-CH₂-C(O)-CH(OAc)-R, R = Ph, 4-MeO-C₆H₄, Bn, 95–99% ee, 69–75% yield	PCL-C, p-chlorophenyl acetate, 60°C, 6 d	Huerta and Bäckvall (2001)
Cl-CH₂-CH(OH)-Ar or Bn, Cl-CH₂-CH(OH)-CH₂-O-Ar	Shvo's catalyst	Cl-CH₂-CH(OAc)-Ar or Bn, Cl-CH₂-CH(OAc)-CH₂-O-Ar, 91–96% ee, 58–91% yield	PCL-C, p-chlorophenyl acetate, 70°C, 3 d	Pàmies and Bäckvall (2002)
OH-CH(R)-CH=CH-Ar	(β5-cyclo-C₅Ph₄NHCHMe₂)RuCl(CO)₂	OAc-CH(R)-CH=CH-Ar, 99% ee, >90% yield	CAL-B, p-isopropenyl acetate, 25°C, 1-7 d	Choi et al. (2002; 2004)
OH-CH(R)-C≡C-Ph	(β5-cyclo-C₅Ph₄NHCHMe₂)RuCl(CO)₂	OAc-CH(R)-C≡C-Ph, 92–99% ee, 74–90% yield	Subtilisin, trifluoromethyl butyrate, 25°C, 3-4 d	Kim et al. (2003)
OH-CH(Ar)-R	(β5-cyclo-C₅Ph₅)RuCl(CO)₂	OAc-CH(Ar)-R, 99% ee, >90% yield	CAL-B, p-isopropenyl acetate, 25°C, 3 h – 3 d	Martin-Matute et al. (2004; 2005)

[a] The racemization catalyst was (β5-Ph₄C₅OH)₂Ru₂H(CO)₄, Shvo's catalyst, typically 2-6 mol% catalyst. [b] The indene catalyst requires one catalyst equivalent of oxygen to oxidize one of the triphenylphosphines to triphenylphosphine oxide and to create an open coordination site for catalysis.

Two key side reactions of the imine intermediate are: water can add and hydrolyze the imine or another amine molecule can add and lead to reductive deamination (Choi et al., 2001). To minimize these side reactions, Choi et al. (2001) used dry reaction conditions and a ketoxime as the starting material. Hydrogen reduced the ketoxime to the amine slowly thus keeping the total concentration of amine low. These researchers tested eight examples 76-89% yield, 94-99% ee. A general dynamic kinetic resolution of amines is still an unsolved problem.

2.1.2.6 Related Strategies

Due to the difficulty of finding reaction conditions where both racemization and the resolution are facile, researchers have sometimes used a stepwise approach. For example, Strauss and Faber (1999) alternately used a *Pseudomonas* lipase-catalyzed acetylation of (±)-mandelic acid to the α-acetoxy acid in organic solvent with a mandelate racemase-catalyzed racemization of the remaining starting material in water (Fig. 21). This reaction must be done stepwise because the acetylation requires nonaqueous conditions, but mandelate racemase is only active in water. Four repetitions of this one-pot sequence yielded (*S*)-α-acetoxy mandelic acid the in 98% ee and 80% yield.

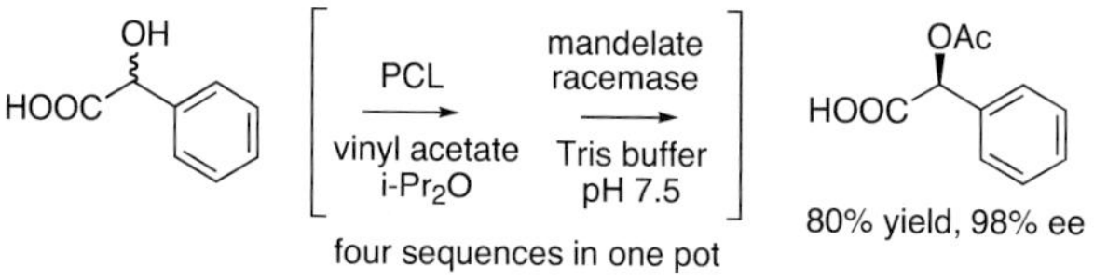

Fig. 21. Resolution of (±)-mandelic acid by a stepwise approach using lipase-catalyzed acetylation in organic solvent and mandelate racemase-catalyzed racemization in water.

Larissegger-Schnell et al. (2005) used a similar approach to prepare (*R*)- and (*S*)-2-hydroxy-4-phenylbutanoic acid, an analog of mandelic acid that is also a substrate for mandelate racemase. (*R*)-2-Hydroxy-4-phenylbutanoic acid (3) is a building block for angiotensin converting enzyme (ACE) inhibitors containing the (*S*)-homophenylalanine moiety.

A related strategy, although it is not a dynamic kinetic resolution, is to invert the configuration of one enantiomer (Mitsuda et al., 1988; Schneider and Goergens, 1992). Vänttinen and Kanerva (1995) resolved α-phenylethanol by PCL-catalyzed acetylation with vinyl acetate yielding a mixture of the (*R*)-acetate and the (*S*)-alcohol. Treating the mixture as shown in Eq. 7 converted the alcohol to the acetate while inverting the configuration. The net reaction was converting a racemic alcohol to the (*R*)-acetate.

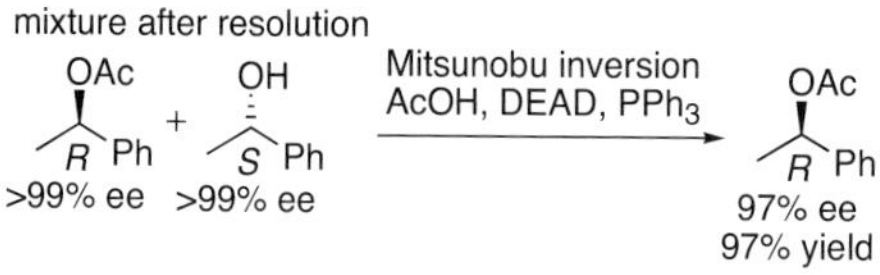

(7)

2.2 Asymmetric Syntheses

Lipase-catalyzed asymmetric syntheses start with *meso* compounds or prochiral compounds and yield chiral products in up to 100% yield. In an asymmetric synthesis the enantiomeric purity of the product remains constant as the reaction proceeds and is given by ee = (E-1)/(E+1), where E is the enantiomeric ratio. For example, an enantioselectivity of 50 yields product with 96% ee. Rearrangement of this equation gives E = (1 + ee)/(1 − ee), useful to calculate the enantioselectivity from the enantiomeric purity of the product.

In practice, however, many lipase-catalyzed asymmetric syntheses undergo a subsequent reaction, a kinetic resolution (Fig. 22). For example, hydrolysis of a *meso* diester first gives the chiral monoester, but this monoester also reacts giving the *meso* diol. Although this overhydrolysis lowers the yield of the monoester, it usually favors the minor enantiomer and thus increases the enantiomeric purity of the monoester by kinetic resolution. For the PPL-catalyzed hydrolysis of 1,5-diacetoxy-*cis*-2,4-dimethylpentane, the enantiomeric ratio for the diacetate to monoacetate hydrolysis was 16 yielding an enantiomeric purity of 88% ee (Wang et al., 1984). The subsequent kinetic resolution with an enantiomeric ratio of 5 increased the enantiomeric purity to 97% ee, but lowered the yield of monoacetate to ~70%. Quantitative analysis of the enantioselectivity in asymmetric syntheses is more difficult than for kinetic resolutions because three variables must be measured: the enantioselectivity of each step and the relative rate of each step. Wang et al. (1984) developed the necessary equations, but most researchers only report the enantiomeric purity and yield of the product. For this reason, we will also report only the enantiomeric purity and yield for asymmetric syntheses in this book.

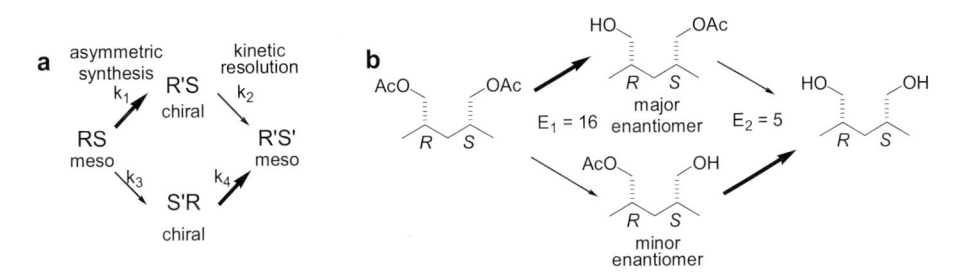

Fig. 22: Asymmetric synthesis are usually coupled to kinetic resolutions. **a** Schematic diagram; **b** PPL-catalyzed hydrolysis of 1,5-diacetoxy-*cis*-2,4-dimethylpentane.

Selected examples of lipase-catalyzed asymmetric syntheses are shown in Fig. 23; more are included in the survey of enantioselectivity in Sect. 5.2.1 and 5.2.2 and in an excellent review (Schoffers et al., 1996). The lipase-catalyzed asymmetric syntheses include a wide range of primary and secondary alcohols, as well as carboxylic acids. One example of the advantage of the combined asymmetric synthesis and kinetic resolution is the PCL-catalyzed acetylation of *cis*-2-cyclohexen-1,4-diol (Harris et al., 1991), a *meso*-secondary alcohol. Although the enantioselectivity for first acetylation (asymmetric synthesis) is only 4 and the enantioselectivity for the second acetylation (kinetic resolution) is only 10, the monoacetate was isolated in moderate yield (51%) and high enantiomeric

2.2 Asymmetric Syntheses

purity (95% ee). Many of the primary alcohol examples are 2-substituted 1,3-propanediols, which are versatile synthetic starting materials. The dihydropyridine example below is a chiral acid, but the acetyloxymethyl group places the stereocenter in the alcohol part of the ester; thus, this prochiral compound can be classified as a chiral alcohol.

meso secondary alcohols

51% yield, 95% ee
PCL, isopropenyl acetate
Harris *et al.* (1991)

PCL, E >100, vinyl acetate
Laumen & Ghisalba (1994)

CAL-B, E >50
vinyl acetate or
hydrolysis of diacetate
Johnson & Bis (1992)

meso primary alcohols

PPL, hydrolysis of diacetate
88 to >96% ee, 65-80% yield
Guanti *et al.* (1990c)

PCL, >98% ee, 92-100% yield
vinyl acetate
Tsuji *et al.* (1989), Itoh *et al.* (1993a)

prochiral 'alcohol' with
remote stereocenter

PFL (Amano AK), 97% ee, 31% yield
PCL, 88% ee, 21% yield
Holdgrün & Sih (1991), Ebiike *et al.* (1991)
Hirose *et al.* (1992, 1995), Salazar & Sih (1995)

PCL, vinyl acetate or
hydrolysis of dibutyrate
>99% ee, 81-94% yield
Tanaka *et al.* (1992)
Mohar *et al.* (1994)
also high E with ROL, CVL, RJL

PCL, 99% ee, 78% yield
vinyl acetate
Gais *et al.* (1992)

meso acid

PPL, 97% ee, 97% yield
hydrolysis of dimethyl ester
Nagao *et al.* (1989)

prochiral acid

PPL, 91% ee, 86% yield
hydrolysis of diester
Tamai *et al.* (1994)

prochiral acid with a remote stereocenter

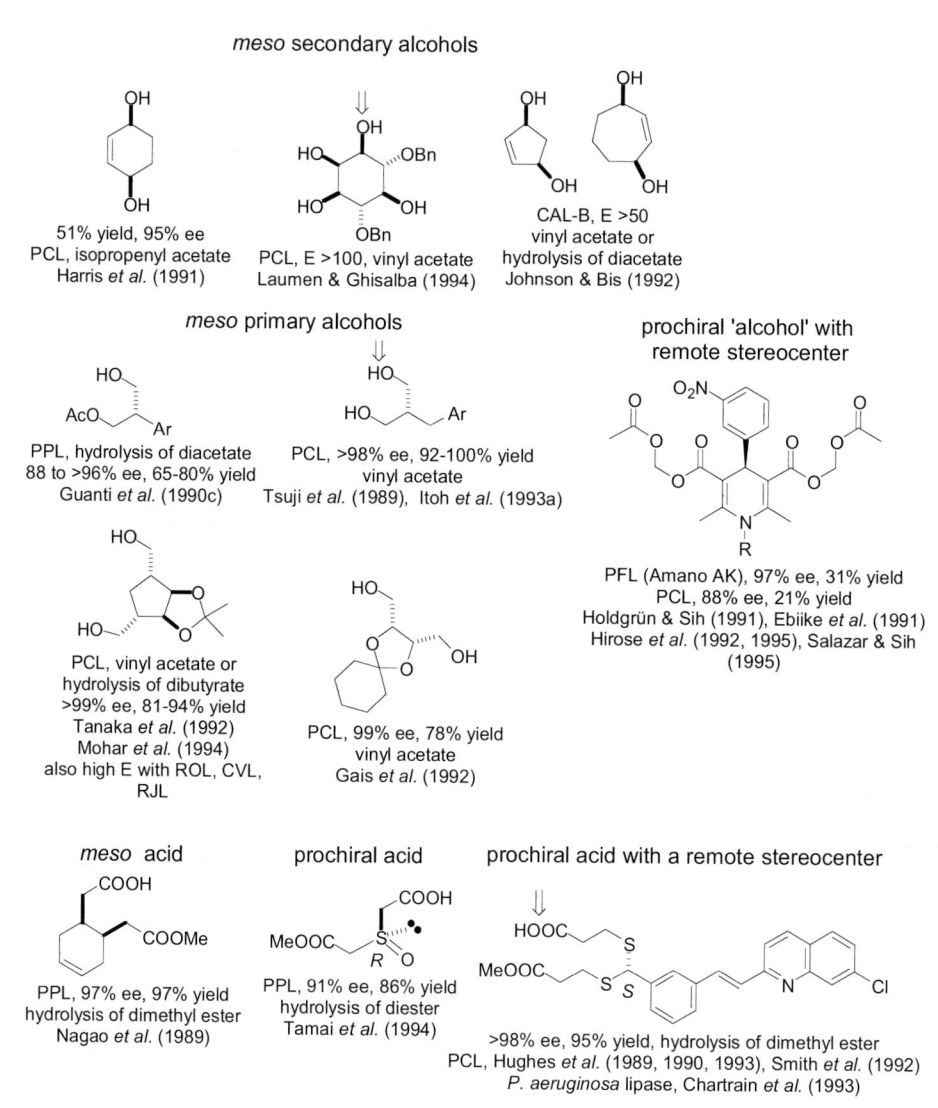

>98% ee, 95% yield, hydrolysis of dimethyl ester
PCL, Hughes *et al.* (1989, 1990, 1993), Smith *et al.* (1992)
P. aeruginosa lipase, Chartrain *et al.* (1993)

Fig. 23. Examples of lipase catalyzed asymmetric syntheses.

3 Choosing Reaction Media: Water and Organic Solvents

3.1 Hydrolysis in Water

The simplest hydrolase-catalyzed reaction is the hydrolysis of substrates in water or biphasic mixtures of water and an organic solvent. Although proteases require a soluble substrate, lipases and esterases do not. A second phase is even desirable because it activates most lipases by 10 to 100-fold, probably due to lid opening as discussed in 5.1.4. Liquid substrate can also act as the organic phase.

A general experimental procedure for a lipase-catalyzed reaction is as follows. Add 100 mg of liquid ester such as acetate or butyrate (or 1 mL of a solution of ester in a water-immiscible solvent such as toluene or ethyl ether) to 5 mL of 50 mM sodium phosphate buffer at pH 7. Monitor reaction by pH stat until reaction reaches ~30% conversion. If a pH stat is not available, monitor by TLC or GC and use 100 mM buffer. Work up reaction, measure enantiomeric purity of both starting material and product, and calculate E using Eq. 1 in Sect. 2.1. Roberts (1999) contains many detailed and tested procedures.

3.2 Transesterifications and Condensations in Organic Solvents

Researchers reported lipase-catalyzed esterifications in organic solvents containing approximately 10% water more than 50 years ago (Sperry and Brand, 1941; Sym, 1936), but most of this work was forgotten. In 1976, Unilever researchers patented a process for cocoa butter equivalent using a lipase-catalyzed transesterification of lipids in hydrocarbon solvents (Coleman and Macrae, 1977). Klibanov's group discovered many other examples of enzyme-catalyzed reactions in organic solvents and further demonstrated that enzymes require only traces of water (Cambou and Klibanov, 1984; Zaks and Klibanov, 1984; 1985; for reviews see: Klibanov, 1989; 1990; Koskinen and Klibanov, 1996). Klibanov's work convinced others that enzyme-catalyzed reactions are not only possible, but also sometimes more convenient, in organic solvents. Today, researchers report slightly more reactions in organic solvents than in water.

One advantage of reactions in organic solvents is the ability to do an esterification reaction instead of hydrolysis. Although lipases favor the same prochiral group in both cases, the two reactions yield opposite enantiomers. For example, acetylation of 2-benzyl glycerol with PCL yields the (*S*)-monoacetate, while hydrolysis of the diacetate with PPL yields the (*R*)-monoacetate. The lipases react at the *pro-R* position in both cases (Fig. 24).

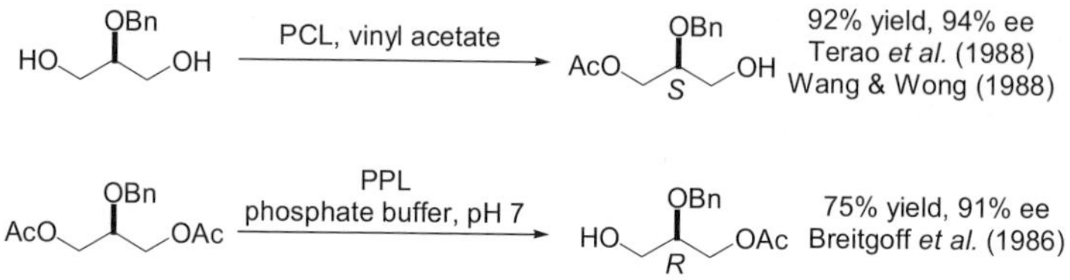

3.2 Transesterifications and Condensation in Organic Solvents

Fig. 24. Although lipases favor the same prochiral group in both cases, acylation of the *meso* alcohol and hydrolysis of the *meso* ester yield opposite enantiomers.

Another advantage of organic solvents is the potential to change the selectivity of the enzyme in different solvents, sometimes called solvent (or medium) engineering. For example, the regioselectivity of the transesterification of a 2-octyl-1,4-dihydroxybenzene butyric acid ester reversed from favoring the 4-position in cyclohexane to the 1-position in acetonitrile (Rubio et al., 1991) (Fig. 25).

Rubio et al. (1991) rationalized the reversal in selectivity according to differences in substrate solvation. Cyclohexane solvates the octyl group well, thus, the ester at the less hindered 4-position reacts. Acetonitrile solvates the octyl group poorly, thus the substrate binds in a manner that places the octyl group within the hydrophobic lipase active site. For this reason, the ester at the 1-position now reacts more rapidly.

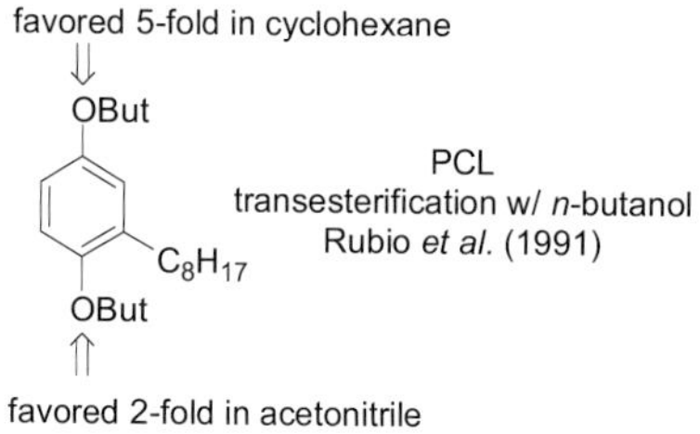

Fig. 25. Regioselectivity changes in different solvents.

Consistent with this explanation, Halling (1990) found that although the observed reaction rate can vary in different solvents, the true specificity constants of the enzyme vary only slightly after correcting for the activity of the substrate in different solvents. Note that substrate solvation changes do not explain changes in *enantio*selectivity, since both enantiomers are solvated equally in achiral solvents. However, solvation of enzyme-enantiomer complexes may differ and these differences may account for changes in enantioselectivity (see Sect. 3.2.2).

Another reason to avoid water is to prevent decomposition of water-sensitive compounds such as organometallics or to simplify the workup of hydrophilic compounds which are difficult to recover from water. Some organic chemists favor reactions in organic solvents simply because they are less familiar with reactions in water.

3.2.1 Increasing the Catalytic Activity in Organic Solvents

Enzymes suspended in organic solvents are less active than enzymes dissolved in water. For crystalline subtilisin, Schmitke et al. (1996) and Klibanov (1997) attributed this drop in activity to approximately equal contributions from (1) changes in the pH optimum, (2) changes in substrate solvation, and (3) low thermodynamic activity of water. Noncrystalline subtilisin is even less active, probably due to denaturation during lyophilization. Noone has done such a careful study for lipases, but additional possibilities for lipases are (1) diffusional limitations (ability of the substrate to reach the active site) and (2) lid orientation. To minimize diffusional limitations, researchers disperse the lipase on supports with high surface area or modify the lipase so that it dissolves in organic solvents. To ensure an open orientation of the lid, researchers add lipids or surfactants to the lipase. Salts hydrates or other techniques can optimize the water activity (Sect. 3.2.3) and organic phase buffers (Blackwood et al., 1994) or solid $KHCO_3$ (Berger et al., 1990) can control the pH. Addition of water in a manner that slows agglomeration of the enzyme particles increased the rate of reaction by approximately a factor of 10 for CRL (Tsai and Dordick, 1996). Heating the reaction mixture also accelerates the reaction. Several groups reported that heating the reaction mixture with microwaves is more effective than simple heating (Carrillo-Munoz et al., 1996; Gelo-Pujic et al., 1996; Parker et al., 1996).

3.2.1.1 Choosing the Best Organic Solvent for High Activity

Finding the best organic solvent for a enzyme-catalyzed resolution is still a trial and error process, but nonpolar solvents are usually better than polar solvents. Laane (Laane, 1987; Laane et al., 1987) divided solvents into three groups according to their $\log P$ value – the logarithm of their partition coefficient between n-octanol and water. Lipases, as well as other biocatalysts, showed low activity in solvents with $\log P$ values less than 2, which includes polar solvents like methanol, acetone, pyridine and diethyl ether. Biocatalytic activity was difficult to predict for solvents of moderate polarity ($2 < \log P < 4$) such as n-octanol, toluene and n-hexane. Biocatalysts are usually active in nonpolar solvents ($\log P > 4$) such as n-decane and diphenyl ether. Good solvents for lipase-catalyzed reactions include n-hexane, vinyl acetate, and toluene. Other researchers correlated activity with solvent parameters such as the dielectric constant or the dipole moment (Bornscheuer et al., 1993; Fitzpatrick and Klibanov, 1991), but $\log P$ gave the best correlation.

Researchers believe that enzymes must retain an essential shell of water to remain active in organic solvents. Nonpolar solvents do not affect this shell of water, while polar solvents strip this essential shell thereby inactivating the biocatalyst. Adding water to the polar solvent helps to retain activity (see Sect. 3.2.3 on optimizing the amount of water in an organic solvent). Clumping or aggregation of the enzyme upon addition of water can lower the activity.

3.2.2 Increasing the Enantioselectivity in Organic Solvents

Lowering the temperatures sometimes increases enantioselectivity. For example, Sakai et al. (1997) increased the enantioselectivity of a PCL-catalyzed acylation in diethyl ether from E = 17 at 30°C to E = 84–99 at –40°C. On the other hand, Yasufuku and Ueji (1995; 1996; 1997) increased the temperature to increase enantioselectivity. The enantioselectivity of a CRL-catalyzed reaction of 2-phenoxypropionic acid increased from E = 5 at 10°C to E = 33 at 57°C.

Changing from water to organic solvent often changes lipase enantioselectivity, as does changing from one organic solvent to another. Many researchers used this 'medium engineering' to optimize reactions in organic solvents. For example, Mori et al. (1987) reported that PPL showed no enantioselectivity in the hydrolysis of seudenol acetate, but Johnston et al. (1991) reported moderate enantioselectivity (E = 17) in the acetylation of seudenol with trifluoroethyl acetate in ethyl ether. The enantioselectivity of the CAL-B-catalyzed acetylation of seudenol with vinyl acetate varied from 8–32 depending on the solvent and the water content. The highest enantioselectivity was in dry benzene (Orrenius et al., 1995). Occasionally, the enantioselectivity even reverses upon changing the solvent. For example, CRL esterified the (R)-enantiomer of a chiral acid, 2-phenoxypropionic acid, with butanol in carbon tetrachloride (E = 16), but the (S)-enantiomer in acetone (E = 1.6) (Ueji et al., 1992). In another example, PCL acetylated the (R)-enantiomer of a secondary alcohol, methyl 3-hydroxyoctanoate, with vinyl acetate in methylene chloride (E = 5), but the (S)-enantiomer in hexane (E = 16) (Bornscheuer et al., 1993). In the most dramatic example, Amano lipase AH hydrolyzed the *pro-R* ester of a dihydropyridine derivative in cyclohexane (E~20), but the *pro-S* ester in diisopropyl ether (E > 100) (Hirose et al., 1992).

Finding the molecular basis of the enantioselectivity changes is difficult because energies involved are small. Rarely does the enantioselectivity change by more than a factor of ten, almost never by a factor of one hundred. A factor of ten corresponds to a $\Delta\Delta G^{\#}$ of 1.4 kcal/mol, a relatively weak interaction. Nevertheless, these enantioselectivity changes are often enough to tip the balance from a useless reaction to useful one. For this reason, many researchers have tried to understand why enantioselectivity changes in different solvents. They proposed at least four different explanations, but none can predict changes in enantioselectivity reliably (reviewed by Carrea et al., 1995).

First, Klibanov proposed that differences in the solvation of the enzyme-substrate transition state complex determine the selectivity (Ke et al., 1996; Rubio et al., 1991).

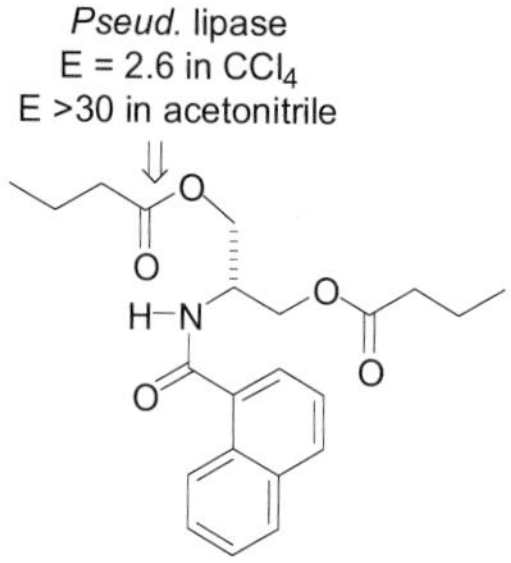

Fig. 26. Enantioselectivity changes in different solvents.

For example, they explained the variation in enantioselectivity for the lipase-catalyzed hydrolysis of a prochiral diester, Fig. 26 (Terradas et al., 1993).

They suggested that the substrate, especially the hydrophobic naphthyl group, binds tightly to the active site in polar solvents giving high enantioselectivity, but binds loosely, or not at all in nonpolar solvents giving low enantioselectvity. To estimate the differences in solvation the authors used the solvent log$P$ and found a good correlation with enantioselectivity. Other resarchers reported similar correlations (recent examples are given by Ema et al., 1996; Hof and Kellog, 1996), but Parida and Dordick (1991) found only partial correlation in a CRL-catalyzed esterification of 2-hydroxy acids and, in this case, the enantioselectivity increased in less polar solvents. Carrea et al. (1995) found no correlation of enantioselectivity and log$P$ in PCL-catalyzed acylation of several secondary alcohols. The lack of correlation suggests that log$P$ maybe a poor measure of enzyme-substrate solvation in these cases. Recently, Ke et al. (1996) used a more sophisticated approach to estimate solvation of the enzyme-substrate complex. They estimated the portion of the substrate bound to the enzyme and calculated the activity coefficient of this fragment. This approach predicts the enantioselectivity of crystalline proteases in organic solvents, but not for amorphous proteases in organic solvents.

Second, researchers suggested that solvent changes the active site by binding in it or near it (Hirose et al., 1992; Nakamura et al., 1991; Secundo et al., 1992). X-ray structures of protease crystals soaked in solvents indeed showed solvent molecules (hexane or acetonitrile) bound to the active site (Fitzpatrick et al., 1993; Yennawar et al., 1995). Ottolina et al. (1994) found that the activities of lipases (PCL, CVL, PPL, CRL, RML) increased by as much as a factor of eight in (R)-carvone as compared to (S)-carvone. The enantiomeric solvents presumably form different complexes with the lipase. However, the enantioselectivity toward secondary alcohols did not differ in the two solvents. Arroyo and Sinisterra (1995) reported small changes in enantioselectivity of CAL-B toward a carboxylic acid (ketoprofen) in (R)- vs. (S)-carvone.

Third, van Tol et al. (1995a; b) suggested that the changes in enantioselectivity are due to a combination of errors in measuring enantioselectivity and changes in solvation of the substrate. The endpoint method for measuring enantioselectivity can give erroneous results when product inhibits the reaction (Sect. 2.1). After correcting for inhibition and the thermodynamic activity of the substrates, van Tol et al. (1995a; b) found that the enantioselectivity of a PPL-catalyzed acylation of glycidol with vinyl butyrate did not change in hexane, diisopropyl ether, tetrachloromethane, and 2-butanone (E = 5.5). Without these corrections, the apparent enantioselectivity varied between 20 and 2.7 in a single reaction.

Fourth, researchers correlated the increased enantioselectivity of subtilisin in different solvents with the increased flexibility of the active site on the nanosecond time scale (Broos et al., 1995b). Noone has yet made such a correlation for lipases.

The enantioselectivity of serine proteases such as subtilisin and chymotrypsin toward amino acid esters drops from 10^3–10^4 in water to less than 10 in organic solvents (Sakurai et al., 1988). Sometimes the enantioselectivity even inverts, but remains low (Broos et al., 1995a). This lowered enantioselectivity permits coupling of D-amino acids.

3.2.3 Water Content and Water Activity

The amount of water in the reaction mixture strongly influences the reaction rate, and to a lesser extent the enantioselectivity, of enzyme-catalyzed reactions in organic solvent. Polar solvents require typically 1–3% added water for optimal activity, while nonpolar solvents require only 0.05–1% added water. Enzymes require this minimum amount of water to maintain their structure and flexibility (Affleck et al., 1992; Broos et al., 1995b; Rupley et al., 1983).

Initially, researchers optimized the amount of water for a lipase-catalyzed reaction in organic solvent by measuring the total water content with Karl-Fischer titration. More recently, researchers found that the thermodynamic water activity (a_w) is a better measure of the amount of water, especially when comparing different reaction conditions (Bell et al., 1995; Halling, 1990; 1994; 1996). For example, the optimum reaction rate for RML in solvents ranging from 3-pentanone to hexane occurred at the same thermodynamic water activity, $a_w = 0.55$, but at widely differing total water content (Valivety et al., 1992a). Similarly, optimal activity of RML immobilized on different supports occurred at $a_w = 0.55$, but at different total water content (Oladepo et al., 1994; 1995). Polar solvents require more added water than nonpolar solvents because polar solvents competed more effectively with the lipase for the available water. Bell et al. (1995) suggested that water activity is like temperature, while water content is like heat content. Two systems may have the same water activity or temperature, but at the same time differ in the water content or heat content.

Of the available methods for controlling the thermodynamic water activity (Tab. 4), the simplest are equilibration of the reaction components with salt solutions of known a_w or pairs of salt hydrate. Equilibration occurs, albeit slower, even without direct contact between the reaction components and the salt solutions or salt hydrates. The pairs of salt hydrates (e.g., $CuSO_4*5\ H_2O$ / $CuSO_4*3\ H_2O$ gives $a_w = 0.32$ at 25°C; $Na_4P_2O_7*10\ H_2O$ /$Na_4P_2O_7$ gives $a_w = 0.56$ at 35°C) act as water buffers taking up or releasing water to the reaction components. Halling (1992) lists water activity values for 48 salt hydrate pairs.

Tab. 4. Methods to Control Water Activity in Organic Media.

Method	Reference
Equilibrate with saturated salt solutions	Adlercreutz (1991); Bloomer et al., (1991); Goderis et al. (1987); Rosell et al. (1996); Valivety et al. (1992a)
Equilibrate with salt hydrates	Halling (1992); Kim and Choi (1995); Kvittingen et al. (1992)
Equilibrate with saturated salt solutions via silicone tubing	Wehtje et al. (1993; 1997)
Equilibrate with wet silica gel	Halling (1994)
Measure a_w with sensor and control drying of headspace	Khan et al. (1990)

It is more difficult to maintain constant water activity in reactions that consume or produce water (e.g., esterification between an acid and alcohol) because water must be removed or added during the reaction. Salt solutions or pairs of salt hydrates can be added directly to the reaction to buffer the water activity. Alternatively, preequilibrated silica particles of known a_w may be used (Halling, 1994). However, these approaches are not practical on a large scale due to cost, low water activity-buffering capacity, and difficulties in recovering the catalyst. One improvement is to add a silicon tube containing a saturated salt solution (e.g., $a_w = 0.75$ for water saturated with NaCl). The circulating salt solution can both take up and release water through the silicon (Wehtje et al., 1993; 1997). On an industrial scale, the best route maybe an a_w sensor (several are commercially available, Halling, 1994) combined with either drying by recirculation of the headspace gases through a drying column or water addition (Khan et al., 1990). These humidity sensors can also monitor the water activity continuously during reactions (Bornscheuer et al., 1993; Goldberg et al., 1988; 1990; Khan et al., 1990; Lamare and Legoy, 1995).

Lipases differ in the amount of water needed to maximize the rate of esterification between decanoic acid and dodecanol in hexane (Valivety et al., 1992b). RML and ROL were most active at low a_w (optimum: $0.32 < a_w < 0.55$), HLL, CRL, and PCL required a_w close to one. Sequence comparison of HLL and ROL suggested that changes in charged residues in the 'hinge and lid' region may be significant in low a_w tolerance. Different reactions may also have different optima. The overall rate of esterification between glycerol and oleic acid using RML did not change significantly with changes in water activity, but the synthesis of diolein from monoolein was fastest at $a_w = 0.5$ and triolein synthesis was fastest at low values of a_w (Dudal and Lortie, 1995). It may also be useful to change a_w as the reaction proceeds. For example, the initial phase of a PCL-catalyzed esterification of decanoic acid and dodecanol proceeds faster at high a_w, but at later stages a lower a_w is useful to obtain higher yields (Svensson et al., 1994).

Water activity also influenced the enantioselectivity of lipase-catalyzed reaction, but not in a consistent manner. The effect varied with solvent and also whether the chiral moiety was in the acyl or alcohol part of the substrate. For the resolution of 2-methyl alkanoic acids with CRL (Eq. 8), the enantioselectivity was higher at higher water activity (Berglund et al., 1994; Högberg et al., 1993). However, in several resolutions of secondary alcohols, water activity had little effect on enantioselectivity of CAL-B (Orrenius et al., 1995; Hansen et al. 1995) or PCL (Bovara et al., 1993). For example, the enantioselectivity in *n*-hexane was $E = 20$ at $a_w < 0.11$, but dropped to less than $E = 10$ at $a_w > 0.75$. However, for dichloromethane and *t*-amyl alcohol the enantioselectivity was independent of water activity and for vinyl acetate and 3-pentanone the enantioselectivity was higher at high water activity. This variation reflects the fact that the effects of different solvents on enantioselectivity are still not well understood (see Sect. 3.2.2).

$$n\text{-}C_6H_{13}\text{-}CH(COOH) + n\text{-}C_{12}H_{25}OH \xrightarrow[\text{salt hydrate pairs}]{\substack{\text{CRL}\\\text{cyclohexane}}} n\text{-}C_6H_{13}\text{-}CH(COOn\text{-}C_{12}H_{25}) + n\text{-}C_6H_{13}\text{-}CH(COOH) \quad (8)$$

a_w	E
no control	7
0.15	23
0.76	95

3.3 Other Reaction Media

3.3.1 Ionic Liquids

Room temperature ionic liquids are salts that do not crystallize at room temperature (review: Welton, 1999). The most common ionic liquids for biocatalysis are the imidazolium-based ionic liquids such as BMIM·BF$_4$, 1-butyl-3-methylimidazolium tetrafluoroborate (Fig. 27). Ionic liquids are possible 'green' replacements for organic solvents. Ionic liquids have no vapor pressure and therefore may be easier to reuse efficiently than organic solvents.

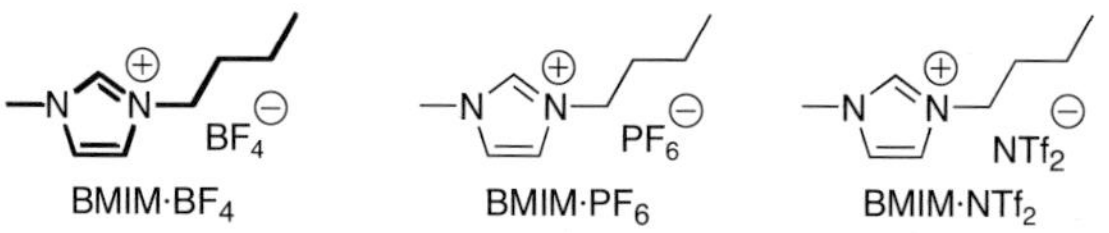

Fig. 27. Common ionic liquids in biocatalysis – 1-butyl-3-methylimidazolium salts.

The 1-alkyl-3-methylimidazolium ionic liquids are polar solvents. They are miscible with polar solvents like methylene chloride and immiscible with *n*-hexane and usually water. Reichardt's dye indicates that the polarity of common ionic liquids is similar to that of methanol, 2-chloroethanol, or *N*-methylformamide. Although these polar organic solvents inactivate enzymes (Chin et al., 1994; Park and Kazlauskas, 2001), the ionic liquids surprisingly do not, even though they have the same polarity. It appears that enzymes that work in organic solvents will also work in ionic liquids (Tab. 5).

The polar nature of ionic liquids increases the solubility of polar substrates, such as glucose, maltose, or ascorbic acid (Park and Kazlauskas, 2001; Park et al., 2003a), leading to faster reactions and changes in selectivity. For example, in CAL-B-catalyzed acylation of ascorbic acid with oleic acid in an ionic liquid, the conversion was higher (83%) when compared to the typical results in organic solvents (50%) (Park et al., 2003a). This improvement is due to the higher solubility of ascorbic acid in the ionic liquid. Several excellent recent reviews are available (Park and Kazlauskas, 2003; van Rantwijk et al., 2003; Kragl et al., 2002; Sheldon et al., 2002).

Other potential advantages of ionic liquids are increased stability or enantioselectivity. For example, α-chymotrypsin (Lozano et al., 2001b), CAL-B (Lozano et al., 2001a), and BSE (Persson and Bornscheuer, 2003) are 17, 3, and 30 times, respectively, more stable in ionic liquids than in organic solvents. Adding substrate further enhanced enzyme stability up to 2300-fold (Lozano et al., 2001a; b). The reasons for this increased stability are unclear. In some cases lipases and proteases were more enantioselective in ionic liquids than organic solvents (Kim et al., 2001; Schöfer et al., 2001; Zhao and Malhotra, 2002). In other cases, there was no change in enantioselectivity (Park and Kazlauskas, 2001; Persson and Bornscheuer, 2003) or even a 1.5- to 4.4-fold decrease (Itoh et al., 2001).

Since ionic liquids are nonvolatile, vacuum can remove volatile products such as alcohol and water to drive the equilibrium toward product formation. Itoh et al. removed methanol from CAL-B-catalyzed transesterification of a secondary alcohol using a methyl ester as an acyl donor (Itoh et al., 2002). Similarly, Bélafi-Bakó et al. (2002) used pervaporation to remove water in the CRL-catalyzed esterification of 2-chloropropionic acid with butanol. Ionic liquids are "tailored solvents" because it is easier to change their structure and thus their solvent properties than it is for normal organic solvents. For example, small changes in the alkyl substituent of the 1-alkyl-3-methylimidazolium cation increased the solubility of the substrate, thereby increasing regioselectivity or conversion (Park and Kazlauskas, 2001; Park et al., 2003a).

Tab. 5. Selected Hydrolases Active in Ionic Liquids.[a]

Enzyme	Reaction	References
Lipases		
CAL-B[b]	transesterification	Park and Kazlauskas (2001); Lau et al. (2000); Lozano et al. (2000); Kim et al. (2001); Itoh et al. (2001); Schöfer et al. (2001)
	perhydrolysis	Lau et al. (2000)
	ammoniolysis	Lau et al. (2000)
PCL[b]	transesterification	Park and Kazlauskas (2001); Kim et al.(2001); Nara et al. (2002)
	polyester synthesis	Nara et al. (2003)
Proteases		
Thermolysin	peptide synthesis	Erbeldinger et al. (2000)
α-Chymotrypsin	transesterification	Lozano et al. (2001b; 2003); Laszlo and Compton (2001); Eckstein et al. (2002)
Esterase		
BSE[b]	transesterification	Persson and Bornscheuer (2003)
Glycosidase		
β-Galactosidase	N-acetyllactosamine synthesis[c]	Kaftzik et al. (2002)

[a]For a more complete list see Kragl et al. (2002). [b]CAL-B, *Candida antarctica* lipase B; PCL, *Pseudomonas cepacia* lipase; BSE, *Bacillus stearothermophilus* esterase. [c]25% ionic liquid in water

As in organic solvents, enzymes did not dissolve in ionic liquids, but remained suspended as a powder. In a few cases, enzymes did dissolve in ionic liquids, but in these cases they were inactive (Erbeldinger et al., 2000; Lozano et al. 2001a).

Unlike organic solvents, which can be purified by distillation, ionic liquids must be purified by other methods. Purity of ionic liquids may be a key issue for biocatalysis. For example, several groups disagreed on whether an enzyme is active in a particular ionic liquid. Schöfer et al. (2001) reported no activity of CAL-B in BMIM·BF_4 or BMIM·PF_6, but other groups reported good activity for transesterification or ammoniolysis in the same ionic liquids (Lau et al., 2000; Kim et al., 2001; Itoh et al., 2001). Halide impurities may cause these inconsistencies. Many procedures start from 1-alkyl-3-methylimidazolium chloride or bromide, in which enzymes are inactive. If halide exchange with the desired anion is incomplete, the remaining halide salt may reduce or eliminate enzymatic activity.

Washing with water followed by vacuum drying purifies water-immiscible ionic liquids. However, the purification of water-miscible ionic liquids such as BMIM·BF_4 involved filtering through silica gel followed by washing with aqueous sodium carbonate solution. These purified ionic liquids worked reliably, but purification of ionic liquids still needs more research.

Since ionic liquids are more expensive than organic solvents (~800 times more in the Fluka catalog), a key to their industrial use will be their efficient recovery, product isolation, and reuse. Ionic liquids are much more viscous when compared to typical organic solvents. For example, BMIM·BF_4 has a viscosity similar to ethylene glycol. This higher viscosity makes handling (filtration, dissolution of solutes, etc.) more difficult. On a laboratory scale, researchers often ignore costs and greenness and simply use the most convenient method, such as extraction with an organic solvent or column chromatography. Several groups have also tested recycling of ionic liquids (Itoh et al., 2003) as well as separation methods that would be suitable on an industrial scale, such as extraction with benign solvents like water or supercritical CO_2 (Reetz et al., 2002), or distillation (Schöfer et al., 2001). The toxicity of ionic liquids has not been carefully investigated yet.

3.3.2 Reverse Micelles

Reverse micelles are the simplest way to run enzyme-catalyzed reactions in almost pure organic solvent, but researchers rarely use reverse micelles for preparative reactions due to difficulties in workup related to the surfactant. Reverse micelles consist of a bulk organic phase containing aqueous droplets stabilized by surfactant. The biocatalyst remains soluble and active in the water, while substrates and products dissolve in the organic phase (Fig. 28). The amount of aqueous phase is small, so that e.g., lipases can catalyze transesterification and ester synthesis reactions under these conditions. A second advantage is a large interfacial area between the micelles and the organic phase, which eliminates mass transfer limitations. These advantages simplify the kinetic analysis of lipases (Han et al., 1987; Stamatis et al., 1995; Walde et al., 1993) and other enzymes (Bommarius et al., 1995). Reverse micelles are also transparent and therefore suitable for spectrophotometric studies. Holmberg (1994) and Ballesteros et al. (1995) reviewed enzymic reactions in microemulsions.

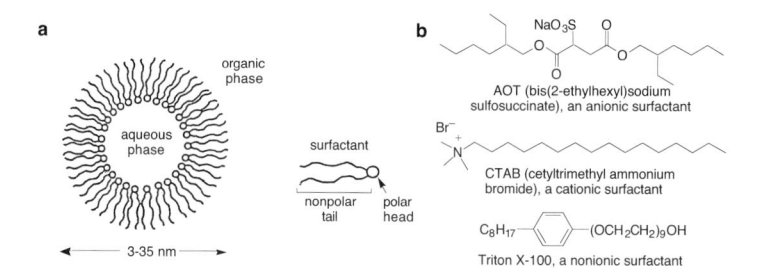

Fig. 28. Reverse micelles or water in oil microemulsions. **a** Reverse micelle contain an inner aqueous phase stabilized by surfactant in a bulk organic phase. **b** Surfactants used to stabilize reverse micelles include anionic (e.g., AOT), cationic (e.g., CTAB) and nonionic (e.g., Triton X-100) surfactants.

Anionic surfactants, in particular AOT, are best for lipase-catalyzed reactions because nonionic surfactants can inhibit lipases and can also react in transesterification reactions when the surfactant contains a free hydroxyl group (Skagerlind et al., 1992). A cationic surfactant (CTAB) decreased the maximal rate of a ROL-catalyzed hydrolysis of triolein by a factor of 50 compared to AOT (Valis et al., 1992).

For preparative work, reverse micelles have several disadvantages. First, recovery of products from surfactant-containing organic solvent can be difficult. To simplify recovery, researchers added gelatin to the aqueous phase. Simple filtration recovers the lipase-containing aqueous phase. Several groups used CVL to make simple esters (Backlund et al., 1995; Rees et al., 1991; 1993; 1995; Uemasu and Hinze, 1994) and CVL or *Pseudomonas* sp. lipase (Genzyme) to resolve secondary alcohols (de Jesus et al., 1995). CRL was inactive under these conditions. Interestingly, CVL within the gel remained active at −20°C (Rees et al., 1991). Another method to recover products from reverse micelles is to disrupt the emulsion with a temperature change into an oil-rich and a water-rich phase (Larsson et al., 1990).

Another disadvantage is that esterifications and transesterifications in reverse micelles often have lower yields than in other systems. For example, Borzeix et al. (1992) compared the RML-catalyzed synthesis of butyl butyrate in hexane, in a two-phase mixture of water-hexane, and in AOT-stabilized reverse micelles in hexane. The rate of ester synthesis was similar in all three systems, but the yield was significantly lower in the reverse micelles. Lipid modification in reverse micelles, especially synthesis of monoglycerides, also yielded less product than other reaction systems (Bornscheuer et al., 1994b; Chang et al., 1991; Hayes and Gulari, 1991; Holmberg et al., 1989; Singh et al., 1994a; b). In addition, the surfactants used to stabilize reverse micelles can also denature lipases, but optimizing the water-to-surfactant ratio can minimize denaturation (Fletcher et al., 1985; Han and Rhee, 1986; Kim and Chung, 1989; Valis et al., 1992).

Although some enzymes show 'superactivity' and changes in selectivity in reverse micelles (Martinek et al., 1982), lipases show only small changes in selectivity. Bello et al. (1987) noted that CRL, which normally shows little fatty acid chain length selectivity, favored longer chain lengths in the transesterification of triglycerides in reverse micelles. Hedström et al. (1993) reported significantly increased enantioselectivity of the CRL-catalyzed esterification of ibuprofen in reverse micelles as compared to hexane ($E > 100$ vs. 3) (Eq. 9). Many other treatments and reaction conditions also increase the enantioselectivity of this reaction (see Sect. 5.2.2.2).

$$\text{ibuprofen (COOH)} \xrightarrow[\text{hexane, E=3}]{\text{CRL, }i\text{-PrOH, AOT/hexane, E>100}} \text{COO}i\text{-Pr product} + \text{COOH product} \quad (9)$$

Lipase from *Penicillium simplicissimum* showed low selectivity (relative initial rates of 6–7) toward menthol enantiomers in reverse micelles (Stamatis et al., 1995).

3.3.3 Supercritical Fluids

As the temperature and pressure of a liquid are raised above the critical point, separate phases of liquid and gas disappear into a single phase called a supercritical fluid. Supercritical fluids have densities and dissolving powers near those of a liquid, but the viscosities near that of a gas. The advantages of supercritical fluids are rapid mass transfer due to the low viscosity, simple downstream processing by evaporation, the elimination of organic solvents, and the ability to change solvation properties by changing the pressure. The disadvantages of supercritical fluids are higher equipment costs and more complex reaction engineering.

Of the several possible supercritical fluids, most researchers use carbon dioxide because it is nonflammable, non-toxic, cheap, and reaches the supercritical state at low temperature (31.1°C). Moreover, its solvating properties are comparable to acetone. To dissolve polar substrates in supercritical carbon dioxide ($SCCO_2$), researchers either added a small amount of polar solvent such as dichloromethane, acetone, or t-butanol (Capewell et al., 1996) or they used techniques developed previously for organic solvents, e.g., complexation of fructose with phenyl boronic acid or immobilization of glycerol on silica gel (Castillo et al., 1994). Kamat et al. (1995) suggested that fluoroform may be a better supercritical fluid for enzyme-catalyzed reaction because carbon dioxide reacts with the lysine residues on an enzyme to make carbamates. In addition, supercritical fluoroform is a better solvent than $SCCO_2$.

Nakamura et al. (1986) first showed that lipases remain active in supercritical fluids. ROL catalyzed the interesterification of triolein and stearic acid to 8% conversion in $SCCO_2$. Since then researchers examined a wide range of reactions, especially lipid modifications (Tab. 6). The observed changes in conversion, enantioselectivity or lipase stability are similar to those in organic solvents. Several reviews of enzyme-catalyzed reactions in supercritical fluids have appeared (Aaltonen and Rantakylae, 1991; Ballesteros et al., 1995; Hammond et al., 1985; Nakamura, 1990).

Most reactions in Tab. 6 used immobilized RML in $SCCO_2$ at ~40°C and ~150 bar in batch systems and the research focused on optimizing the activity and stability of the lipase. For example, Marty et al. (1992) varied the water content to maximize the rate of esterification of oleic acid with ethanol. The optimum water amount increased upon addition of a small amount of ethanol to the reaction. As expected, Marty et al. (1992) observed no diffusion limitations, nor did Miller et al. (1990) in a similar reaction, but Bernard and Barth (1995) observed a partial diffusion-limitation. Conversion and residual activity of PCL were improved by adding molecular sieves to the reaction (Capewell et al., 1996), maximum conversion were influenced by pressure and temperature (Chi et al., 1988; Nakamura et al., 1986), initial rates were twice higher in $SCCO_2$ compared to n-hexane, which was attributed in part due to different solubility of the substrates in the two solvents (Marty et al., 1990; 1992).

Most comparisons suggest that the enantioselectivity of lipases in $SCCO_2$ is similar or slightly lower than in organic solvents (Capewell et al., 1996; Martins et al., 1992; Michor et al., 1996).

Pressure changes the solvating power of a supercritical fluid and several groups found that pressure changes the selectivity of a lipase. Ikushima et al. (1995) found that the enantioselectivity of a CRL-catalyzed acetylation of ($\pm$)-citronellol in $SCCO_2$ varied with pressure and suggested that pressure changes may change the conformation of the lipase. On the other hand, Rantakylae and Aaltonen (1994) found no changes in enantioselectivity for the RML-catalyzed esterification of ibuprofen with *n*-propanol. Chaudhary et al. (1995) controlled the molecular weight of polyester formed in a PPL-catalyzed transesterification of 1,4-butanediol and bis(2,2,2-trichloroethyl)adipate by changing the pressure of supercritical fluoroform. As the pressure increased, supercritical fluoroform dissolved longer polymer chains and the molecular weight of the product increased.

Batch supercritical reactors allow analysis only at the end of the reaction. To monitor while the reaction is in progress, Marty et al. (1990; 1992) used a reactor with a sampling loop and a saphire window for visual monitoring. To avoid taking samples, Bornscheuer et al. (1996) monitored formation of acetaldehyde in the acylation of a 3-hydroxy ester with vinyl acetate through a high-pressure flow-through cell at 320 nm. The on-line data agreed with off-line values up to 60% conversion.

Tab. 6. Examples of Lipase-Catalyzed Reactions in Supercritical Carbon Dioxide.

Lipase	Reaction	Process conditions[a]	References
ROL	triolein/stearic acid	B, 35β50°C, 150 bar	Chi et al. (1988); Nakamura et al. (1986)
RML	ethyl acetate/iso-amyl alcohol	C, 35β80°C, 80β140 bar, IPI	Doddema (1990); Janssens et al. (1992)
ROL	trilaurin/palmitic acid	B, 40°C, 90β290 bar	Erickson et al. (1990)
RML	oleic acid/ethanol	C, 33β50°C,110β170 bar,	Marty et al. (1990; 1992)
RML	myristic acid/ethanol	B, 50°C, 150 bar	Dumont et al. (1992)
ROL	trilaurin/myristic acid	SβC, 35°C, 79β107 bar	Miller et al. (1990)
RML	trilaurin/oleic acid methylester	C, 40°C, 100 bar, IPI	Adshiri et al. (1992)
RML	myristic acid/ethanol	B, 50°C, 125 bar	Bernard and Barth (1995)
RML, ANL	oleic acid/oleyl alcohol	B, 40°C, 84β167 bar	Knez and Habulin (1992)
RML	myristic acid/ethanol	B, 50°C, 125 bar	Bernard et al. (1992)
PPL	butyric acid/glycidol	B, 35°C, 140 bar	Martins et al. (1992)
RML	ethyl acetate/nonanol	C, 60°C, 125β200 bar	Vermuë et al. (1992)
Several	methacrylate/2-ethyl hexanol	B, 40β50°C, 107 bar[b]	Kamat et al. (1992; 1993)
RML	propyl acetate/geraniol	B, 40°C, 140 bar	Chulalaksananukul et al. (1993)
RML	e.g., fructose-PBA/oleic acid	B, 40°C, 150 bar	Castillo et al. (1994)
PPL	adipate/1,4-butanediol	B, 50°C, 60β340 bar[c]	Chaudhary et al. (1995)
CRL	oleic acid/citronellol	B, 31–40°C, 76β193 bar	Ikushima et al. (1993)
PCL	3-hydroxyester/vinyl acetate	B, 36β70°C, 100β170 bar	Bornscheuer et al. (1996); Capewell et al. (1996)
PCL	2°-alcohols/acylation	B, 40°C, 200 bar	Cernia et al. (1994a; b)
CRL, PME	menthol or citronellol /IA, TA	B, 35β70°C, 150 bar	Michor et al. (1996)
RML	ibuprofen	B, 50°C, 100β150 bar	Rantakylae and Aaltonen (1994)
CAL-B	randomization of e.g., palm olein	B, 65°C, 200β345 bar	Jackson et al. (1997)
CAL-B	soybean oil/glycerol or MeOH	B, 40β70°C, 200β345 bar	Jackson and King (1997)

[a] B: batch reactor; C: continuous reactor; S-C: semi-continuous reactor; IPI: integrated product isolation; PBA: phenyl boronic acid; IA: isopropenyl acetate; TA: triacetylglycerol. [b] Ethane, ethylene, fluoroform, propane, sulfur hexafluoride also. [c] Fluoroform also.

3.4 Immobilization

3.4.1 Introduction

Even if an enzyme is identified to be useful for a given reaction, its application is often hampered by instability under process conditions or difficult recovery and recycling. Immobilization can often overcome these limitations. Immobilization can enhance stability, enable repeated or continuous use, ease separation from the reaction mixture and modulate catalytic properties. Since the first uses of biocatalysts in organic synthesis, researchers have linked enzymes to solid carriers. Several recent reviews document examples for a broad range of enzymes for both aqueous solutions and organic solvents (reviews: Boller et al., 2002; Lalonde and Margolin, 2002; Bornscheuer 2003; a book: Cao, 2005). Despite many methodologies, a general, broadly applicable procedure for enzyme immobilization still needs to be discovered. The four most frequently used immobilization categories are: (1) non-covalent adsorption or deposition, (2) covalent attachment, (3) entrapment into a polymeric gel and (4) cross-linking of an enzyme. All these approaches are a compromise between achieving the advantages given above, while maintaining high catalytic activity.

3.4.1.1 Increasing the Surface Area to Increase Catalytic Activity

Enzyme powders are insoluble in organic solvents and can be recovered by simple filtration at the end of the reaction. Unfortunately, even after optimizing the solvent and the water content, catalysis is often thousands of times slower than in water or water-organic solvent mixtures. One reason for the drop in activity is diffusional limitations, that is, the substrate cannot reach the enzyme molecules in the center of the particle. Another reason for the drop in activity is denaturation of the biocatalyst during lyophilization. The simplest solution to both these problems is adsorbing the enzyme on an insoluble support such as Celite. Adsorption both increases the surface area and also avoids lyophilization of the biocatalyst. Most examples given below deal with the immobilization of lipases. However, the principles and most findings also hold true for other hydrolases.

Adsorption and entrapment
In a typical procedure, PCL (0.4 g) was dissolved in buffer (15 mL), mixed with insoluble support (4.0 g) and dried at room temperature (Bianchi et al., 1988; Inagaki et al., 1992; Kanerva and Sundholm, 1993). Reactions catalyzed by PPL absorbed on Celite were seven to twenty times faster than for crude PPL (Banfi et al., 1995). Sugars added to the buffer further increased the activity of immobilized PCL by a factor of 2–3. Dabulis and Klibanov (1993) also found similar rate increases for PCL, while Sanchez-Montero et al. (1991) found that the rate of heptyl oleate formation catalyzed by CRL depended on the type of carbohydrate. Rates increased upon adding lactose, but decreased upon adding fructose, glucose, sucrose or sorbitol. Sanchez-Montero et al. (1991) suggested these differences may be due to the ability of sugars to change the activity of water. Indeed, many commercial samples of lipases contain large amounts of

inert materials such as sugars or Celite (sometimes > 95 wt%), so the available surface area is already large. CRL adsorbed on Celite showed increased stability toward acetaldehyde, a product of esterifications with vinyl esters (Kaga et al., 1994). Adsorption of PCL on an acrylic resin (Amberlite XAD-8) also increased the catalytic activity > 200-fold (Hsu et al., 1990). Several adsorption-immobilized lipases are commercially-available from Novozymes: CAL-B immobilized by adsorption on macroporous acrylic resin and RML immobilized by adsorption on microporous phenolic anion exchange resin. Adsorption-immobilized lipases may desorb from the support in water; thus, covalently immobilized lipases (see below) should be used in water.

Because lipases often show higher catalytic activity in the presence of insoluble organic substrates, several groups have adsorbed or entrapped lipases in hydrophobic matrices. For example, hydrolysis of mixtures of tetramethoxysilane and alkyltrimethoxysilanes, $RSi(OCH_3)_3$, in the presence of lipases forms sol gel-entrapped lipases. Lipases entrapped in hydrophobic sol-gels also showed up to 100-fold increased activity in organic solvents (Reetz, 1997; Reetz et al., 1995; 1996a, b). Enantioselectivities remained unchanged in most cases. The sol-gel entrapped lipases are also easily recovered and reused with no loss in activity. Fluka awarded the Reagent-of-the-year-1997 prize to Manfred Reetz for his discovery of the sol-gel immobilized lipases. Other workers found smaller improvements in similar systems (Kawakami and Yoshida, 1995; Sato et al., 1994). Sol-gel immobilized lipases also work in aqueous solutions (Reetz et al., 1995). The mechanism for activation is probably lid opening as suggested for the lipid- or surfactant-coated lipases.

A simple and elegant immobilization method is the formation of protein-coated microcrystals (PCMC's, Kreiner et al., 2001). A solution of an aqueous protein solution and an excipient such as a salt, a sugar or an amino acid is added drop-wise to a rapidly mixing water-miscible organic solvent. This addition dilutes the water and co-precipitates the protein with excipient. These PCMC's can be used efficiently in non-aqueous media. For PCMC from subtilisin Carlsberg with excipient K_2SO_4 precipitated in ethanol a 1000-fold rate improvement in the transesterification of N-AcTyrOET with 1-propanol was observed. For lipase from *Pseudomonas* sp. a 200-fold and for CAL-A a 60-fold enhancement were reported without changes in enantioselectivity.

Covalent Immobilization

Covalent immobilization creates a more stable link between the enzyme and the support, but requires more effort than adsorption (reviews: Akita, 1996; Balcao et al., 1996). The most common method is the cross-linking of adsorbed biocatalysts with glutaraldehyde. Several enzymes from the Chirazyme series were immobilized by this method. Other examples include linking to an epoxy-containing resin (Berger and Faber, 1991), to polystyrene via a cysteinyl-*S*-ethyl spacer (Stranix and Darling, 1995) or via a polyethylene glycol linker (Ampon et al., 1994), or entrapping the lipase in urethane prepolymers (Koshiro et al., 1985). Covalent immobilization often increases the thermal or operational stability of the enzyme, but does not activate it.

For example, lipase from *Pseudomonas fluorescens* was immobilized on four different carriers (Fernández-Lafuente et al., 2001). The native enzyme and two carrier-linked lipase preparations show no or only modest changes in activity and enantioselectivity in the kinetic resolution of a racemic carboxylic acid ethylester. However, two immobilisates exhibited substantially altered properties (Tab. 7). Specific activity was increased 10-fold and enantioselectivity increased from $E = 7$ to $E = 86$ for the lipase adsorbed on decaoctyl-sepharose. The authors claim, that during this (also much more rapid) immobilization procedure, the lipase underwent a conformational change from the closed to an open structure, as the hydrophobic 'lid' – known to be present in most lipases – moves aside by an interfacial activation caused by the carrier and immobilization procedure, providing enhanced substrate access to the active site residues. With a similar strategy, the same group also reported modulation of the properties of penicillin acylases from three different species, which also underwent conformational changes upon binding of the acyl donor substrate (Terreni et al, 2001).

Tab. 7. Comparison of different 'controlled' immobilization techniques for a lipase from *Ps. fluorescens* in the resolution of a racemic carboxylic acid ester (Fernández-Lafuente et al., 2001).

Carrier	Activity [U/mg][a]	Enantiomeric excess[b] [% ee$_S$]	E [ ]
None	0.5	59	7
Dextran-agarose	0.45	58	6.9
Glyoxyl-agarose	0.35	74	20
IHA-Octyl-agarose[c]	3.5	92	79
IHA-Decaoctyl-sepharose[c]	3.4	92.5	86

[a] μmol substrate hydrolyzed per minute per mg of immobilized protein
[b] of remaining (R)-ester
[c] IHA, interfacial hydrophobic adsorption

Cross-Linked Enzyme Crystals – CLEC's

Cross-linked enzyme crystals are crystals (typically 1–100 μm diameter) of pure enzyme cross-linked with glutaraldehyde (Margolin, 1996; St.Clair and Navia, 1992). Several hydrolases including two lipase CLECs were commercially-available from Altus Biologics Inc. (Cambridge, MA, USA): lipase from *Candida rugosa* (CLEC-CR) and *Pseudomonas cepacia* (CLEC-PC) (Lalonde, 1995; Lalonde et al., 1995). Lipase CLEC remain active and insoluble in organic solvents and, unlike adsorbed lipases, in water-organic solvent mixtures. In addition, CLEC lipases are more stable to high temperatures and retain their activity over many cycles of use. For example, Lalonde et al., 1995 reused CLEC-CR eighteen times (recovering 20% of the activity), whereas crude CRL lost virtually all activity after one cycle. Mechanical losses and lipase inactivation contributed approximately equally to the loss in CLEC-CR activity. CLEC-CR also showed improved enantioselectivity toward 2-arylpropionic acids; for ketoprofen the enantioselectivity increased from 5.2 to 64 (Lalonde et al., 1995; Persichetti et al., 1996). Lalonde et al., 1995 attributed the increase to the removal of a nonselective esterase during purification, but a conformational change may also contribute. CLEC's prepared from different conformations of CRL (open vs. closed) also differed in their enantioselectivity. Coating the crystals with surfactants increased the activity of the lipase in organic solvents (Khalaf et al., 1996). After taking into account the protein content, the activity was 2–90 times greater than crude lipase. The surfactant may maintain the water balance or it may facilitate transfer of hydrophobic substrates through the tightly bound layer of water. Two protease CLEC's were also available from Altus Biologics Inc.: CLEC-BL, derived from subtilisin A and CLEC-TR, derived from thermolysin. However, CLEC's are no longer available.

Cross-Linked Enzyme Aggregates – CLEA's

A much simpler and cost-effective alternative to CLEC's is the cross-linking of enzyme aggregates. In this procedure, the enzyme is precipitated from an aqueous solution by adding a salt or a water-miscible organic solvent or polymer (e.g., polyethylene glycol). In a subsequent step, the physical aggregates are cross-linked with a bifunctional agent such as glutaraldehyde. A major advantage is, that – in contrast to other methods using carrier-bound enzymes – the dilution of catalytic activity resulting from the introduction of a large proportion of non-catalytic mass (often 90–99% of total mass) is avoided. Thus, higher volumetric activity and space-time yields are possible as the molecular weight of the bifunctional cross-linking agent is negligible. CLEA's from penicillin G acylase had the same activity in the synthesis of ampicillin as CLEC's from the same enzyme, but the synthesis over hydrolysis ratio was better. In addition, the CLEA also catalyzed the reaction in a broad range of organic solvents (Cao et al., 2000; 2001). The same concept was later successfully applied to various other enzymes (Mateo et al., 2004) and the physical structure of the CLEA's was determined (Schoevaart et al, 2004). For lipases it could be shown that careful choice of the precipitant can be crucial to achieve high activity. Precipitation of lipases from *Thermomyces lanuginosa* and RML with ammonium sulfate in the presence of SDS and cross-linking with glutaraldehyde gave CLEA's with 3- to 2-fold enhanced hydrolytic activity compared to the native enzymes. Some preparations gave 10-fold enhanced activity in organic media (López-Serrano et al., 2002). Further applications have been published by Wilson et al. (2004a; b).

Covalently-Modified Enzymes Soluble in Organic Solvents

Another way to increase the activity in organic solvents is to modify the enzyme so that it dissolves in the organic solvent. Kikkawa et al. (1989) coupled polyethylene glycol (PEG) to the free amino groups of PCL and lipase from *Pseudomonas fragi*. The modified lipases were soluble in benzene, toluene and chlorinated hydrocarbons and catalyzed the formation of lactones from ethyl 16-hydroxyhexadecanoate and the resolution of 2-phenylethanol. Addition of hexane precipitated the PEG-lipase thereby simplifying recovery (Kodera et al., 1994). Using the same principle, Kodera et al. (1994) coupled PFL to a comb-shaped polymer yielding a more stable and more active lipase. PEG-modified CRL was significantly more stable in organic solvents than unmodified CRL (Basri et al., 1995; Hernáiz et al., 1996). Only PEG-modified CAL-B was found to be active in sugar ester synthesis in pure ionic liquids, commercial immobilisate gave no conversion (Ganske and Bornscheuer, 2005).

4 Protein Sources and Optimization of Biocatalyst Performance

4.1 Accessing Biodiversity

The traditional method to identify new enzymes is screening of environmental samples (e.g. soil samples) or strain collections by enrichment culture (Asano, 2002; Ogawa and Shimizu, 2002). Once a suitable biocatalyst is identified, strain improvement as well as cloning and expression of the encoding gene enables production on a large scale. Unfortunately, only 0.001-1% of the total number of microorganisms grow under common culture conditions (Lorenz and Eck, 2004; Lorenz et al., 2003; Miller, 2000). Thus, more than 99% of the biodiversity escaped using these approaches.

Two new strategies seek to include these unculturable microorganisms in the search for biocatalysts: (i) the metagenome approach and (ii) sequence-based discovery.

The metagenome approach directly extracts, clones and expresses DNA from uncultivated microbial consortia such as soil samples. Microbial cells in the sample are lysed to yield high molecular weight DNA, which is purified and cloned by standard procedures into cultivable host cells like *Escherichia coli*. Growing these host cells to express the cloned proteins is followed by screening or selection to identify distinct enzymatic activities (Handelsman, 2004; 2005; Lorenz and Eck, 2004; Short, 1997; Uchiyama et al., 2005). The major advantage of this approach is the ability to discover new subclasses of enzymes. The classes can show a broad evolutionary diversity thereby increasing the chances to find biocatalysts with unique properties. In addition, the identified enzymes are already recombinantly expressed and thus in principle available at large scale. The disadvantage is that the host cells may not express some of the biocatalysts and thus these will be missed.

One impressive example of the metagenome approach is the discovery of >130 novel nitrilases from more than 600 environmental DNA libraries obtained from diverse biotopes (Robertson et al., 2004). This number is much larger than the twenty previously known nitrilases, which were isolated by classical cultivation methods (see Sect. 9.4).

Another alternative is sequence-based discovery. This is increasingly attractive as data from sequencing singular genes, whole genomes and even biotopes are collected in public databases. Searching these databases with known (nucleotide or amino acid) sequences encoding the enzyme of interest or even using consensus sequence parts only results in identification of homolog proteins. Once new sequences are identified, the cloning of the encoding genes is straight-forward either by a PCR-based approach amplifying known open reading frames or by the introduction of necessary mutations in already cloned homologous enzyme genes. Useful databases can be found at PubMed for sequence retrieval (*http://www.ncbi.nlm.nih.gov/entrez/query.fcgi*) and alignment (*http://www.ncbi.nlm.nih.gov/BLAST/*). A specific database (Pleiss et al., 2000) has been developed (*http://www.led.uni-stuttgart.de*) for lipases, esterases, epoxide hydrolases and dehalogenases.

For instance, a gene (*yvak*) encoding an esterase from *Bacillus subtilis* showing 74% identity and 95% homology to an esterase from *B. stearothermophilus* was discovered improved variants. For non-recombining mutagenesis of an enzyme of 200 amino acids (N = 200), the number of possible variants is $19^M[N!/(N-M)!M!]$, where M is the number of substitutions. Thus, two random mutations create 7 million variants: with three or more substitutions, the creation and screening of a library becomes very challenging (Tab. 8).

Tab. 8. Sequence Space of Possible Variants for a Protein Consisting of 200 Amino Acids at a Given Number of Substitutions.

Substitutions (M)	Number of variants (sequence length N=200)
1	3 800
2	7 183 900
3	9 008 610 600
4	8 429 807 368 950

The most prominent method for the creation of libraries is the error-prone polymerase chain reaction (epPCR), which introduces approximately one mutation per 1000 base pairs (Cadwell and Joyce, 1992). This is achieved by changing the reaction conditions, i.e. use of Mn^{2+} salts instead of Mg^{2+} salts (the polymerase is magnesium-dependent), use of the *Taq* polymerase from *Thermomyces aquaticus*, and variations in the concentrations of the desoxynucleotides. Another approach utilizes mutator strains, e.g. the *Escherichia coli* derivative *Epicurian coli* XL1-Red, lacking DNA repair mechanisms (Bornscheuer et al., 1998; Greener et al., 1996). Introduction of a plasmid bearing the gene encoding the protein of interest leads to mutations during replication. Both methods introduce point mutations and several iterative rounds of mutation followed by identification of best variants are usually required to obtain a biocatalyst with desired properties.

Alternatively, methods of recombination (also referred to as sexual mutagenesis) can be used. The first example was the DNA (or gene-) shuffling developed by Stemmer, in which DNase I degrades the gene followed by recombination of the fragments using PCR with and without primers (Stemmer, 1994a; b). This process mimics natural recombination and has been proven in various examples as a very effective tool to create desired enzymes. More recently, this method was further refined and termed DNA family shuffling or molecular breeding enabling the creation of chimeric libraries from a family of genes.

Several other groups also developed recombination methods of mutagenesis. The Arnold laboratory developed the staggered extension process (StEP). This method uses a PCR protocol with short reaction times for annealing and polymerization. Truncated oligomers dissociate from the template and anneal randomly to different templates leading to recombination. Several repetitions allow the formation of full-length genes (Zhao et al., 1998). Other methods are ITCHY (incremental truncation for the creation of hybrid enzymes) and related approaches (Lutz et al., 2001; Ostermeier et al., 1999). Tab. 9 provides an overview of methods, more details and comparisons of different strategies for the creation of mutant libraries can be found in reviews (Kurtzman et al., 2001; Neylon, 2004).

Tab. 9. Selected Methods to Create Mutant Libraries for Directed Evolution (Kurtzman et al., 2001; Neylon, 2004).

Method	Pros	Cons	References
Error-prone PCR (ep-PCR)	easy to perform, mutation rate adjustable	only point mutations accessible	Cadwell and Joyce (1992)
Mutator strains	easy to perform	entire organism / plasmid is mutated only point mutations accessible	Bornscheuer et al. (1998); Greener et al. (1996)
DNA-shuffling	modest sequence homology sufficient several parent genes can be used creation of chimeras possible useful mutations are combined, harmful ones lost	requires sequence homology	Crameri et al. (1998); Ness et al. (1999a); Stemmer (1994a)
StEP	similar to DNA-shuffling, more simple no fragment purification necessary	requires sequence homology PCR protocol must be specifically adapted	Zhao et al. (1998)
SHIPREC	no sequence homology required	low diversity library in single round (might be repeated) limited to two parents of similar length deletions / duplications possible	Sieber et al. (2001)
ITCHY	similar to SHIPREC	similar to SHIPREC	Ostermeier et al. (1999)
THIO-ITCHY	similar to ITCHY, but more efficient / easier	similar to ITCHY	Lutz et al. (2001)
GSSM	all single amino acid substitutions are covered	technically out of reach for most researchers	DeSantis et al. (2003)
SeSaM	Complete coverage at selected sites	Sites to be saturated should be known	Wong et al. (2004)

4.2.1.2 Assay Systems

The major challenge in directed evolution is the identification of desired variants within the mutant libraries. Suitable assay methods should enable a fast, very accurate and targeted identification of desired biocatalysts out of libraries comprising 10^4-10^6 mutants. In principle, two different approaches can be applied: screening or selection.

Selection

Selection-based systems have been used traditionally to enrich certain microorganisms. For in vitro evolution, selection methods are less frequently used as they usually can only be applied to enzymatic reactions, which occur in the metabolism in the host strain. On the other hand, selection-based systems allow a considerably higher throughput compared to screening systems (see below). Often, selection is performed as a complementation, e.g., an essential metabolite is produced only by a mutated enzyme. For instance, a growth assay was used to identify active esterase variants (Fig. 30). Mutants of an esterase from *Pseudomonas fluorescens* (PFE) produced by directed evolution using the mutator strain *Epicurian coli* XL1-Red were assayed for altered substrate specificity using a selection procedure (Bornscheuer et al., 1999).

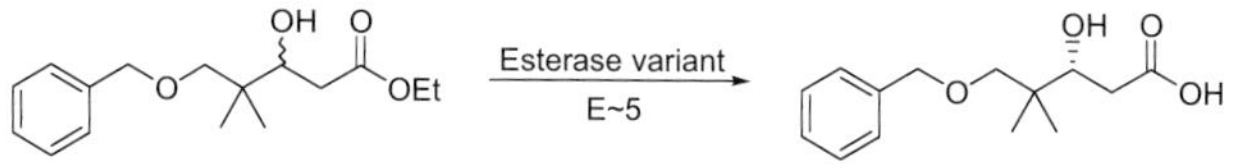

Fig. 30. Expanding the substrate range of an esterase from *Pseudomonas fluorescens* by using a mutator strain.

Key to the identification of improved variants acting on a sterically-hindered 3-hydroxy ester – which was not hydrolyzed by the wild-type esterase – was an agar plate assay system based on pH-indicators, thus leading to a change in color upon hydrolysis of the ethyl ester. Parallel assaying of replica-plated colonies on agar plates supplemented with the glycerol derivative of the 3-hydroxy ester was used to refine the identification, because only *E. coli* colonies producing active esterases had access to the carbon source glycerol, thus leading to enhanced growth and in turn larger colonies. By this strategy, a double mutant was identified, which efficiently catalyzed hydrolysis.

Stemmer's group subjected four genes of cephalosporinases from *Enterobacter*, *Yersinia*, *Citrobacter* and *Klebsiella* species to error-prone PCR or DNA-shuffling. Libraries from four generations (a total of 50 000 colonies) were assayed by selection on agar plates with increasing concentrations of Moxalactam (a β-lactam antibiotic). Only those clones could survive, which were able to hydrolyze the β-lactam antibiotic. The best variants from epPCR gave only an 8-fold increased activity, but the best chimeras from multiple gene-shuffling showed 270-540 fold resistance to Moxalactame (Crameri et al., 1998). Sequencing of a mutant revealed low homology compared to the parental genes and a total of 33 amino acid substitutions and seven crossovers were found. These changes would have been impossible to achieve using epPCR and single-gene shuffling only and the work demonstrates the power of DNA-shuffling.

Another method is in vitro compartmentalization (IVC), which can be extended to a selection approach. IVC is based on water-in-oil emulsions, where the water phase is dispersed in the oil phase to form microscopic aqueous compartments. Each droplet contains, in average, a single gene, and serves as an artificial cell allowing for transcription, translation, and the activity of the resulting proteins, to take place within the compartment. The droplet volume (~5 femtoliter) enables a single DNA molecule to be transcribed and translated (Griffiths and Tawfik, 1998), as well as the detection of single enzyme molecules (Griffiths and Tawfik, 2003). The high capacity of the system ($>10^{10}$ in 1 ml emulsion), the ease of preparing emulsions, and their high stability over a broad range of temperatures, render IVC an attractive system for enzyme HTS. Other applications of IVC can be found in a recent review (Aharoni et al., 2005).

Screening

Much more frequently used are screening-based systems (not to be confused with the use of the term 'screening' for the identification of microorganisms). Due to the very high number of variants generated by directed evolution, common analytical tools like gas chromatography and HPLC are less useful, as they are usually too time-consuming. Also, high-throughput GC-MS or NMR techniques have been described, but these require the availability of rather expensive equipment and in case of screening for enantioselective biocatalysts also the use of deuterated substrates. In addition phage display, ribosome display and FACS have been used to screen within mutant libraries. Although they allow to screen mutant libraries in the order of $>10^6$ variants, they are hardly generally applicable.

The most frequently used methods are based on photometric and fluorimetric assays performed in microtiter plate (MTP)-based formats in combination with high-throughput robot-assistance (see also Sect. 5.1.7). They allow a rather accurate screening of several 10 000 variants within reasonable time and provide sufficient information about the enzymes investigated, i.e. activity by determining inital rates or endpoints and stereoselectivity by using both enantiomers of the compound of interest. One versatile example is the use of umbelliferone derivatives (Fig. 31). Esters or amides of umbelliferone are rather unstable, especially at extreme pH and at elevated temperatures. The ether derivatives shown in Fig. 31 are very stable as the fluorophore is linked to the substrate via an ether bond. Only after enzymatic reaction and treatment with sodium periodate and bovine serum albumin (BSA), the fluorophore is released (Reymond and Wahler, 2002).

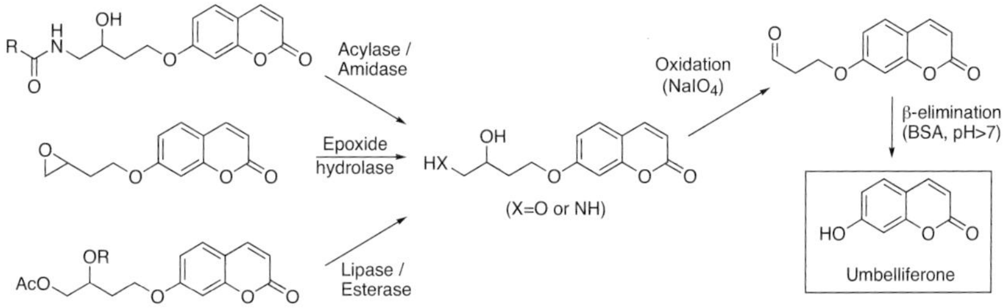

Fig. 31. Fluorogenic assay based on umbelliferone derivatives. Enzyme activity yields a product, which upon oxidation with sodium periodate and treatment with bovine serum albumin (BSA) yields umbelliferone (Reymond and Wahler, 2002).

Another alternative is the recently described 'Surface Enhanced Resonance Raman Scattering', which was shown to enable a rapid and highly sensitive identification of lipase activity and enantioselectivity on dispersed silver nanoparticles (Bornscheuer, 2004; Moore et al., 2004). Janes et al. (1999) developed the Quick E screening method, which uses pH indicators to monitor the rate of ester hydrolysis. The main advantage is that any ester substrate can be used, not just chromogenic substrates (review: Kazlauskas, 2005).

A variety of further assay methods can be found in a number of recent reviews (Bornscheuer, 2001; Goddard and Reymond, 2004; Reetz, 2002; Wahler and Reymond, 2001).

4.2.1.3 Selected Examples

Reetz, Jäger and coworkers turned a non-enantioselective (2% ee, E = 1.1) lipase from *Pseudomonas aeruginosa* (PAL) into a variant with very good selectivity (E>51, >95% ee) in the kinetic resolution of 2-methyl decanoate (Fig. 32). Identification of variants was based on optically pure (*R*)- and (*S*)-*p*-nitrophenyl esters of 2-methyl decanoate in a spectrophotometric screening. In the first step, the wild-type lipase gene was subjected to several rounds of random mutagenesis by epPCR leading to a variant with an E = 11 (81% ee) followed by saturation mutagenesis (E = 25).

Fig. 32. Model reaction to improve and alter enantioselectivity of a lipase from *Pseudomonas aeroginosa* (Reetz et al., 1997).

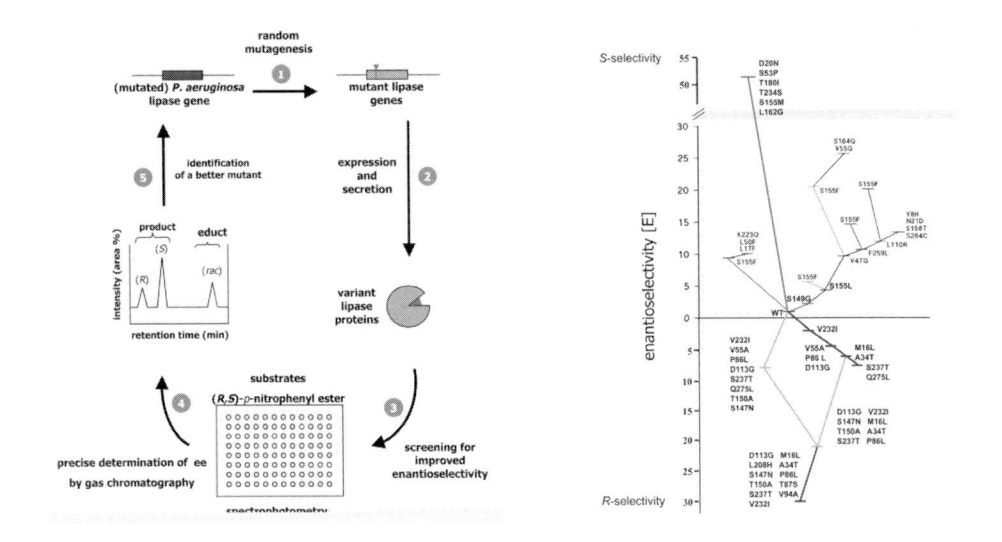

Fig. 33. **Left:** Overview of the directed evolution of a lipase from *Pseudomonas aeruginosa* for the enantioselective resolution of 2-methyl decanoate. In the first step (1), the lipase gene was subjected to random mutagenesis, next the mutated genes were expressed and secreted (2). Screening for improved enantioselectivity was based on a spectrophotometric assay using optically pure (*R*)- or (*S*)-*p*-nitrophenyl esters of the substrate (3). Hit mutants with improved enantioselectivity were then verified by gas chromatography (4). The cycle was repeated several times to identify best mutants (5). **Right:** Changes in enantioselectivity of a lipase from *Ps. aeruginosa* using the strategy shown in the left panel. Starting from the non-selective wild-type (WT, E = 1.1), the combination of various genetic tools led to the creation and identification of variants with high (*S*)-selectivity (E = 51) and with good (*R*)-selectivity (E = 30) (Reetz et al., 2001).

Key to further doubling of enantioselectivity was a combination of DNA-shuffling, combinatorial cassette mutagenesis and saturation mutagenesis which led to a maximal recombination of best variants.The best mutant (E>51) contained six amino acid substitutions and a total of approximately 40 000 variants were screened (Reetz et al., 2001). The overall strategy the changes in enantioselectivity using the combination of different approaches for random mutagenesis are summarized in Fig. 33.

The Arnold group reported the inversion of enantioselectivity of a hydantoinase from D-selectivity (40% ee) to moderate L-preference (20% ee at 30% conversion) by a combination of epPCR and saturation mutagenesis. Only one amino acid substitution was sufficient to invert enantioselectivity. Thus production of L-methionine from D,L-5-(2-methylthioethyl)hydantoin in a whole-cell system of recombinant *E. coli* containing also a L-carbamoylase and a racemase at high conversion became feasible (May et al., 2000).

Even if a biocatalyst with proper substrate specificity (and stereoselectivity) is already identified, requirements for a cost-effective process are not always fulfilled. Enzyme properties such as pH-, temperature and solvent stability are very difficult to improve by 'classical' methods like immobilization techniques or site-directed mutagenesis. Again, directed evolution has been shown to be a versatile tool to meet this challenge.

For instance, an esterase from *Bacillus subtilis* hydrolyses the *p*-nitrobenzyl ester of Loracarbef, a Cephalosporin antibiotic. Unfortunately, the wild-type enzyme was only weakly active in the presence of dimethylformamide (DMF), which must be added to dissolve the substrate. A combination of epPCR and DNA-shuffling led to the generation of a variant with 150 times higher activity in 15% DMF compared to the wild-type (Moore and Arnold, 1996). Later, the thermostability of this esterase could also be increased by ~14°C by directed evolution. In a similar manner, performance of subtilisin E in DMF was improved 470-fold.

A combination of error-prone PCR and DNA-shuffling led to the generation of a more stable and active variant of an esterase from *Bacillus subtilis* (Moore and Arnold, 1996). This enzyme hydrolyzes the *p*-nitrobenzyl ester of Loracarbef, a cephalosporin antibiotic, with 150 times higher activity compared to the wild-type in 15% DMF (Fig. 34) (Arnold and Moore, 1997). In another paper, the thermostability of this esterase could also be increased by ~14°C by directed evolution (Giver et al., 1998).

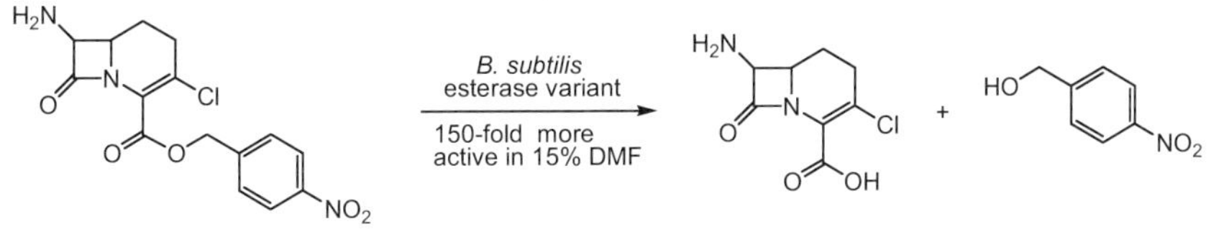

Fig. 34. A combination of error-prone PCR and DNA-shuffling led to a variant of an esterase from *Bacillus subtilis*, which exhibits 150-fold higher activity in 15% DMF compared to the wild-type in the cleavage of the *p*-nitrobenzyl ester of Loracarbef. For practical reasons, the corresponding *p*-nitrophenyl ester was used in the assays.

It could also been shown, that it is possible to increase the thermostability of a cold-adapted protease to 60°C while maintaining high activity at 10°C (Miyazaki et al., 2000). The best psychrophilic subtilisin S41 variant contained only seven amino acid substitutions resembling only a tiny fraction of the usual 30–80% sequence difference found between psychrophilic enzymes and mesophilic counterparts.

Subtilisin is a useful catalyst for organic synthesis, particularly in the presence of organic solvents (see Sect. 6.2.3). However, wild-type subtilisin E is unstable when researchers add a cosolvent like dimethylformamide (DMF) to dissolve the substrate. Directed evolution of subtilisin E using error-prone PCR and a *Bacillus subtilis-Escherichia coli* shuttle vector led to a variant which was 471 times more active in 60% DMF than the wild-type protease in the hydrolysis of succinyl-Ala-Ala-Pro-Phe-*p*-nitroanilide. Even in the absence of DMF, the mutant was ca. 15-fold more active (You and Arnold, 1994). In another example, the half-life of subtilisin E at 65°C was increased 50 times through recombination of two subtilisin E variants using the StEP-method (Zhao et al., 1998). Nakano et al. (1999) increased the stability of lipase from *Pseudomonas* sp. KWI-56 in DMSO by subjecting the gene to error-prone PCR. The best variant containing four mutations was ~40% more stable in 80% DMSO compared to the wild-type enzyme.

In another example, researchers at Maxygen (USA) and Novozymes (Denmark) simultaneously screened for four properties in a library of family-shuffled subtilisins: activity at 23°C, thermostability, organic-solvent tolerance and pH-profile and reported variants with considerably improved characteristics for all parameters (Ness et al., 1999b).

Thus, directed evolution was also used to solve problems related to the technical application of biocatalysts, such as sufficient stability under process conditions with respect to pH profile, temperature activity and stability and solvent tolerance. These deficits are extremely difficult to overcome by site-directed mutagenesis and directed evolution can contribute to the fast generation and identification of improved enzymes.

Despite the methods developed so far and the successful examples given above, several problems have to be solved to allow a broader application of this method. This includes the further optimization of methods for the generation of mutants and enzyme libraries and the development of highly efficient assay systems.

4.2.2 Focused Directed Evolution

Surprisingly, many of the mutations identified by directed evolution were far from the substrate-binding site - so far that the mutated residue did not directly contact the substrate (Park et al., 2005a; Morley and Kazlauskas, 2005). This preference for distant mutations stems from a combination of 1) larger number of distant amino acids and thus, distant mutations and 2) incomplete formation or screening of libraries.

There are more amino acid residues far from the active site than there are close to the active site, Fig. 35. Random mutagenesis methods that target the entire protein create more mutations far from the active site than close to the active site. One can imagine an enzyme like a matryoshka (nesting Russian) doll with multiple shells of amino acid residues surrounding an active site. As one moves away from the active site the shells are larger and contain more amino-acid residues. For a typical enzyme, >50% of the amino acids lie 13–22 Å from the active site and thus, >50% of the random mutations occur in this region.

In spite of this bias toward distant mutations, if one generates and screens all the mutants, one will find the best ones. In practice, it is easier to generate large numbers of mutants than it is to screen them and many experiments involve incomplete screening. These experiments are likely to discover distant mutations and indeed a recent survey showed that most directed evolution experiments discover distant mutations (Morley and Kazlauskas, 2005).

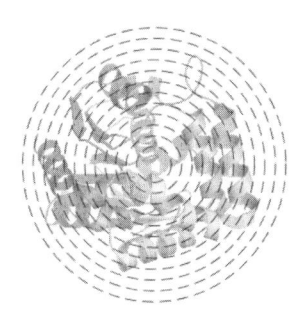

Fig. 35. The spherical shape of enzyme means that there are more amino acids farther from the center than close to the center.

A recognition of this bias prompted many researchers to focus random mutagenesis to the substrate-binding site where mutations are likely to be more effective. This approach can be called focused directed evolution. For example, Park et al. (2005a) focused single mutations in the active site of esterase from *Pseudomonas fluorescens* and screened for increased enantioselectivity. They found several single mutants with up to five-fold increased enantioselectivity toward methyl 3-bromo-2-methylpropanoate (from $E = 12$ to $E = 60$). Earlier work, where mutations throughout the enzyme were permitted, found a mutation outside the substrate-binding site that increased enantioselectivity only 1.5-fold (from $E = 12$ to $E = 19$).

Single mutations have the disadvantage that they create only nineteen different mutants. To get greater numbers of mutants, several groups simultaneously mutated several amino acid residues near the active site. One of the most dramatic successes of this approach – ~500-fold – altered the substrate specificity of organophosphorus hydrolase (Hill et al., 2003). Most organophosphorus nerve agents (chemical weapons) contain P–F, P–CN, or P–S bonds and are poor substrates for organophosphorus hydrolase. Mutagenesis of a pair of residues gave 400 possible mutants, from which Hill et al. identified mutant His254Gly/His257W with ~100-fold increased activity for a 4-nitrophenolate analog of a nerve agent, which simplified screening of the mutants. A subsequent single mutagenesis created 19 additional mutants from which they identified a triple mutant which showed ~500-fold increased activity over wild type for the nerve-agent analog.

Several other groups have also varied several amino acids in the active site of a hydrolase to alter the substrate specificity. For example, Hubner et al. (1999) varied six amino acid residues to increase the relative guanine/adenine specificity of ribonuclease, Antikainen et al. (2003) varied three amino acids in the headgroup binding pocket of phospholipase C to alter the favored headgroup from phosphatidyl choline to ethanolamine or serine, Cheon et al. (2004) varied two amino acids in substrate-binding site of hydantoinase to increase the activity toward larger substrates such as hydroxyphenylhydantoin, and Reetz et al. (2005) varied two amino acids in the substrate-binding site of *Pseudomonas aeruginosa* lipase to increase the activity toward esters with larger acyl groups.

4.3 Catalytic Promiscuity in Hydrolases

Catalytic promiscuity in enzymes is the ability of enzyme active sites to catalyze distinct chemical transformations (reviews: O'Brian and Herschlag (1999), Copley (2003), Kazlauskas and Bornscheuer (2004)). The chemical transformations may differ in the functional group involved, that is, the type of bond formed or cleaved during the reaction and/or may differ in the catalytic mechanism or path of bond making and breaking. Most examples of catalytic promiscuity include both changes. For example, Sect. 10.3.2 on diisopropyl fluorophosphatases mentions two peptidases, which normally catalyze hydrolysis of amide bonds (a C-N link), but also catalyzes hydrolysis of organophosphorus compounds containing a P-F or P-O link. In addition to the different type of bond cleaved, the transition state and reaction mechanism must differ for these two substrates since amides contain a trigonal carbonyl carbon, while phosphate triesters contain a four-coordinate phosphorus. Another example is a catalytic antibody that catalyzes both ester hydrolysis and decarboxylation (Backes et al., 2003). The mechanisms of these reactions differ, but both have anionic transition states that may be stabilized by an arginine and a histidine residue in the active site.

Previously, researchers discovered examples of catalytic promiscuity mainly by serendipity, but with a growing understanding of catalytic mechanisms and an awareness of the possibility of catalytic promiscuity some deliberately search for catalytic promiscuity. In most cases, the promiscuous activity is much lower than the normal activity and therefore too slow for large-scale preparative applications. One exception is pyruvate decarboxylase, whose promiscuous catalysis of an acyloin condensation makes a key intermediate for L-ephedrine manufacture (Goetz et al., 2001). Catalytic promiscuity is likely one mechanism by which new enzymes evolve in nature and a similar evolution in laboratory may create preparatively useful catalysts.

The examples below focus on hydrolases and are divided into reactions involving functional group analogs, aldol and Michael additions, and new reactions with modified hydrolases.

4.3.1 Reactions Involving Functional Group Analogs

The most common examples of catalytic promiscuity in hydrolases are hydrolyses of functional group analogs. For example, many proteases also catalyze ester hydrolysis. The bonds broken in the two cases (C–N vs. C–O) differ, but the catalytic mechanism is likely very similar. On a commercial scale, BASF uses a lipase, which normally cleaves C–O bonds in triglycerides, to resolve amines by enantioselective acylation, which forms a C–N bond (review: Breuer et al., 2004). Similarly, several esterases and lipases cleave the C-N bond in β-lactams: pig liver esterases (Jones et al., 1991), *Pseudomonas fluorescens* lipase (Brieva et al., 1993), or *Candida antarctica* lipase B (Adam et al., 2000; Forró et al., 2003; Park et al., 2003b).

The protease pepsin can cleave the S–O bond in sulfites (Reid and Fahrney, 1967) and subtilisin Carlsberg cleaves the S-N bond in sulfinamides (Mugford et al., 2005). In the sulfinamide case, the substrate also contained a carboxamide link, but subtilisin Carlsberg favored hydrolysis of the unnatural sulfinamide link. Asparaginase, which cleaves the primary amide in the side chain of asparagine, also cleaves a nitrile in an analogous substrate, β-cyanoalanine (Jackson and Handschumacher, 1970). Alkaline phosphatase also catalyzes sulfate ester hydrolysis (O'Brian and Herschlag, 1998).

Lipase (Nishino et al., 2002) or trypsin (Bassindale et al., 2003) catalyze the condensation of silanols or alkoxysilanes, which involves formation of a Si–O–Si bond. Trypsin catalyzed the hydrolysis and condensation of trimethylethoxysilane to hexamethylsiloxane in water, Fig. 36. Although silanols and alkoxysilanes are inherently reactive and can undergo spontaneous condensation or peptide promoted condensation, the trypsin-catalyzed reaction was at least ten times faster than the spontaneous reaction. The condensation involves the trypsin active site because addition of trypsin-specific inhibitors eliminated catalysis and because not all trypsins catalyze this reaction – porcine trypsin was effective, but not trypsin from Atlantic cod.

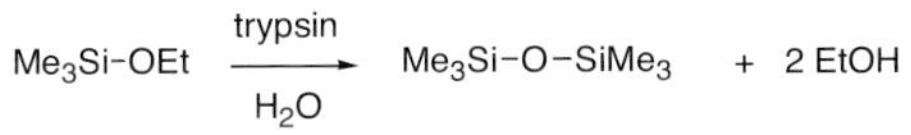

$$\text{Me}_3\text{Si-OEt} \xrightarrow[\text{H}_2\text{O}]{\text{trypsin}} \text{Me}_3\text{Si-O-SiMe}_3 + 2 \text{ EtOH}$$

Fig. 36. Trypsin-catalyzed hydrolysis and condensation of trimethylethoxysilane to hexamethyldisiloxane in water.

4.3.1.1 Perhydrolases

Some esterases and lipases also catalyze perhydrolysis - the reversible formation of a peroxycarboxylic acid from a carboxylic acid and hydrogen peroxide, Fig. 37. For example, esterase from *Pseudomonas fluorescens* (Pelletier and Altenbuchner, 1995), a lactonase (Kataoka et al., 2000), and many lipases (Björkling et al., 1992, Kirk and Conrad, 1999), show low perhydrolase activity in the presence of a carboxylic acid and hydrogen peroxide, see also Sect. 5.3.4.

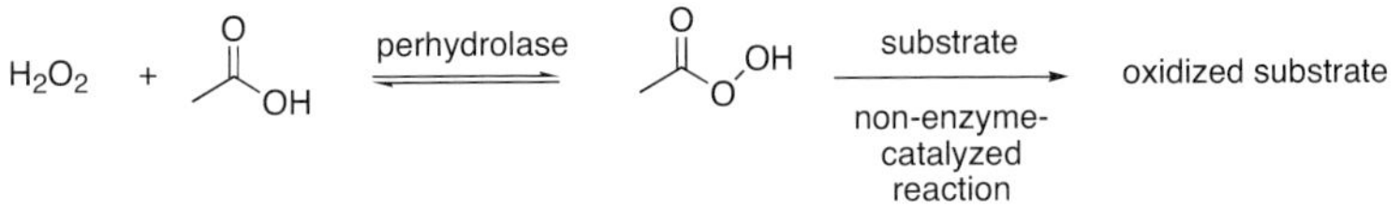

Fig. 37. Perhydrolases catalyze the reversible formation of a peroxycarboxylic acid from hydrogen peroxide and a carboxylic acid, often acetic acid, using an esterase-like mechanism. Peroxycarboxylic acids are more reactive than hydrogen peroxide and can oxidize substrates in a non-enzyme-catalyzed reaction.

The catalytic mechanism of perhydrolysis is analogous to that for hydrolysis, see Sect. 5.1.3. The carboxylic acid reacts with the active site serine to form an acyl enzyme intermediate. Reaction of this acyl enzyme with water regenerates the starting carboxylic acid, while reaction with hydrogen peroxide forms the peroxycarboxylic acid. The change from water to hydrogen peroxide as a nucleophile could be considered as a change in substrate selectivity, but the very different chemical reactivity of the products (carboxylic acid vs. peroxycarboxylic acid) makes this also an example of an alternate catalytic activity.

Another class of enzymes – non-heme haloperoxidases, or more accurately, perhydrolases – also catalyzes perhydrolysis as well as hydrolysis (Picard et al., 1997). However, esterases and lipases are more efficient at ester hydrolysis, while perhydrolases are more efficient at generating peroxycarboxylic acids. Recently, Bernhardt et al. (2005) identified a difference between these two classes of enzymes: perhydrolases contain a key proline residue that orients a main-chain carbonyl group to make a hydrogen bond with hydrogen peroxide, but not with water.

4.3.2 Aldol and Michael additions Catalyzed by Hydrolases

Several hydrolases catalyze the hydrolysis of vinylogous β-keto compounds analogs similar to a Claisen reaction. For example, dienelactone hydrolase (Beveridge and Ollis, 1995) and 2-hydroxy-6-keto-nona-2,4-diene-1,9-dioic acid 5,6-hydrolase (MhpC hydrolase) (Li et al., 2005; Dunn et al., 2005) catalyze the hydrolyses shown below in Fig. 38.

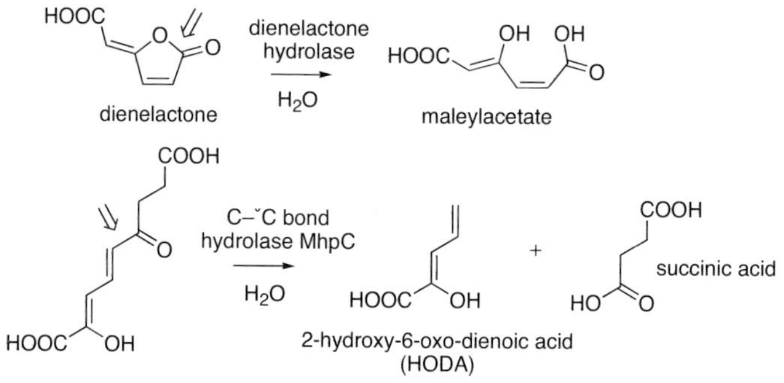

Fig. 38. C-C bond hydrolases catalyze the hydrolysis of vinylogous β-keto compounds.

These hydrolases are involved in the degradation of aromatic compounds and likely activate the carbonyl by hydrogen bonding and promote nucleophilic attack by either water or the active site nucleophile. They do not involve chiral starting materials or products and have not found synthetic applications. Although they are not examples of catalytic promiscuity, their structural similarity to lipases suggests that lipases may show catalytic promiscuity in similar carbon-carbon forming reactions that involve activation of the carbonyl group. Indeed, researchers have discovered two such reactions – lipase-catalyzed aldol addition and Michael additions.

4.3.2.1 Aldol Additions

Lipase B from *Candida antarctica* (CAL-B), a carboxyl ester hydrolase, catalyzed the formation of a carbon-carbon bond in the aldol addition of hexanal in cyclohexane (Branneby et al., 2003), Fig. 39. This catalytically promiscuous reaction is $>10^5$ times slower than the normal hydrolysis of triglycerides, but is at least ten times faster than aldol additions catalyzed by an aldolase catalytic antibody. Although the reaction was not enantioselective, the diastereoselectivity differed from the spontaneous reaction. The authors hypothesized that the aldol addition did not require the active site serine and indeed, replacement with alanine (Ser105Ala) increased the aldol addition approximately two-fold. Molecular dynamics simulations further suggested that the substrate orientation was not ideal and adjusting this orientation might significantly increase the reaction rate (Branneby et al., 2004).

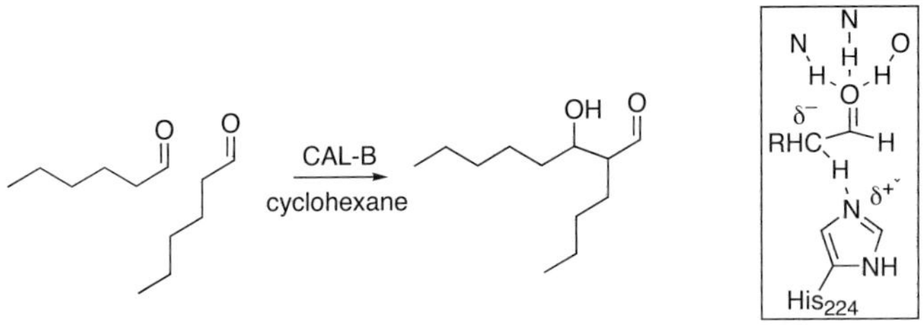

Fig. 39. Lipase B from *Candida antarctica* catalyzes an aldol addition of hexanal. The box shows the calculated transition state for enolate formation.

4.3.2.2 Michael-Type Additions

Four groups reported slow hydrolase-catalyzed Michael-type additions, Fig. 40. The first was a CRL-catalyzed a Michael addition of *o*-aminophenol to 2-(trifluoromethyl) propenoic acid (Kitazume et al., 1986; 1988), Fig. 40. The overall reaction also included formation of an amide link, but the order of the two reactions was not established. Subtilisin catalyzed the addition of imidazole to acrylates at 50°C in pyridine (Cai et al., 2004; Yao et al., 2004). CAL-B catalyzed the addition of secondary amines to an α,β-unsaturated nitrile (acrylonitrile) (Torre et al., 2004) or the addition of thiols or secondary amines to α,β-unsaturated aldehydes (Carlqvist et al., 2005). The apparent k_{cat} for the addition of thiols to α,β-unsaturated aldehydes was 0.01 to 4 min^{-1}, which is similar to the k_{cat} values of aldolase antibodies (0.001 to 1 min^{-1}).

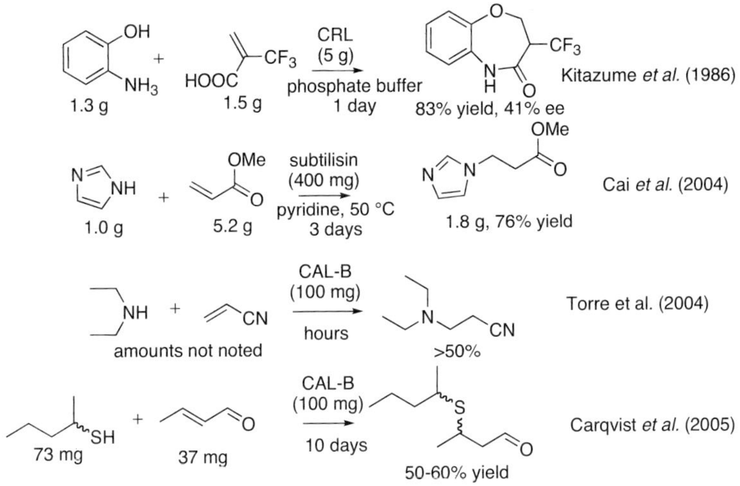

Fig. 40. Examples of hydrolase-catalyzed addition of nucleophiles to α,β-unsaturated carbonyl compounds. In the first example, condensation to form an amide bond also occurs.

Quantum-modeling of this addition (Carlqvist et al., 2005) suggests that the oxyanion hole in CAL-B activates the aldehyde for addition, while the active site histidine acts as a base, but the active site serine is not involved in the reaction. Consistent with this prediction, replacing serine with alanine increased catalysis six-fold for the addition of 2-

pentanethiol to 2-butenal. None of the purely Michael addition reactions were enantioselective, but the combined reaction (CRL-catalyzed amide link formation plus Michael addition) showed moderate enantioselectivity. The enantioselectivity may come from amide link reaction or from the use of a different hydrolase – CRL vs. CAL-B or subtilisin in the other reactions.

4.3.3 Modifications to Introduce New Reactivity in Hydrolases

Chemical modification of enzymes can introduce new functional groups in the active site and thus new chemical reactivity (review: Qi et al., 2001). Several examples involving hydrolases are summarized below.

One simple way to change the catalytic activity of a metalloenzyme is to change the metal ion. For example, replacement of the active site Zn^{2+} in a carboxypeptidase with Cu^{2+} converted this peptidase into a slow oxidase (Yamamura and Kaiser, 1976). More recently, replacement of the active site Zn^{2+} ion in thermolysin with much larger ions such as tungstate, molybdate, or selenate created enzymes that catalyze oxidation of thioethers to sulfoxides with hydrogen peroxide (Bakker et al., 2002).

4.3.3.1 Enantioselective Reduction of Hydroperoxides with Selenosubtilisin

Wu and Hilvert (1989) chemically converted subtilisin to selenosubtilisin, where an – SeH replaces the active site serine –OH. This conversion involved sulfonylation of serine 221 in subtilisin Carlsberg with phenyl methanesulfonyl fluoride followed by displacement with hydrogen selenide. Häring and Scheier (1998a; b) extended this preparation to a 10 g scale and to cross-linked enzyme crystals by directly modifying the crystals. Selenosubtilisin does not catalyze peptide hydrolysis, but does catalyze transfer of the cinnamoyl group from cinnamoyl imidazole to butyl amine and shows 10^4-fold higher selectivity for acylation (acyl transfer to butyl amine) over hydrolysis (acyl transfer to water) as compared to native subtilisin.

Oxidation of selenosubtilisin with hydrogen peroxide yields the seleninate (–Se(O)O$^-$). The X-ray structure of this form shows an intact catalytic triad in spite of the added negative charge in the seleninate form (Syed et al., 1993). Thiols reduce the seleninate form back to selenosubtilisin selenol (–SeH) form. This redox activity allows selenosubtilisin to catalyze the reduction of hydroperoxides with thiols (Wu and Hilvert, 1990; Bell et al., 1993), Fig. 41. This activity mimics the activity of glutathione peroxidases, which also contains a naturally occurring selenocysteine residue at the active site. Glutathione peroxidase is approximately 10^5-fold more efficient than selenosubtilsin in the reduction of hydroperoxides (Bell et al., 1993), but this comparison is for different substrates and reaction conditions.

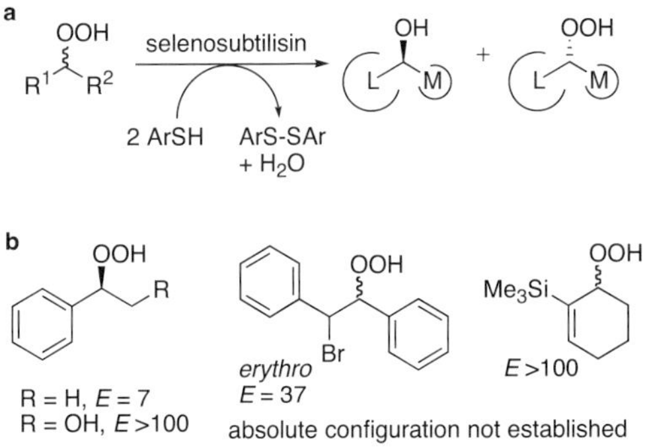

4 Protein Sources and Optimizaton of Biocatalyst Performance

Fig. 41. Selenosubtilisin catalyzes the enantioselective reduction of hydroperoxides with thiols. a) The reduction often favors the enantiomer with the shape shown where L represents a large substituent and M represents a medium-sized substituent. b) Examples of hydroperoxides resolved by selenosubtilisin.

The reduction of hydroperoxides with thiols catalyzed by selenosubtilisin is enantioselective and can be used on a preparative scale (Häring et al., 1999). Typical reactions require a mass of selenosubtilisin approximately equal to the mass of substrate. The favored enantiomer follows an extension of a rule for secondary alcohols as shown in Fig. 53. Savile and Kazlauskas (2005) recently revised this rule for reactions in water (see Sect. 5.2.1.1). This revision and strategies to increase enantioselectivity would also likely apply to the hydroperoxides since these reactions also occur in water.

4.3.3.2 Vanadate-Modified Phosphatases as Peroxidases

Another example of overlapping catalytic activity are acid phosphatases and vanadate-dependent haloperoxidases (Neuwald, 1997; Hemrika et al., 1997; Littlechild et al., 2002). The amino acid sequence, three-dimensional structure and active site are similar in both classes of enzymes. Vanadate binds to the same site as a phosphate ester presumably because it readily adopts a five-coordinate structure that resembles the transition state for phosphate ester hydrolysis. The vanadate ion catalyzes peroxidation by binding peroxide to the vanadium center thereby increasing its electrophilicity. Further support for the similarity of the two active sites is the ability of vanadate to inhibit phosphatases and the ability of phosphate to inactivate vanadate-dependent haloperoxidases by displacing the vanadate. This exchange of active sites also exchanges the catalytic activity of these two classes of enzymes. Several acid phosphatases show low haloperoxidase activity upon addition of vanadate (Tanaka et al., 2002) and conversely, apo-haloperoxidases show low phosphatase activity (Renirie et al., 2000). Sheldon and co-workers reported an enantioselective oxidation of sulfides to sulfoxides using a vanadate-substituted phytase (van de Velde et al., 1998; 2000). However, the altered enzymes were much less effective catalysts than the true enzyme: the turnover numbers were 10^3-10^4 times lower for the haloperoxidase activity of a vanadate-containing phosphatase as compared to a true haloperoxidase or for the phosphatase activity of apo-haloperoxidase as compared to a true phosphatase. This large difference shows that each enzyme is optimized for its catalytic activity, but even with x-ray crystal structures, it is not clear which structural features favor the different activities.

5 Lipases and Esterases

5.1 Availability, Structures and Properties

Both lipases (EC 3.1.1.3) and esterases (EC 3.1.1.1) catalyze the hydrolysis of esters, but lipases preferentially catalyze hydrolysis of water-insoluble esters such as triglycerides. For example, lipases catalyze the hydrolysis of triolein to diolein (Eq. 10).

$$\text{triolein (a triglyceride)} + H_2O \xrightarrow[\text{pH 7}]{\text{lipase}} \text{1,2- or 2,3-diolein (a diglyceride)} + {}^-O\text{-}C(=O)\text{-}R + H^+ \text{ (oleic acid, a fatty acid)} \quad (10)$$

$$R = (CH_2)_7CH=CH(CH_2)_7CH_3$$

In addition, lipases also catalyze the hydrolysis of a broad range of natural and unnatural esters, while retaining high enantio- or regioselectivity. This combination of broad substrate range and high selectivity makes lipases an ideal catalyst for organic synthesis. Chemists use lipase-catalyzed biotransformations to prepare enantiomerically-pure pharmaceuticals and synthetic intermediates (Sect. 5.2.4), to protect and deprotect synthetic intermediates (Sect. 5.3.1) as well as for more specialized uses. A survey of these reactions is the main focus of this book.

Besides high selectivity and broad substrate range, another major advantage of lipases for synthetic reactions is that they act efficiently on water-insoluble substrates. Lipases need this ability because the natural substrates of lipases – triglycerides – are insoluble in water. Lipases bind to the water-organic interface and catalyze hydrolysis at this interface. This binding not only places the lipase close to the substrate, but also increases the catalytic power of the lipase, a phenomenon called interfacial activation. Most lipases are poor catalysts in the absence of an interface such as an organic droplet or a micelle. A conformational change in the lipase probably causes the interfacial activation (see Sect. 5.1.4). In contrast, efficient reactions with proteases often require chemical modification of the substrate to increase water solubility.

Cheese manufacturers use lipase-catalyzed hydrolysis of milk fat to enhance flavors, accelerate cheese ripening and to manufacture cheese-like products (for reviews containing sections on cheese making and detergents see: Berry and Paterson, 1990; Cheetham, 1993; 1997; Haas and Joerger, 1995; Vulfson, 1994). Traditional cheese-making adds extracts containing lipases to the raw cheeses to impart characteristic flavors. For example, extracts of the pregastric gland of a calf imparts a buttery and slightly peppery flavor, while a kid extract imparts a sharp flavor and a lamb extract imparts a strong 'dirty sock' flavor. In addition, microbes responsible for cheese ripening secrete lipases. For example, lipase from *Penicillium roquefortii* liberates short and medium chain fatty acids which add flavor both directly and by serving as precursors for δ-lactones and methylketones. Modern cheese-makers can substitute commercial lipases (e.g., lipases from *Aspergillus niger* or *Rhizomucor miehei*) for the pregastric gland extracts and for the microbes. Also, addition of lipases to cow's milk can mimic the flavor of goat's or sheep's milk. Addition of lipases to cheese followed by incubation at high temperature yields a concentrated cheese flavor that can be used to flavor sauces and other prepared foods.

Some detergents include microbial lipases (e.g., lipase from *Humicola lanuginosa*) to aid removal of fat stains, but the advantage accumulates only after multiple washing. The wash cycle is too short for significant hydrolysis, but the lipase remains on the fat in the subsequent drying where it hydrolyzes the fats. The next wash cycle removes these fats. Lipases may also prevent redeposition of fats on textiles. Recently, Novozymes introduced a new lipase preparation (LipoPrime^{TM}) optimized by protein engineering proven to be stable during the washing process in the presence of protease, high ionic strength, bleach, chlorinated water and within a broad range of water hardness. Moreover, the enzyme is active in the temperature range of 10–50°C, which meets the washing conditions in many countries.

Using lipases for biotransformations is a smaller market, so biotransformations often use lipases that were originally developed for other uses. Two exceptions are lipase from *Rhizomucor miehei*, which Novozymes developed specifically for lipid modification and lipase B from *Candida antarctica*, which Novozymes produces for applications in organic syntheses.

A number of books on lipases, or with large sections on lipases and esterases, and extensive reviews are available (Alberghina et al., 1991; Boland et al., 1991; Borgstrom and Brockman, 1984; Collins et al., 1992; Drauz and Waldmann, 1995; Faber, 1997; Gandhi, 1997; Jaeger et al., 1994; Jaeger and Reetz, 1998; Kazlauskas and Bornscheuer, 1998; Poppe and Novak, 1992; Roberts, 1999; Schmid and Verger, 1998; Sheldon, 1993; Theil, 1997; Wong and Whitesides, 1994; Woolley and Petersen, 1994). More specialized reviews will be cited in the appropriate sections.

5.1.1 Lipases

Lipases occur in plants, animals and microorganism where the biological role of lipases is probably digestive. Most biotransformations use commercial lipases, about 70 of which are available. Tab. 10 lists the most popular of these. Pancreatic cholesterol esterase is included with the lipases because its sequence and biochemical properties are identical to bile-salt stimulated lipase (Hui and Kissel, 1990; Nilsson et al., 1990).

Lipases are usually named according to the (micro)organism that produces the lipase. The classification, and thus name, of a microorganism can change as researchers learn more about it. Likewise, the name of the lipase sometimes changes which can be frustrating to organic chemists accustomed to molecules whose name rarely changes. For example, Amano researchers first classified the microorganism that produces 'Amano P' (ATCC 21808) as *Pseudomonas fluorescens*, but have since reclassified it as *P. cepacia*. For this reason pre-1990 papers on this lipase refer to it as *P. fluorescens* lipase. Confusingly, some researchers continue to refer to this lipase as *P. fluorescens* lipase and Fluka sells SAM-II under the name of lipase from *Pseudomonas fluorescens*. Unfortunately, ATCC 21808 has been again renamed to *Burkholderia cepacia*. For this book, we will continue to use *Pseudomonas cepacia*. Furthermore, a lipase isolated from a strain initially designated as *Pseudomonas* sp. KWI-56 (Iizumi et al., 1991) was reclassified as *Burkholderia cepacia*. The enzyme was later produced by Boehringer Mannheim (now Roche) and sold as Chirazyme L-1. However, all enzymes from the Chirazyme series are no longer available from Roche.

A recent reclassification of the *Rhizopus* fungi renamed *R. niveus*, *R. delemar* and *R. javanicus* all as *Rhizopus oryzae* (review: Haas and Joerger, 1995). Consistent with this reclassification, the lipases isolated from *R. delemar*, *R. javanicus* and *R. niveus* have identical amino acid sequences and lipase from *R. oryzae* (ROL) differs only by two conservative substitutions (His134 is Asn and Ile234 is Leu in ROL). In spite of these similarities, Amano sells three different lipases from this group, and they show slightly different selectivities, perhaps due to cleaving the prolipase at different positions (Beer et al., 1998; Uyttenbröck et al., 1993). The prolipase contains extra amino acid residues to guide folding and secretion of the lipase. After folding, proteases cleave the extra amino acid residues to give the mature lipase. This cleavage may not always occur at the same amino acid residue. Furthermore, the prolipases ProROL and PreProROL had considerably higher thermostability and it could be shown that the natural leader sequence of ROL is able to inhibit the folding supporting properties of the prosequence, resulting in a retardation of folding (Beer et al., 1998). Expression of ROL in *E. coli* proceeds with formation of inclusion bodies, which have to be refolded in order to get active enzyme. This could be circumvented by expression of the ROL gene in the yeasts *Saccharomyces cerevisiae* (Takahashi et al., 1998) or *Pichia pastoris* (Minning et al., 1998), which also facilitated secretion of active enzyme into the culture supernatant.

Note that even when the microorganism classification is settled, the same species may produce different lipases. Amano sells lipase AH from *Pseudomonas cepacia* which differs from lipase P in the amino acid sequence in 16 of 320 residues. These two lipases had opposite selectivity for a dihydropyridine substrate (Sect. 3.2.2) (Hirose et al., 1995).

For the purposes of this book, we will simplify the lipase names. The properties of all commercial preparations of lipase from *Candida rugosa* seem similar, for this reason we will refer to all of them as CRL. Amano P, purified forms of this lipase, and SAM-II (Fluka) all come from microorganism ATCC 21808. We will refer to all of these as PCL, even if the authors did not. The amino acid sequence and biochemical properties of lipase from *Pseudomonas glumae* and lipase from *Chromobacterium viscosum* are identical (Lang et al., 1996; Taipa et al., 1995), and we will refer to both of these as CVL. We will refer to all the *Rhizopus* lipases as ROL. For the other lipases, we will use the abbreviations shown in Tab. 10 or the full name.

Tab. 10. Selected Examples of Commercially-Available Lipases.

Abbreviation	Origin of Lipase	Other Names	Commercial Source and Name
	Mammalian Lipases		
PPL	Porcine pancreas		Amano, Fluka, Sigma
	Fungal Lipases		
CRL	Candida rugosa	Candida cylindracea	Amano (lipase AY), Meito Sangyo (lipase MY, lipase OF-360)
HLL	Humicola lanuginosa	Thermomyces lanuginosa	Novozymes (TL IM, SP 524, Lipolase®)
PcamL	Penicillium camembertii	Penicillium cyclopium	Amano (lipase G)
RJL	Rhizomucor javanicus	Mucor javanicus	Amano (lipase M)
RML	Rhizomucor miehei	Mucor miehei	Amano (MAP), Novozymes (RM IM, Lipozyme®), Fluka
ROL	Rhizopus oryzae	R. javanicus, R. delemar, R. niveus[a]	Amano (lipase F), Amano (lipase D), Amano (lipase N), Fluka, Sigma, Seikagaku Kogyo Co. (Japan)
CAL-A	Candida antarctica A		Novozymes (SP 526)[b]
CAL-B	Candida antarctica B		Novozymes (SP 525 or Novozym435)[b] Sigma
ANL	Aspergillus niger		Amano (lipase A, AP), Röhm, Novozymes (Palatase®)
CLL	Candida lipolytica		Amano (lipase L)
ProqL	Penicillium roquefortii		Amano (lipase R)
	Bacterial Lipases[c]		
PCL	Pseudomonas cepacia	Burkholderia cepacia[c]	Amano (P, P-30, PS, LPL-80, LPL-200S), Fluka, Sigma
PCL-AH	Pseudomonas cepacia		Amano (lipase AH)
PFL	Pseudomonas fluorescens		Amano (lipase AK), Amano (lipase YS), Biocatalysts Ltd.
PfragiL	Pseudomonas fragi		lipase B, Wako Pure Chemical (Osaka)
CVL[d]	Chromobacterium viscosum	Pseudomonas glumae	Sigma, Genzyme, Asahi Chemical, Biocatalysts Ltd.
	Pseudomonas sp.		Amano (K-10)
	Alcaligenes sp.		Meito Sangyo (lipase QL)

[a] The amino acid sequences of lipases R. delemar, R. javanicus and R. niveus are identical and differ from ROL by only two substitutions. [b] SP 525 is a powder containing 40 wt% protein, while Novozym 435 is the same enzyme immobilized on macroporous polypropylene (1 w/w% protein). [c] Lipase from microorganism ATCC21808. Early reports classified this microorganism as Ps. fluorescens, later as Ps. cepacia, then as Burkholderia cepacia. [d] The amino acid sequence and biochemical properties of lipase from Ps. glumae and lipase from Chromobacterium viscosum are identical (Lang et al., 1996; Taipa et al., 1995).

5.1.1.1 Classification of Lipases

Naming lipases according to their microbial source sometimes obscures structural similarities. A better classification uses protein sequence alignments (Tab. 11), which is also consistent with the 3-D structures of lipases (see Sects. 5.1.3 and 5.1.5). The mammalian (pancreatic) lipases form one group, the fungal lipases form two – the *Candida rugosa* and the *Rhizomucor* families – and the bacterial lipases also form two – the *Pseudomonas* and the *Staphylococcus* families. The *Candida rugosa* family includes CRL, GCL and, even though it is a mammalian lipase, pancreatic cholesterol esterase. These lipases are large (60–65 kDa). Note that *Candida antarctica* lipase B does not belong to this family, even though it comes from a *Candida* yeast. The *Rhizomucor* family includes lipases from a wide range of fungi: the *Rhizopus* lipases, the *Rhizomucor* lipases, *Penicillium camembertii* lipase, HLL, CAL-B. These lipases are all small (30–35 kDa). The *Pseudomonas* lipases are also small and include all the *Pseudomonas* lipases and CVL. The *Staphylococcus* lipases are medium-sized (40–45 kDa), but none are commercially-available. One lipase in this group, a thermostable lipase from *Bacillus thermocatenulatus* (BTL2), was available as Chirazyme enzyme from Roche. A number of lipases remain unclassified. For some, e.g., ANL, the amino acid sequence is not known, for others, e.g., CAL-A, the sequence is known (Hoegh et al., 1995), but it shows little similarity to the other lipases. The Lipase Engineering Database (*http://www.led.uni-stuttgart.de*) facilitates comparison of lipases – but also of other hydrolases – as it allows sequence retrieval, alignment and classification (Pleiss et al., 2000). Readers are also referred to a review (Arpigny and Jäger, 1999).

The most useful lipases for organic synthesis are: porcine pancreatic lipase (PPL), lipase from *Pseudomonas cepacia* (Amano lipase PS, PCL), lipase from *Candida rugosa* (CRL), and lipase B from *Candida antarctica* (CAL-B). For lipid modification, lipase from *Rhizomucor miehei* (RML) is the most important. For this reason, we emphasize these five lipases in this book. Note that the synthetically useful lipases include examples from all the classifications in Tab. 11 except the *Staphyloccocus* family. Two examples – RML and CAL-B – come from the *Rhizomucor* family.

Tab. 11. Classification of Commercial Lipases According to Similarities in Protein Sequence[a]

Classification	Characteristics	Examples
Mammalian (pancreatic) lipases	50 kDa	PPL
Fungal lipases		
Candida rugosa family	60–65 kDa	CRL, GCL, CE
Rhizomucor family	30–35 kDa	CAL-B, RML, ROL, HLL, PcamL
Unclassified		ANL, CAL-A, CLL
Bacterial lipases		
Pseudomonas family	30–35 kDa	PCL, PFL, CVL
Staphylococcus family	40–45 kDa	BTL2

[a] Classification according to Cygler et al. (1993) and Svendsen (1994) with some additions.

5.1.1.2 General Features of PPL, PCL, CRL, CAL-B, and RML

Researchers use crude, rather than purified lipases, in most biocatalytic applications for two reasons. First, crude enzymes are less expensive. Microbes secrete lipases into the growth medium. To isolate the lipase, manufacturers simply remove the cells and concentrate. Crude preparations often contain other proteins, but they usually contain only one hydrolase. The second reason researchers often use crude preparations is that they often work better than purified enzymes. The crude preparations contain sugars and other inert carriers which increase the surface area and stabilize the lipases, especially for reactions in organic solvents. A bound calcium ion stabilizes the 3-D structure of PCL, for this reason, crude preparation of PCL often contain added calcium salts. Because of the ill-defined nature of the product, most commercial lipases remain proprietary products. Suppliers sometimes create different preparations of the same lipase intended for different applications. In addition, lipases from different suppliers may be identical due to cross-licensing agreements or may be different due to separate patents on different strains of the same species. Two values for wt% protein in crude lipases are listed below. The higher value is the Lowry assay on the crude sample. This assay overestimates protein content due to interferences in the Lowry assay by sugars and other additives. The lower value refers to the Lowry assay after precipitation of the proteins with trichloroacetic acid (Weber et al., 1995b). This assay will underestimate the protein content if the proteins do not precipitate completely.

Crystallographers have solved the X-ray crystal structures of all five lipases (see below).

PPL

Porcine pancreatic lipase has a molecular weight of 50 kDa. PPL from Sigma contains 8–20 wt% protein (Weber et al., 1995b). Of all the commonly-used lipases for synthesis, PPL is the least pure. Microbial lipases, even when they are not recombinant lipases, are purer because microbes secrete the lipase into the medium. Removal of the cells and precipitation of the lipase yields much purer lipase. In contrast, PPL must be isolated from pancreas or bile which contains numerous hydrolases. SDS gel electrophoresis of crude PPL from Sigma shows four or five major proteins. Cholesterol esterase, trypsin, and chymotrypsin are likely contaminating hydrolases. Several groups reported increased enantioselectivity upon purification of PPL (e.g., Cotterill et al., 1991; Quartey et al., 1996; Ramos-Tombo et al., 1986, see also: Bornemann et al., 1992).

CRL

Commercial samples contain 2–11 wt% protein (Weber et al., 1995b), the rest is sugars and inert carriers. Gel electrophoresis shows a single protein with a molecular weight of 63 kDa when stained with Coomassie blue, but more sensitive staining reveals small amounts of other proteins. Molecular biologists have cloned five different isozymes of CRL from the *Candida rugosa* yeast (Lotti et al., 1993), which all have similar molecular weights. However, heterologous expression of these clones failed because *Candida rugosa* uses an unusual codon for serine which leads to incorrect translation into leucine in other microorganisms. Recently, this was overcome by designing a synthetical gene (*lip1*, 1647 bp) encoding the most prominent isozyme. This enabled functional expression in the yeasts *Saccharomyces cerevisae* and *Pichia pastoris*. Furthermore, expression in the methylotrophic yeast *P. pastoris* allows secretion of active (150 U/ml) and highly pure CRL into the cultivation medium facilitating downstream processing (Brocca et al., 1998).

Commercial samples of CRL are non-recombinant enzymes and contain several isozymes. SDS gel electrophoresis of commercial samples of CRL shows a major band at an apparent molecular weight of 60.1 kDa and a minor one at 58.4 kDa. Lip A (major band) has an isoelectric point of 4.8, while Lip B (minor band) has an isoelectric point of 5.5 (Chang et al., 1994; Rúa et al., 1993). Lotti et al. (1993) identified five isozymes according to their amino acid sequence. All have 85-90% sequence identity and the major differences are the positions and amount of *N*-glycosylation. The major band corresponds to Lip 1 (57.2 kDa according to the amino acid sequence), while the minor band is a mixture of of isozymes Lip 2 - Lip 5 (56.9-57.7 kDa) according to amino acid sequence). Grochulski et al. (1994) solved several x-ray crystal structures of Lip1, while Ghosh et al. (1995) solved the structure of Lip3. In addition, commercial samples of CRL contain a small amount of contaminating protease (Lalonde et al., 1995). Some purification procedures appear to change the conformation of the lipase (Colton et al., 1995; Wu et al., 1990). In spite of this complexity, commercial CRL is a useful and reproducible biocatalyst. A detailed study using purified isozymes of CRL in kinetic resolutions of secondary alcohols showed that they differ in their reactivity and enantioselectivity, but exhibited the same enantiopreferences (Lundell et al., 1998).

RML

RML has a molecular weight of 33 kDa (Huge-Jensen et al., 1987) and commercial material contains 25–57 wt% protein (Weber et al., 1995b). RML is a recombinant lipase produced in *Aspergillus* fungus (Huge-Jensen et al., 1989).

CAL-B

CAL-B has a molecular weight of 33 kDa and commercial material contains 16–51 wt% protein (Weber et al., 1995b). CAL-B is a recombinant protein produced in *Aspergillus* fungus (Hoegh et al., 1995). CAL-B shows little or no interfacial activation and hydrolyzes long chain triglycerides only slowly. For this reason, it may be better classified as an esterase. It shows very high activity and high enantioselectivity toward a wide range of alcohols. Its enantioselectivity is usually low toward carboxylic acids. The application of CAL-B in organic synthesis has been reviewed (Anderson et al., 1998).

PCL

PCL is 320 amino acids long with a molecular weight of 33 kDa. Amano lipase P or PS is the industrial grade which contains 1–25 wt% protein as well as diatomaceous earth, dextran, and $CaCl_2$. LPL-80 and LPL-200S are diagnostic grades that contain glycine. LPL-200S contains no detectable amounts of any other proteins. SAM-II from Fluka differs from lipase P or PS only in the purification method. Four groups have cloned and expressed PCL starting from different *Pseudomonas* strains, but the amino acid sequences of all four are very similar (for original references see: Hom et al., 1991; Iizumi et al., 1991; Jorgensen et al., 1991; Nakanishi et al., 1991; for reviews see: Gilbert, 1993; Svendsen et al., 1995). Expression of active lipase required stoichiometric amounts of an additional protein which guides the proper folding of the prolipase (Hobson et al., 1993;

Quyen et al., 1999). Commercial PCL is probably not a recombinant protein. PCL shows interfacial activation with an increase in activity of ~25 in the presence of an interface (Curtis and Kazlauskas, unpublished data). A single step purification yields crystalline PCL, but this pure material is no longer active in organic solvents (Bornscheuer et al., 1994a). Xie (1991) reviewed the application of this lipase in organic synthesis.

5.1.2 Esterases

Esterases (Carboxylester hydrolases, EC 3.1.1.1) catalyze like lipases the hydrolysis of carboxylic acid esters and can be isolated from the same sources. Esterases and lipases show many similarities with respect to their biochemical and structural properties. All esterases, from which the structures are known have the characteristic α/β-hydrolase fold (Sect. 5.1.3, Fig. 42) and a similar catalytic triad.

The physiological role of most esterases is still unknown. The only exceptions are acetyl- and butyryl choline esterases, both hydrolyze *in vivo* these neutrotransmitters. Some acetyl- and cinnamic acid esterases are involved in metabolic pathways giving access to carbon sources through degradation of hemicelluloses (Dalrymple et al., 1996). Esterases also catalyze the detoxification of biocides. For instance, an insectizide resistance was related to an amplification of esterase genes (Blackman et al., 1995) and an esterase from *Bacillus subtilis* is capable of cleaving the phytotoxin Brefeldin A (Wie et al., 1996). Also, an esterase was described, which converts heroin into morphin with high specificity (Rathbone et al., 1997). Esterases might also be involved in the formation of ω-hydroxy acids from lactones (Griffin and Trudgill, 1976; Khalameyzer et al., 1999; Onakunle et al., 1997), which are produced *in vivo* in an enzymatic Baeyer-Villiger oxidation (Kelly et al., 1998; Roberts and Wan, 1998; Taschner and Black, 1988). This might enable growth on carbon sources such as cyclic alkanes or cyclic alkanones or is required for the production of flavor lactones.

In contrast to lipases, only a few esterases have practical use in organic synthesis. The most widely used mammalian esterase is isolated from pig liver (Sect. 5.4.1), examples for the use of other mammalian esterases appear to a much lesser extent and are summarized in Sect. 5.4.2 (acetylcholine esterase, AChE) and Sect. 5.4.3. The use of microbial esterases is reviewed in Sect. 5.4.4.

5.1.3 Lipases and Esterases are α/β Hydrolases

Although lipases differ significantly in their amino acid sequences, all 11 lipases whose structures have been solved show similar 3-D structures (Tab. 12) for reviews see: Cambillau and Tilbeurgh, 1993; Cygler et al., 1992; Derewenda et al., 1994b; Derewenda, 1994; Dodson et al., 1992; Ransac et al., 1996). This fold, called the α/β-hydrolase fold (Ollis et al., 1992), consists of a core of eight mostly parallel β-sheets, which are surrounded on both sides by α-helices. The connectivity of the sheets and helices is the same in all α/β-hydrolases (Fig. 42). A similar structural pattern was found for several esterase from *Pseudomonas fluorescens* (Kim et al., 1997a, Cheeseman et al., 2004).

α/β-hydrolase fold

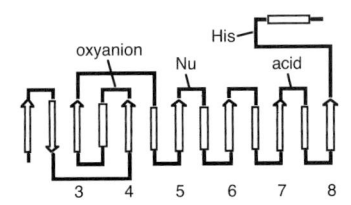

Fig. 42 Schematic diagram of the α/β-hydrolase fold. Oxyanion: residues that stabilize the oxyanion, Nu: nucleophilic residue; for lipases, esterases, and proteases this is a serine; α-helices are shown as rectangles, β-sheets as arrows.

Lipases and esterases are serine hydrolases. The catalytic machinery consists of a triad – Ser, His, and Asp(Glu) – and several oxyanion-stabilizing residues. These residues occur in the same order in all lipase amino acid sequences and orient in the same three-dimensional way in all the structures as shown schematically in Fig. 42. One characteristic of lipase is the location of the serine a tight γ-turn. Most turns are β-turns where at least four Cα are involved in the bend. However, the γ-turn is tighter where only three Cα are involved in the bend. This tight turn also explains the conserved –G-X-S-X-G– motif around the active site serine in lipases. The glycines are small enough to allow this sharp turn to form.

The 3-D orientation of the catalytic machinery is approximately the mirror image of that in the subtilisin and chymotrypsin families of proteases (see Sect. 6.3.1).

The catalytic mechanism for lipase-catalyzed hydrolysis is similar to that for serine proteases (Dodson and Wlodawer, 1998). First, the ester binds to the lipase and the catalytic serine attacks the carbonyl forming a tetrahedral intermediate (Fig. 43).

Fig. 43. Hydrolysis of a butyric acid ester catalyzed by lipase or esterase involves an acyl enzyme intermediate and two different tetrahedral intermediates. Formation of the acyl enzyme involves the first tetrahedral intermediate, T_d1. Alcohol is released in this step, thus, this step determines the selectivity of lipases toward alcohols. Release of the acyl enzyme involves the second tetrahedral intermediate, T_d2. When deacylation limits the rate, this step determines the selectivity of the lipase toward acids. The amino acid numbering corresponds to the active site of lipase from *Candida rugosa*, CRL.

Collapse of this tetrahedral intermediate releases the alcohol and leaves an acyl enzyme intermediate. In a hydrolysis reaction, water attacks this acyl enzyme to form a second tetrahedral intermediate. Collapse of this intermediate releases the acid. Alternatively, another nucleophile such as an alcohol can attack the acyl enzyme thereby yielding a new ester (a transesterification reaction). In most cases, it appears that formation of the acyl enzyme is fast; thus, deacylation is the rate-determining step.

5.1.4 Lid or Flap in Interfacial Activation of Lipases

The x-ray structures of lipases usually show the 'closed' conformation where a lid or flap (a helical segment) blocks the active site. However, x-ray structures of lipases containing bound transition state analogs or bound lipids show the 'open' conformation where the lid is opened to permit access to the active site. For this reason researchers believe a lipid-induced change in the lid orientation causes interfacial activation. Lipases show poor activity toward soluble substrates in aqueous solution because the lid is closed. Upon binding to a hydrophobic interface such as a lipid droplet, the lid opens and the catalytic activity of the lipase increases. In addition, the opening of the lid places one of the oxyanion-stabilizing residues into the catalytic orientation. Cutinase and acetylcholine esterase, which show no interfacial activation, lack a lid and contain a pre-formed oxyanion hole (Martinez et al., 1992; 1994). However, the interfacial activation mechanism may be more complex. A number of lipases (for example, lipase from *Pseudomonas aeruginosa*, CVL, and CAL-B) do not show interfacial activation even though they contain a (small) lid. Lipase from *Staphylococcus hyicus* shows interfacial activation with some substrates, but not with others (for reviews see: Ransac et al., 1996; Verger, 1997).

Tab. 12. X-Ray Crystal Structures of Selected Lipases.

Lipase	Comments	pdb Code[a]	References
Mammalian pancreatic lipases			
humanPL	with colipase & phospholipid	1lpa, 1n8s	Tilbeurgh et al. (1992; 1993)
humanPL	with colipase & phosphonate	1lpb	Egloff et al. (1995a; b)
humanPL	closed form	none	Winkler et al. (1990)
horsePL	closed form	1hpl	Bourne et al. (1994)
pigPL	with colipase & surfactant	1eth	Hermoso et al. (1996)
CE (BSSL)[b]	open form with bile salts	1akn, 1aqn, 1f6w, 1jmy, 2bce	Wang et al. (1997); Chen et al. (1998); Terzyan et al. (2000); Moore et al. (2001)
***Candida rugosa* family**			
CRL	closed form	1trh	Grochulski et al. (1993)
CRL	open form	1crl	Grochulski et al. (1994)
CRL[c]	linoleate complex	1cle, 1lle	Ghosh et al. (1995); Pletnev et al. (2003)
CRL	sulfonate complexes	1lpn, 1lpo, 1lpp	Grochulski et al. (1994)
CRL	phosphonate complexes	1lpm, 1lps	Cygler et al. (1994); Grochulski et al. (1994)
GCL	closed form	1thg	Schrag et al. (1991)
***Rhizomucor* family**			
CAL-B	open form	1tca, 1tcb, 1tcc	Uppenberg et al. (1994)
CAL-B	phosphonate complex	1lbs	Uppenberg et al. (1995)
CAL-B	Tween 80 complex	1lbt	Uppenberg et al. (1995)
RML	closed form	3tgl	Brady et al. (1990)
RML	phosphonate complex	4tgl, 5tgl	Brzozowski et al. (1991); Derewenda et al. (1992)
PcamL		1tia	Derewenda et al. (1994a)
HLL	disordered lid	1tib	Derewenda et al. (1994a; b); Brzozowski et al. (2000)
ROL	closed and partially open forms	1tic, 1lgy	Derewenda et al. (1994a; b); Kohno et al. (1996)
***Pseudomonas* family**			
CVL (PGL)	closed form	1tah, 1cvl	Lang et al. (1996); Noble et al. (1993; 1994)
PCL	open form	1oil, 2lip, 3lip	Kim et al. (1997); Schrag et al. (1997)
PCL	phosphonate complex	1hqd, 1lys1, 1lys2	Luic et al. (2001); Mezzetti et al. (2005)
PAL	phosphonate complex	1ex9	Nardini et al. (2000)

[a] Accession code for the Brookhaven protein data bank. [b] Human pancreatic cholesterol esterase is identical to human bile salt stimulated lipase in milk. [c] Isozyme of CRL.

5.1.6 Designing Reactions with Lipases and Esterases

5.1.6.1 Acyl Donor for Acylation Reactions

The ideal acyl donor would be inexpensive, acylate quickly and irreversibly in the presence of lipase, and be completely unreactive in the absence of lipase. No acyl donor fulfills all three criteria. For transformations of inexpensive chemicals (for example, modified lipids, Sect. 5.3.2), cost is most important, so researchers use acids and simple esters (e.g., methyl, glyceryl). Acylations with these donors are often slow and reversible with an equilibrium constant near one. To drive reactions to completion, researchers removed the water or alcohol by evaporation (Björkling et al., 1989), azeotropic distillation (Bloomer et al., 1992), microwave heating (Carrillo-Munoz et al., 1996) or chemical drying agents such as molecular sieves or inorganic salts (Kvittingen et al., 1992). In other cases, crystallization of the product drives the reaction (Cao et al., 1996; 1997; McNeill et al., 1991).

For resolution reactions of fine chemicals, researchers use activated acyl donors (Fig. 46). Lipase-catalyzed acylations with these donors are one to two orders of magnitude faster than with acids or simple esters. In addition, activated acyl donors shift the equilibrium constant in favor of acylation.

activated esters

C_3H_7 — trifluoroethyl butyrate

C_7H_{15} — S-ethyl thiooctanoate

biacetyl monooxime acetate

enol esters

R = H, vinyl acetate
R = CH_3, isopropenyl acetate
R = OEt, 1-ethoxyvinyl acetate

diketene

anhydrides

acetic acid anhydride

succinic acid anhydride

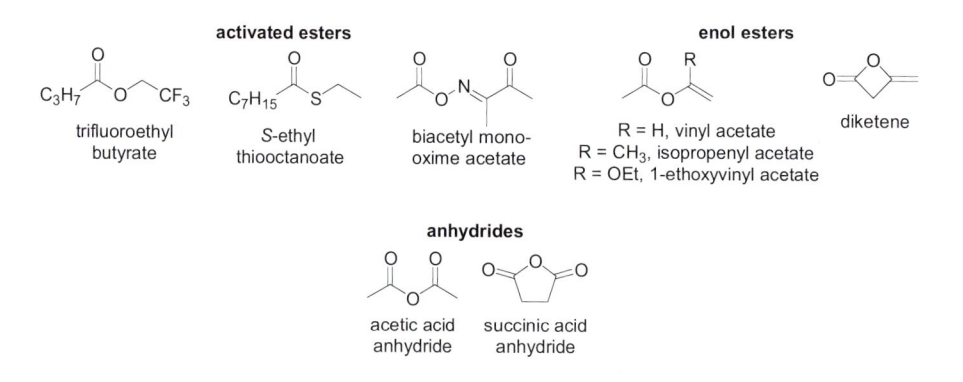

Fig. 46. Examples of activated acyl donors for irreversible acylation of alcohols.

In the case of enol esters and acid anhydrides, acylation is practically irreversible. An irreversible reaction is important for kinetic resolution of alcohols because the reverse reaction degrades the enantiomeric purity of the remaining starting material thereby lowering the efficiency of the resolution (see Sect. 2.1).

Researchers first used activated esters where the alcohol is a better leaving group. For example, Stokes and Oehlschlager (1987) acylated sulcatol with trifluoroethyl laurate and recovered the unreacted alcohol in 97% ee (Eq. 11).

$$\text{R} \overset{O}{\underset{}{\parallel}} \text{O-CF}_3 \quad + \quad \overset{OH}{\underset{}{\diagup\!\diagdown}} \quad \xrightarrow[E=80]{\text{PPL, Et}_2\text{O}} \quad \overset{OH}{\underset{}{\diagup\!\diagdown}} \quad + \quad \overset{O}{\underset{}{\parallel}}\text{R} \quad (11)$$

R = n-C$_{11}$H$_{23}$

97% ee
(S)-(+)-sulcatol
insect pheromone

90% ee

However, for another secondary alcohol, De Amici et al. (1989) could not recover the unreacted starting material in high enantiomeric purity even though the transesterification was enantioselective. They attributed this difficulty to the reversibility even for this activated ester. The less expensive trichloroethyl esters are less convenient because the product trichloroethanol is difficult to remove (bp. 151°C). The thioester S-ethyl thiooctanoate drives the reaction both because the thiol is a good leaving group and because the ethanethiol is easily removed by evaporation (Frykman et al., 1993; Öhrner et al., 1992), but working with volatile thiols requires extra care. The oxime esters react faster than simple esters and even enol esters (Ghogare and Kumar, 1989; 1990), but the nonvolatile oxime may complicate separations. Several other leaving groups (not shown) are less useful: Cyanomethyl esters release the toxic formaldehyde cyanohydrin while 2-chloroethyl esters do not activate the ester enough.

The most useful activated acyl donors are enol esters, such as vinyl acetate or isopropenyl acetate. The product alcohol tautomerizes to a carbonyl compound, thereby driving the reaction and eliminating potential product inhibition. The first reports of lipase-catalyzed acylations with enol esters appeared in 1986–1987 (Degueil-Castaing et al., 1987; Sweers and Wong, 1986) and the first enantioselective acylations appeared in 1988 (Laumen et al., 1988; Terao et al., 1988; Wang et al., 1988; Wang and Wong, 1988). Hoechst AG patented the resolution of alcohols using vinyl esters in 1988. Since that time researchers have resolved hundreds of alcohols using this method. For example, Berkowitz et al. (1992) efficiently resolved glycals for the synthesis of artificial oligosaccharides (Eq. 12).

$$(12)$$

>97% ee   >97% ee

Most lipases except CRL and GCL tolerate the liberated acetaldehyde. Acetaldehyde slowly inactivates these two probably by formation of a Schiff base with Lys residues (Weber et al., 1995a). Acetone from isopropenyl acetate is less reactive and may not inactivate CRL or GCL, but this has not been investigated. Another alternative is 1-ethoxyvinyl acetate which liberates ethyl acetate (Kita et al., 1996; Schudok and Kretzschmar, 1997).

Although the acylation of alcohols by enol esters, such as vinyl acetate, is indeed irreversible, another equilibrium can cause reversibility and lower the enantiomeric purity of the remaining alcohol (Lundh et al., 1995). Since the reaction mixture contains small amounts of water, the lipase can catalyze hydrolysis of the product acetate ester to the alcohol plus acetic acid. Hydrolysis of the faster-reacting acetate lowers the enantiomeric purity of the remaining alcohol. To minimize this hydrolysis, Lundh et al. (1995) recommend dry conditions and an excess of vinyl acetate. In addition, stopping the reaction before 50% conversion, separating the ester and alcohol and subjecting the alcohol to a second esterification will also minimize hydrolysis.

Acylation with diketene, a cyclic enol ester, is fast and has the advantage that it produces no by-products (Balkenhohl et al., 1993; Jeromin and Welsch, 1995; Suginaka et al., 1996). However, the reported enantioselectivity was slightly lower than that for vinyl acetate, possibly due to nonenzymic acylation. For example, the acylation of α-phenylethanol with diketene showed an enantioselectivity of 12 to 80 (Eq. 13), while vinyl acetate showed an enantioselectivity > 100 with the same enzyme (Laumen et al., 1988; Nishio et al., 1989).

$$\underset{Ph}{\overset{OH}{\bigwedge}} + \overset{O}{\underset{}{\bigtriangleup}} \xrightarrow[E = 12 - 80]{PCL} \underset{Ph}{\overset{O \quad O}{\bigwedge\bigwedge\bigwedge}} + \underset{Ph}{\overset{OH}{\bigwedge}} \quad (13)$$

Acid anhydrides also irreversibly acylate alcohols (Bianchi et al., 1988c), but the release of carboxylic acid may decrease the enantioselectivity of the reaction. For example, CRL-catalyzed acetylation of a bicyclic secondary alcohol with acetic anhydride was moderately enantioselective (E = 20), but in the presence of solid potassium bicarbonate the enantioselectivity increased dramatically to E = 240 (Eq. 14) (Berger et al., 1990). In polar solvents, uncatalyzed acylation by acid anhydrides can lower the overall selectivity.

$$(14)$$

base	E
none	20
KHCO$_3$	240

In some cases, acid anhydrides inactivate lipases probably by depleting essential water. Xu et al. (1995a; b) minimized this inactivation in a CRL-catalyzed resolution of menthol by controlling the rate of anhydride addition. Fast addition promoted hydrolysis of anhydride and decreased the available amount of water, while slow addition promoted reaction between free acid and alcohol which released water. Controlled addition of anhydride kept the water constant at 2–4 mM and the reactor was stable for over two months.

Terao et al. (1989) used succinic acid anhydride to both acylate an alcohol and to simplify the separation of unreacted alcohol and ester. Simple extraction separated the neutral alcohol from the charged succinate half ester.

For resolution of amines, researchers avoid chemical acylation by either using less reactive acyl donors (e.g., ethyl methoxyacetate (Balkenhohl et al., 1997)) or minimizing the contact time between acyl donor and amine (Gutman et al., 1992). An additional problem sometimes encountered with amines is that the resulting amide can be difficult to cleave. To solve this problem, Wong's group developed two acyl donors that yield readily cleaved carbamates or amides (Fig. 47) (Orsat et al., 1996; Takayama et al., 1996).

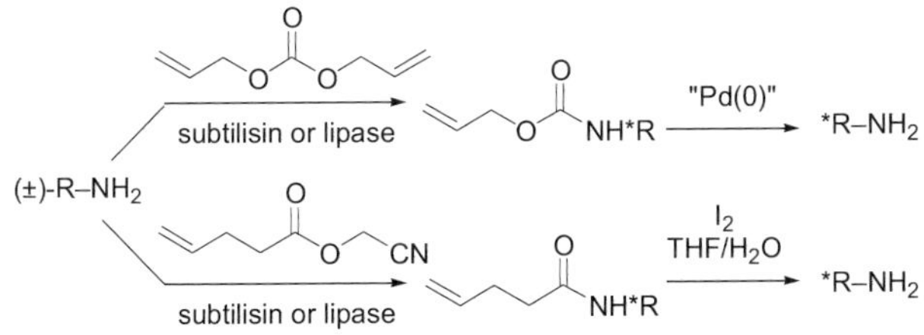

Fig. 47. Special acyl donors that yield carbamates or amides that can be readily cleaved to the free amine.

Kanerva and Sundholm (1993b) compared the enantioselectivity of PCL-catalyzed acylation with butyric acid anhydride, vinyl butyrate, and trifluoroethyl butyrate. All showed similar enantioselectivity, but for one substrate the rate of reaction with trifluoroethyl butyrate was very slow. On the other hand, the enantioselectivity of a CAL-B-catalyzed acylation of several secondary alcohols was highest with 2-chloroethyl butyrate (Hoff et al., 1996). Vinyl butyrate, butyric acid anhydride, and 2,2,2-trichloroethyl butyrate showed 5 to 35 times lower enantioselectivity. In most cases, the key to high enantioselectivity is to avoid chemical acylation. Sometimes acyl donors with longer chains (butyrates and above) show higher enantioselectivity than those with shorter chains like acetate (for examples see: Ema et al., 1996; Guo et al., 1990; Holmberg et al., 1989; Sonnet, 1987; Stokes and Oehlschlager, 1987; Yamazaki and Hosono, 1990).

5.1.7 Assays for Lipases and Esterases

5.1.7.1 Requirements for a Suitable Assay

The common way to determine the activity of lipases and esterases is by means of a pH-stat assay – the hydrolysis of a triglyceride in an emulsion system with continuous titration of released fatty acids with sodium hydroxide solution – or using p-nitrophenyl esters (e.g. p-nitrophenyl acetate, pNPA) with spectrophotometric quantification of the liberated p-nitrophenolate at 405-410 nm. In addition, substrates and products are usually analyzed by means of GC or HPLC methods, especially if chiral compounds are investigated and optical purities must be determined. All of these methods are useful, if a few enzymes need to be characterized, but can not be applied to the screening of many biocatalysts and especially not for the high-throughput screening (HTS) of enzyme libraries derived from directed evolution experiments (see below and for a more detailed overview Sect. 4.2.1.2).

One unit of lipase activity is generally defined as the amount of enzyme liberating 1 μmol fatty acid per min under assay conditions. Common methods for the determination of lipase activity are either based on a titrimetric (e.g., pH-stat) or spectrophotometrical (e.g., hydrolysis of p-nitrophenyl ester) assay. However, no standard protocol has defined so far and as a consequence each research group or supplier of lipases uses a different procedure. A number of lipase assays including detailed protocols are described by Vorderwülbecke et al. (1992), who compared the activity of a wide range of commercial lipases. Care should be taken when comparing lipases from commercial suppliers, because most enzymes are sold as crude preparations, which might contain several hydrolytic enzymes including isozymes of lipases. In addition, different protein content, additives or carriers for immobilization affect specific activity, which is usually expressed as U/mg protein (Soumanou et al., 1997; Vorderwülbecke et al., 1992). Representative procedures for the determination of lipase activity are:

pH-Stat assay

Triglyceride (5% v/v) and gum arabic (2% w/v) (or other appropriate detergents) are emulsified with water using an ultraturrax for 3 min. 20 ml of this emulsion are added to the reaction chamber of the pH-stat equipment, thermostated to 37°C and pH is adjusted to pH 7.0. Then a known amount of lipase (solid or dissolved in buffer) is added to the vigorously stirred emulsion and pH is kept constant by continuous addition of NaOH solution. Specific activity is then calculated from the initial rate of NaOH consumption. If no pH-stat is available, one can perform an end-point titration by addition of a stop solution (EtOH:acetone, 1:1) after 10–20 min followed by titration of the total amount of released fatty acids with NaOH. However, this method is less accurate due to a significant pH-drop during hydrolysis causing enzyme inactivation and inhibition of lipase by the fatty acids released.

Spectrophotometric and fluorimetric assays

Common substrates are *p*-nitrophenyl esters such as *p*-nitrophenyl palmitate (pNPP) or acetate (pNPA). 100 µl pNPP (10 mM dissolved in DMSO or isopropanol) solution and 900 µl lipase solution are mixed in a cuvette and the amount of *p*-nitrophenol released is determined by recording the absorbance at 410 nm at 25°C using a spectrophotometer. One unit is defined as the amount of lipase releasing 1 µmol *p*-nitrophenol per min under assay conditions.

More sophisticated assays take into account the phenomenon of interfacial activation (see Sect. 5.1.4) by e.g., monolayer techniques using Langmuir-Blodget film balance, however these methods require pure enzyme and are not of practical importance. Other assays allow the determination of the regioselectivity of lipases (Farias et al., 1997; Jaeger et al., 1994) by using artificial triglycerides or the direct measurement of enantioselectivity (Janes and Kazlauskas, 1997b; Janes et al., 1998) (see Quick E below).

If a large number of enzymes or variants obtained from directed evolution experiments need to be analyzed, HTS assays are the method of choice and in the past few years, tremendous progress has been made in HTS as outlined below. A suitable HTS should allow a rapid and reliable determination of enzyme properties with high accuracy and reproducibility, especially if 10^3->10^4 samples have to be tested (often per day). In the ideal case, the results of a HTS assay provides true information about the enzyme kinetics, activities and (enantio-)selectivities and can be directly transferred, e.g. to a kinetic resolution of the compound of interest. In reality, HTS assays at least allow to considerably reduce the number of enzyme variants to be investigated with common methodologies (GC/HPLC) and also save time. Further requirements of a suitable HTS method are cheap and readily available screening substrates, no need for expensive instruments and easy performance.

Quick E

The first HTS method described for the determination of activity and enantioselectivity of lipases and esterases was the 'Quick E' test. Kazlauskas and coworkers used a pH indicator to follow enzymatic hydrolysis. To provide high accuracy in the determination of E-values, initial rates of hydrolysis of pure enantiomers (e.g., 4-nitrophenyl 2-phenyl propionate or Ibuprofen-4-nitrophenyl ester) were measured in the presence of an achiral reference compound (resorufin tetradecanoate). Hydrolysis of the chiral substrates yields the *p*-nitrophenolate ion, monitored at 404 nm; the hydrolysis of the reference compound resorufin is detected at 572 nm (Janes and Kazlauskas, 1997b). Advantages are the short measurement time of only 1 min and the need of very small amounts of hydrolases. Disadvantages are the requirement of pure enantiomers, transparent solutions and that only chromogenic 'surrogate' substrates can be used. Later, this format was extended to the use of *p*-nitrophenol serving itself as pH-indicator (Janes et al., 1998). Enzymatic hydrolysis of an ester releases a proton, which causes a pH-change. Working in a buffer such as BES ([*N,N*-bis(2-hydroxyethyl)-2-aminoethanesulfonic acid) with a pKa=7.15 (close to that of *p*-nitrophenol; pKa=7.20) allows quantification of the pH change. This format was verified for solketal butyrate using 27 commercial enzymes and optically pure enantiomers of this ester. This lead to the identification of horse liver esterase as the most enantioselective (E_{app}=15) enzyme. The assay also works in the presence of organic cosolvents and the true substrate can be employed. Disadvantages are the need of pure enantiomers and that the assay conditions must match the pH optimum of the enzymes to be investigated.

Another HTS assay was the adaption of the well-established *p*-nitrophenyl assay (see above). Reetz and coworkers used optically pure (*R*)- or (*S*)-2-methyl decanoate *p*-nitrophenyl esters to screen for enantioselective variants of a lipase from *Pseudomonas aeruginosa* (Reetz et al., 1997), see Sect. 4.2.1.2, Fig. 32.

Acetic acid assay

In this assay hydrolysis of an acetate (e.g. of a chiral secondary alcohol) catalyzed by an esterase or lipase releases acetic acid, which is used to determine the initial rate of the reaction (Baumann et al., 2001). Acetic acid is converted in the presence of ATP and coenzyme A (CoA) by acetyl-CoA-synthetase (ACS) to acetyl-CoA. Next, acetyl-CoA and oxaloacetate react by a citrate synthase to citrate. Oxaloacetate in turn is produced by a L-malate-dehydrogenase (L-MDH) catalyzed oxidation of malate with stoechiometric consumption of NAD^+ yielding NADH (Fig. 48).

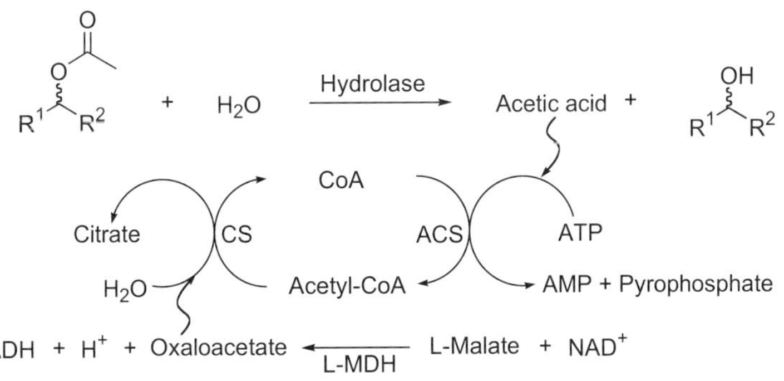

Fig. 48. Principle of the acetic acid test kit format (Baumann et al., 2001).

Thus, the initial rate of acetate hydrolysis can be measured spectrophotometrically at 340 nm using this enzyme cascade via the increase in NADH.

The assay is very reliable and fast, and the exact determination of activity and enantioselectivity (by using the corresponding (*R*)- and (*S*)-enantiomers of the chiral acetate in separate wells) is possible within minutes. In addition, the acetic acid kit is commercially available from R-Biopharm GmbH (Darmstadt, Germany) (initially produced and distributed by Roche Diagnostics, Penzberg, Germany) and can be easily applied following a slightly modified manufacturers protocol.

The acetic acid assay was used to identify enzyme variants obtained by epPCR from an esterase of *Bacillus subtilis* (BsubE), a para-nitrobenzoyl-esterase from *Bacillus subtilis* (BsubpNBE), esterases from *Bacillus stearothermophilus* (BsteE), *Streptomyces diastatochromogenes* (SDE) and PFE, with higher enantioselectivity towards esters of secondary alcohols. The enzymes of these mutant libraries were produced in microtiter plates and split into two wells. After addition of the acetic acid kit solution, the optically pure (*R*)- and (*S*)-acetates of chiral secondary alcohols were added and E_{app}-values were determined from initial rate measurements.

Fluorescence assays have the advantage, that they are very sensitive and that they are less susceptible to compounds causing background signals. For instance, enzymatic activity can be measured in crude cell lysates and even in turbid solutions. As a disadvantage – which holds true for most chromophoric/fluorophoric assays – a non-natural substrate is used and the bulky chromophoric group can lead to erroneous results. In addition, the esters serving as substrates are usually not stable at high or low pH and/or elevated temperatures and only work in aqueous solutions, which is especially true for simple umbelliferyl esters to assay lipases or esterases (Demirjan et al., 1999).

Reymond and coworkers developed an elegant variation of the umbelliferyl test. Initially, they used a chiral secondary alcohol linked to umbelliferyl via an ether bond. Oxidation to the corresponding ketone by an alcohol dehydrogenase, followed by treatment with bovine serum albumin (BSA) releases the fluorophore via a β-elimination. This concept was later adapted (the intermediate is first treated with $NaIO_4$ before addition of BSA) for a variety of enzyme substrates to screen for lipases and esterases, phosphatases, epoxide hydrolases and Baeyer-Villiger-monooxygenases (see Sect. 4.2.1.2, Fig. 31). In case of lipases/esterases, they tested 25 different enzymes and could detect the activity as a time-dependent increase in fluorescence, whereby sodium periodate and BSA did not effect the assay. In addition, the use of optically pure substrates allowed to determine the apparent enantioselectivity (Badalassi et al., 2000). The major advantage of this approach is, that the substrates are highly stable at elevated temperatures and over a broad range of pH-values. The method was recently shown to establish fingerprints of different enzymes, including lipases and esterases, using fluorogenic and chromogenic substrate arrays (Grognux and Reymond, 2004; Wahler et al., 2001). A further variation uses acyloxymethylethers of umbelliferone (Leroy et al., 2003).

Another example is a resorufin-based fluorescence assay, which was used to identify more enantioselective mutants of an esterase from *Pseudomonas fluorescens* (PFE), First, esterase variants obtained from epPCR or using a mutator strain were produced via induction with L-rhamnose in a microtiter plate. Next, the crude lysate of the enzyme solution was split into two wells, diluted with buffer and the assay solution consisting of optically pure (*R*)- and (*S*)-3-phenylbutyric acid resorufin esters were added. The initial rates of PFE-variants catalyzed hydrolysis of the substrate was then determined by measurement of the fluorescence of released resorufin (excitation 544 nm, emission 590 nm) separately for each enantiomer (Fig. 49). The apparent enantioselectivity (E_{app})

was calculated by comparing the rates of hydrolysis for both enantiomers. For mutants exhibiting higher selectivities than the wild-type, the true enantioselectivity (E_{true}) was determined for the hydrolysis of the corresponding racemic methyl ester of 3-phenyl butyric acid. This assay is very sensitive and allowed to identify mutants with enhanced enantioselectivity with good agreement between E_{app} and E_{true}-values (Henke and Bornscheuer, 1999). As a disadvantage, it is only applicable to (chiral) carboxylic acids as substrates.

Fig. 49. Fluorophoric assay based on resorufin esters of chiral carboxylic acids. To determine apparent enantioselectivites of a lipase/esterase, both enantiomers have to be used in separate wells of a microtiter plate (Henke and Bornscheuer, 1999).

"Adrenaline test"

Reymond and coworkers described an "endpoint" assay, the so-called "adrenaline test for enzymes", which is based on the principle of back-titration (Wahler and Reymond, 2002). A periodate-resistant substrate is hydrolyzed by the enzyme to yield a periodate-sensitive product, which is then oxidized by periodate (Fig. 50). The hydrolysis is proportional to a decrease of the deep red-colored adrenochrome, which is also formed by the oxidation of adrenaline with periodate within seconds. First, they add a known amount of sodium periodate and then a chromogenic or fluorogenic reagent and detect the adrenochrome at 490 nm.

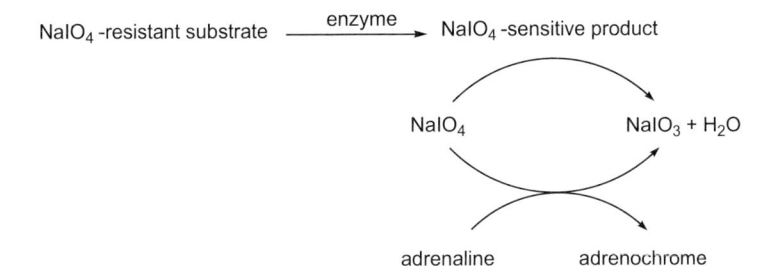

Fig. 50. 'Adrenaline' test for enzymes (Wahler and Reymond, 2002).

This format was used to determine lipase and esterase activity towards vegetable oils such as olive oil, sunflower oil, and rape seed oil etc and provided more reliable results compared to a simple pH indicator test for these substrates, because the long chain fatty acids might remain in the oil phase and do not provide a pH effect. Other advantages are that the reagents for this assay are inexpensive and easily commercially available.

Mass spectroscopy based assay

In a very elegant assay, Reetz and co-workers enable the direct determination of true enantioselectivities (Reetz et al., 1999). One enantiomer of the acetate of a chiral secondary alcohol is deuterated, the other is not. Products of enzymatic hydrolysis of this *pseudo*-racemic mixture are then analyzed by automated electrospray ionization mass spectroscopy (ESI-MS). From the data derived, true enantioselectivity values can be easily calculated at high accuracy. However, this format requires the synthesis of deuterated enantiomers and especially investments into the ESI-MS device capable of handling samples from microtiterplates.

Synthesis activity assay

Many biocatalytic reactions with esterases and lipases are performed as transesterifications in the absence of water in organic solvents. The reaction is shifted from hydrolysis to synthesis with the advantages, that conditions can be widely influenced by the choice of organic solvent, enantioselectivity is often altered and organic chemists are more used to organic solvents rather than water. To find suitable enzymes, an assay was developed, which allows the direct determination of synthesis activity in a microtiter plate format (Konarzycka-Bessler and Bornscheuer, 2003). This allows a considerably higher throughput compared to (trans)esterification assays used by Novozymes (Denmark) (incorporation of decanoic acid into *sn*-1 and *sn*-3 positions of triglycerides of high oleic sunflower oil) or by Roche (acetylation of α-phenylethanol with vinyl acetate in *n*-hexane) as both methods require time-consuming GC or HPLC analysis.

The HTS method by Konarzycka-Bessler and Bornscheuer is based on the transesterification between an alcohol and a vinyl ester of a carboxylic acid. Acetaldehyde generated from the vinyl alcohol by keto-enol tautomerization reacts with a hydrazine (NBD-H) to produce the corresponding hydrazone, which is then quantified by fluorimetric measurement (Fig. 51). This principle allows the rapid identification of active enzymes, which could be demonstrated for a range of hydrolases in a microtiter plate format using a broad range of solvents (e.g. isooctane, toluene, hexane, ether). The enzymatic reaction is detected in real time and no interference between the derivatization agent and the enzymatic reaction was observed.

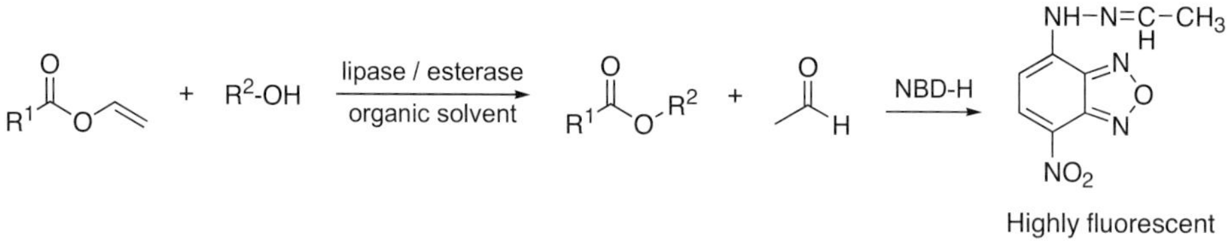

Fig. 51. Assay to determine the synthetic activity of lipases and esterases by *in-situ* derivatization of acetaldehyde with a hydrazine (NBD-H) yielding a fluorescent hydrazone (Konarzycka-Bessler and Bornscheuer, 2003).

Zymogram

This activity staining allows identification, calculation of molecular weight and purity of lipases and esterases on polyacrylamide gels. Proteins are first denatured with sodium dodecyl sulfate (SDS) at 90°C to obtain monomeric proteins, followed by separation using polyacrylamide gel electrophoresis. SDS is removed by incubation of the gel in a detergent solution (e.g., 0.5% w/v Triton-X-100 in Tris/HCl buffer (100 mM, pH 7.5)) and then stained by adding a 1:1 mixture of solutions A and B (see below). The gel is then incubated until a red-colored band indicates proteins with hydrolytic activity due to formation of a complex between α-naphthol and Fast Red. Solution A: 20 mg α-naphhtyl acetate (dissolved in 5 ml acetone and 45 ml Tris/HCl buffer (100 mM, pH 7.5)), solution B: 50 mg Fast Red (Sigma) in 50 ml of the same buffer. Both solutions must be prepared directly prior to use.

Fig. 52 shows an SDS-PAGE of samples taken during cultivation of *E. coli* harboring the gene encoding an esterase from *Pseudomonas fluorescens* (PFE). Induction was performed by adding L-rhamnose. Proteins were stained with coomassie brillant blue (left gel) or solutions A and B (right gel).

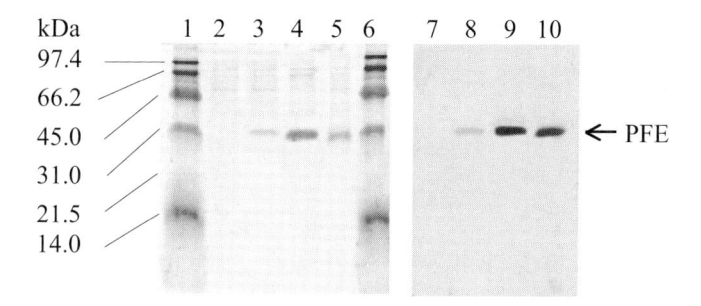

Fig. 52. Separation of proteins (*left*) and identification of the esterase by activity staining with α-naphthyl acetate and Fast Red (*right*). Lanes 1 and 6: low molecular weight standard, lanes 2 and 7: before induction, lanes 3 and 8: 1 h after induction, lanes 4 and 9: 3.5 h after induction, lanes 5 and 10: crude cell extract after cultivation (Krebsfänger et al., 1998a).

Screening of new microbial lipases or esterases can be performed by an agar plate assay. Tributyrin (1% v/v) is added to the agar nutrient media before sterilization and mixed with an Ultraturrax. Lipase or esterase producing microorganims are identified by the formation of clear zones surrounding the colonies caused by release of free fatty acids. The assay becomes more sensitive by addition of rhodamin B creating a fluorescent complex.

5.1.7.2 How to Distinguish Between Lipase, Esterase, and Protease

The far most reliable feature to distinguish these enzymes is their substrate specificity. Proteases usually cleave peptide or amide bonds. Lipases preferentially hydrolyze triglycerides composed of long chain fatty acids and esterases usually only accept water-soluble esters or short-chain fatty acid triglycerides, such as tributyrin. For lipases and esterases, a similar observation can be made in the hydrolysis of *p*-nitrophenyl esters: lipases hydrolyze *p*-nitrophenyl palmitate, esterases do not, but both hydrolyze *p*-nitrophenyl acetate. This substrate specificity might be expanded to water-soluble (lipases and esterases) vs. water-insoluble compounds (lipases), but this is not valid for all enzymes. Pig liver esterase mainly accepts methyl esters of carboxylic acids and acetates of alcohols. Proteases usually do not accept any of these typical lipase or esterase substrates. In contrast, some lipases are capable to convert amines or amides (Sect. 5.2.1.3).

Another distinction can be made on the basis of the protein structure. Most lipases possess a lid covering the active site, esterases do not have a lid (Sect. 5.1.4). Exceptions are lipases from *Candida antarctica* B (CAL-B) and Cutinase from *Fusarium solani pisi*, which have only a small or no lid. A further criteria is the interfacial activation phenomenon observed for most lipases. Measurement of enzyme activity at different substrate concentrations should reveal, whether the enzyme is a lipase or an esterase. However, no interfacial activation was found for CAL-B. Esterases also obey normal Michaelis-Menten kinetics. Proteases and some esterases are inhibited by phenylmethylsulfonyl fluoride (PMSF), but not lipases.

5.2 Survey of Enantioselective Lipase-Catalyzed Reactions

5.2.1 Alcohols

A number of reviews also include surveys of lipase enantioselectivity (Boland et al., 1991; Chen and Sih, 1989; Faber and Riva, 1992; Gais and Elferink, 1995; Kazlauskas et al., 1991; Klibanov, 1990; Margolin, 1993a; Mori, 1995; Santaniello et al., 1992; Schoffers et al., 1996; Sih and Wu, 1989; Theil, 1995; Xie, 1991). The survey below includes only representative examples to give the reader a feel for the type and range of molecules that undergo enantioselective reactions with lipases.

5.2.1.1 Secondary Alcohols

Overview and Models

Although lipases show high enantioselectivity toward a wide range of substrates, the most common substrates are secondary alcohols and their derivatives. Researchers have resolved hundreds of secondary alcohols using lipases. Selected examples, including asymmetric syntheses of secondary alcohols, are collected below.

Based on the observed enantioselectivity of lipases, researchers proposed a rule to predict which enantiomer reacts faster in lipase-catalyzed reactions (Fig. 53; Tab. 13). This rule is based on the size of the substituents and suggests that lipases distinguish between enantiomeric secondary alcohols primarily by comparing the sizes of the two substituents.

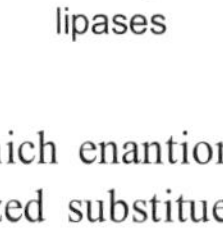

Fig. 53. An empirical rule to predict which enantiomer of a secondary alcohol reacts faster in lipase-catalyzed reactions. M, medium-sized substituent, e.g., methyl. L, large substituent, e.g., phenyl. In acylation reactions, the enantiomer shown reacts faster; in hydrolysis reactions, the ester of the enantiomer shown reacts faster.

Indeed, a number of researchers increased the enantioselectivity of lipase-catalyzed reactions by modifying the substrate to increase the size of the large substituent (for examples see: Adam et al., 1997b; Gupta and Kazlauskas, 1993; Johnson et al., 1991; Kazlauskas et al., 1991; Kim and Choi, 1992; Rotticci et al., 1997; Scilimati et al., 1988). Similarly, Shimizu et al. (1992a) reversed the enantioselectivity by converting the medium substituent into the large one.

To add more detail to this model, many groups tried to more precisely define the size limits of the medium and large substituents (PFL: Burgess and Jennings, 1991; Naemura et al., 1994; 1995; lipase QL: Naemura et al., 1996; PCL: Lemke et al., 1997; Theil et al., 1995), while others have tried to include electronic effects (PCL: Hönig et al., 1994).

Tab. 13. Sized-Based Rules Similar to Those in Fig. 53 Proposed for Different Lipases.

Lipase	Reference	Comments
CAL-B	Orrenius et al. (1995b)	8 substrates
CRL	Kazlauskas et al. (1991)	86 substrates; reliable for cyclic, but not acyclic, substrates
PAL	Kim and Cho (1992)	28 substrates
PCL	Laumen (1987)	tried to also include primary alcohols and acids
PCL	Xie et al. (1990)	6 substrates
PCL	Kazlauskas et al. (1991)	64 substrates
PFL	Burgess and Jennings (1991)	31 substrates
PFL	Naemura et al. (1993b; 1995)	27 substrates
PPL	Janssen et al. (1991b)	23 substrates
PPL	Lutz et al. (1992)	21 substrates
RML	Roberts (1989)	6 substrates
CE	Kazlauskas et al. (1991)	15 substrates
lipase QL	Naemura et al. (1996)	27 substrates

Although steric effects are the most important determinant of lipase enantiopreference, electronic effects also contribute. For example, the CAL-B shows high enantioselectivity toward 3-nonanol ($E > 300$), but low enantioselectivity toward 1-bromo-2-octanol ($E = 7.6$) under the same conditions (Fig. 54). Both an ethyl and a $-CH_2Br$ group are similar in size, so the difference suggests that an electronic effect lowers the enantioselectivity.

E >300   E = 7.6
CAL-B, acylation w/ S-ethyl thiooctanoate
Orrenius et al. (1995b), Rotticci et al. (1997)

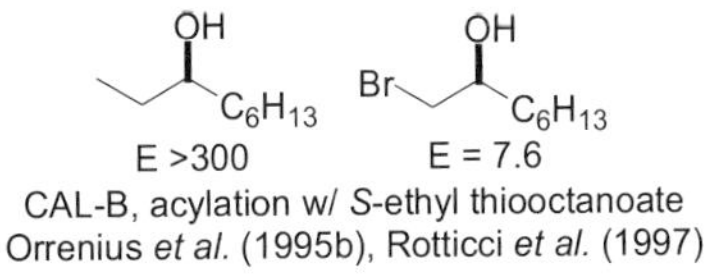

Fig. 54. Electronic effects also change enantioselectivity.

X-ray structures of transition state analogs containing a secondary alcohol, menthol, bound to CRL identified the alcohol binding pocket (Cygler et al., 1994). This pocket indeed resembled the empirical rule: a large hydrophobic pocket and a smaller pocket for the medium-sized substituent (Fig. 55). A comparison of the structures of the fast- and slow-reacting enantiomers of menthol showed that in both cases the large substituent binds in the large hydrophobic pocket and the medium substituent binds in the smaller pocket. In the slow enantiomer the alcohol oxygen points away from the catalytic histidine, thus, this transition state analog lacks a key hydrogen bond. This observation suggests that enantiomers differ mainly in their rate of reaction, not in their relative affinity to the lipase. Consistent with this idea, Nishizawa et al. (1997) measured the kinetic constants with PCL for two enantiomers of a secondary alcohol and found similar values for the apparent K_m, but very different values for k_{cat}. However, modeling of the transition state for ester hydrolysis in CAL-B suggested another explanation. Uppenberg et al. (1995) found that transition states for hydrolysis of the enantiomers of 1-phenylethanol showed different binding. In the fast enantiomer the large substituent binds in the large pocket, and medium in the medium pocket, but in the slow enantiomer, the large substituent binds in the medium pocket and medium substituent in the large pocket.

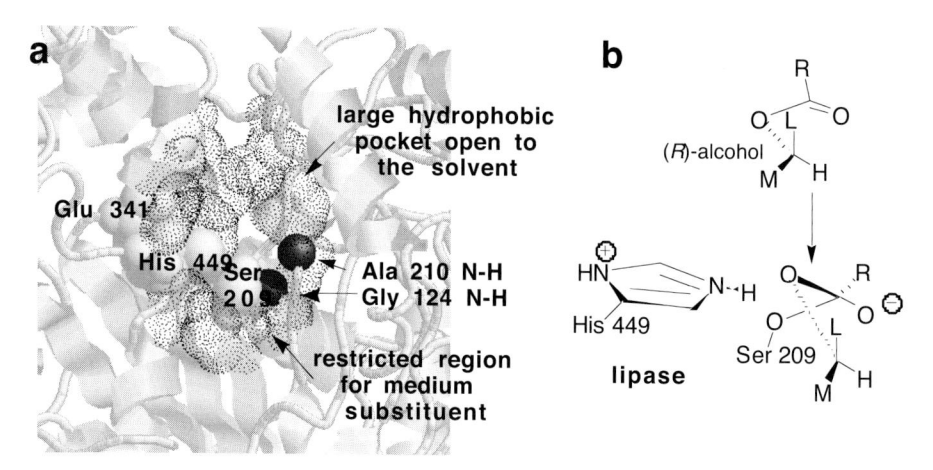

Fig. 55. Proposed binding site for secondary alcohols in CRL. **a** X-ray structure of CRL highlighted to show the catalytic machinery (Ser 209, His 449, Glu 341 and the N-H groups of Ala 210 and Gly 124) and the alcohol binding site. **b** Schematic of the first step of hydrolysis of an ester of a secondary alcohol. The alcohol oxygen orients to form a hydrogen bond with the catalytic His, while the large and medium substituents orient in their respective pockets.

Further support for the proposed alcohol binding site comes from variations in the amino acids within the pocket for the medium substituent (Tab. 14). These variations are consistent with differences in the selectivity of lipases. For example, smaller amino acids line the M region of CRL (Glu, Ser, Gly) than in PCL and CVL (=PGL) (His, Leu, Gly). If the backbone lies in the same place for both lipases, then the smaller side chains in CRL create a larger binding site. Consistent with this suggestion, CRL catalyzes the hydrolysis of esters of large alcohols (esters of norborneols (Oberhauser et al., 1987)) and esters of tertiary alcohols (O'Hagan and Zaidi, 1992), while the *Pseudomonas* lipases do not. Using substrate mapping Exl et al. (1992) found that CRL had a larger alcohol binding site than PCL. Further, CRL shows low enantioselectivity toward esters of primary alcohols, while the *Pseudomonas* lipases show moderate enantioselectivity. All of these characteristics are consistent with a larger binding site in CRL.

Because the same size rule works for all lipases, Cygler et al. (1994) suggested that structures common to all lipases cause this enantiopreference. Indeed, all lipases follow the α/β-hydrolase fold and have a similar catalytic machinery. On the basis of x-ray crystal structures of chiral transition state analogs bound to the active site of CRL, Cygler et al. (1994) proposed that the loops that assemble the catalytic machinery also assemble an alcohol binding site that is similar to the rule in Fig. 53. A large hydrophobic pocket open to the solvent can bind the large substituent, while a restricted pocket near the catalytic machinery can bind the medium substituent (Fig. 55).

Although all lipases favored the same enantiomer of a secondary alcohol, subtilisin favors the opposite enantiomer and the enantioselectivity is usually lower (Kazlauskas and Weissfloch, 1997). One way to reverse the enantiopreference of CAL-B toward secondary alcohols is to put the alcohol in a different place within the binding site. For example, aminolysis of an allyl carbonate derivative (Eq. 15), replaces the allyl group (Pozo and Gotor, 1993b). In this reaction, RCHMeOC(O)– behaves as the acid portion of an ester; thus, the alcohol stereocenter probably binds in the acid binding site. Of course, the secondary alcohol rule no longer applies to this reaction. Pozo and Gotor (1993b) found that CAL-B favored the alcohol enantiomer opposite to the one predicted in Fig. 53.

$$\text{(structure)} \xrightarrow{\text{CAL-B, NH}_2\text{Bn}} \text{(structure)} \quad (15)$$

R = *n*-hexyl, Ph, Et; E >50

5.2 Survey of Enantioselective Lipase-Catalyzed Reactions

Fig. 58. Selected examples of acyclic secondary alcohols resolved by CAL-B.

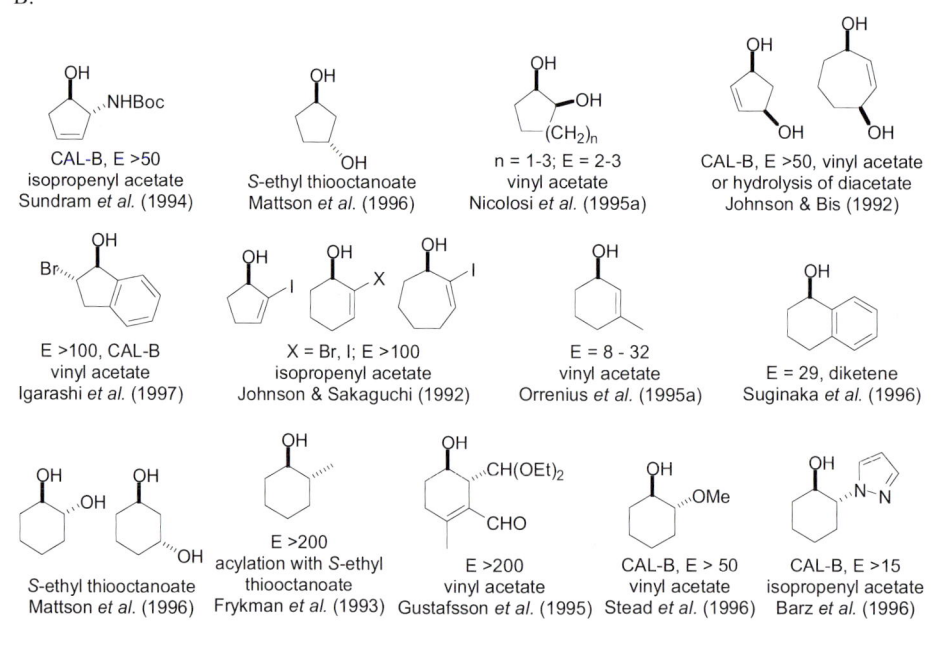

Fig. 59. Selected examples of cyclic secondary alcohols resolved by CAL-B.

Ohtani et al. (1998) compared the enantioselectivity of CAL-B toward twelve secondary alcohols with that of CRL and several *Pseudomonas* lipases (PCL, PFL, PAL). In general, CAL-B showed the highest enantioselectivity followed by the *Pseudomonas* lipases while CRL showed the lowest enantioselectivity. However, PCL showed the highest enantioselectivity for one alcohol, in another case, CRL showed higher enantioselectivity than PCL.

Candida rugosa *Lipase*

In a previous survey of secondary alcohols resolved by CRL, Kazlauskas et al. (1991) found that the secondary alcohol rule in Fig. 53 is *not* reliable for acyclic alcohols. Out of 31 examples, only 14 followed the rule. Since there are only two choices in predicting the fast reacting enantiomer, even guessing yields 50% correct predictions. Thus the rule is little better than guessing. More recent examples (Fig. 60), include seven examples that follow the rule, four that do not and two with either an uncertain absolute configuration or large and small substituents with similar sizes.

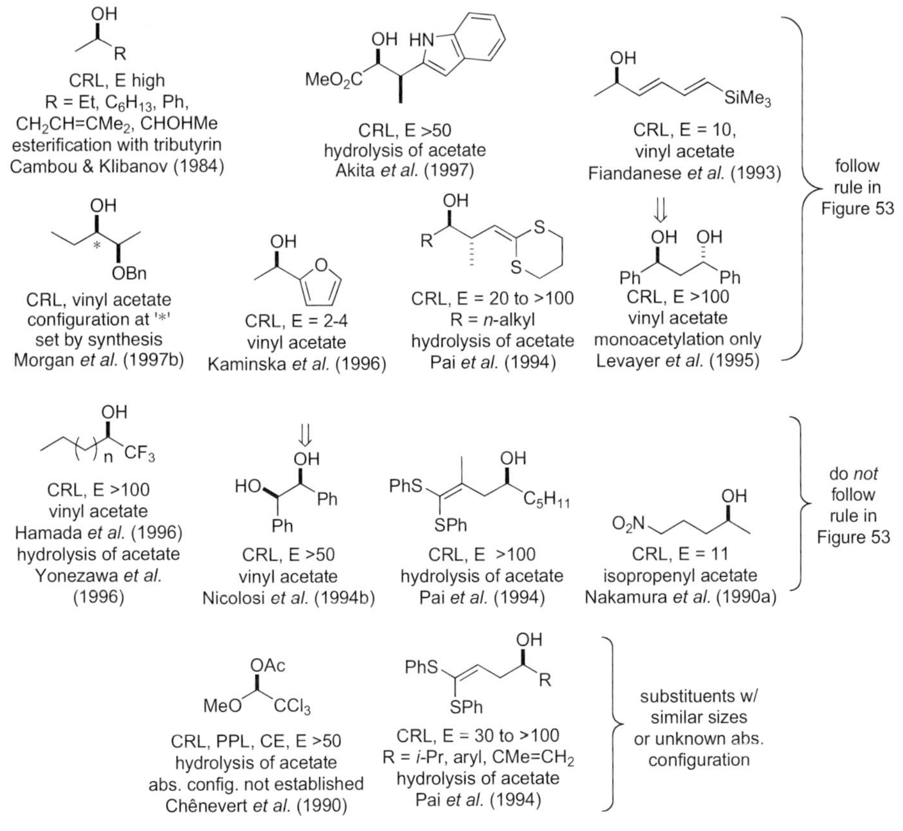

Fig. 60. Selected examples of acyclic secondary alcohols resolved by CRL. Note that several examples do not follow the empirical rule in Fig. 53. This rule is not reliable for CRL-catalyzed reactions of *acyclic* secondary alcohols.

5.2 Survey of Enantioselective Lipase-Catalyzed Reactions

Compared to other lipases (especially CAL-B, PCL and RML) CRL accepts larger substrates and the x-ray crystal structures show a larger active site. This larger active site may allow acyclic substrates to react in several productive conformations. Some of these conformations may favor the opposite enantiomers.

In contrast, cyclic secondary alcohols reliably follow the rule in Fig. 53. Of the 55 substrates in an earlier survey, 51 followed the rule (Kazlauskas et al., 1991). All but two of the selected examples in Fig. 61 follow the rule.

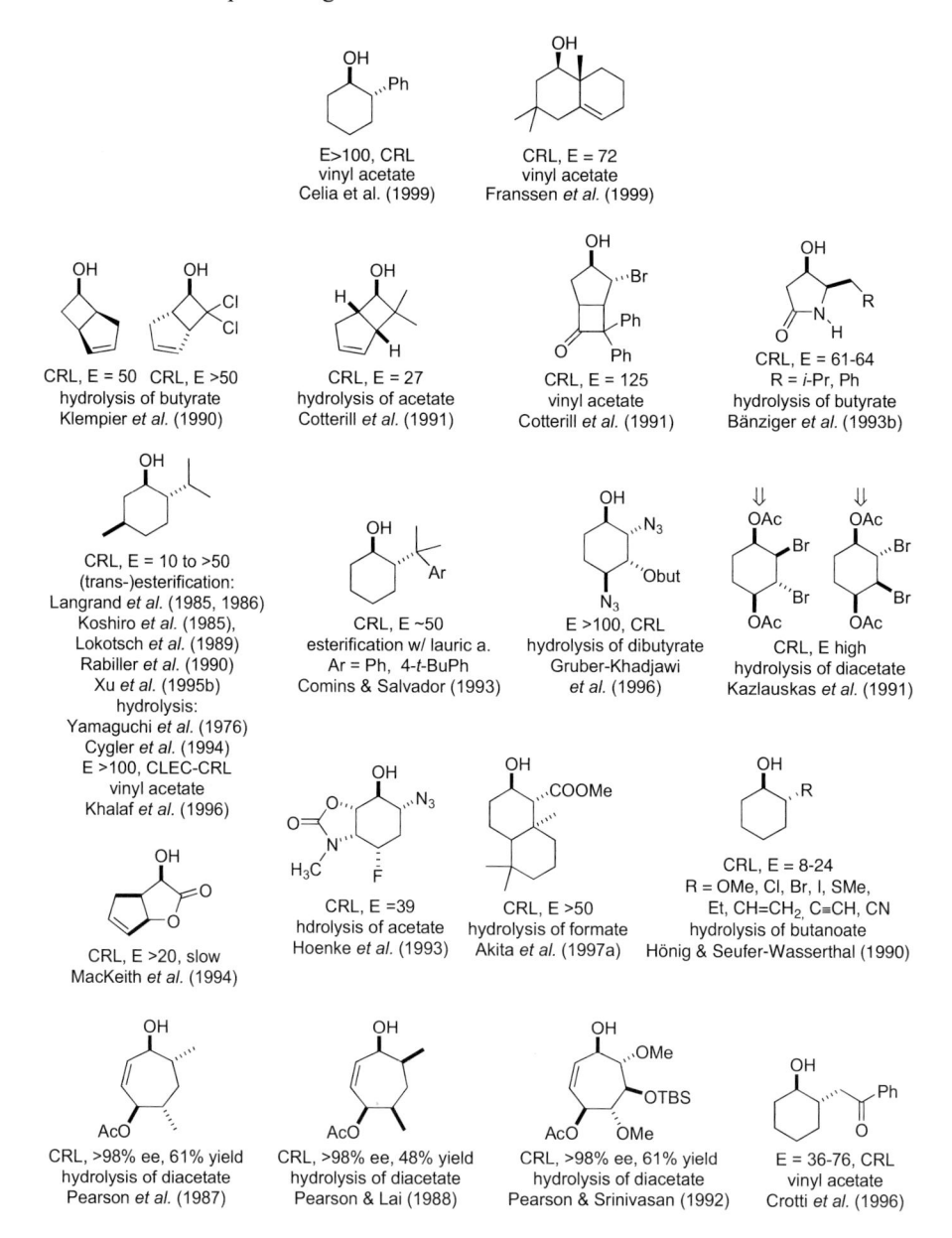

E>100, CRL
vinyl acetate
Celia et al. (1999)

CRL, E = 72
vinyl acetate
Franssen et al. (1999)

CRL, E = 50   CRL, E >50
hydrolysis of butyrate
Klempier et al. (1990)

CRL, E = 27
hydrolysis of acetate
Cotterill et al. (1991)

CRL, E = 125
vinyl acetate
Cotterill et al. (1991)

CRL, E = 61-64
R = i-Pr, Ph
hydrolysis of butyrate
Bänziger et al. (1993b)

CRL, E = 10 to >50
(trans-)esterification:
Langrand et al. (1985, 1986)
Koshiro et al. (1985),
Lokotsch et al. (1989)
Rabiller et al. (1990)
Xu et al. (1995b)
hydrolysis:
Yamaguchi et al. (1976)
Cygler et al. (1994)
E >100, CLEC-CRL
vinyl acetate
Khalaf et al. (1996)

CRL, E ~50
esterification w/ lauric a.
Ar = Ph, 4-t-BuPh
Comins & Salvador (1993)

E >100, CRL
hydrolysis of dibutyrate
Gruber-Khadjawi
et al. (1996)

CRL, E high
hydrolysis of diacetate
Kazlauskas et al. (1991)

CRL, E >20, slow
MacKeith et al. (1994)

CRL, E =39
hdrolysis of acetate
Hoenke et al. (1993)

CRL, E >50
hydrolysis of formate
Akita et al. (1997a)

CRL, E = 8-24
R = OMe, Cl, Br, I, SMe,
Et, CH=CH$_2$, C≡CH, CN
hydrolysis of butanoate
Hönig & Seufer-Wasserthal (1990)

CRL, >98% ee, 61% yield
hydrolysis of diacetate
Pearson et al. (1987)

CRL, >98% ee, 48% yield
hydrolysis of diacetate
Pearson & Lai (1988)

CRL, >98% ee, 61% yield
hydrolysis of diacetate
Pearson & Srinivasan (1992)

E = 36-76, CRL
vinyl acetate
Crotti et al. (1996)

Fig. 61. Selected examples of cyclic secondary alcohols resolved by CRL. Two examples do not follow the rule in Fig. 53: MacKeith et al. (1994) and Hoenke et al. (1993).

5 Lipases and Esterases   93

Porcine Pancreatic Lipase

All examples of PPL-catalyzed reactions of secondary alcohols follow the rule in Fig. 53. The examples are divided into 2-alkanols in Fig. 62, other acyclic secondary alcohols in Fig. 63, and cyclic secondary alcohols in Fig. 64. Note in Fig. 62 that the *cis* vs. *trans* configuration of the double bond in the large substituent strongly influenced the enantioselectivity.

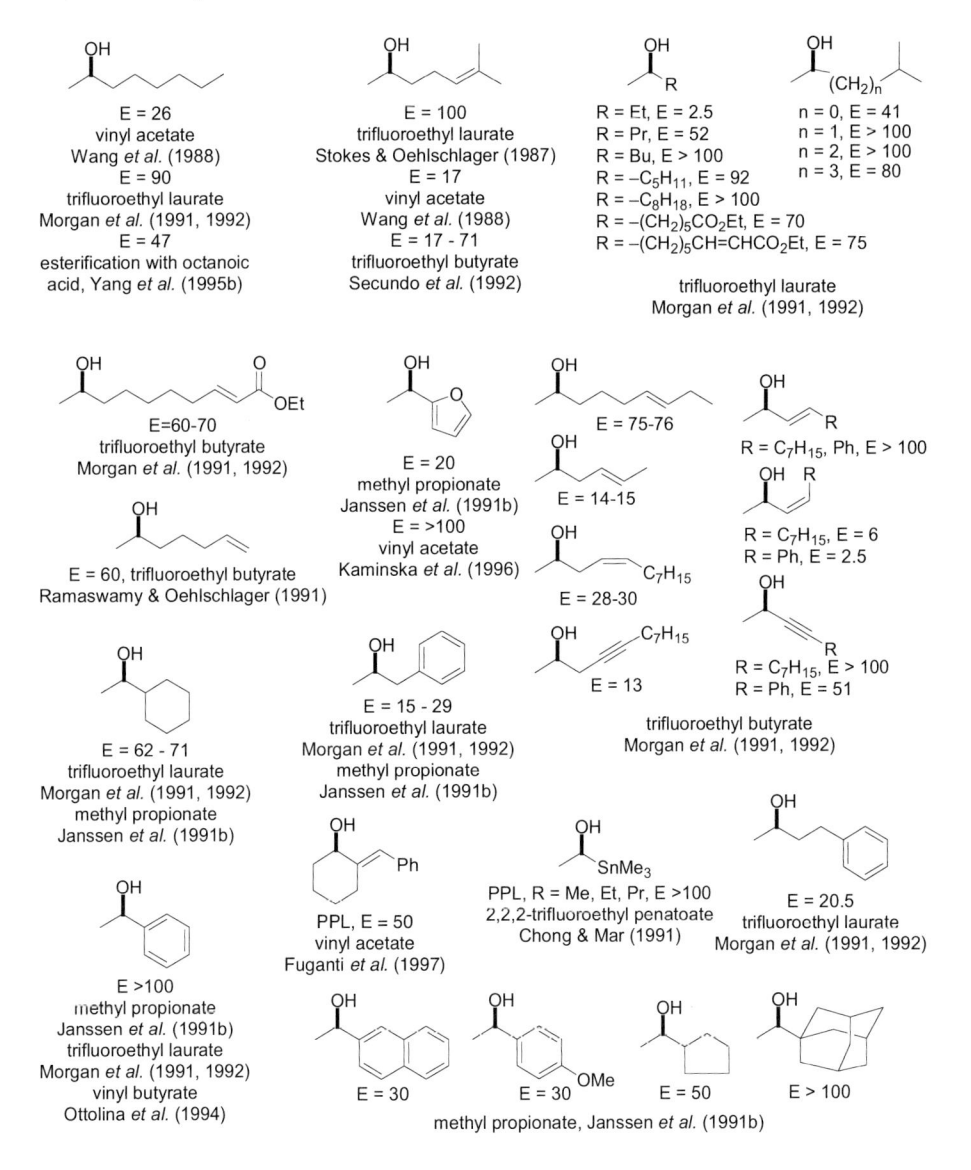

E = 26
vinyl acetate
Wang et al. (1988)
E = 90
trifluoroethyl laurate
Morgan et al. (1991, 1992)
E = 47
esterification with octanoic acid, Yang et al. (1995b)

E = 100
trifluoroethyl laurate
Stokes & Oehlschlager (1987)
E = 17
vinyl acetate
Wang et al. (1988)
E = 17 - 71
trifluoroethyl butyrate
Secundo et al. (1992)

R = Et, E = 2.5
R = Pr, E = 52
R = Bu, E > 100
R = $-C_5H_{11}$, E = 92
R = $-C_8H_{18}$, E > 100
R = $-(CH_2)_5CO_2Et$, E = 70
R = $-(CH_2)_5CH=CHCO_2Et$, E = 75

trifluoroethyl laurate
Morgan et al. (1991, 1992)

n = 0, E = 41
n = 1, E > 100
n = 2, E > 100
n = 3, E = 80

E=60-70
trifluoroethyl butyrate
Morgan et al. (1991, 1992)

E = 60, trifluoroethyl butyrate
Ramaswamy & Oehlschlager (1991)

E = 20
methyl propionate
Janssen et al. (1991b)
E = >100
vinyl acetate
Kaminska et al. (1996)

E = 75-76

E = 14-15

E = 28-30

E = 13

R = C_7H_{15}, Ph, E > 100

R = C_7H_{15}, E = 6
R = Ph, E = 2.5

R = C_7H_{15}, E > 100
R = Ph, E = 51

trifluoroethyl butyrate
Morgan et al. (1991, 1992)

E = 62 - 71
trifluoroethyl laurate
Morgan et al. (1991, 1992)
methyl propionate
Janssen et al. (1991b)

E = 15 - 29
trifluoroethyl laurate
Morgan et al. (1991, 1992)
methyl propionate
Janssen et al. (1991b)

PPL, E = 50
vinyl acetate
Fuganti et al. (1997)

PPL, R = Me, Et, Pr, E >100
2,2,2-trifluoroethyl penatoate
Chong & Mar (1991)

E = 20.5
trifluoroethyl laurate
Morgan et al. (1991, 1992)

E >100
methyl propionate
Janssen et al. (1991b)
trifluoroethyl laurate
Morgan et al. (1991, 1992)
vinyl butyrate
Ottolina et al. (1994)

E = 30

E = 30
methyl propionate, Janssen et al. (1991b)

E = 50

E > 100

Fig. 62. Examples of 2-alkanols resolved by porcine pancreatic lipase.

94   5.2 Survey of Enantioselective Lipase-Catalyzed Reactions

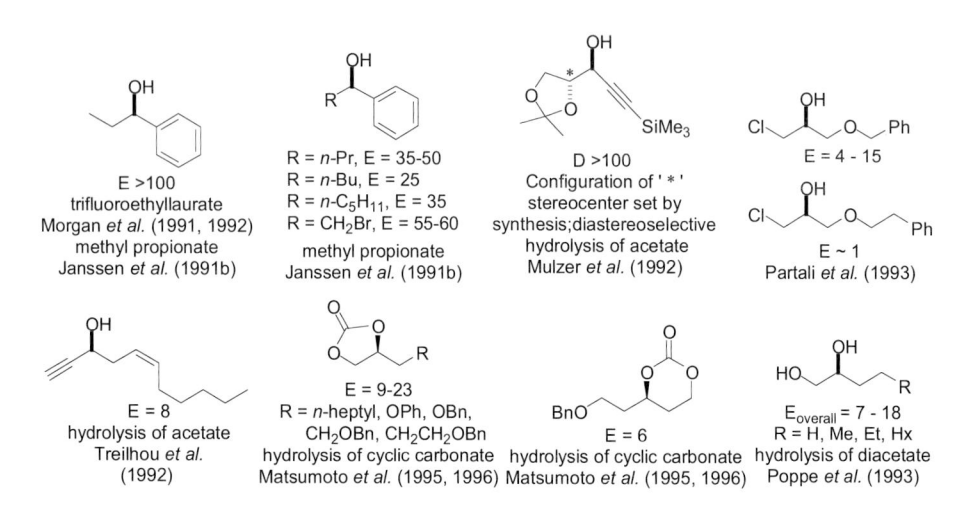

Fig. 63.   Examples of other acyclic secondary alcohols resolved by porcine pancreatic lipase.

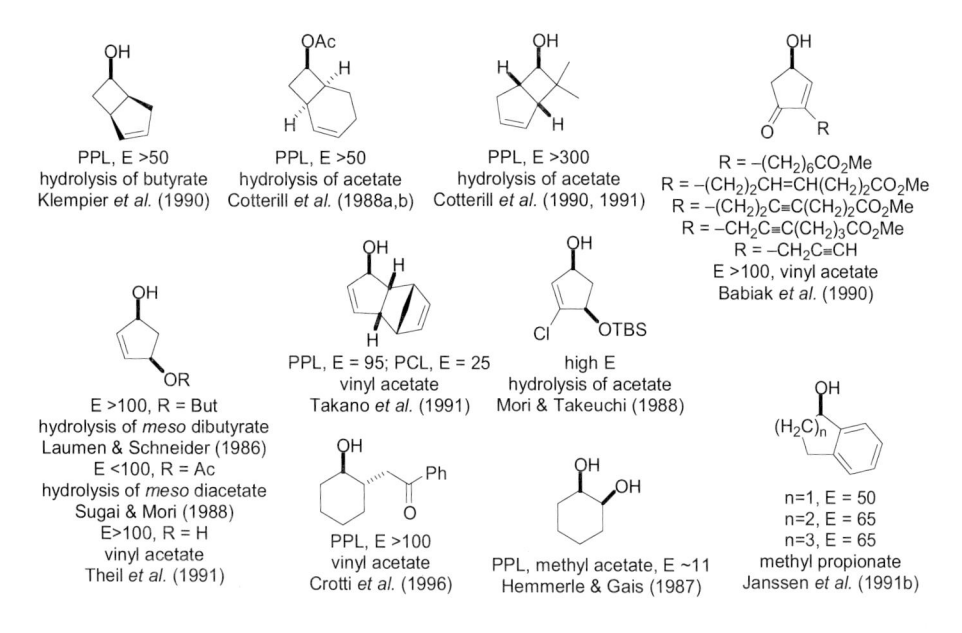

Fig. 64.   Examples of cyclic secondary alcohols resolved by porcine pancreatic lipase.

Pseudomonas *Lipases*

The *Pseudomonas* lipases show high enantioselectivity toward a wide range of secondary alcohols, Figs. 65–67. A previous survey also includes 64 secondary alcohols (Kazlauskas et al., 1991).

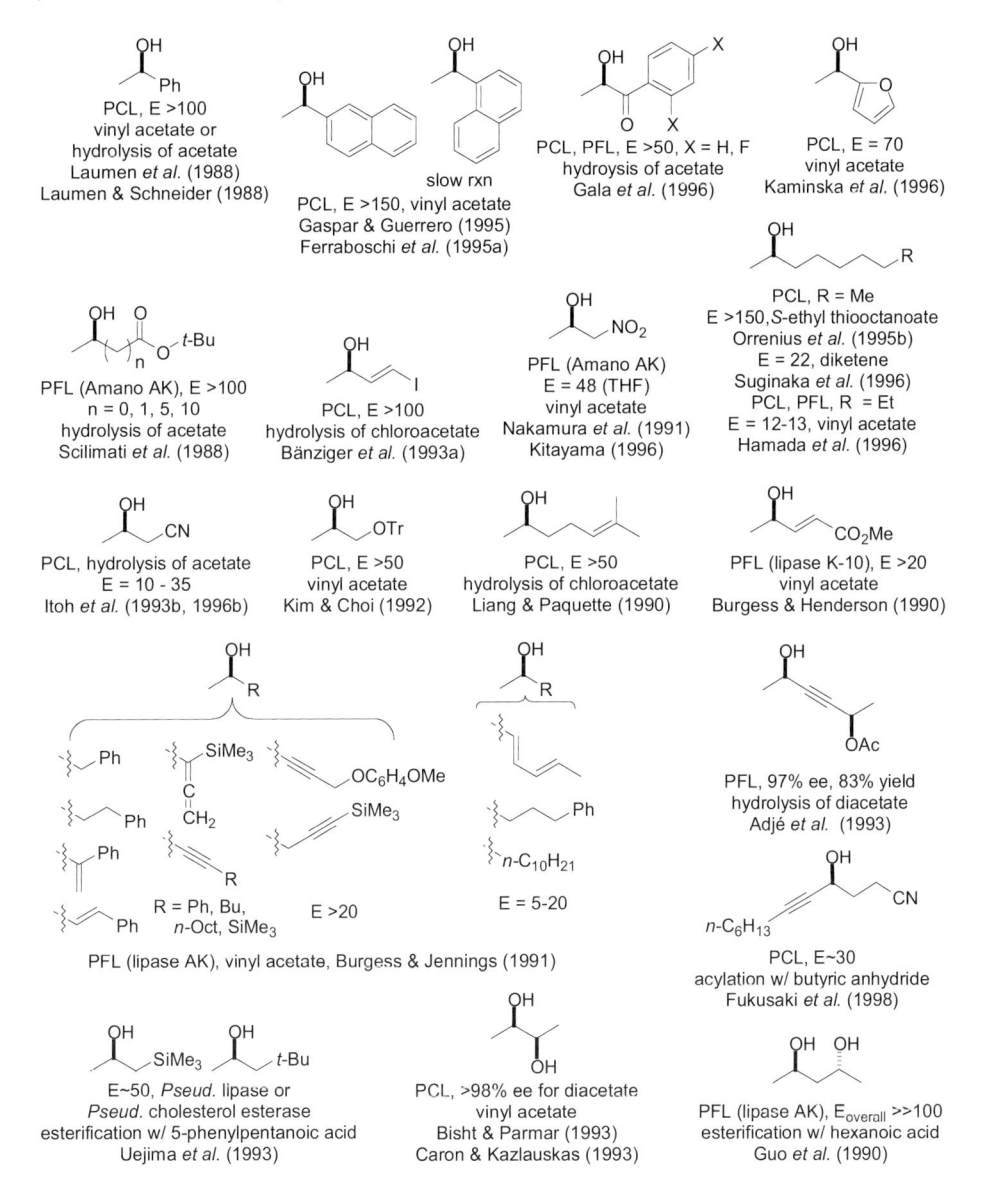

Fig. 65. Selected examples of 2-alkanols resolved by *Pseudomonas* lipases.

5.2 Survey of Enantioselective Lipase-Catalyzed Reactions

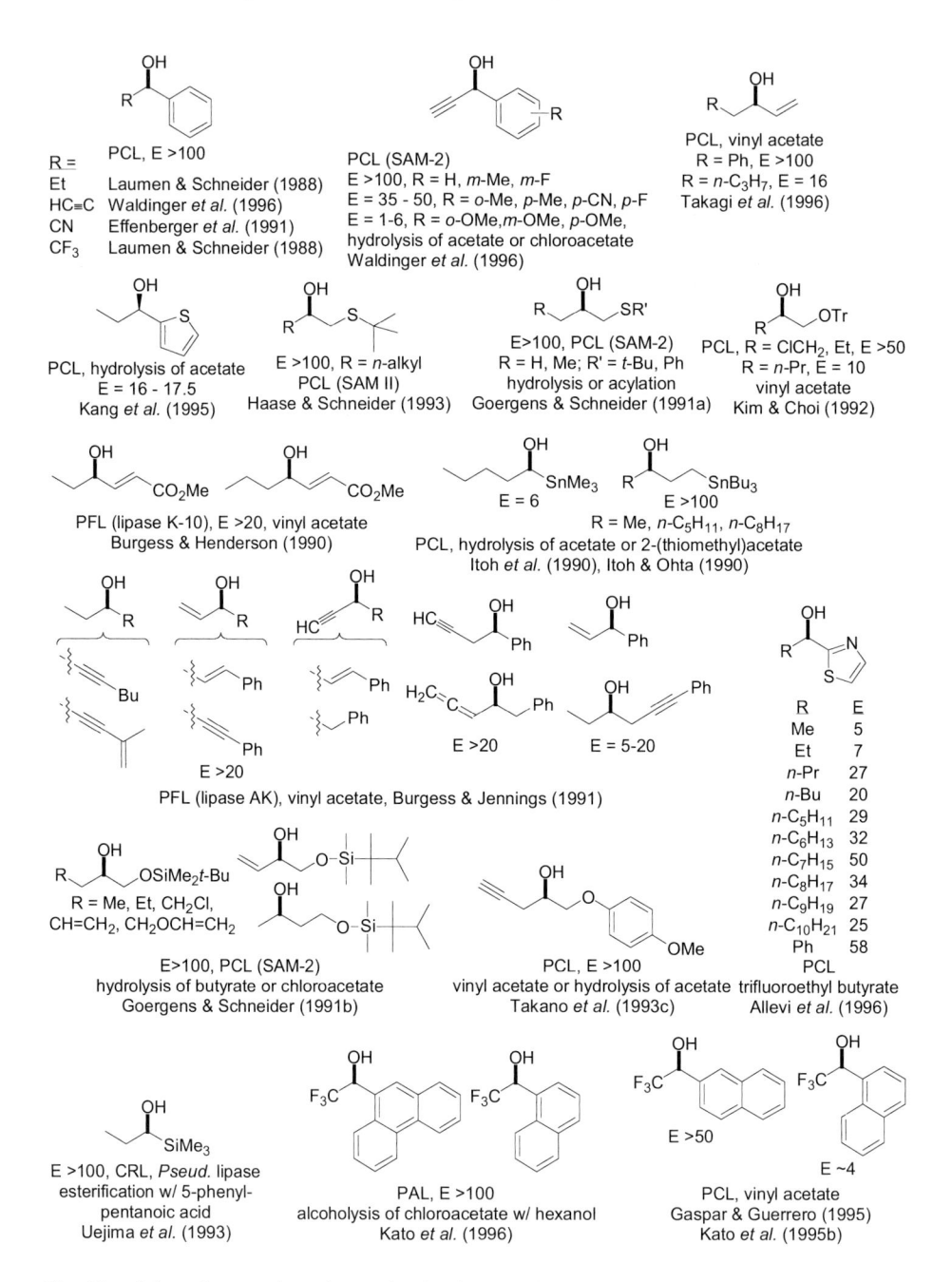

PCL, E >100
R =	
Et	Laumen & Schneider (1988)
HC≡C	Waldinger et al. (1996)
CN	Effenberger et al. (1991)
CF$_3$	Laumen & Schneider (1988)

PCL (SAM-2)
E >100, R = H, m-Me, m-F
E = 35 - 50, R = o-Me, p-Me, p-CN, p-F
E = 1-6, R = o-OMe, m-OMe, p-OMe,
hydrolysis of acetate or chloroacetate
Waldinger et al. (1996)

PCL, vinyl acetate
R = Ph, E >100
R = n-C$_3$H$_7$, E = 16
Takagi et al. (1996)

PCL, hydrolysis of acetate
E = 16 - 17.5
Kang et al. (1995)

E >100, R = n-alkyl
PCL (SAM II)
Haase & Schneider (1993)

E>100, PCL (SAM-2)
R = H, Me; R' = t-Bu, Ph
hydrolysis or acylation
Goergens & Schneider (1991a)

PCL, R = ClCH$_2$, Et, E >50
R = n-Pr, E = 10
vinyl acetate
Kim & Choi (1992)

PFL (lipase K-10), E >20, vinyl acetate
Burgess & Henderson (1990)

E = 6 / E >100
R = Me, n-C$_5$H$_{11}$, n-C$_8$H$_{17}$
PCL, hydrolysis of acetate or 2-(thiomethyl)acetate
Itoh et al. (1990), Itoh & Ohta (1990)

E >20 / E >20 / E = 5-20

PFL (lipase AK), vinyl acetate, Burgess & Jennings (1991)

R	E
Me	5
Et	7
n-Pr	27
n-Bu	20
n-C$_5$H$_{11}$	29
n-C$_6$H$_{13}$	32
n-C$_7$H$_{15}$	50
n-C$_8$H$_{17}$	34
n-C$_9$H$_{19}$	27
n-C$_{10}$H$_{21}$	25
Ph	58

R = Me, Et, CH$_2$Cl, CH=CH$_2$, CH$_2$OCH=CH$_2$
E>100, PCL (SAM-2)
hydrolysis of butyrate or chloroacetate
Goergens & Schneider (1991b)

PCL, E >100
vinyl acetate or hydrolysis of acetate
Takano et al. (1993c)

PCL
trifluoroethyl butyrate
Allevi et al. (1996)

E >100, CRL, Pseud. lipase
esterification w/ 5-phenyl-pentanoic acid
Uejima et al. (1993)

PAL, E >100
alcoholysis of chloroacetate w/ hexanol
Kato et al. (1996)

E >50
PCL, vinyl acetate
Gaspar & Guerrero (1995)

E ~4
Kato et al. (1995b)

Fig. 66. Selected examples of enantioselective reactions of *Pseudomonas* lipases with other racemic acyclic alcohols. Note that substituent type and substituent location in the aromatic ring strongly influences the enantioselectivity in some cases.

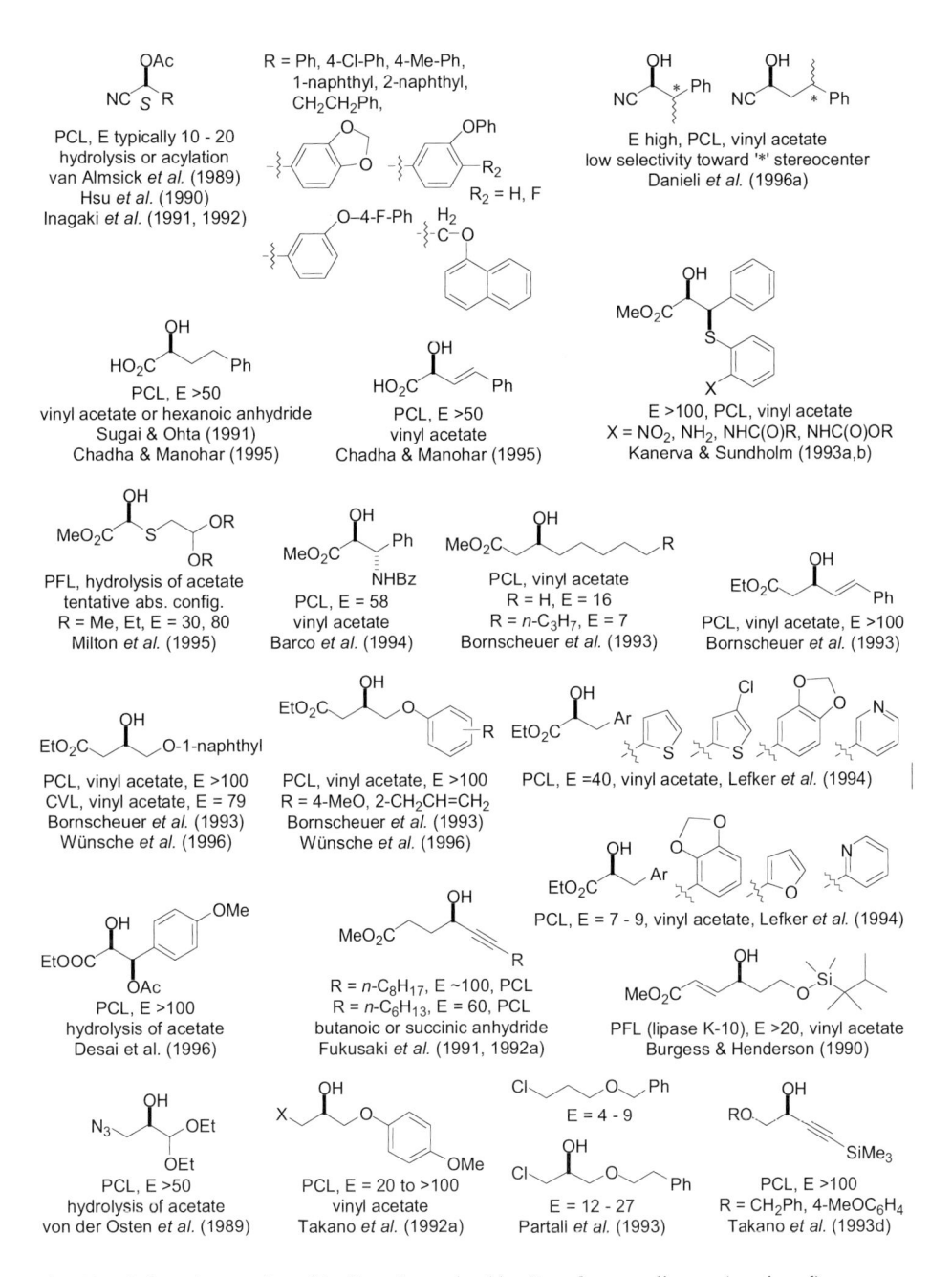

Fig. 66. Selected examples of 2-alkanols resolved by *Pseudomonas* lipases (continued).

5.2 Survey of Enantioselective Lipase-Catalyzed Reactions

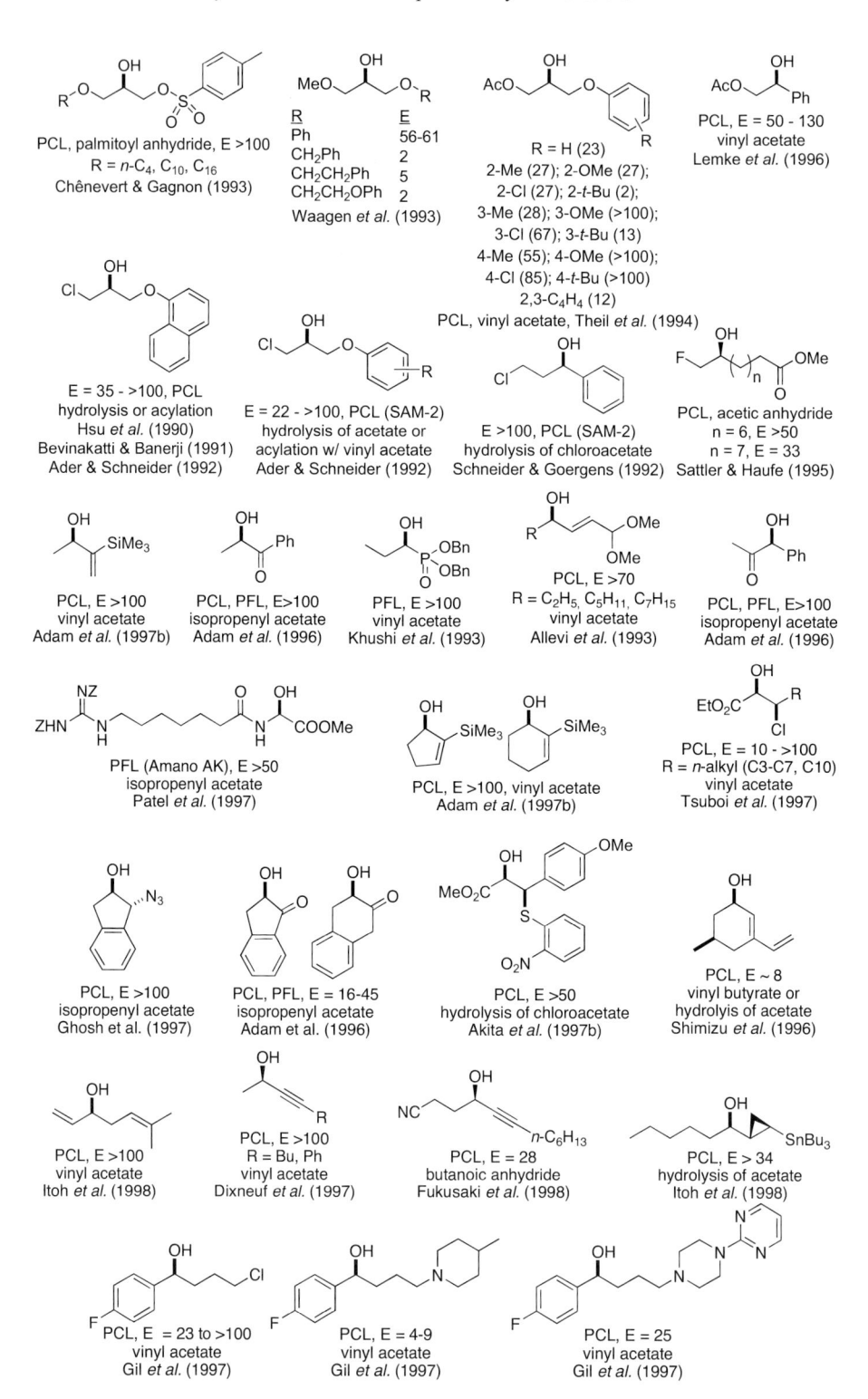

Fig. 66. Selected examples of 2-alkanols resolved by *Pseudomonas* lipases (continued).

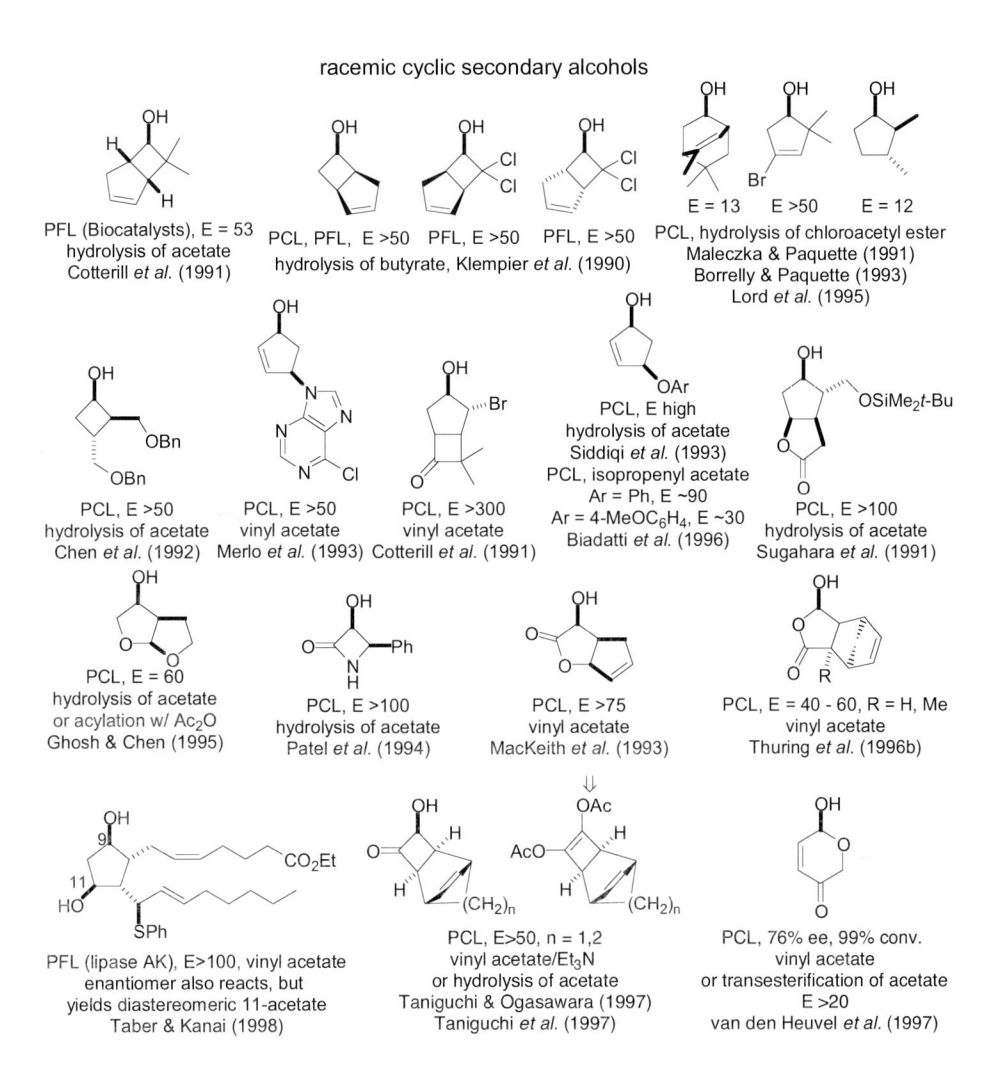

Fig. 67. Selected examples of enantioselective reactions of cyclic secondary alcohols catalyzed by *Pseudomonas* lipases.

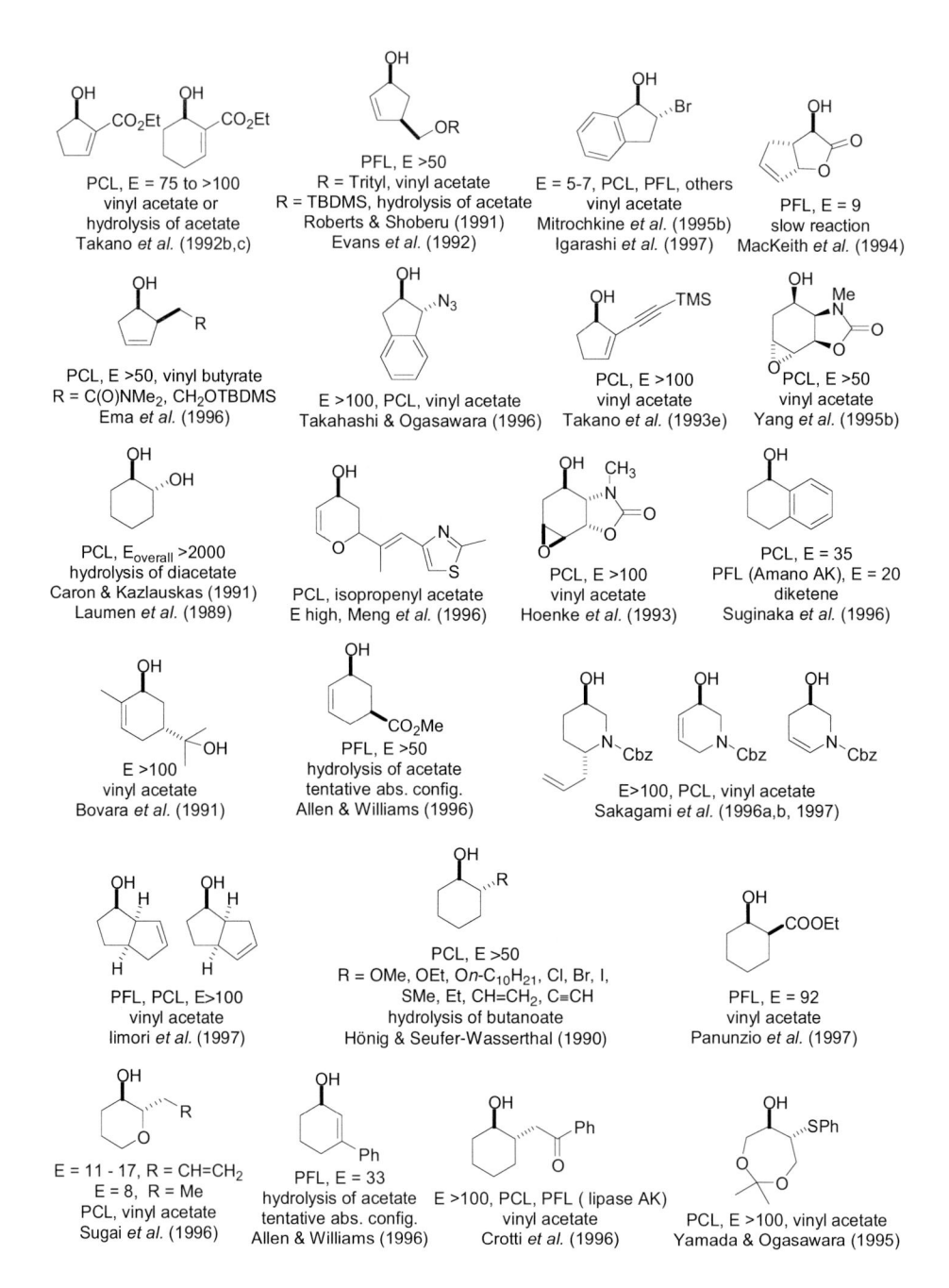

Fig. 67. Selected examples of enantioselective reactions of cyclic secondary alcohols catalyzed by *Pseudomonas* lipases (continued).

meso cyclic secondary alcohols

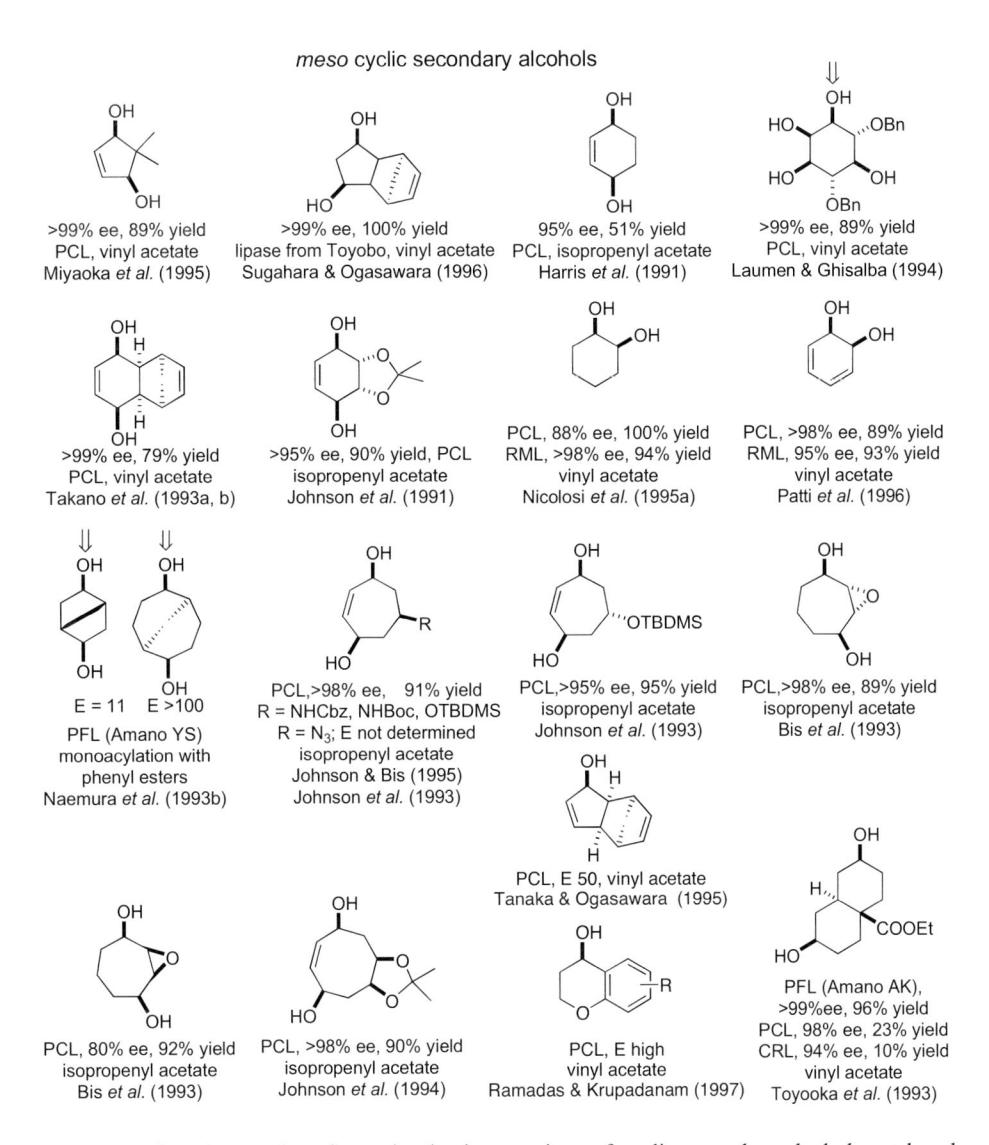

>99% ee, 89% yield
PCL, vinyl acetate
Miyaoka et al. (1995)

>99% ee, 100% yield
lipase from Toyobo, vinyl acetate
Sugahara & Ogasawara (1996)

95% ee, 51% yield
PCL, isopropenyl acetate
Harris et al. (1991)

>99% ee, 89% yield
PCL, vinyl acetate
Laumen & Ghisalba (1994)

>99% ee, 79% yield
PCL, vinyl acetate
Takano et al. (1993a, b)

>95% ee, 90% yield, PCL
isopropenyl acetate
Johnson et al. (1991)

PCL, 88% ee, 100% yield
RML, >98% ee, 94% yield
vinyl acetate
Nicolosi et al. (1995a)

PCL, >98% ee, 89% yield
RML, 95% ee, 93% yield
vinyl acetate
Patti et al. (1996)

E = 11   E >100
PFL (Amano YS)
monoacylation with
phenyl esters
Naemura et al. (1993b)

PCL, >98% ee, 91% yield
R = NHCbz, NHBoc, OTBDMS
R = N_3; E not determined
isopropenyl acetate
Johnson & Bis (1995)
Johnson et al. (1993)

PCL, >95% ee, 95% yield
isopropenyl acetate
Johnson et al. (1993)

PCL, >98% ee, 89% yield
isopropenyl acetate
Bis et al. (1993)

PCL, E 50, vinyl acetate
Tanaka & Ogasawara (1995)

PCL, 80% ee, 92% yield
isopropenyl acetate
Bis et al. (1993)

PCL, >98% ee, 90% yield
isopropenyl acetate
Johnson et al. (1994)

PCL, E high
vinyl acetate
Ramadas & Krupadanam (1997)

PFL (Amano AK),
>99%ee, 96% yield
PCL, 98% ee, 23% yield
CRL, 94% ee, 10% yield
vinyl acetate
Toyooka et al. (1993)

Fig. 67. Selected examples of enantioselective reactions of cyclic secondary alcohols catalyzed by *Pseudomonas* lipases (continued).

Except for increasing the difference in size of the substituents (see Sect. 5.2.1.1), or lowering the temperature (for an example see: Sakai et al., 1997), no general method exists to increase the enantioselectivity of PCL-catalyzed reactions. Longer esters or acylating agents (butyrates and above) sometimes give higher enantioselectivity than acetates. Hydrolyses of β-(phenylthio)- or β-(methylthio)acetoxy groups increased enantioselectivity ten-fold (e.g., from 6 to 55) as compared to hydrolyses of acetates or valerates (Itoh et al., 1991). Thiocrown ethers (e.g., 1,4,8,11-tetrathiacyclotetradecane) increased the enantioselectivity of PCL four-fold (from 9 to 27–37 and from 100 to 400) in the resolution of several secondary alcohols (Itoh et al., 1996b; Takagi et al., 1996). Itoh et al. (1993b) observed smaller increases with simple crown ethers.

Rhizomucor Lipases

Researchers have resolved fewer secondary alcohols using RML (Fig. 68). In triglycerides, RML selectively hydrolyzes esters at the primary alcohol positions, but the examples below show that RML can also hydrolyze esters of secondary alcohols.

2-alkanols

R = Bu, E = 9.5
R = Hx, E > 50
R = C_8H_{17}, E = 7.7
R = $C_{10}H_{21}$, E = 15
R = C≡C$(CH_2)_7CH_3$, E >50
R = CH_2CHMe_2, E = 24
R = c-Hx, E > 50
R = Ph, E = 42
RML, esterification with hexanoic or octanoic acid
Sonnet (1987)

RML, E = 19 (45°C), 170 (7°C)
vinyl butyrate
Noritomi et al. (1996)

RML, E = 50 - 106
vinyl acetate
Kaminska et al. (1996)

other acyclic secondary alcohols

RML, E = 10
hydrolysis of acetate
Chan et al. (1990)

RML, E = 12
hydrolysi of butyrate
Partali et al. (1993)

RML, E = 5
hydrolysis of butyrate
Waagen et al. (1993)

cyclic secondary alcohols

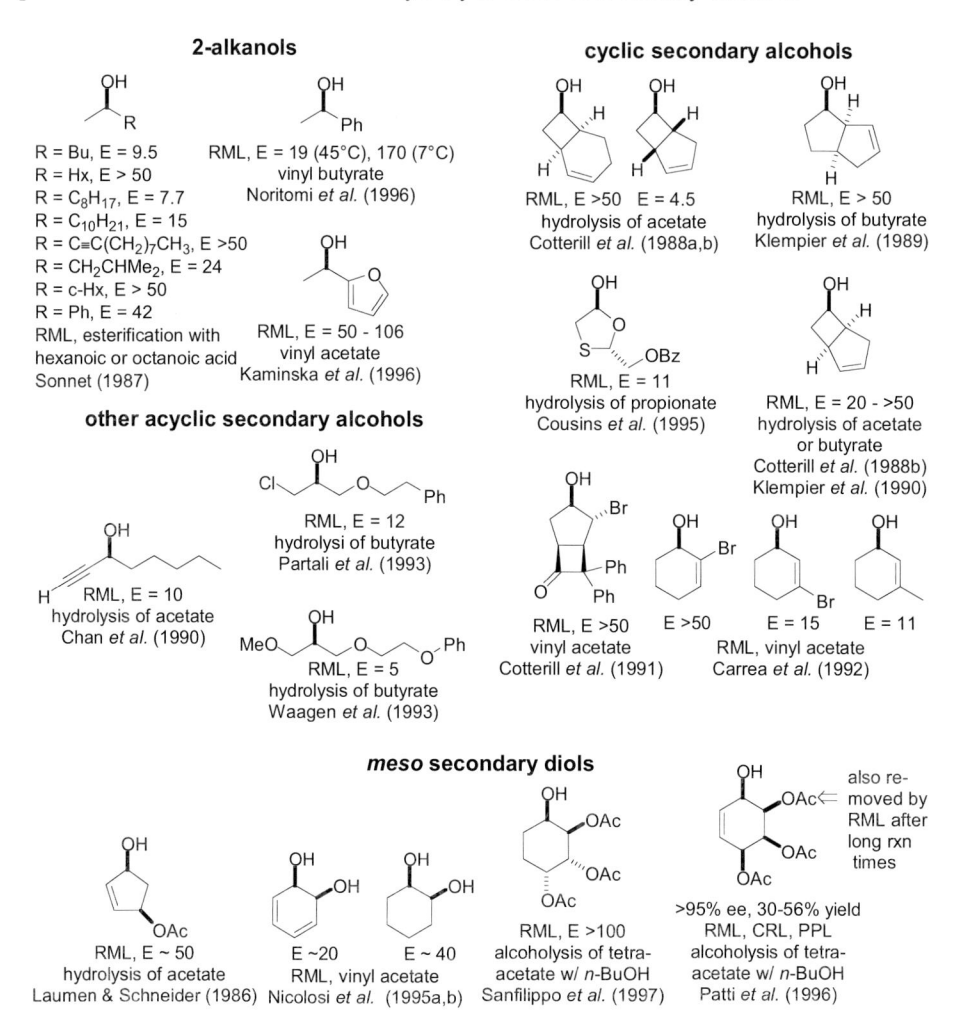

RML, E >50   E = 4.5
hydrolysis of acetate
Cotterill et al. (1988a,b)

RML, E > 50
hydrolysis of butyrate
Klempier et al. (1989)

RML, E = 11
hydrolysis of propionate
Cousins et al. (1995)

RML, E = 20 - >50
hydrolysis of acetate or butyrate
Cotterill et al. (1988b)
Klempier et al. (1990)

RML, E >50   E >50
vinyl acetate
Cotterill et al. (1991)

E = 15   E = 11
RML, vinyl acetate
Carrea et al. (1992)

meso secondary diols

RML, E ~ 50
hydrolysis of acetate
Laumen & Schneider (1986)

E ~20   E ~ 40
RML, vinyl acetate
Nicolosi et al. (1995a,b)

RML, E >100
alcoholysis of tetraacetate w/ n-BuOH
Sanfilippo et al. (1997)

also removed by RML after long rxn times

>95% ee, 30-56% yield
RML, CRL, PPL
alcoholysis of tetraacetate w/ n-BuOH
Patti et al. (1996)

Fig. 68. Examples of *Rhizomucor miehei* lipase-catalyzed enantioselective reactions of secondary alcohols.

Noritomi et al. (1996) increased the enantioselectivity of an RML-catalyzed acylation of α-phenylethanol by lowering the temperature. Acylation with vinyl butyrate in dioxane gave E = 19 at 45°C, but E = 170 at 7°C. In other solvents (pyridine, THF, triethylamine) temperature did not affect E.

Other Lipases

Other lipases are also enantioselective toward secondary alcohols and usually follow the rule in Fig. 53. Selected examples are in Fig. 69.

Naemura et al. (1996) surveyed the enantioselectivity of a lipase from *Alcaligenes* sp. and found good enantioselectivity toward a range of secondary alcohols (27 examples, mostly MeCH(OH)aryl). The favored enantiomer was the one predicted by the empirical rule in Fig. 53.

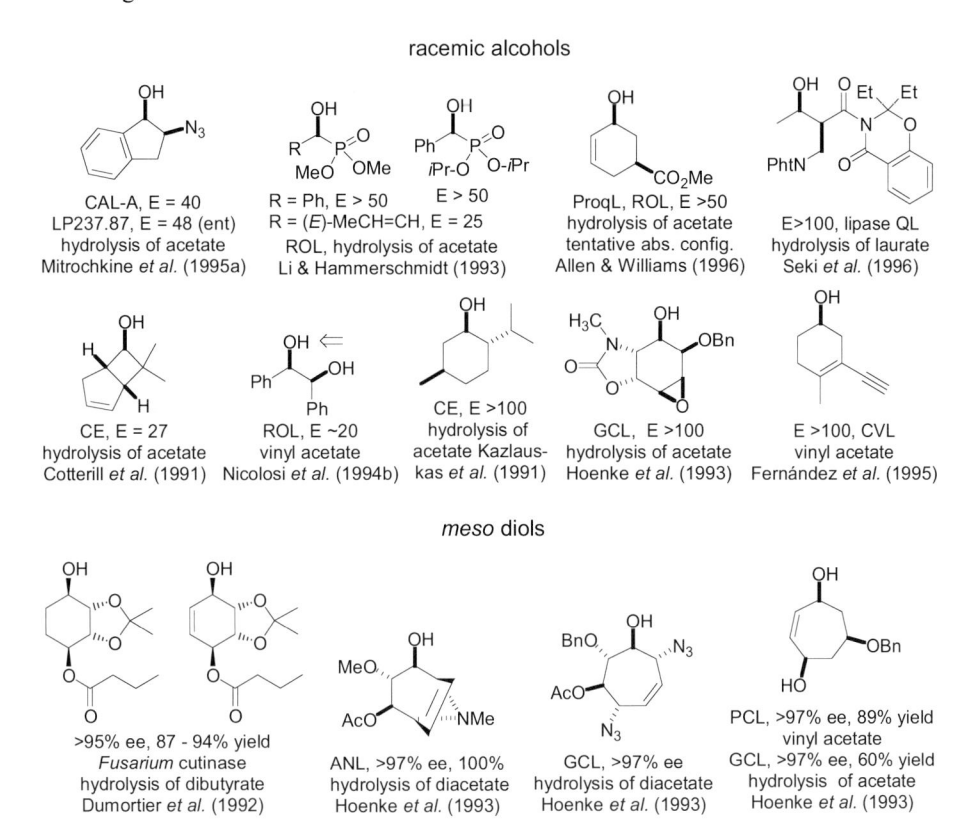

Fig. 69. Selected examples of enantioselective reactions catalyzed by other lipases.

5.2 Survey of Enantioselective Lipase-Catalyzed Reactions

Choosing the Best Route

The best route to a particular compound is rarely a straightforward choice. In addition to several lipase-catalyzed routes, researchers consider other chemical and biochemical routes. The 'best' is often an individual decision which depends on the intended next steps and available starting materials. The two examples below summarize only the lipase-catalyzed routes to these targets.

Inositols. Researchers have found a number of different routes to enantiomerically pure inositol derivatives. Starting from the achiral *myo*-inositol, researchers added protective groups to increase the size of one substituent. Protection yields either a racemate or a *meso* derivative. Several different lipases showed excellent enantioselectivity. The asymmetric synthesis starting from *meso* derivatives is probably the best route since it gives both high yield and high enantioselectivity. However, in some cases another route may fit better with subsequent synthetic steps (Fig. 70).

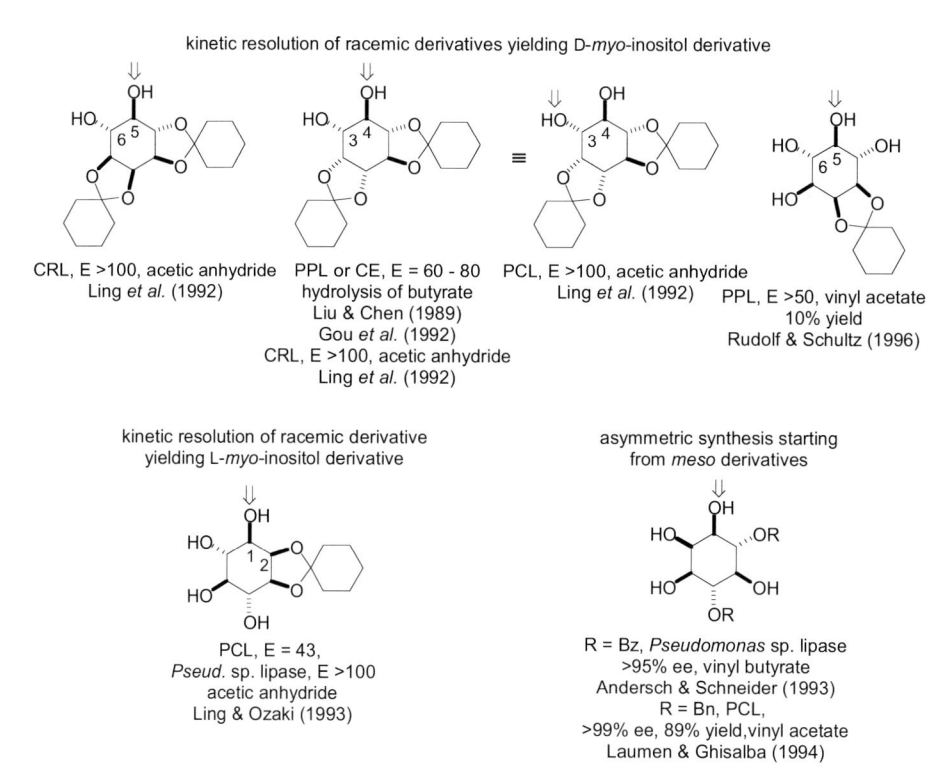

Fig. 70. Lipase-catalyzed enantioselective reactions of *myo*-inositol derivatives.

β-Blockers. Most β-adrenergic antagonists (β-blockers), used for the treatment of hypertension and angina pectoris, are 3-aryloxy-2-propanolamines which contain a secondary alcohol stereocenter. The (*S*)-enantiomer is usually more active than (*R*), e.g., one hundred times more active in the case of propranolol. For this reason, chemists have developed routes to the (*S*)-enantiomer, including lipase-catalyzed reactions (Fig. 71) (reviewed by Kloosterman et al., 1988; Sheldon, 1993). None of these routes have been commercialized. In most examples in Fig. 71 the aryl group is the large substituent, thus the (*R*)-enantiomer reacts faster. Resolution by acylation of the alcohol is preferred over hydrolysis of the acetate because acylation yields the desired (*S*)-alcohol as the unreacted starting material. A resolution by hydrolysis requires an extra step to hydrolyze the unreacted ester.

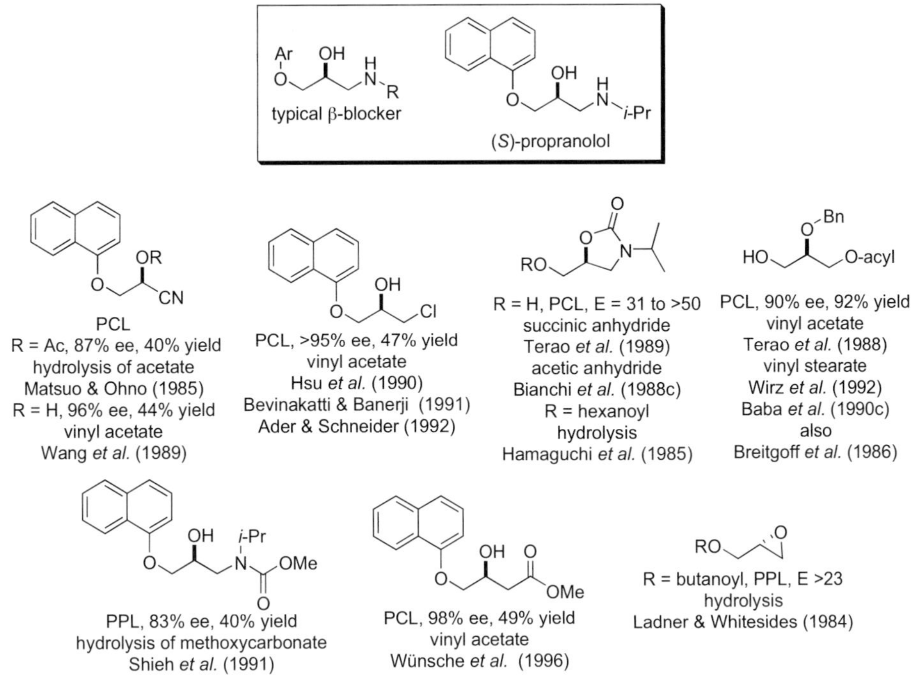

Fig. 71. Lipase-catalyzed routes to enantiomerically pure precursors of propranolol.

5.2.1.2 Primary Alcohols

Lipases usually show lower enantioselectivity toward primary alcohols than toward secondary alcohols. Only PPL and PCL show high enantioselectivity toward a wide range of primary alcohols.

Pseudomonas *Lipases*

An empirical rule can predict some of the enantiopreference of PCL toward primary alcohols (Weissfloch and Kazlauskas, 1995). Like the secondary alcohol rule above, the primary alcohol rule is based on the size of the substituents, but, surprisingly the sense of enantiopreference toward it is opposite. That is, the –OH of secondary alcohols and the –CH_2OH of primary alcohols point in opposite directions. One way to rationalize this opposite enantiopreference is to assume the extra CH_2 in primary alcohols introduces a kink between the stereocenter and the oxygen as suggested in Fig. 72. In this manner the large and medium substituents bind in the same enzyme pockets in both cases. Another possibility is a different binding mode for primary alcohols. Indeed, modeling suggests that the large substituent of primary alcohols does not bind in the same pocket as secondary alcohols (Tuomi and Kazlauskas, 1999).

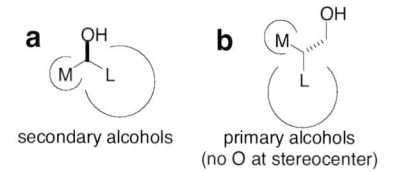

Fig. 72. Empirical rules summarize the enantiopreference of PCL toward primary and secondary alcohols. **a** Shape of the favored enantiomer of secondary alcohols. **b** Shape of the favored enantiomer of primary alcohols. This rule is reliable only when there is no oxygen directly bonded at the stereocenter. Computer modeling (and the drawing above) suggest that the large substituent, L, binds to different regions of the active site. Note that PCL shows an opposite enantiopreference toward these two classes of alcohols.

Not all primary alcohols fit this rule. In particular, primary alcohols that have an oxygen at the stereocenter (e.g., glycerol derivatives) do not fit this rule. Of the remaining primary alcohol examples, the rule showed an 89% reliability (correct for 54 of 61 examples).

An example of how the empirical rule does not apply to primary alcohols with an oxygen at the stereocenter are the γ-butyrolactones in Fig. 73. For the *trans* isomer, PCL favors one enantiomer, but for the *cis* isomer PCL favors the other (Ha et al., 1998).

R = H, E = 30
R = Me, E = 38
R = Et, E = 16

E >100

PCL, hydrolysis of acetate
Ha et al. (1998)

Fig. 73. Examples which do not follow the empirical rule for primary alcohols.

For secondary alcohols, increasing the difference in the size of the substituents often increases the enantioselectivity of PCL and other lipases. Indeed, researchers often introduce a large protective group to increase the enantioselectivity (Sect. 5.2.1.1). However, for primary alcohols this strategy is not reliable. Upon adding large substituents the enantioselectivity sometimes increased (Lampe et al., 1996), sometimes decreased, and sometimes remained unchanged (Weissfloch and Kazlauskas, 1995).

Selected examples of primary alcohols that are resolved or desymmetrized by PCL are shown in Fig. 74. More examples are found in a survey (Weissfloch and Kazlauskas, 1995).

One group of popular substrates are the *meso*-1,3-diols which can be asymmetrized either by hydrolysis of the diester or acylation of the diol. For hydrolysis reactions, Liu et al. (1990) noted that acyl migration in the monoester was fast enough in aqueous solution at pH 7 and above to lower the enantiomeric purity of the product. Acyl migration slows considerably in organic solvents (benzene, chloroform, THF). For this reason, asymmetrization of a diol by acylation can give higher enantiomeric purity.

racemic primary alcohols

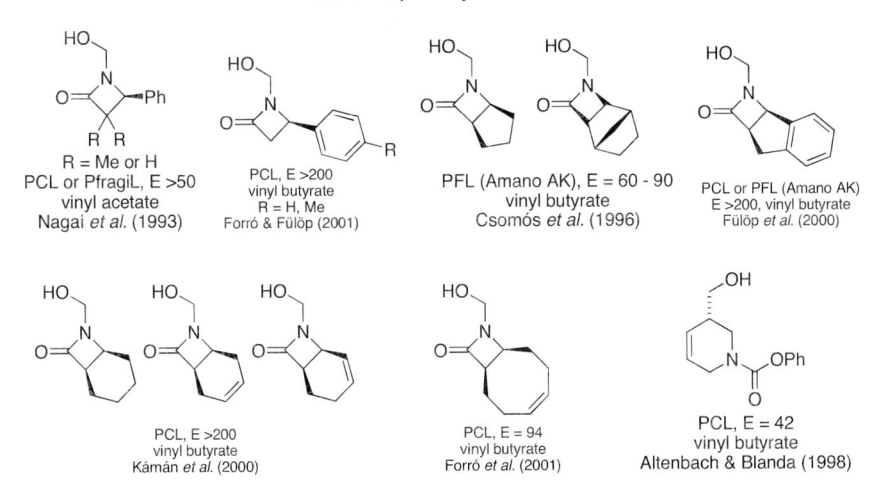

R = Me or H
PCL or PfragiL, E >50
vinyl acetate
Nagai et al. (1993)

PCL, E >200
vinyl butyrate
R = H, Me
Forró & Fülöp (2001)

PFL (Amano AK), E = 60 - 90
vinyl butyrate
Csomós et al. (1996)

PCL or PFL (Amano AK)
E >200, vinyl butyrate
Fülöp et al. (2000)

PCL, E >200
vinyl butyrate
Kámán et al. (2000)

PCL, E = 94
vinyl butyrate
Forró et al. (2001)

PCL, E = 42
vinyl butyrate
Altenbach & Blanda (1998)

Fig. 74. Selected examples of PCL-catalyzed enantioselective reactions of primary alcohols.

108   5.2 Survey of Enantioselective Lipase-Catalyzed Reactions

racemic primary alcohols

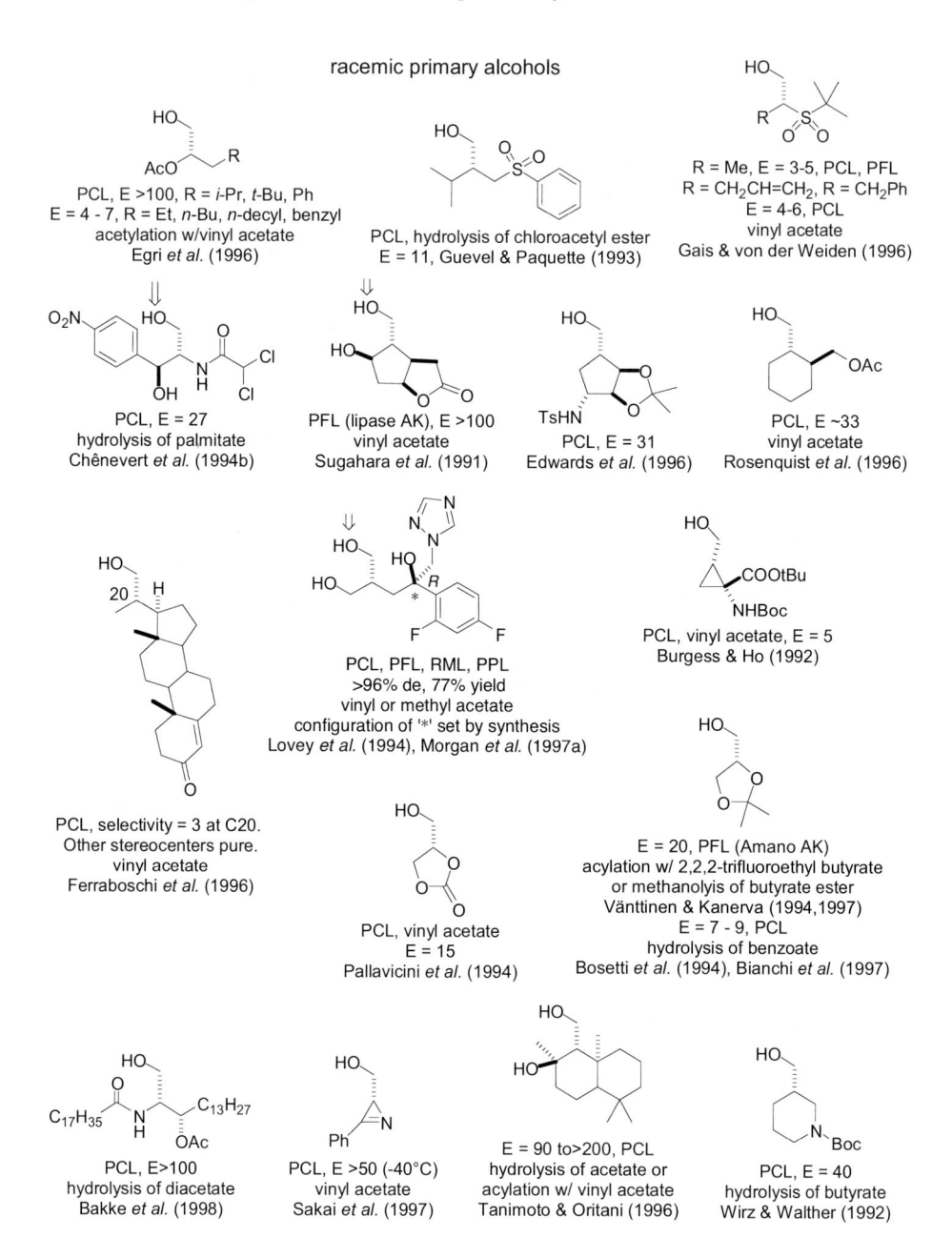

PCL, E >100, R = *i*-Pr, *t*-Bu, Ph
E = 4 - 7, R = Et, *n*-Bu, *n*-decyl, benzyl
acetylation w/vinyl acetate
Egri *et al.* (1996)

PCL, hydrolysis of chloroacetyl ester
E = 11, Guevel & Paquette (1993)

R = Me, E = 3-5, PCL, PFL
R = CH$_2$CH=CH$_2$, R = CH$_2$Ph
E = 4-6, PCL
vinyl acetate
Gais & von der Weiden (1996)

PCL, E = 27
hydrolysis of palmitate
Chênevert *et al.* (1994b)

PFL (lipase AK), E >100
vinyl acetate
Sugahara *et al.* (1991)

PCL, E = 31
Edwards *et al.* (1996)

PCL, E ~33
vinyl acetate
Rosenquist *et al.* (1996)

PCL, selectivity = 3 at C20.
Other stereocenters pure.
vinyl acetate
Ferraboschi *et al.* (1996)

PCL, PFL, RML, PPL
>96% de, 77% yield
vinyl or methyl acetate
configuration of '*' set by synthesis
Lovey *et al.* (1994), Morgan *et al.* (1997a)

PCL, vinyl acetate, E = 5
Burgess & Ho (1992)

PCL, vinyl acetate
E = 15
Pallavicini *et al.* (1994)

E = 20, PFL (Amano AK)
acylation w/ 2,2,2-trifluoroethyl butyrate
or methanolyis of butyrate ester
Vänttinen & Kanerva (1994,1997)
E = 7 - 9, PCL
hydrolysis of benzoate
Bosetti *et al.* (1994), Bianchi *et al.* (1997)

PCL, E>100
hydrolysis of diacetate
Bakke *et al.* (1998)

PCL, E >50 (-40°C)
vinyl acetate
Sakai *et al.* (1997)

E = 90 to>200, PCL
hydrolysis of acetate or
acylation w/ vinyl acetate
Tanimoto & Oritani (1996)

PCL, E = 40
hydrolysis of butyrate
Wirz & Walther (1992)

Fig. 74. Selected examples of PCL-catalyzed enantioselective reactions of primary alcohols (continued).

meso and prochiral primary alcohols

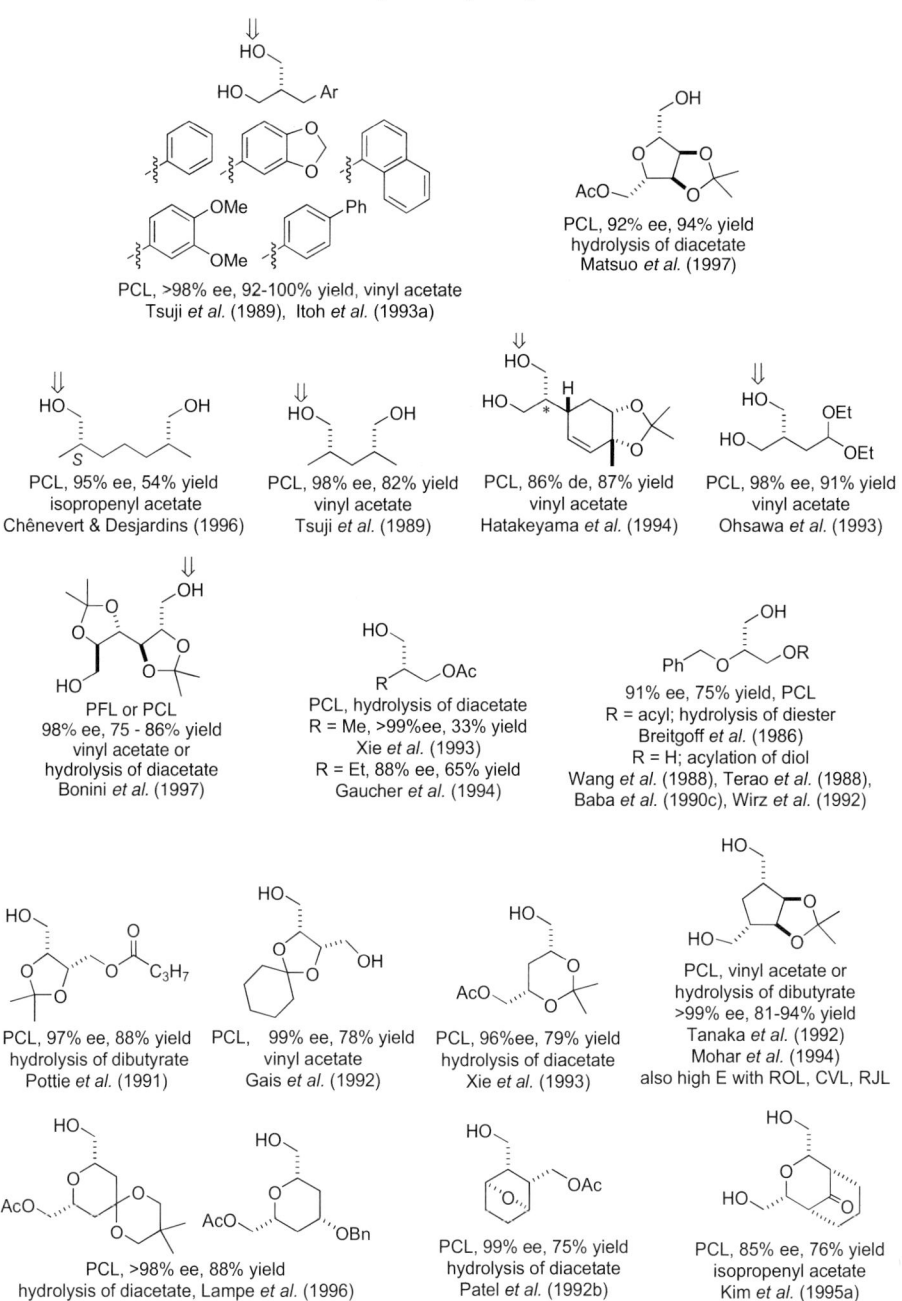

PCL, >98% ee, 92-100% yield, vinyl acetate
Tsuji *et al.* (1989), Itoh *et al.* (1993a)

PCL, 92% ee, 94% yield
hydrolysis of diacetate
Matsuo *et al.* (1997)

PCL, 95% ee, 54% yield
isopropenyl acetate
Chênevert & Desjardins (1996)

PCL, 98% ee, 82% yield
vinyl acetate
Tsuji *et al.* (1989)

PCL, 86% de, 87% yield
vinyl acetate
Hatakeyama *et al.* (1994)

PCL, 98% ee, 91% yield
vinyl acetate
Ohsawa *et al.* (1993)

PFL or PCL
98% ee, 75 - 86% yield
vinyl acetate or
hydrolysis of diacetate
Bonini *et al.* (1997)

PCL, hydrolysis of diacetate
R = Me, >99%ee, 33% yield
Xie *et al.* (1993)
R = Et, 88% ee, 65% yield
Gaucher *et al.* (1994)

91% ee, 75% yield, PCL
R = acyl; hydrolysis of diester
Breitgoff *et al.* (1986)
R = H; acylation of diol
Wang *et al.* (1988), Terao *et al.* (1988),
Baba *et al.* (1990c), Wirz *et al.* (1992)

PCL, 97% ee, 88% yield
hydrolysis of dibutyrate
Pottie *et al.* (1991)

PCL, 99% ee, 78% yield
vinyl acetate
Gais *et al.* (1992)

PCL, 96%ee, 79% yield
hydrolysis of diacetate
Xie *et al.* (1993)

PCL, vinyl acetate or
hydrolysis of dibutyrate
>99% ee, 81-94% yield
Tanaka *et al.* (1992)
Mohar *et al.* (1994)
also high E with ROL, CVL, RJL

PCL, >98% ee, 88% yield
hydrolysis of diacetate, Lampe *et al.* (1996)

PCL, 99% ee, 75% yield
hydrolysis of diacetate
Patel *et al.* (1992b)

PCL, 85% ee, 76% yield
isopropenyl acetate
Kim *et al.* (1995a)

Fig. 74. Selected examples of PCL-catalyzed enantioselective reactions of primary alcohols (continued).

5.2 Survey of Enantioselective Lipase-Catalyzed Reactions

Porcine Pancreatic Lipase

No generally applicable rule to predict the fast-reacting enantiomer in PPL catalyzed reactions of primary alcohols exists. Researchers have proposed several rules, but none are satisfactory (Ehrler and Seebach, 1990; Guanti et al., 1992; Hultin and Jones, 1992; Wimmer, 1992). Two rules even predict opposite enantiomers. An example of the difficulties is shown in Fig. 75. Enantiopreference of PPL reversed upon changing from a *trans* to a *cis* configuration of the double bond in the 2-substituted 1,3-propandiol derivatives below. PPL favored the (*S*)-enantiomer with high enantioselectivity for the *trans* isomer, the (*S*) enantiomer with moderate enantioselectivity for the saturated analog, but the (*R*)-enantiomer with low to moderate enantioselectivity for the *cis* isomer.

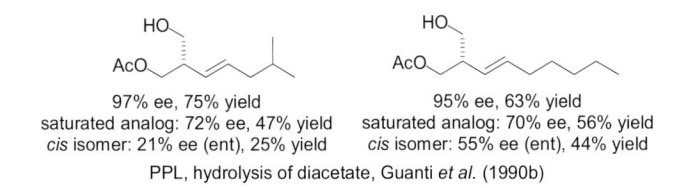

97% ee, 75% yield
saturated analog: 72% ee, 47% yield
cis isomer: 21% ee (ent), 25% yield

95% ee, 63% yield
saturated analog: 70% ee, 56% yield
cis isomer: 55% ee (ent), 44% yield

PPL, hydrolysis of diacetate, Guanti et al. (1990b)

Fig. 75. The presence and configuration of a double bond change and even reverse the enantioselectivity.

This reversal is difficult to explain using only the relative sizes of the substituents. Note that for secondary alcohols, the enantioselectivity also varied with the configuration of double bonds in the large substituent, but the enantiopreference remained the same (Morgan et al., 1992). Selected examples of PPL-catalyzed resolutions of primary alcohols are shown in Fig. 76.

kinetic resolution of racemic primary alcohols

PPL
R = Me, Et, E=17-21
R = Pr, Bu, E >50
ethyl acetate
Fernández et al. (1992)

PPL, E = 13
hydrolysis of butyrate
Ladner & Whitesides
(1984)

R = n-C$_6$H$_{13}$ or i-Pr
E = 13 to >50, PPL
hydrolysis or acylation
Bianchi et al. (1988a)
Fukusaki et al. (1992b)

purified PPL, E=17-24
hydrolysis of butyrate
Quartey et al. (1996)

PPL, E >50
vinyl acetate or hydrolysis of acetate
Serebryakov et al. (1995)
Gamalevich et al. (1998)

PPL, E = 30
hydrolysis of acetate
Ley et al. (1996)

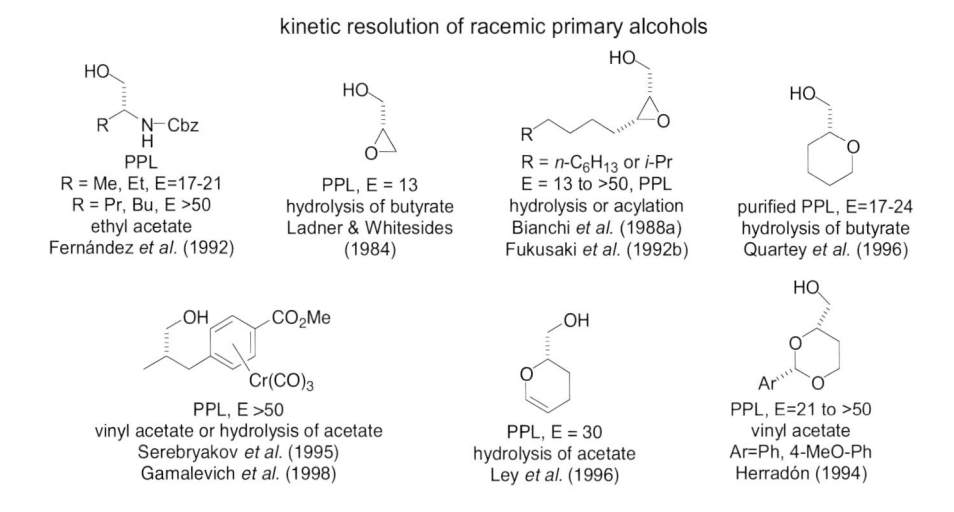

PPL, E=21 to >50
vinyl acetate
Ar=Ph, 4-MeO-Ph
Herradón (1994)

Fig. 76. Selected examples of PPL-catalyzed enantioselective reactions of primary alcohols.

Although there is no general method to increase the enantioselectivity of PPL-catalyzed reactions of primary alcohols, Serebryakov et al. (1995) dramatically increased the enantioselectivity of PPL toward 3-aryl-2-methyl propanol from E = 2 to E > 50 by coordinating a chromium tricarbonyl group to the aryl. After resolution, oxidation with iodine removed the chromium tricarbonyl group. Large differences in the size of the substituents are not always necessary. Ley et al. (1996) efficiently resolved pyrans having minimal differences in the sizes of the substituents (Fig. 76).

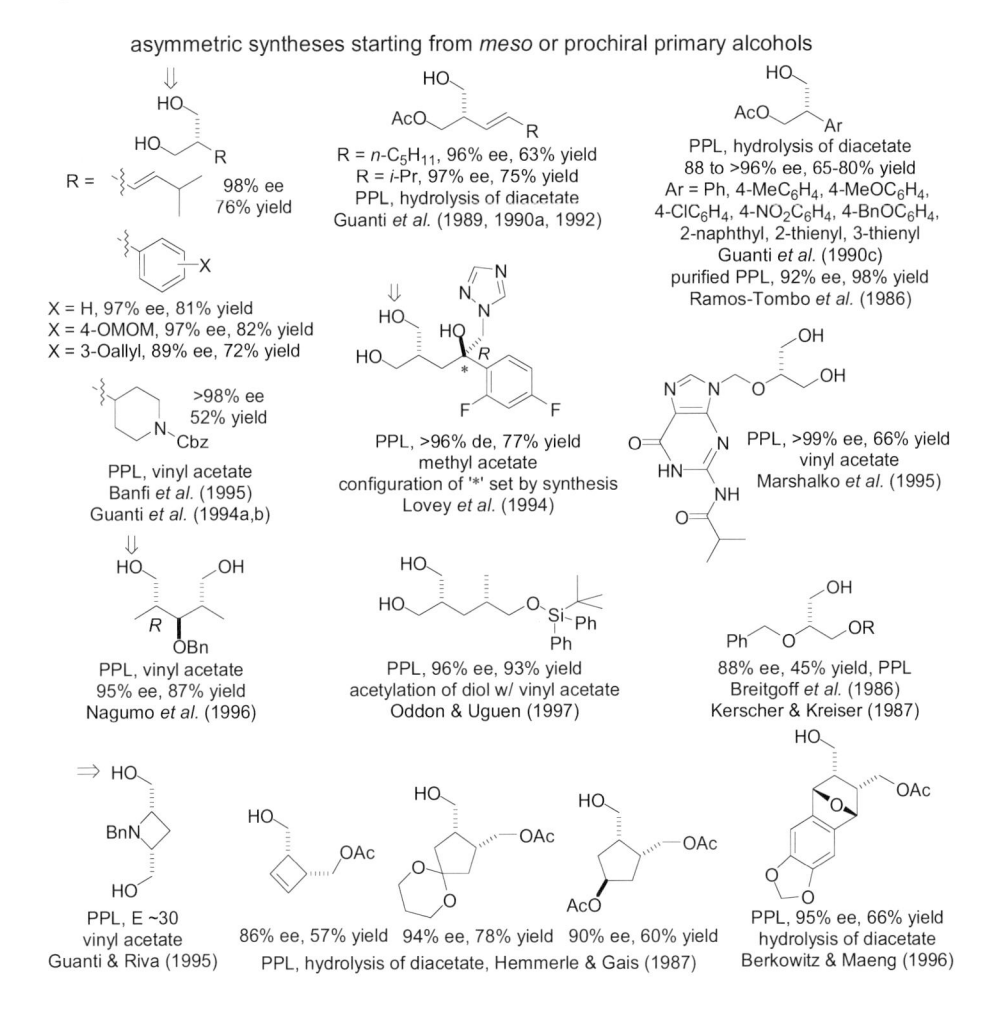

Fig. 76. Selected examples of PPL-catalyzed enantioselective reactions of primary alcohols (continued).

5.2 Survey of Enantioselective Lipase-Catalyzed Reactions

Other Lipases

Only a handful of primary alcohols have been either resolved or desymmetrized by other lipases. Selected examples are shown in Fig. 77.

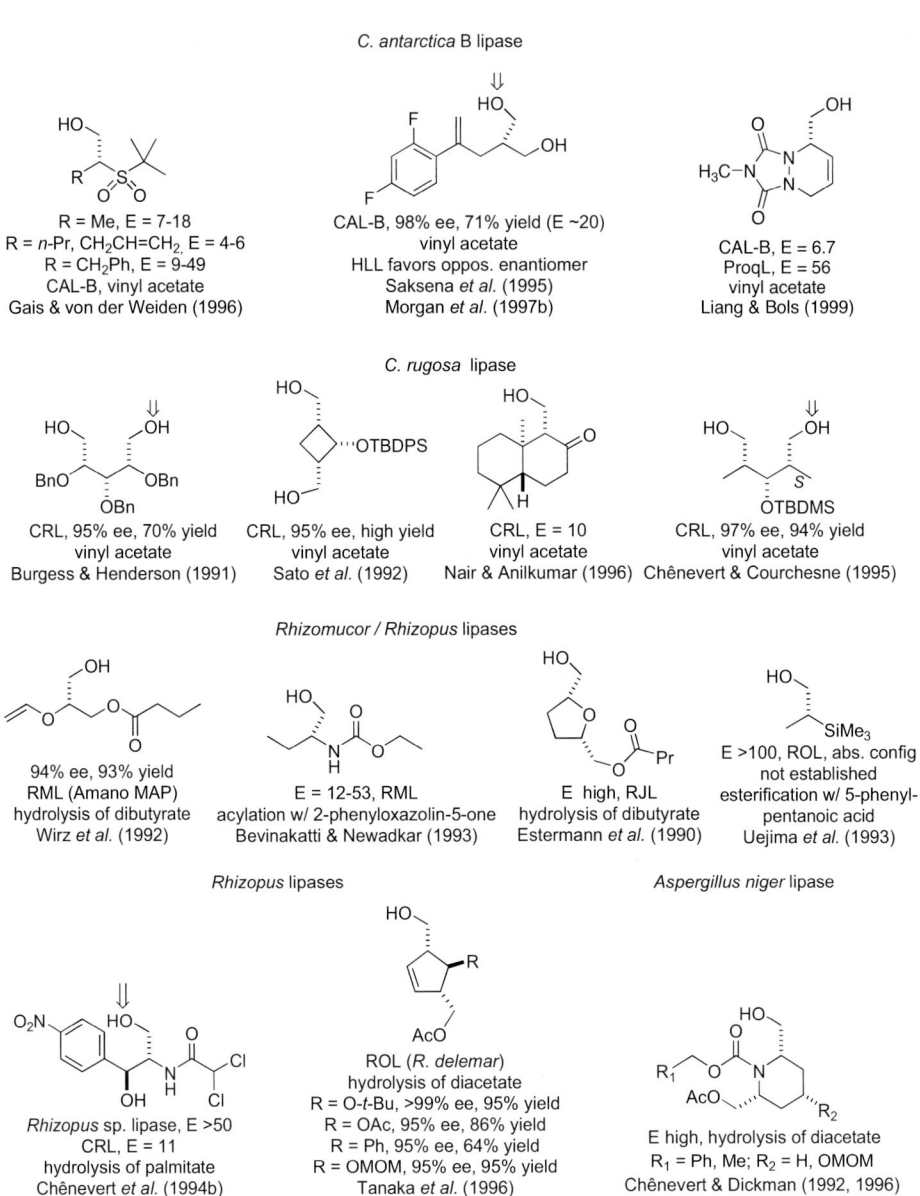

C. antarctica B lipase

R = Me, E = 7-18
R = n-Pr, CH$_2$CH=CH$_2$, E = 4-6
R = CH$_2$Ph, E = 9-49
CAL-B, vinyl acetate
Gais & von der Weiden (1996)

CAL-B, 98% ee, 71% yield (E ~20)
vinyl acetate
HLL favors oppos. enantiomer
Saksena et al. (1995)
Morgan et al. (1997b)

CAL-B, E = 6.7
ProqL, E = 56
vinyl acetate
Liang & Bols (1999)

C. rugosa lipase

CRL, 95% ee, 70% yield
vinyl acetate
Burgess & Henderson (1991)

CRL, 95% ee, high yield
vinyl acetate
Sato et al. (1992)

CRL, E = 10
vinyl acetate
Nair & Anilkumar (1996)

CRL, 97% ee, 94% yield
vinyl acetate
Chênevert & Courchesne (1995)

Rhizomucor / Rhizopus lipases

94% ee, 93% yield
RML (Amano MAP)
hydrolysis of dibutyrate
Wirz et al. (1992)

E = 12-53, RML
acylation w/ 2-phenyloxazolin-5-one
Bevinakatti & Newadkar (1993)

E high, RJL
hydrolysis of dibutyrate
Estermann et al. (1990)

E >100, ROL, abs. config
not established
esterification w/ 5-phenyl-
pentanoic acid
Uejima et al. (1993)

Rhizopus lipases

Rhizopus sp. lipase, E >50
CRL, E = 11
hydrolysis of palmitate
Chênevert et al. (1994b)

ROL (*R. delemar*)
hydrolysis of diacetate
R = O-t-Bu, >99% ee, 95% yield
R = OAc, 95% ee, 86% yield
R = Ph, 95% ee, 64% yield
R = OMOM, 95% ee, 95% yield
Tanaka et al. (1996)

Aspergillus niger lipase

E high, hydrolysis of diacetate
R$_1$ = Ph, Me; R$_2$ = H, OMOM
Chênevert & Dickman (1992, 1996)

Fig. 77. Selected examples of enantioselective reactions of primary alcohols catalyzed by other lipases.

5 Lipases and Esterases

Enantioselectivity of Lipases Toward Triglycerides

Triglycerides are presumably the natural substrates of lipases, so many researchers have investigated the enantioselectivity of lipases toward triglycerides. Triglycerides with identical functional groups at the two primary alcohols (*sn*-1 and *sn*-3) are prochiral. The stereochemical numbering for triglycerides and other glycerol derivatives starts with a Fischer projection of the triglyceride with the central hydroxyl group positioned to the left. Numbering the carbons *sn*-1, *sn*-2, and *sn*-3 from top to bottom uniquely identifies each position. Enantioselectivity towards triglycerides is based on the ability of a lipase to discriminate between the *sn*-1 and *sn*-3 position. Enantiomers thus formed are 1,2- or 2,3-diglycerides or 1- or 3- monoglycerides (Fig. 78).

```
                    sn-1 position
                    ↙
                 CH₂OAc
            AcO──┼──H
                 CH₂OAc
                    ↖
                    sn-3 position
```

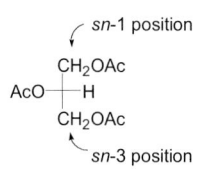

Fig. 78.   Stereochemical numbering of triglycerides.

Rogalska et al. (1993) surveyed the enantioselectivity of 25 lipases in triglyceride monolayers. Some lipases (e.g., from *Pseudomonas* sp., RML, CRL) showed high selectivity toward the *sn*-1 position, but the stereoselectivity of other lipases varied with the triglyceride: CAL-B showed *sn*-3 selectivity with trioctanoin, but *sn*-1 selectivity with triolein. Selectivity also varied with interfacial tension. Stadler et al. (1995) found large changes as well as reversals in selectivity using analogs of triglycerides with ether or alkyl groups at the *sn*-2 position. For ROL, Holzwarth et al. (1997) rationalized these changes using computer modeling. In recent publications it was proposed that the flexibility of the substituent in *sn*-2 position as well as the geometry of the substrate docked to the active site of ROL determines enantioselectivity. If the torsion angle Φ_{O3-C3} of glycerol is > 150°, the substrate will be hydrolyzed preferentially in *sn*-1 position, if Φ_{O3-C3} is < 150°, ROL is *sn*-3 selective. Moreover, enantioselectivity could be altered by site-directed mutagenesis (Scheib et al., 1998; 1999).

5.2.1.3   Other Alcohols, Amines, and Alcohol Analogs

Tertiary Alcohols and Other Quaternary Stereocenters

Lipase-catalyzed reactions involving tertiary alcohols are slow, presumably due to steric hindrance. O'Hagan and Zaidi (1992; 1994b) resolved several acetylenic alcohols with CRL (Fig. 79). O'Hagan and Zaidi (1992, 1994b) suggested that the acetylenic moiety in these tertiary alcohols occupies the same site as a hydrogen in secondary alcohols because CRL showed low enantioselectivity toward alcohols containing both a hydrogen and a –C≡CH substituent at the stereocenter. No CRL-catalyzed hydrolysis occurred when O'Hagan and Zaidi (1992, 1994b) replaced the acetylenic moiety with Me, vinyl, or nitrile. The ability of CRL to catalyze reactions involving tertiary alcohols is consistent with a large alcohol binding site as suggested above in Sect. 5.2.1.1. Surprisingly, RML also catalyzed hydrolysis of an acetate of a tertiary alcohol.

Brackenridge et al. (1993) used an oxalate ester to introduce a less hindered ester group. PPL-catalyzed hydrolysis of the mixed oxalate ester showed moderate enantioselectivity, although the actual site of reaction was not determined.

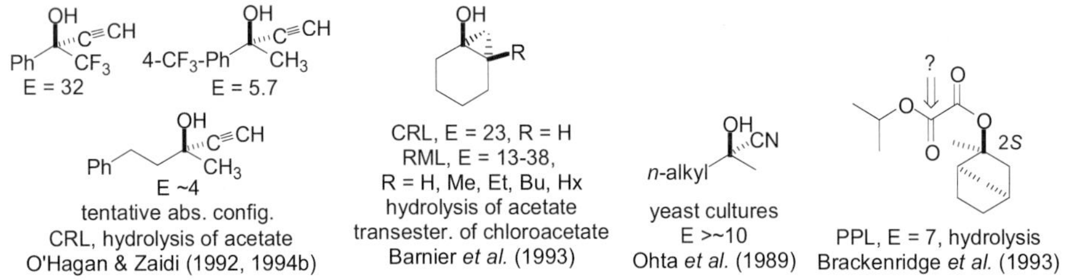

Fig. 79. Lipase catalyzed enantioselective reactions of tertiary alcohols.

More recently, it was discovered that a certain amino acid motif (GGG(A)X-motif, in single-letter amino acid code where G denotes glycine and X denotes any amino acid; in a few enzymes, one glycine is replaced by an alanine, A) located in the oxyanion binding pocket of lipases and esterases determines activity towards tertiary alcohols (Henke et al., 2002). All enzymes investigated bearing this motif (e.g. lipase from *Candida rugosa*, pig liver esterase, acetyl choline esterases and a *p*-nitrobenzyl esterase from *Bacillus subtilis* (BsubpNBE)) were active towards several acetates of tertiary alcohols, while enzymes bearing the more common GX-motif (Pleiss et al., 2000) did not hydrolyze the model compounds linalyl acetate, methyl-1-pentin-1-yl acetate and 2-phenyl-3-butin-2-yl acetate. Consequently, the search for a hydrolase active towards the sterically demanding tertiary alcohol moiety is considerably facilitated, as it can be simply based on the existence of the GGG(A)X-motif.

The rather low enantioselectivity of all enzymes was later improved based on computer modeling. One point mutation (Gly105Ala) in BsubpNBE increased the E-value from $E = 6$ to $E = 19$ in the kinetic resolution of 2-phenyl-3-butin-2-yl acetate (Henke et al., 2003).

In primary alcohols with quaternary stereocenter, the hindered stereocenter lies further from the reactive hydroxyl, so lipase-catalyzed reactions remain fast. For diols of the type RR'C(OH)CH$_2$OH, two groups proposed models to predict the fast reacting enantiomer in PFL-catalyzed reactions (Fig. 80) (Chen and Fang, 1997; Hof and Kellog, 1996a). Hof and Kellog (1996a) proposed a flat pocket for one substituent, while Chen and Fang proposed a simplified version. Fig. 81 shows more examples of primary alcohols with quaternary stereocenters. Many of these also follow the model, even though the reactions use lipases other than PFL. Other examples of quaternary stereocenters are in Sect. 5.2.2.4 on acids and Sect. 5.2.1.3 on alcohols with remote stereocenters.

Fig. 80. Model and several examples of primary alcohols with a quaternary stereocenter resolved by PFL (lipase AK).

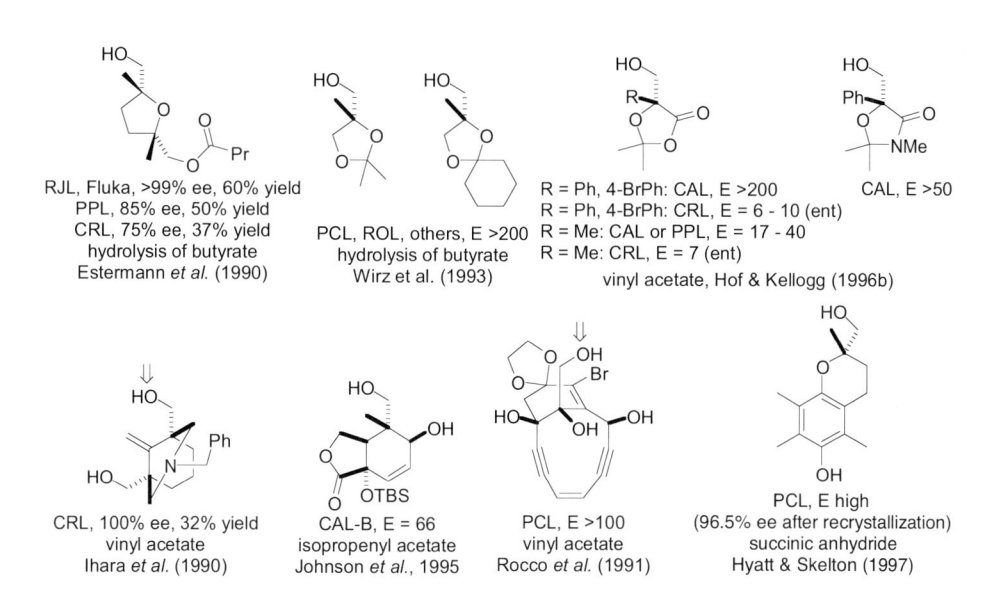

Fig. 81. Lipase-catalyzed enantioselective reactions of primary alcohols with quaternary stereocenters.

116   5.2   Survey of Enantioselective Lipase-Catalyzed Reactions

⇓

HO
HO
MeOOC   n-$C_{14}H_{29}$

PFL (Fluka), E = 15-19
PCL, E = 14
vinyl acetate
Jiménez et al. (1997)

HO
HO
I
R

PPL, vinyl acetate
E >100, R = Ph, SiMe$_3$, t-Bu
E = 24, R = CH_2OCH_2Ph
E = 10 - 47, R = n-Bu, n-Hex
E = 6.5, R = CH_2CH_2Ph
Chen & Fang (1997)

HO
HO
Cl
Ph

PPL, vinyl acetate
E = 72
Chen & Fang (1997)

⇓

MeO
S
HO
HO
F

CRL, E = 45, vinyl acetate or
hydrolysis of acetate
PCL, E >100, vinyl acetate
Khlebnikov et al. (1995)

HO   OBn
OH
OMe
R
N
COO-n-Pr

PCL, E = 32 - 68
isopropenyl acetate
Sih (1996)
Henegar et al. (1997)

N
N
HO
HO
F
F

PPL, E ~13
methyl acetate
Lovey et al. (1994)

HO
NC
Ph

CRL, E =13.5
vinyl acetate
abs. config. not established
Cheong et al. (1996)

HO
BzHN   R

R = i-Pr; GCL, E = 9
R = Bn, CH_2Ar; PFL
(Amano AK), E = 4-5
vinyl acetate
Berkowitz et al. (1994)

HO
F
R   Ph

PCL, acetic anhydride
R = Me, E = 21- 38
R = Et, n-Pr, n-Bu, E >50
Goj et al. (1997)

HO
O
R

R = Bn, E >100
R = C_9H_{19}, E = 82

HO
O
E >100

HO
O

PCL, E >50
vinyl acetate
Ferraboschi et al. (1993)

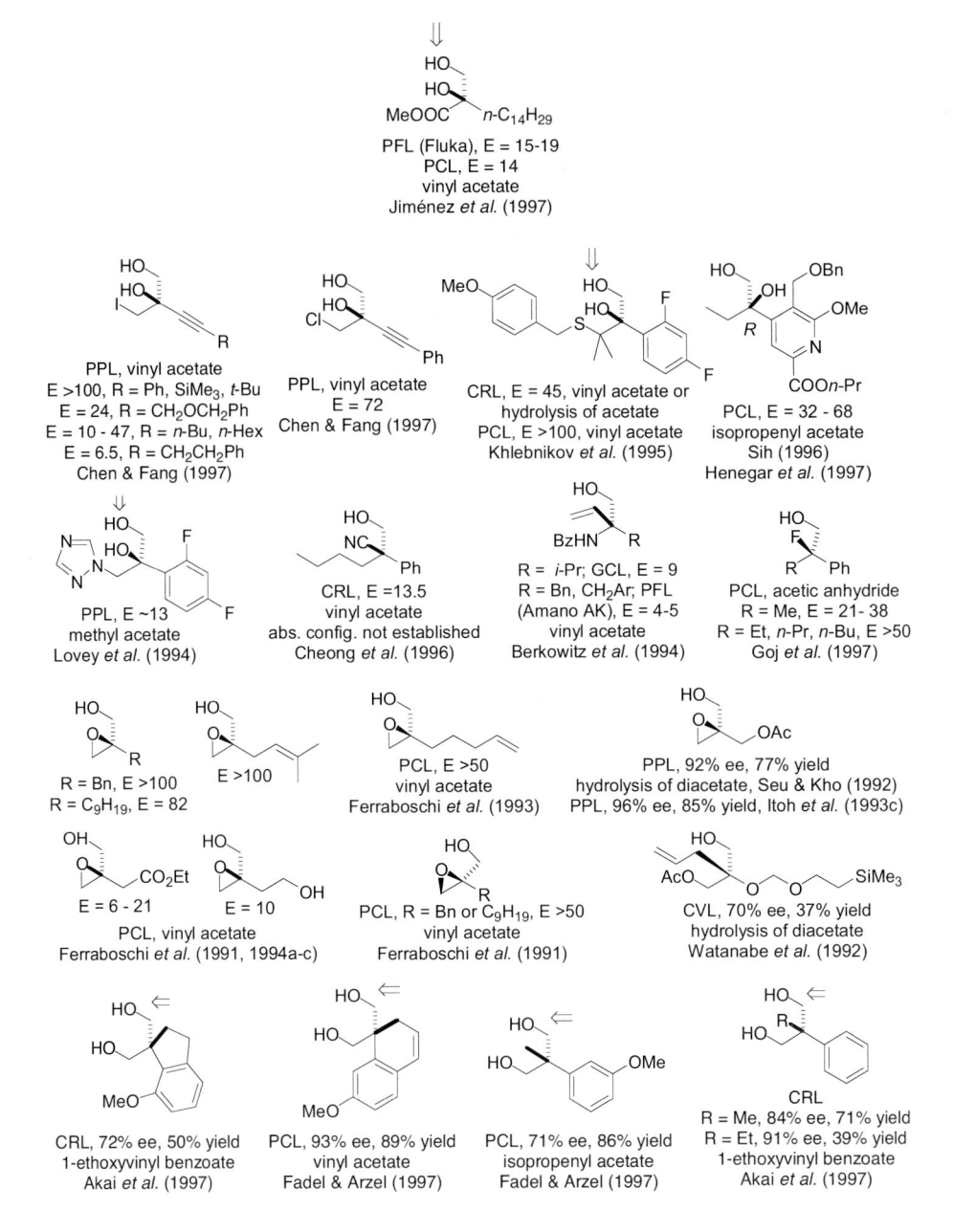

HO
O   OAc

PPL, 92% ee, 77% yield
hydrolysis of diacetate, Seu & Kho (1992)
PPL, 96% ee, 85% yield, Itoh et al. (1993c)

OH
O CO_2Et

E = 6 - 21

HO
O   OH

E = 10

PCL, vinyl acetate
Ferraboschi et al. (1991, 1994a-c)

HO
O   R

PCL, R = Bn or C_9H_{19}, E >50
vinyl acetate
Ferraboschi et al. (1991)

HO
AcO   O   O   SiMe$_3$

CVL, 70% ee, 37% yield
hydrolysis of diacetate
Watanabe et al. (1992)

HO ⇐
HO
MeO

CRL, 72% ee, 50% yield
1-ethoxyvinyl benzoate
Akai et al. (1997)

HO ⇐
HO
MeO

PCL, 93% ee, 89% yield
vinyl acetate
Fadel & Arzel (1997)

HO ⇐
HO   OMe

PCL, 71% ee, 86% yield
isopropenyl acetate
Fadel & Arzel (1997)

HO ⇐
R
HO

CRL
R = Me, 84% ee, 71% yield
R = Et, 91% ee, 39% yield
1-ethoxyvinyl benzoate
Akai et al. (1997)

Fig. 81.   Lipase-catalyzed enantioselective reactions of primary alcohols with quaternary stereocenters (continued).

Alcohols with Axial Chirality or Remote Stereocenters

Pure enantiomers of axially-disymmetric and spiro compounds are often difficult to make using traditional chemical methods, so lipase-catalyzed reactions are often the best route to these compounds (Fig. 82).

Fig. 82. Selected examples of lipase-catalyzed enantioselective reactions of axially-disymmetric and spiro alcohols.

Other difficult-to-resolve compounds are those with a stereocenter remote from the reaction site. Nevertheless, lipases often showed high enantioselectivity toward these compounds (reviewed by Mizuguchi et al., 1994). Selected examples involving alcohols are summarized in Fig. 83.

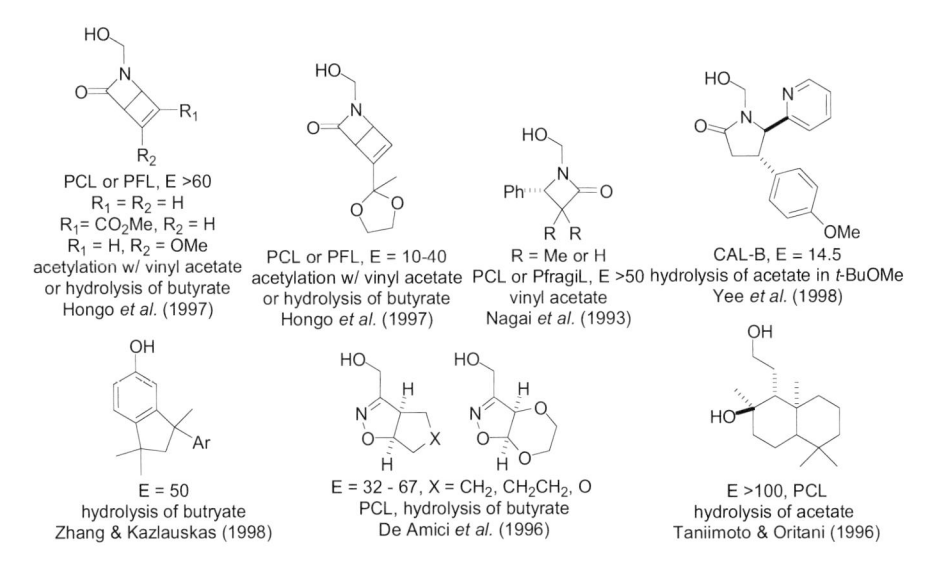

Fig. 83. Lipase-catalyzed enantioselective reactions of alcohols with remote stereocenters.

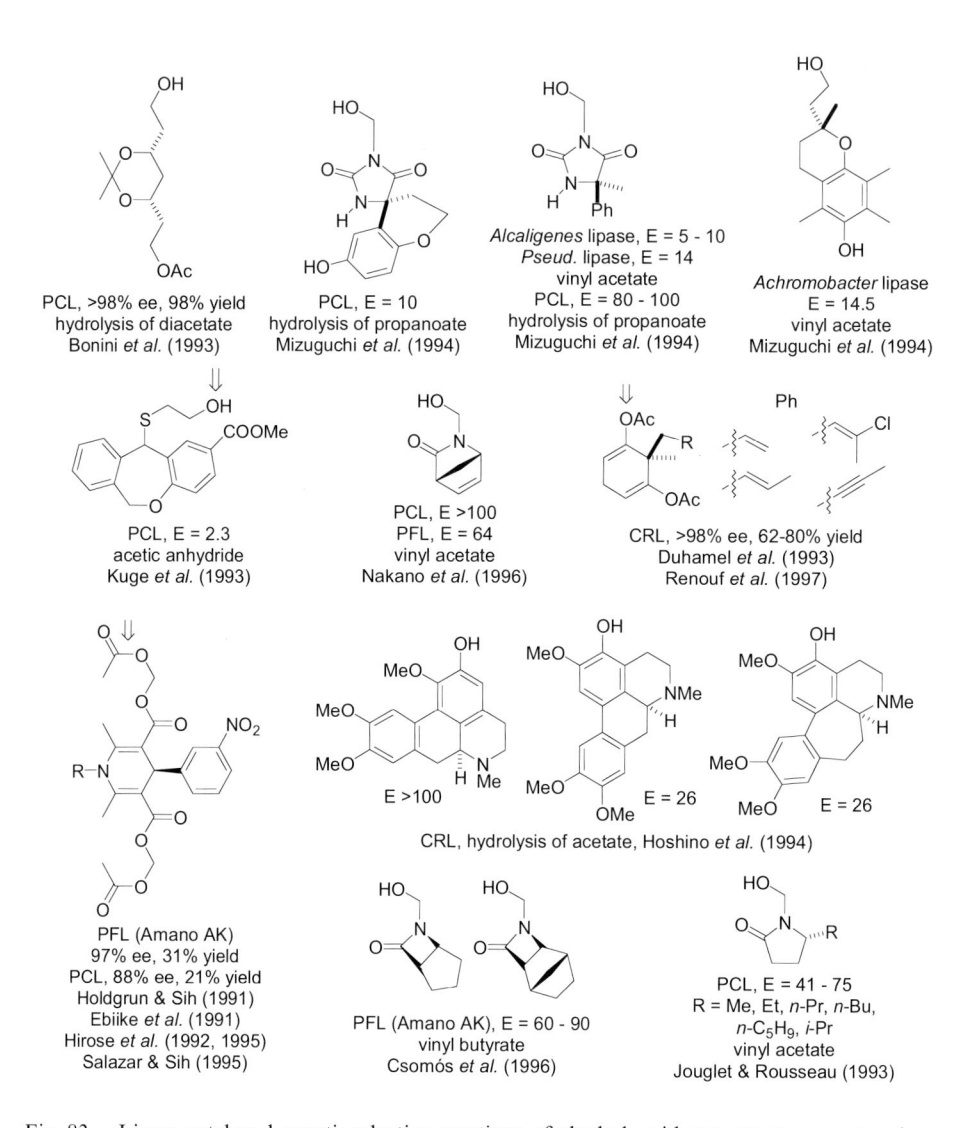

Fig. 83. Lipase-catalyzed enantioselective reactions of alcohols with remote stereocenters (continued).

The prochiral dihydropyridine (Fig. 83, first example in the last row), is a chiral acid, but is included with the chiral alcohols due to the acetyloxymethyl ester. Lipases do not catalyze hydrolysis of simple esters of these dihydropyridines presumably due to a combination of steric hindrance and lower reactivity (this carbonyl is a vinylogous carbamate). Researchers used the acetyloxymethyl group to introduce a more reactive and more accessible carbonyl group. This strategy places the stereocenter to the alcohol portion of the ester.

Alcohols with Non-Carbon Stereocenters

The first examples of lipase-catalyzed resolutions involved secondary alcohols where the organometallic was the large substituent (Boaz, 1989; Chong and Mar, 1991; Izumi et al., 1992; Wang et al., 1988). Acylation in organic solvent was crucial to the success of the resolution of the ferrocenyl derivatives because the corresponding acetate reacts readily with water (Fig. 84).

vinyl acetate
PPL, E = 9 - 20, Wang et al. (1988)
PCL (SAM-2), E >100, Boaz (1989)

CRL, PCL, E >100
PPL, RML, E ~20
Izumi et al. (1992)

PPL, E >100
trifluoroethylvalerate
Chong & Mar (1991)

Fig. 84. Secondary alcohols containing an organometallic substituent.

Later examples of lipase-catalyzed resolutions involved organometallics with planar chirality. For the (arene)chromium tricarbonyl complexes (Fig. 85), the *Pseudomonas* lipases were the most enantioselective. CRL showed opposite, but low E (Uemura et al., 1994). PCL also resolved several other metal carbonyl complexes (Fig. 86), and ferrocenes (Fig. 87). Surprisingly, the shape of the favored enantiomers in PCL catalyzed reactions is similar for all the (arene)chromium tricarbonyl complexes, but is opposite for most of the ferrocenes.

Pseudomonas lipases favor enantiomer shown.
CRL favors opposite enantiomer.

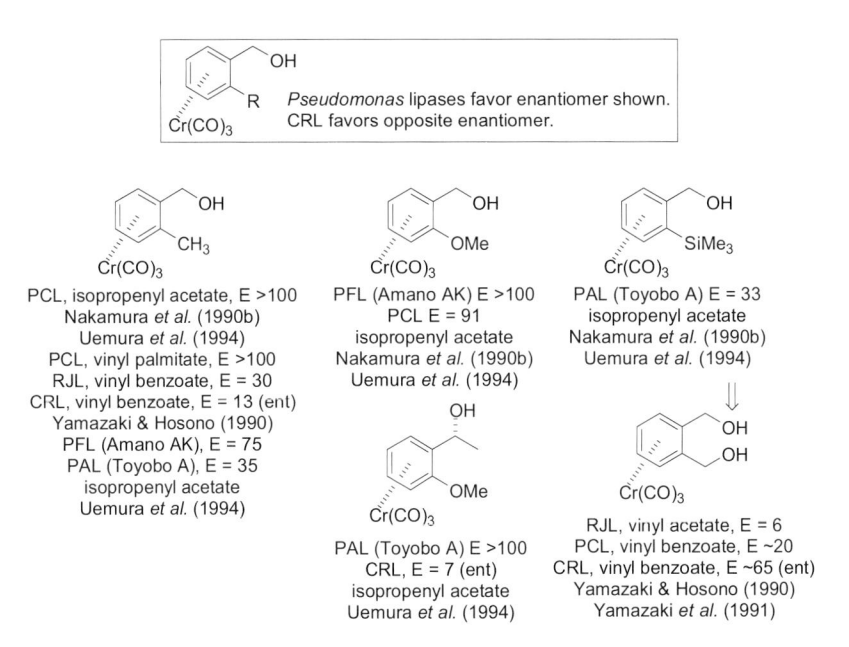

R = CH₃, Cr(CO)₃
PCL, isopropenyl acetate, E >100
Nakamura et al. (1990b)
Uemura et al. (1994)
PCL, vinyl palmitate, E >100
RJL, vinyl benzoate, E = 30
CRL, vinyl benzoate, E = 13 (ent)
Yamazaki & Hosono (1990)
PFL (Amano AK), E = 75
PAL (Toyobo A), E = 35
isopropenyl acetate
Uemura et al. (1994)

R = OMe, Cr(CO)₃
PFL (Amano AK) E >100
PCL E = 91
isopropenyl acetate
Nakamura et al. (1990b)
Uemura et al. (1994)

R = SiMe₃, Cr(CO)₃
PAL (Toyobo A) E = 33
isopropenyl acetate
Nakamura et al. (1990b)
Uemura et al. (1994)

OMe, Cr(CO)₃
PAL (Toyobo A) E >100
CRL, E = 7 (ent)
isopropenyl acetate
Uemura et al. (1994)

OH, OH, Cr(CO)₃
RJL, vinyl acetate, E = 6
PCL, vinyl benzoate, E ~20
CRL, vinyl benzoate, E ~65 (ent)
Yamazaki & Hosono (1990)
Yamazaki et al. (1991)

Fig. 85. Favored enantiomer in the lipase-catalyzed reaction of *ortho*-substituted hydroxymethyl benzene chromium (0) tricarbonyl complexes. The *Pseudomonas* lipases favor the enantiomer with the general structure shown for five examples, while CRL favors the opposite enantiomer in three examples.

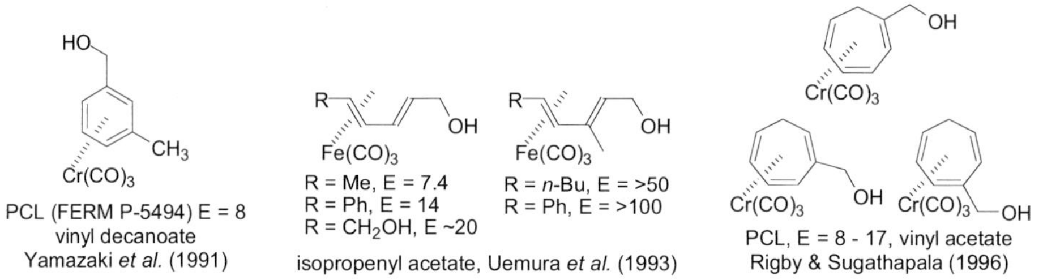

120   5.2   Survey of Enantioselective Lipase-Catalyzed Reactions

Fig. 86.   Other examples of metal carbonyl complexes resolved by lipases.

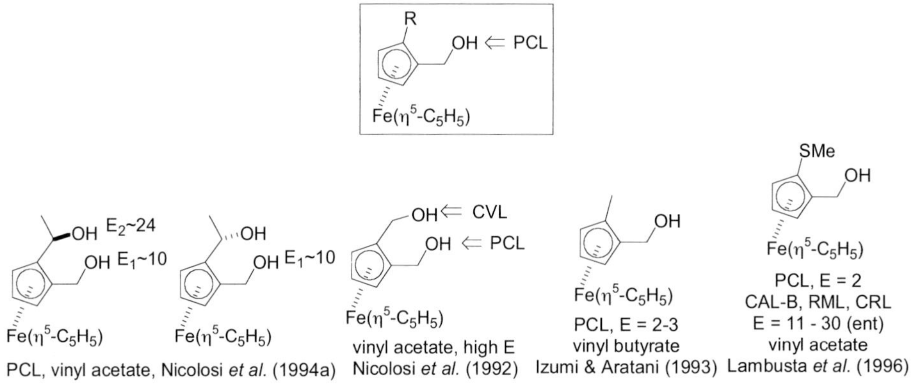

Fig. 87. Favored enantiomer in the lipase-catalyzed reaction of 1,2-disubstituted ferrocenes. For the hydroxymethyl substituted ferrocenes, PCL favors the general structure shown. Note the absolute configuration of the favored enantiomer of the ferrocenes differs from the benzene tricarbonyls in Fig. 85.

CRL and CE were the most enantioselective lipases toward phenols containing sulfur or phosphorus stereocenters (Fig. 88). Examples of acids containing phosphorus or sulfur stereocenters are summarized in Sect. 5.2.2.4.

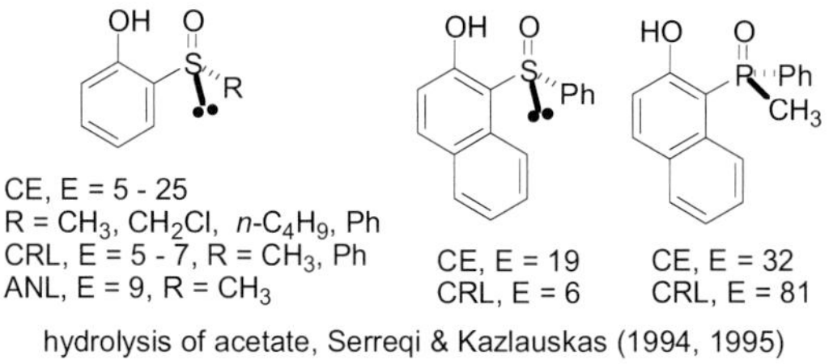

Fig. 88. Lipase-catalyzed enantioselective reactions of alcohols containing phosphorus or sulfur stereocenters.

CRL and CVL catalyzed the enantioselective acylation of prochiral 2-sila-1,3-propanediols (Fig. 89), but favored opposite enantiomers (Djerourou, 1991). Unfortunately, several closely related compounds reacted very slowly and with low enantioselectivity (Aouf et al., 1994).

CRL or CVL (favor opposite enantiomers)
R = Ph, 70% ee, 50-80% yield
R = n-octyl, 75% ee, 63-70% yield
acylation with isobutyrate esters
absolute configuration not established
Djerourou & Blanco (1991)

Fig. 89. CRL and CVL catalyzed enantioselective acylation of prochiral silanes.

Analogs of Alcohols: Amines, Thiols, and Hydroperoxides

Amines. Lipases also catalyze the enantioselective acylation of amines, although reactions are slower than for alcohols. CAL-B is the most popular lipase, but PCL and lipase from *Pseudomonas aeruginosa* also showed high enantioselectivity. Most researchers used less reactive acylating agents (e.g., esters or carbonates) to avoid chemical acylation of the more nucleophilic amines. Primary amines of the type NH_2CHRR' are isosteric with secondary alcohols. Smidt et al. (1996) proposed extending the secondary alcohol rule to primary amines for CAL-B and indeed all of the amines below fit this rule (Fig. 90).

rule to predict favored enantiomer

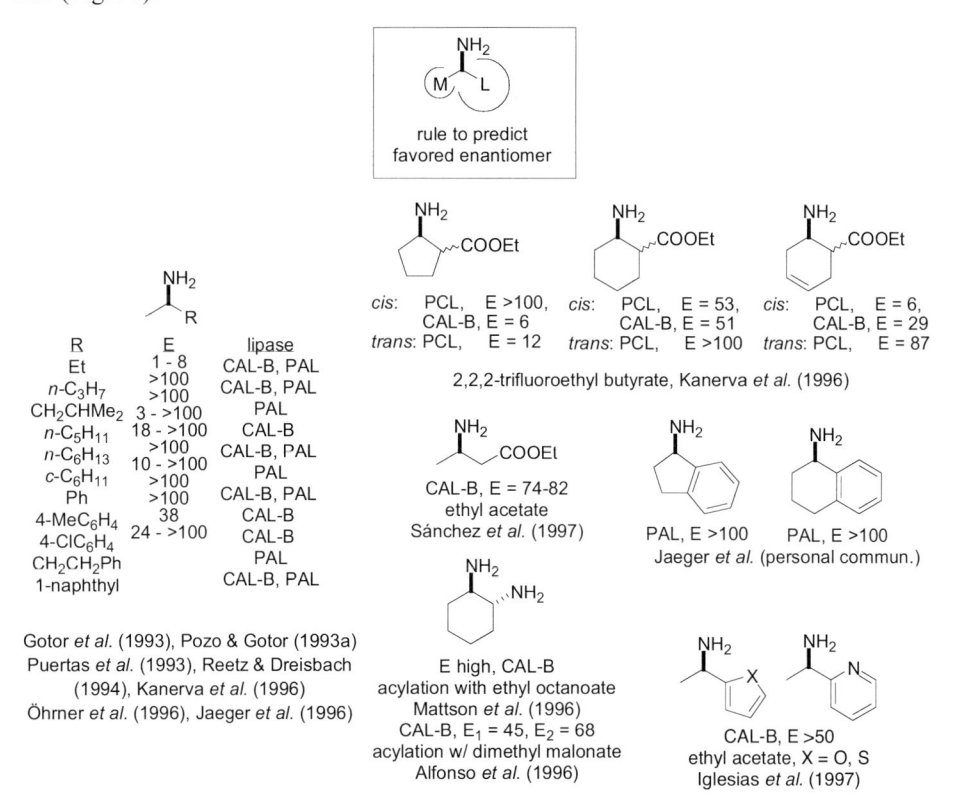

R	E	lipase
Et	1 - 8	CAL-B, PAL
n-C_3H_7	>100	CAL-B, PAL
	>100	
CH_2CHMe_2	3 - >100	PAL
n-C_5H_{11}	18 - >100	CAL-B
n-C_6H_{13}	>100	CAL-B, PAL
	10 - >100	PAL
c-C_6H_{11}	>100	
Ph	>100	CAL-B, PAL
4-MeC_6H_4	38	CAL-B
4-ClC_6H_4	24 - >100	CAL-B
CH_2CH_2Ph		PAL
1-naphthyl		CAL-B, PAL

Gotor et al. (1993), Pozo & Gotor (1993a)
Puertas et al. (1993), Reetz & Dreisbach (1994), Kanerva et al. (1996)
Öhrner et al. (1996), Jaeger et al. (1996)

cis: PCL, E >100, CAL-B, E = 6
trans: PCL, E = 12

cis: PCL, E = 53, CAL-B, E = 51
trans: PCL, E >100

cis: PCL, E = 6, CAL-B, E = 29
trans: PCL, E = 87

2,2,2-trifluoroethyl butyrate, Kanerva et al. (1996)

CAL-B, E = 74-82
ethyl acetate
Sánchez et al. (1997)

PAL, E >100   PAL, E >100
Jaeger et al. (personal commun.)

E high, CAL-B
acylation with ethyl octanoate
Mattson et al. (1996)
CAL-B, E_1 = 45, E_2 = 68
acylation w/ dimethyl malonate
Alfonso et al. (1996)

CAL-B, E >50
ethyl acetate, X = O, S
Iglesias et al. (1997)

Fig. 90. Examples of lipase-catalyzed resolutions of amines.

5.2 Survey of Enantioselective Lipase-Catalyzed Reactions

BASF AG commercialized the resolution of primary amines by a *Pseudomonas* lipase-catalyzed acylation (Balkenhohl et al., 1997). A key discovery was the acylation reagent ethyl methoxyacetate. Activated acyl donors react chemically, while lipase-catalyzed reactions with simple esters or carbonates are usually slow. Acylation with ethyl methoxyacetate proceeds at least 100 times faster than acylation with ethyl butyrate. The reason for this acceleration is not known, but one possibility is that the oxygen may help deprotonate the amine (Lavandera et al., 2005). The hydrolysis of the corresponding methoxyacetate esters of amines also proceeded significantly faster compared to hydrolysis of simple acetates, however here CAL-B was superior compared to *Pseudomonas* sp. lipases (Wagegg et al., 1998).

Researchers also resolved several amines that do not resemble secondary alcohols. Orsat et al. (1996) resolved a secondary amine with CRL and Morgan et al. (2000) resolved an atropisomeric secondary amine. Yang et al. (1995a) resolved a primary amine that is isosteric with a primary alcohol (Fig. 91).

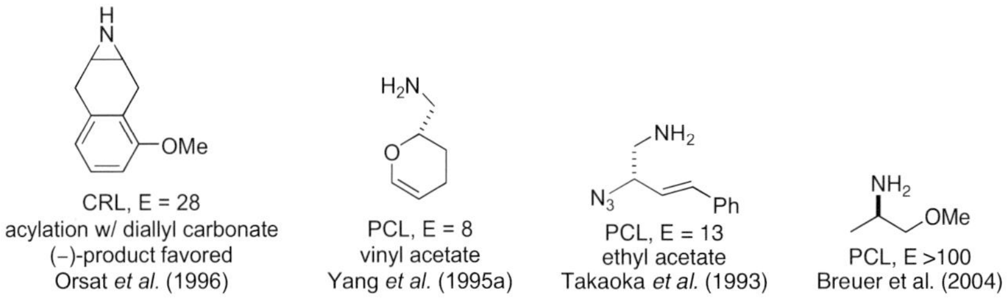

Fig. 91. Other amines resolved by lipases.

Lipases usually do not catalyze hydrolysis of amides. One exception is the CAL-B-catalyzed hydrolysis of *N*-acetyl 1-arylethylamines, but reaction times were a week or longer (Smidt et al., 1996). Chapman et al. (1996) found an indirect way to resolve amides using oxalamic esters. CAL-B catalyzed hydrolysis of the ester group and showed high enantioselectivity toward the remote stereocenter. The stereocenter now lies in the acid portion of the reacting ester, so the rule above no longer applies. Coincidentally, this reaction also favors the same amine enantiomer predicted above (Fig. 92). Hu et al. (2005) extended this oxalamic ester approach to the resolution of secondary amines using proteases, see Sect. 6.4.1.2.

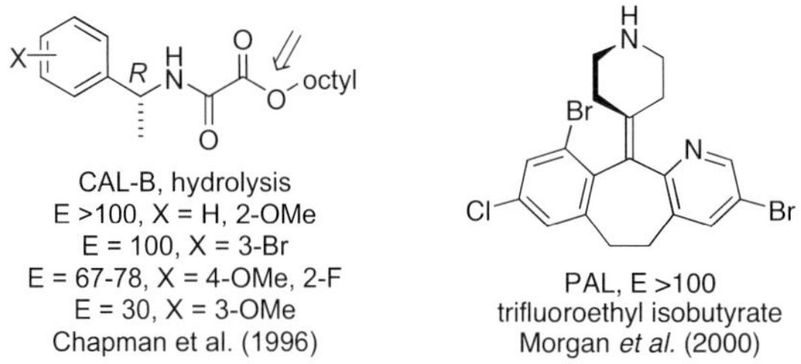

Fig. 92. Resolution of amines as oxalamic esters.

Thiols. Researchers resolved several thiols, which are the simplest analogs of alcohols (Fig. 93). Resolutions used either hydrolysis or alcoholysis of thiol esters and examples include primary and secondary thiols and even an axially chiral thiol. Enantioselectivities were similar to those for the corresponding alcohols. Baba et al. (1990a) and Öhrner et al. (1996) noted that acylation of thiols gave no reaction. The acyl enzyme intermediate may not be a strong enough acyl donor to acylate thiols.

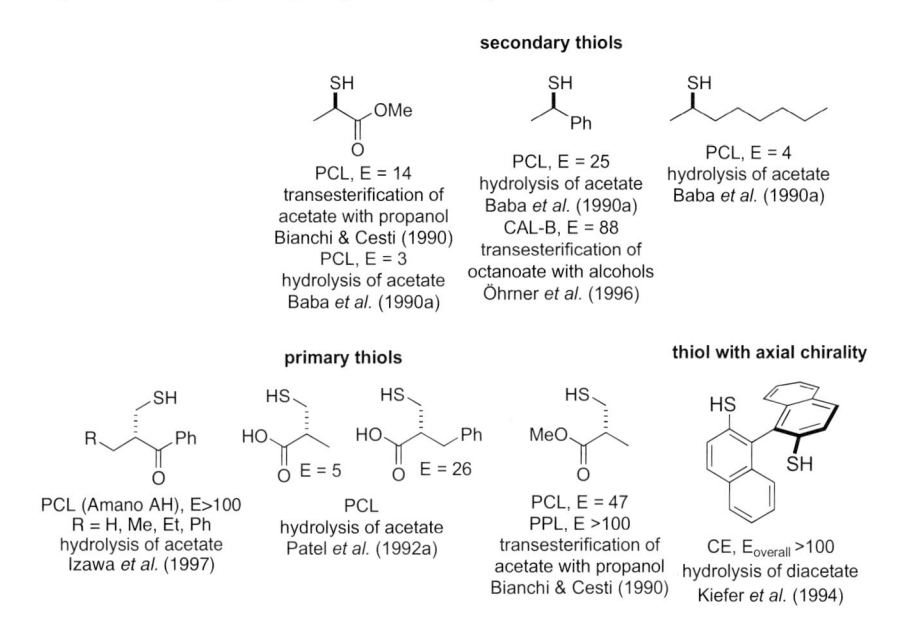

Fig. 93. Selected examples of lipase-catalyzed enantioselective reactions of thiols.

Peroxides. Lipases also discriminate between enantiomers of alkyl peroxides, which resemble primary alcohols. Enantioselective acylation of alkyl peroxides yielded unreacted starting material in high enantiomeric excess, but the produced peroxyesters decomposed under the reaction conditions to ketones. The structures in Fig. 94 show the fast-reacting enantiomer.

Fig. 94. Alkyl peroxides resolved by PCL-catalyzed acylations.

5.2.2 Carboxylic Acids

5.2.2.1 General Considerations

There are fewer examples of lipase-catalyzed enantioselective reactions of carboxylic acids (reviewed by Haraldsson, 1992). In water, lipases catalyze hydrolyses of various carboxylic acid esters, while in organic solvents, lipases catalyze esterification of acids, transesterification of esters and aminolysis of esters. Reactions of chiral anhydrides and lactones are discussed below in Sects. 5.2.2.6 and 5.2.3.

5.2.2.2 Carboxylic Acids with a Stereocenter at the α-Position

Candida antarctica *Lipase B*

In contrast to its high enantioselectivity toward alcohols, CAL-B usually shows low to moderate enantioselectivity toward carboxylic acids (Fig. 95). Preparation of enantiomerically-pure 2-arylpropionic acids, a class of non-steroidal anti-inflammatory drugs, required two sequential resolutions (Morrone et al., 1995; Trani et al., 1995). Starting from 300 g of racemic ibuprofen (Ar = 4-*i*-BuC$_6$H$_4$) Trani et al. (1995) used two sequential esterifications to make 38 g of (*S*)-ibuprofen with 97.5% ee.

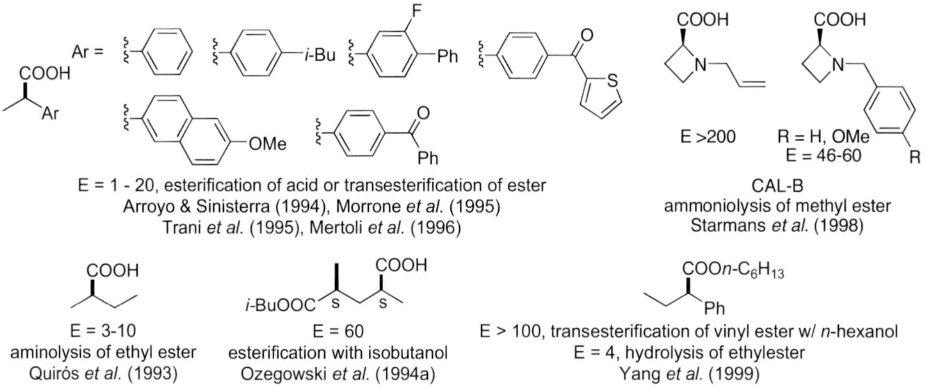

Fig. 95. CAL-B-catalyzed enantioselective reaction of carboxylic acids with a stereocenter at the α-position.

The acyl binding site of CAL-B is a shallow crevice. It is likely that the lower enantioselectivity toward stereocenters in the acyl part of an ester stems from fewer and/or weaker contacts between the acyl part and its binding site. In contrast, the alcohol binding site appears to engulf the alcohol.

Candida rugosa *Lipase*

In contrast to CAL-B, CRL shows high enantioselectivity toward many carboxylic acids (Fig. 96), and a rule can predict the enantiopreference of CRL-catalyzed reactions of carboxylic acids with a stereocenter at the α-position. Recent reviews also contain a few more examples of acids resolved by CRL (Ahmed et al., 1994; Franssen et al., 1996).

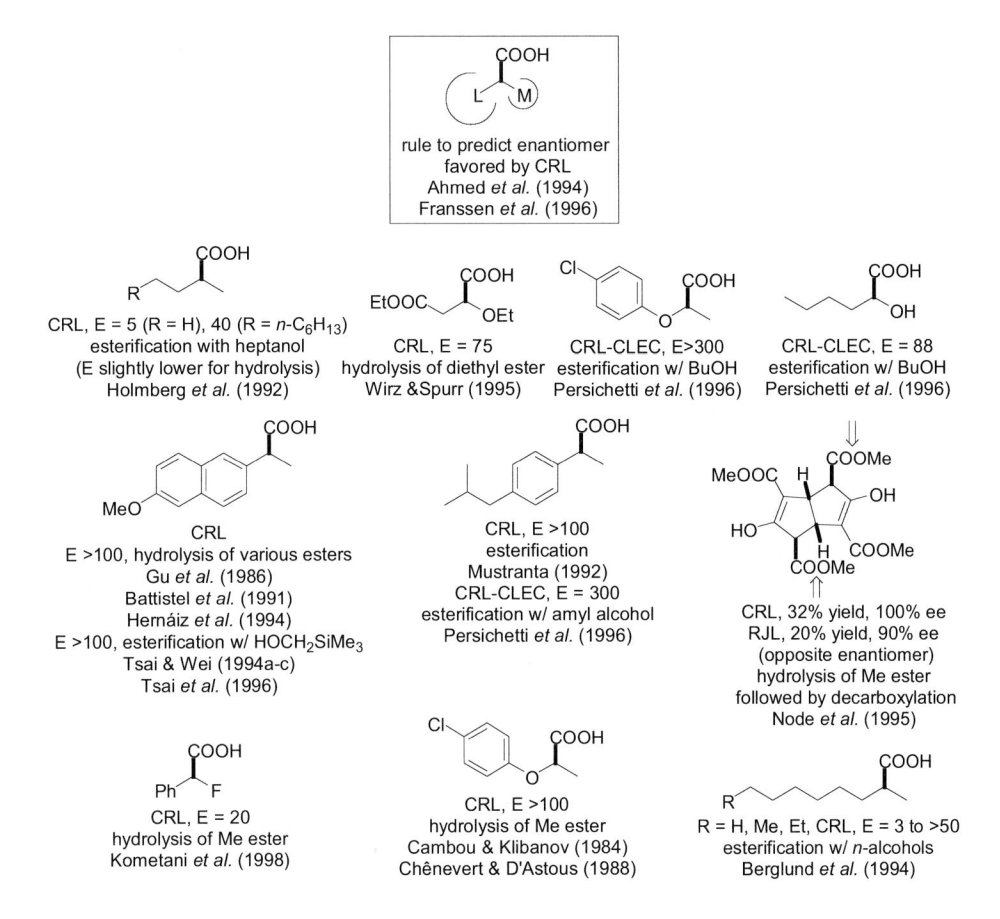

Fig. 96. Selected examples of CRL-catalyzed enantioselective reactions of carboxylic acids with a stereocenter at the α-carbon.

One important commercial target are pure (S)-enantiomers of 2-arylpropionic acids. Although CRL shows high enantioselectivity toward these acids, reaction rates are too slow for commercial use. Chirotech (now Dowpharma Chirotech Technology Ltd.) in the UK developed a resolution of naproxen using a *Bacillus* esterase (Quax and Broekhuizen, 1994, see also Sect. 5.4.4.1). This process produced 13 tons of (S)-naproxen in 1996 (Stinson, 1997).

Comparing the above empirical rule to the X-ray structure of CRL suggests that the large substituent, L, binds in a tunnel, while the stereocenter lies at the mouth of this tunnel. Indeed, molecular modeling supports this proposal (Botta et al., 1997; Holmquist et al., 1996). Further modeling rationalized some known exceptions to the empirical rule (Holmquist et al., 1996). When the large substituent is extensively branched, it no longer fits in the tunnel. An alternate binding mode with the substrate outside the tunnel favors the opposite enantiomer.

126   5.2   Survey of Enantioselective Lipase-Catalyzed Reactions

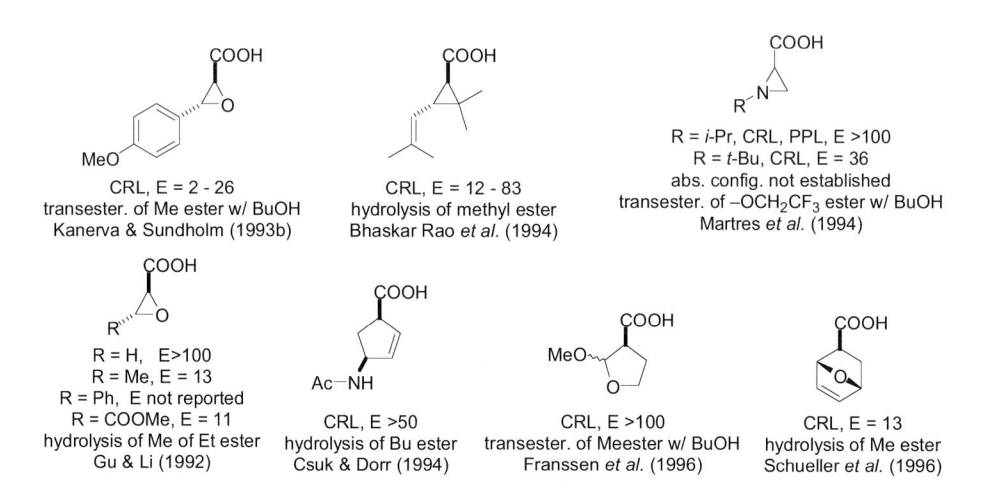

Fig. 96. Selected examples of CRL-catalyzed enantioselective reactions of carboxylic acids with a stereocenter at the α-carbon (continued).

Researchers increased the enantioselectivity of CRL-catalyzed resolution of chiral acids using a number of different methods. Most examples involve either 2-arylpropionic acids or 2-aryloxypropionic acids, a class of herbicides. For example, Guo and Sih increased the enantioselectivity of a CRL-catalyzed hydrolysis of the 2-chloroethyl ester of 2-(3-benzoyl)phenylpropanoic acid by adding a chiral amine, dextromethorphan (Guo and Sih, 1989; Sih et al., 1992). Kinetic analysis showed that dextromethorphan increased the enantioselectivity by inhibiting the hydrolysis of the slow-reacting enantiomer. In another example, Colton et al. (1995) increased the enantioselectivity of a CRL-catalyzed hydrolysis of the methyl ester of 2-(4-chloro)phenoxypropanoic acid by a purification procedure that involved treating the enzyme with isopropanol (Fig. 97).

Fig. 97. Several methods increase the enantioselectivity of CRL toward acids.

Other researchers have increased the enantioselectivity of CRL toward 2-aryl- or 2-aryloxypropionic acids by changing the solvent (Miyazawa et al., 1992), temperature (Yasufuku and Ueji, 1995) or pH, by carrying out the reaction in a microemulsion (Hedström et al., 1993), by adding (S)-2-amino-4-methylthio-1-butanol (Itoh et al., 1991) or Triton X-100 (a surfactant) (Bhaskar-Rao et al., 1994), by linking the ε-amino group of lysine residues to a solid support (Sinisterra et al., 1994), by nitration of tyrosyl residues (Gu and Sih, 1992), by purification and cross-linking of crystals of CRL (Lalonde et al., 1995; Persichetti et al., 1996), by purification (Allenmark and Ohlsson, 1992a; b; Wu et al., 1990), and by careful addition of water to avoid clumping in organic solvents (Tsai and Dordick, 1996).

Chemists do not know how these changes increase enantioselectivity on a molecular level, but two possibilities are most reasonable. First, the treatments may remove or inactivate a contaminating hydrolase with low or opposite enantioselectivity. Indeed, Lalonde et al. (1995) reported a contaminating esterase in commercial CRL. Second, the treatments may change the conformation of CRL. Crystallographers solved the structures for both an 'open' and a 'closed' form of CRL which differed in the orientation of the lipase lid (a surface helical region). Indeed, Lalonde et al. (1995) found that crystals of the 'open' and 'closed' forms differed in their enantioselectivity. These two possibilities do not exclude each other, so both effects may contribute to the increased enantioselectivity.

Pseudomonas *Lipases*

PCL also catalyzes enantioselective reactions of acids (Fig. 98). The relative sizes of the substituents cannot account for the enantiopreference, but note that all but two examples have a similar orientation of a heteroatom (S, O, N, X) at the stereocenter.

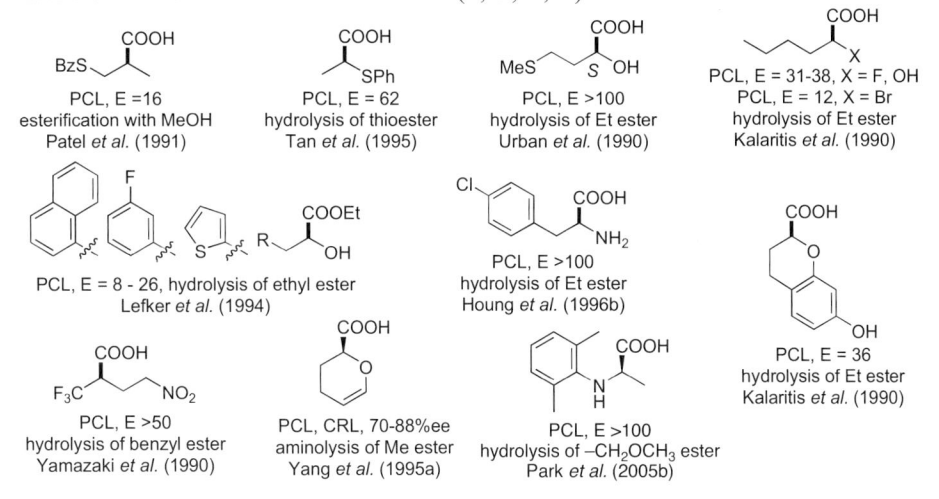

Fig. 98. Selected examples of PCL-catalyzed enantioselective reactions of carboxylic acids with a stereocenter at the α-carbon.

O'Hagan and Rzepa (1994) suggested that the high enantioselectivity of PCL toward acids with a fluorine substituent at the α-position may be due to a stereoelectronic effect. In nonenzymic reactions, nucleophilic attack at a carbonyl favors an *anti* orientation of an electron withdrawing substituent at the α-position. A similar preference in the active site of PCL may also account for the observed enantioselectivity.

Other Lipases

Fig. 99 summarizes selected examples of enantioselective reactions involving other lipases. The PPL-catalyzed resolution of amino acid esters (Houng et al., 1996a; b) used crude enzyme which contains protease contaminants. Researchers observed high enantioselectivity (E > 100) only for amino acids where the alkyl group is $-CH_2-$aryl. These amino acid esters are good substrates for chymotrypsin, thus chymotrypsin, a likely contaminant, may contribute to the observed selectivity.

128   5.2   Survey of Enantioselective Lipase-Catalyzed Reactions

A survey of the enantioselectivity of ANL toward carboxylic acids identified α-amino acids as the best resolved class of carboxylic acids (Janes and Kazlauskas, 1997a). Replacement of the positively charged $-NH_3^+$ substituent with $-OH$ or $-CH_3$ lowered the enantioselectivity drastically. Note that CRL and RML favor opposite enantiomers of 2-arylpropionic acids.

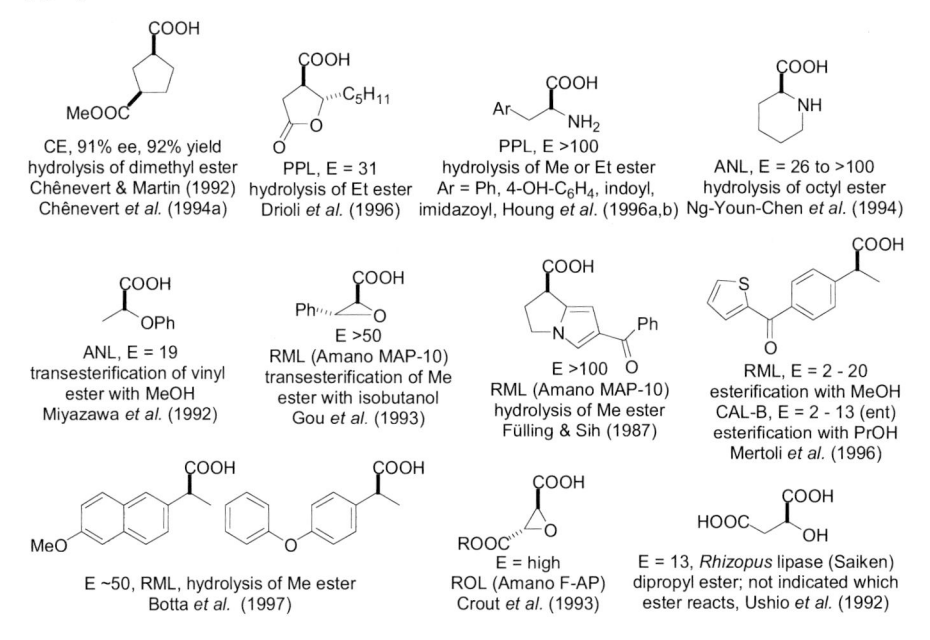

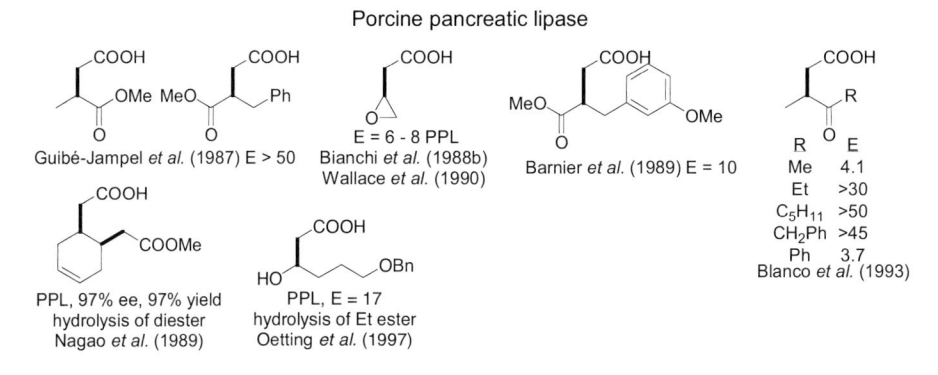

Fig. 99. Selected examples of lipase-catalyzed enantioselective reactions of carboxylic acids with a stereocenter at the α-carbon.

5.2.2.3   Carboxylic Acids with a Stereocenter at the β-Position

Even though the stereocenter is further away, a number of lipases also catalyze enantioselective reactions of carboxylic acids with a stereocenter at the β-position (Fig. 100).

Fig. 100. Selected examples of lipase-catalyzed enantioselective reactions of carboxylic acids with a stereocenter at the β-position.

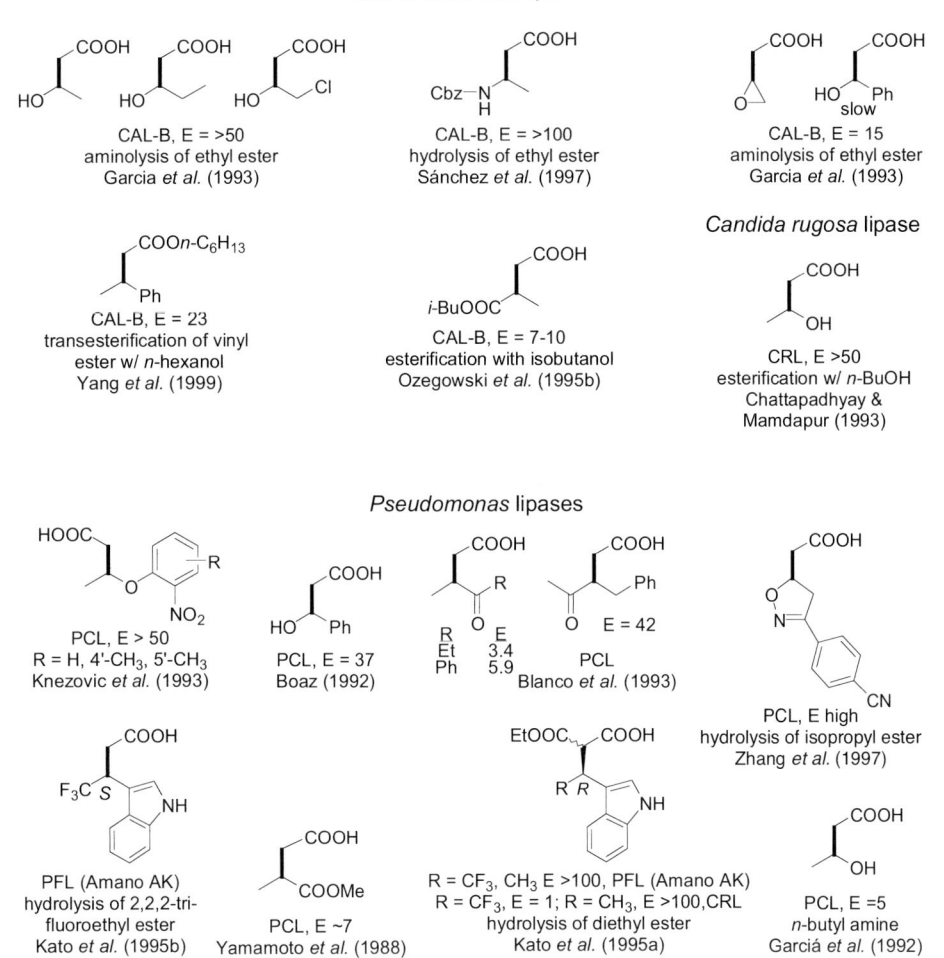

Fig. 100. Selected examples of lipase-catalyzed enantioselective reactions of carboxylic acids with a stereocenter at the β-position (continued).

5.2.2.4 Other Carboxylic Acids

Quaternary Stereocenters

Several examples are shown in Fig. 101.

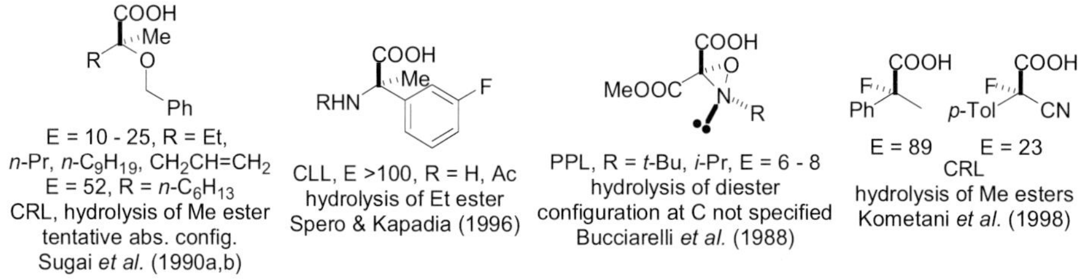

Fig. 101. Selected examples of lipase-catalyzed enantioselective reactions of carboxylic acids with quaternary stereocenters.

Sulfur Stereocenters

Several examples are shown in Fig. 102.

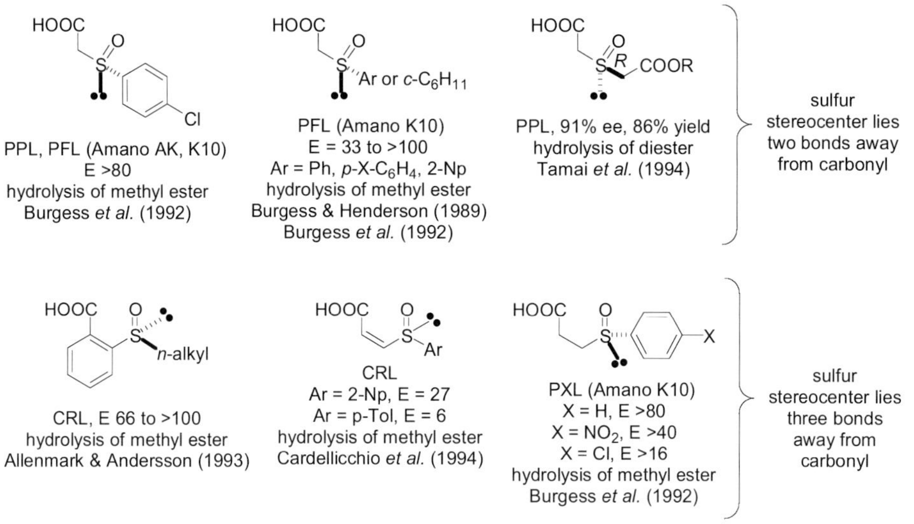

Fig. 102. Selected examples of lipase-catalyzed enantioselective reactions of carboxylic acids with sulfur stereocenters.

Remote Stereocenters

Lipases occasionally show high enantioselectivity toward carboxylic acids with stereocenters far from the carbonyl (Fig. 103). For example, researchers at Merck Research Laboratories (Rahway, NJ, USA) used *Pseudomonas* lipases to enantioselectively hydrolyze the *pro-R* ester in the dithioacetal (Fig. 103) yielding the (*S*)-monoester in enantiomeric purity even though the sterecenter lies four bonds from the carbonyl (Chartrain et al., 1993; Hughes et al., 1989; 1990; 1993; Smith et al., 1992). The enantioselectivity dropped for analogues where the stereocenter lies either three or five bonds from the carbonyl. Bhalerao et al. (1991) found that CRL showed surprisingly high enantioselectivity toward several carboxylic acids where the stereocenter lies eight or nine bonds (but not seven bonds) from the carbonyl. The x-ray crystal structures of CRL suggest that the carboxylic acid binds in a tunnel which bends approximately at C-9 of a fatty acid. The observed enantioselectivity toward stereocenters in this region may be due to this bend. Researchers also resolved several other synthetic intermediates with remote stereocenters. CAL-B shows moderate enantioselectivity toward the phenol below (Kawasaki et al., 1999). The stereocenter lies four bonds away from the reacting carbonyl.

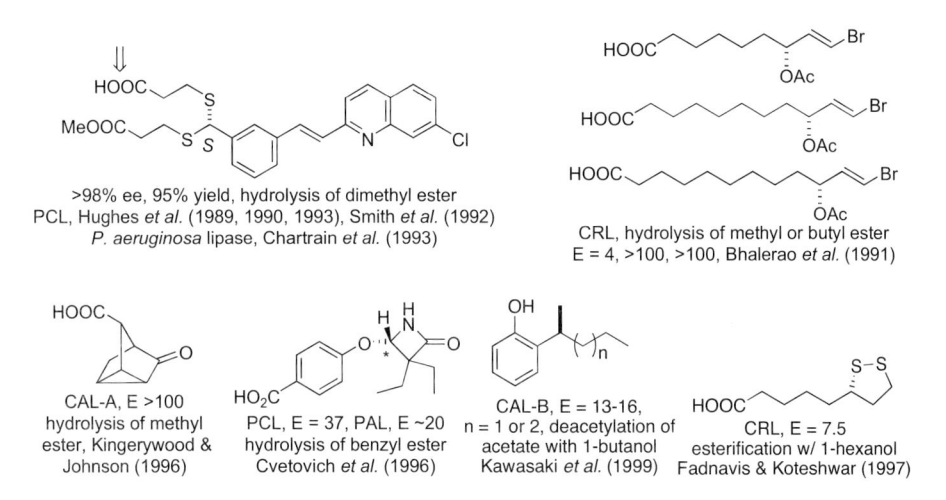

Fig. 103. Selected examples of lipase-catalyzed enantioselective reactions of carboxylic acids with remote stereocenters.

5.2.2.5 Double Enantioselection

The term double enantioselection refers to the synthesis of esters from racemic (or prostereogenic) carboxylic acids and alcohols. However, in order to achieve an efficient reaction, the following criteria must be fulfilled: (1) the enzyme should be highly enantioselective towards both substrates and (2), the reaction equilibrium must be shifted towards ester synthesis in order to obtain high conversion of both enantiomers. The first problem might be solved by investigating different hydrolases, whereas the low reaction rate might be enhanced by using activated acyl donors (see Sect. 5.1.6.1).

To date, double enantioselection has been investigated only to a limited extent (Fig. 104). For instance, Theil and coworkers used *rac*-2,2,2-trifluoroethyl-2-chloropropanoate in the resolution of a *meso*-diol in THF. However, the reaction proceeded only with moderate selectivity and best results were achieved using a lipase from *Candida* sp. yielding a diastereomeric excess of 52% de (Theil et al., 1992). Direct esterification of e.g., *rac*-α-phenoxypropionic acid with *rac*-α-phenylethanol in *n*-hexane was much slower (only 15% conversion after 122 h) and enantioselectivity was very low (E = 10.4 (alcohol) and 12.5 (acid) (Chen et al., 1993).

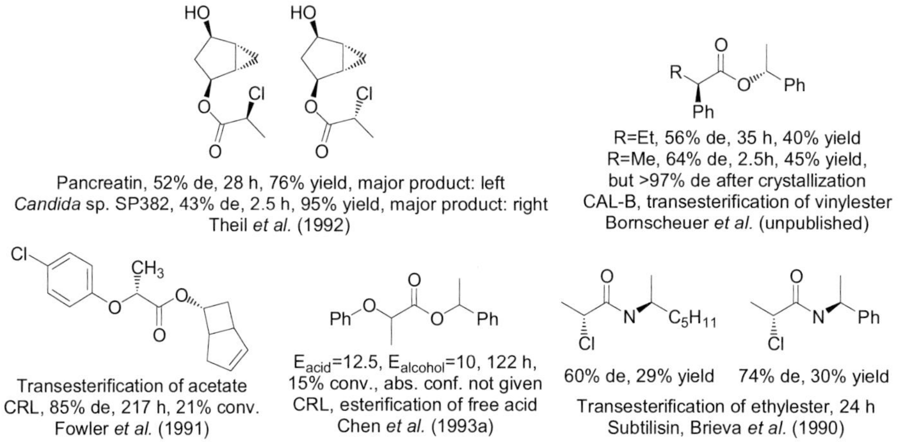

Pancreatin, 52% de, 28 h, 76% yield, major product: left
Candida sp. SP382, 43% de, 2.5 h, 95% yield, major product: right
Theil et al. (1992)

R=Et, 56% de, 35 h, 40% yield
R=Me, 64% de, 2.5h, 45% yield,
but >97% de after crystallization
CAL-B, transesterification of vinylester
Bornscheuer et al. (unpublished)

Transesterification of acetate
CRL, 85% de, 217 h, 21% conv.
Fowler et al. (1991)

E_{acid}=12.5, $E_{alcohol}$=10, 122 h,
15% conv., abs. conf. not given
CRL, esterification of free acid
Chen et al. (1993a)

60% de, 29% yield

74% de, 30% yield
Transesterification of ethylester, 24 h
Subtilisin, Brieva et al. (1990)

Fig. 104. Examples for lipase- or subtilisin-catalyzed double enantioselections.

5.2.2.6 Anhydrides

PCL catalyzed the regioselective ring opening of 2-substituted succinic and glutaric acid anhydrides, but without enantioselectivity (Hiratake et al., 1989). Reaction occurred at the less hindered carbonyl (Fig. 105).

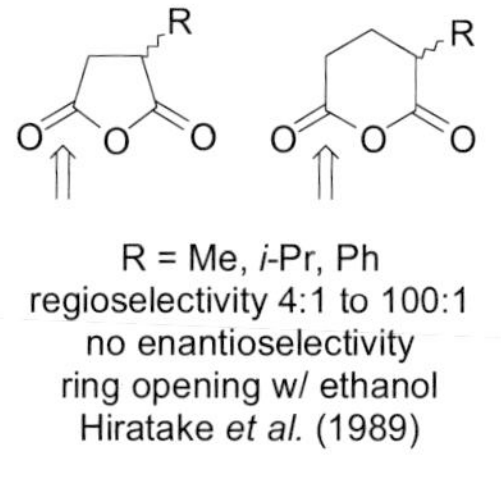

R = Me, i-Pr, Ph
regioselectivity 4:1 to 100:1
no enantioselectivity
ring opening w/ ethanol
Hiratake et al. (1989)

Fig. 105. Regioselective ring opening of anhydrides occurred at the less hindered carbonyl.

However, CAL-B was both regio- and enantioselective (Ozegowski et al., 1994b; b). For 2-methyl glutaric acid anhydride, the (R)-enantiomer reacted at the more hindered carbonyl, while the (S)-enantiomer reacted at the less hindered carbonyl (Eq. 16). Similarly, the (2R)-enantiomer of 2,3-dimethylglutarates reacted at the more hindered carbonyl while the (2S)-enantiomer reacted at the less-hindered carbonyl (Eqs. 17 and 18) (Ozegowski et al., 1996). A similar reaction with the five-membered succinic anhydride was less enantioselective (not shown).

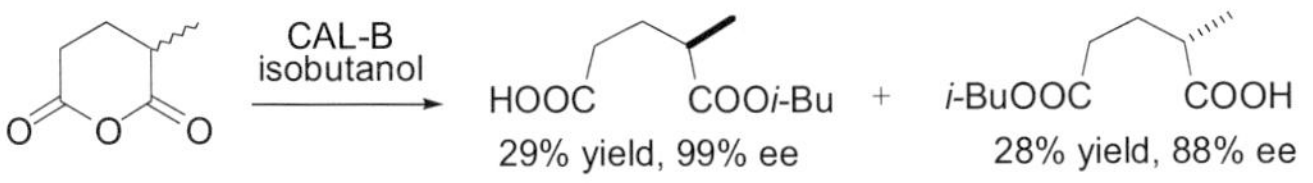

(16) CAL-B, isobutanol → HOOC–COOi-Bu (29% yield, 99% ee) + i-BuOOC–COOH (28% yield, 88% ee)

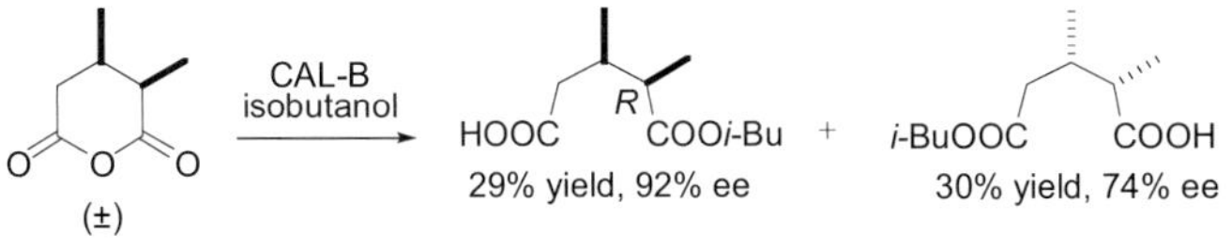

(17) (±) CAL-B, isobutanol → HOOC–(R)–COOi-Bu (29% yield, 92% ee) + i-BuOOC–COOH (30% yield, 74% ee)

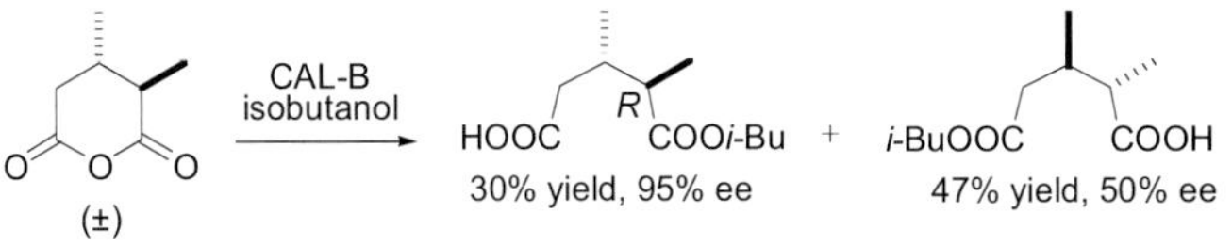

(18) (±) CAL-B, isobutanol → HOOC–(R)–COOi-Bu (30% yield, 95% ee) + i-BuOOC–COOH (47% yield, 50% ee)

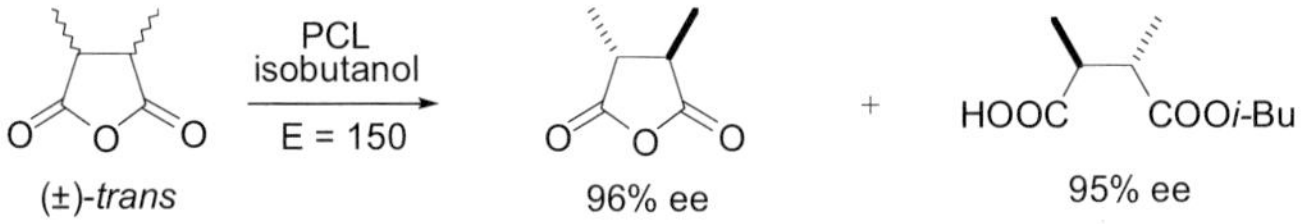

(19) (±)-trans, PCL isobutanol, E = 150 → anhydride (96% ee) + HOOC–COOi-Bu (95% ee)

For resolution of *syn*-2,3-dimethylbutandioic anhydride, PCL was the most enantioselective enzyme (Eq. 19) (Ozegowski et al., 1995a).

Lipase-catalyzed ring-opening of prochiral and *meso* anhydrides also proceeded with good enantioselectivity (Fig. 106).

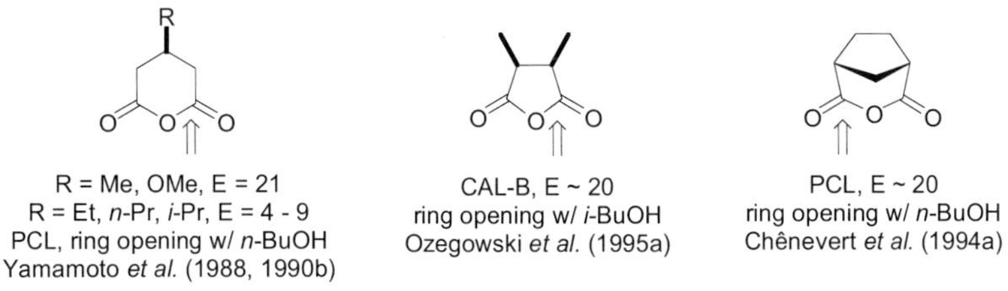

R = Me, OMe, E = 21
R = Et, n-Pr, i-Pr, E = 4 - 9
PCL, ring opening w/ n-BuOH
Yamamoto et al. (1988, 1990b)

CAL-B, E ~ 20
ring opening w/ i-BuOH
Ozegowski et al. (1995a)

PCL, E ~ 20
ring opening w/ n-BuOH
Chênevert et al. (1994a)

Fig. 106. Lipase-catalyzed ring opening of prochiral and *meso* anhydrides.

5.2.3 Lactones

Lactones are important flavor compounds and synthetic intermediates. Researchers have used lipases for the synthesis of enantiomerically pure lactones either directly using reactions involving the lactone link, or indirectly using other reactions that eventually lead to enantiomerically pure lactones. Another application of lipases is the selective formation of macrolides or diolides (cyclic dimers) from hydroxyacids. Without a lipase catalyst, oligomers are the major product.

Enantioselective reactions involving the lactone link are summarized in Fig. 107. PPL catalyzed the enantioselective lactonization of a wide range of γ-hydroxyesters to the five-membered γ-lactones (Eq. 20). Researchers used hydroxy esters, not hydroxy acids, as the starting materials to avoid spontaneous lactonization. The enantioselectivity was moderate to good, but the reaction times were often several days. Hydrolysis of the γ-lactones was less enantioselective than lactonization. Resolutions using esterases are covered in Sect. 5.4.4.2.

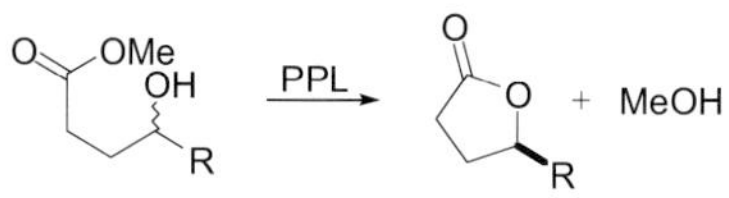
OMe
OH
R
PPL
O
+ MeOH
R

(20)

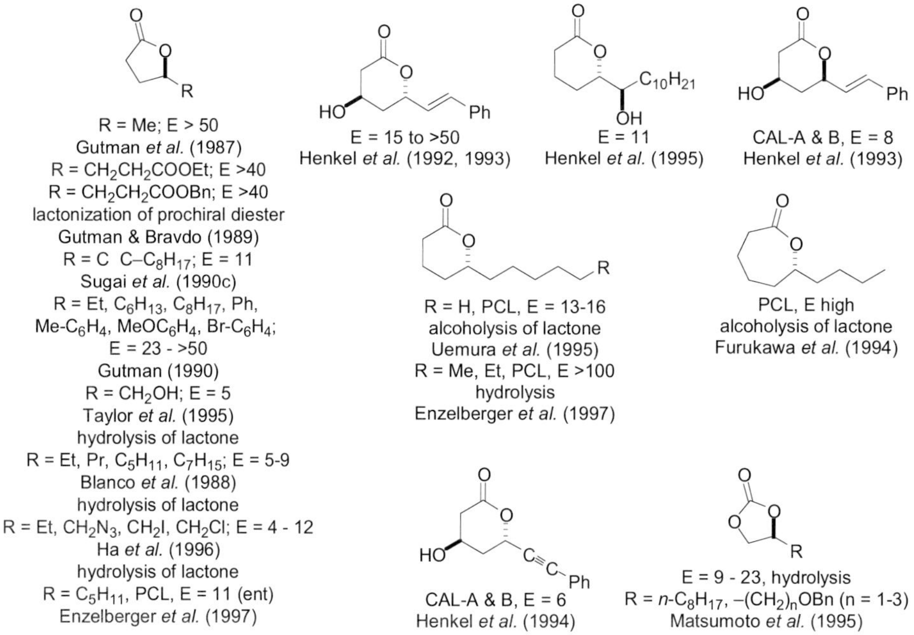
R
R = Me; E > 50
Gutman *et al.* (1987)
R = CH_2CH_2COOEt; E >40
R = CH_2CH_2COOBn; E >40
lactonization of prochiral diester
Gutman & Bravdo (1989)
R = C C–C_8H_{17}; E = 11
Sugai *et al.* (1990c)
R = Et, C_6H_{13}, C_8H_{17}, Ph,
Me-C_6H_4, MeOC_6H_4, Br-C_6H_4;
E = 23 - >50
Gutman (1990)
R = CH_2OH; E = 5
Taylor *et al.* (1995)
hydrolysis of lactone
R = Et, Pr, C_5H_{11}, C_7H_{15}; E = 5-9
Blanco *et al.* (1988)
hydrolysis of lactone
R = Et, CH_2N_3, CH_2I, CH_2Cl; E = 4 - 12
Ha *et al.* (1996)
hydrolysis of lactone
R = C_5H_{11}, PCL, E = 11 (ent)
Enzelberger *et al.* (1997)
HO
Ph
E = 15 to >50
Henkel *et al.* (1992, 1993)
$C_{10}H_{21}$
OH
E = 11
Henkel *et al.* (1995)
HO
Ph
CAL-A & B, E = 8
Henkel *et al.* (1993)
R
R = H, PCL, E = 13-16
alcoholysis of lactone
Uemura *et al.* (1995)
R = Me, Et, PCL, E >100
hydrolysis
Enzelberger *et al.* (1997)
PCL, E high
alcoholysis of lactone
Furukawa *et al.* (1994)
HO
C≡C
Ph
CAL-A & B, E = 6
Henkel *et al.* (1994)
R
E = 9 - 23, hydrolysis
R = *n*-C_8H_{17}, –$(CH_2)_n$OBn (n = 1-3)
Matsumoto *et al.* (1995)

Fig. 107. Lipase-catalyzed enantioselective reactions involving the lactone ring. Unless otherwise noted, all reactions refer to the lactonization of the ester catalyzed by PPL. In all cases, the faster-reacting enantiomer is shown. The cyclic carbonate resembles a lactone so it is included in this list.

PPL also catalyzed the formation of δ-lactones, but with lower enantioselectivity. Surprisingly, the fast-reacting enantiomer differed for the γ- and δ-lactones. Although the alcohol portion of γ- and δ-lactones is a secondary alcohol, the secondary alcohol rule cannot be used here because the stereocenter lies in a different position as shown in Fig. 108.

Attempts to form four- or seven-membered lactones yielded oligomers and polymers as discussed in Sect. 5.3.3. In some cases, oligomeric side products also formed during the lactonization of six-membered rings. However, ring-opening alcoholysis can resolve seven- (Furukawa et al., 1994) or four-membered lactones (Adam et al., 1997a; Xu et al., 1996) (Fig. 109).

Fig. 108. The secondary alcohol rule cannot be used for lactones because the stereocenter lies in a different position. Acyclic esters adopt a *syn* conformation along the carbonyl C–alcohol–O-bond. The crystal structure of transition state analogs bound to lipases suggest that this conformation persists in the active site. On the other hand, the lactone ring forces an *anti* conformation along the carbonyl C–alcohol–O-bond which places the stereocenter in a different part of the enzyme. In particular, the lactone stereocenter appears to lie entirely within the L-pocket of the alcohol binding crevice. Indeed, many of the lactone examples in this section do not follow the secondary alcohol rule.

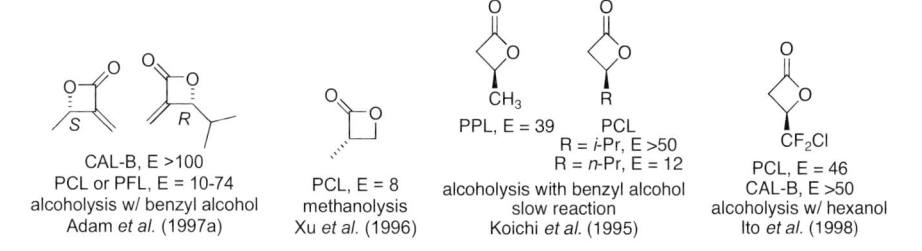

Fig. 109. Seven- and four-membered ring lactones resolved by lipase-catalyzed alcoholysis.

A more common route to enantiomerically pure lactones is to prepare a precursor (usually an efficiently resolved secondary alcohol) and convert it using lipase to the desired lactone. Selected examples are shown in Fig. 110 (see also: Sugai et al., 1990b). Another special case are lactones with additional alcohol or acid functional groups. Researchers resolved several such lactones without affecting the lactone ring; more examples are included in the surveys above. The advantages of these indirect methods are higher enantioselectivity and faster reaction times.

5.2.4 Commercial Enantioselective Reactions

5.2.4.1 Enantiomerically-Pure Chemical Intermediates

DSM-Andeno (Netherlands) produce (R)-glycidol butyrate using a PPL-catalyzed resolution, but they did not reveal details of the process (Kloosterman et al., 1988; Ladner and Whitesides, 1984). Several groups have since studied this reaction and its scale-up in more detail (van Tol et al., 1995a; b; Walts and Fox, 1990; Wu et al., 1993) (Eq. 21).

$$\text{(±)-glycidyl butyrate} \xrightarrow{\text{PPL, E=13}} \text{(R)-glycidyl butyrate} + \text{(R)-glycidol} \quad (21)$$

BASF produces enantiomerically-pure amines using a *Pseudomonas* lipase-catalyzed acylation (Balkenhohl et al., 1997). A key part of the commercialization of this process was the discovery that methoxyacetate esters reacted much fast than simple esters. Activated esters are not suitable due to a competing uncatalyzed acylation (Eq. 22).

$$\text{racemate} \xrightarrow[\text{Pseudomonas sp. lipase}]{\text{MeO-CH}_2\text{-CO-O-Et}} \quad (22)$$

R = H, 4-Me, 3-OMe

Chiroscience (now Celltech, Cambridge, UK) has scaled-up the dynamic kinetic resolution of (S)-*tert*-leucine, an intermediate for the synthesis of conformationally-restricted peptides and chiral auxiliaries (McCague and Taylor, 1997; Turner et al., 1995).

5.2.4.2 Enantiomerically-Pure Pharmaceutical Intermediates

Both DSM-Andeno (Netherlands) and Tanabe Pharmaceutical (Osaka, Japan) in collaboration with Sepracor (Marlborough, MA) have commercialized lipase-catalyzed resolutions of (+)-(2S,3R)-MPGM, a key precursor to diltiazem (Furui et al., 1996; Hulshof and Roskam, 1989; Matsumae et al., 1993; 1994). The DSM-Andeno process uses RML, while the Tanabe process uses a lipase secreted by *Serratia marcescens* Sr41 8000. In both cases the lipase catalyzed hydrolysis of the unwanted enantiomer with high enantioselectivity (E > 100). The resulting acid spontaneously decomposed to an aldehyde (Fig. 113).

In the Tanabe process, a membrane reactor and crystallizer combine hydrolysis, separation, and crystallization of (+)-(2R,3S)-MPGM. Toluene dissolves the racemic substrate in the cystallizer and carries it to the membrane containing immobilized lipase. The lipase catalyzes hydrolysis of the unwanted (−)-MPGM to the acid, which then passes through the membrane into an aqueous phase. Spontaneous decarboxylation of the acid yields an aldehyde which reacts with the bisulfite in the aqueous phase. In the absence of bisulfite, this aldehyde deactivates the lipase. The desired (+)-MPGM remains in the toluene phase and circulates back to the cystallizer where it crystallizes.

Lipase activity drops significantly after eight runs and the membrane must be recharged with additional lipase. Although the researchers detected no lipase-catalyzed hydrolysis of (−)-MPGM, chemical hydrolysis lowered the apparent enantioselectivity to E = 135 under typical reaction conditions. The yield of crystalline (+)-(2R,3S)-MPGM is > 43% with 100% chemical and enantiomeric purity.

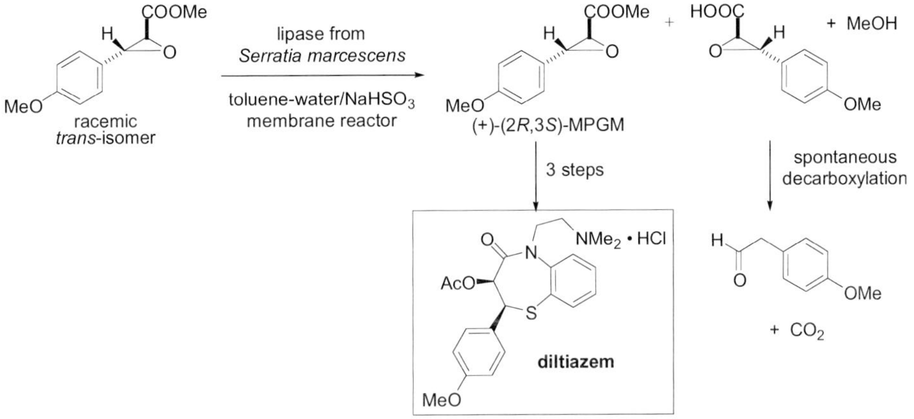

Fig. 113. Commercial synthesis of diltiazem by Tanabe Pharmaceutical uses a kinetic resolution catalyzed by lipase from *Serratia marcescens*.

Glaxo resolves (1S,2S)-*trans*-2-methoxycyclohexanol, a secondary alcohol, on a ton scale for the synthesis of a tricyclic β-lactam antibiotic (Stead et al., 1996). The slow reacting enantiomer needed for synthesis is recovered in 99% ee from an acetylation of the racemate with vinyl acetate in cyclohexane. Immobilized CAL-B and PFL (Biocatalysts, Ltd.) both showed high enantioselectivity, but CAL-B was more stable over multiple use cycles. Other workers had resolved this alcohol by hydrolysis of its esters with PCL, CRL, or pig liver acetone powder (Basavaiah and Krishna, 1994; Hönig and Seufer-Wasserthal, 1990; Laumen et al., 1989), but Glaxo chose resolution by acylation of the alcohol because it yields the required slow-reacting alcohol directly (Fig. 114).

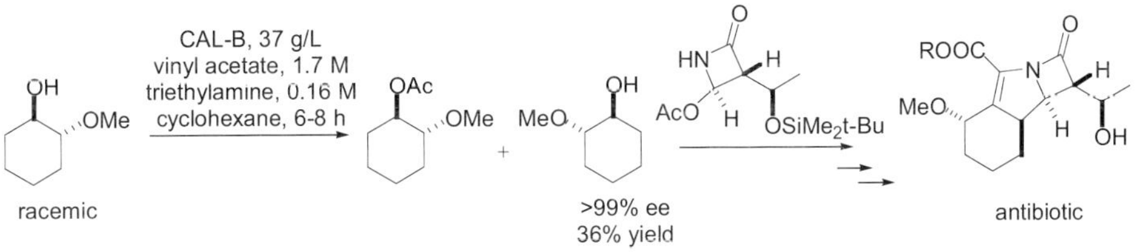

Fig. 114. Glaxo resolves a building block for antibiotic synthesis.

A multitonne scale process using a γ-lactamase has been developed at Dowpharma Chirotech Technology Ltd. (formerly Chirotech, UK) for the enantioselective synthesis of carbocyclic nucleoside precursors. Although the γ-lactamase from *Comamonas acidovorans* showed sufficient enantioselectivity (E = 94), the process was hampered by the

5.2 Survey of Enantioselective Lipase-Catalyzed Reactions

need of approx. 1 kg cells per kg product. Functional expression of the γ-lactamase in *E. coli* enabled a process based on isolated enzyme.

Fig. 115. Dowpharma resolves a building block for carbocyclic nucleoside synthesis.

Researchers reported a number of other kilogram scale routes to pharmaceutical precursors that involve lipases. Selected examples are summarized in Fig. 116.

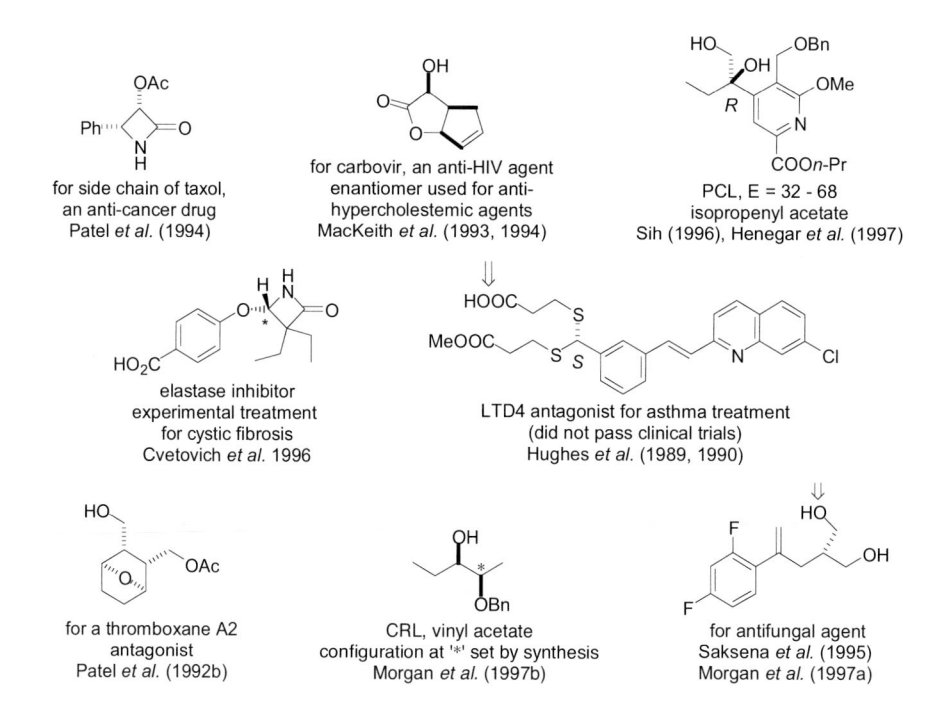

Fig. 116. Kilogram-scale routes to pharmaceutical precursors involving lipases.

5.3 Chemo- and Regioselective Lipase-Catalyzed Reactions

5.3.1 Protection and Deprotection Reactions

5.3.1.1 Hydroxyl Groups

The most difficult part of carbohydrate chemistry is the selective protection and deprotection of the various hydroxyl groups. The difficulty stems from their similar chemical reactivity, so researchers have searched for enzymic methods to simplify this problem. For example, Fink and Hay (1969) investigated the selective deprotection of peracylated sugars thirty years ago, but only more recently have researchers found enzymes and reaction conditions sufficiently selective for synthetic use (for reviews see: Bashir et al., 1995; Riva, 1996; Thiem, 1995; Waldmann and Sebastian, 1994; Wong, 1995).

The simplest examples are sugars with a single ester group. For example, PPL-catalyzed hydrolysis of the esters in the glucopyranose and -furanose shown in Fig. 117 (Kloosterman et al., 1987). However, lipases provide little advantage over chemical methods in these reactions.

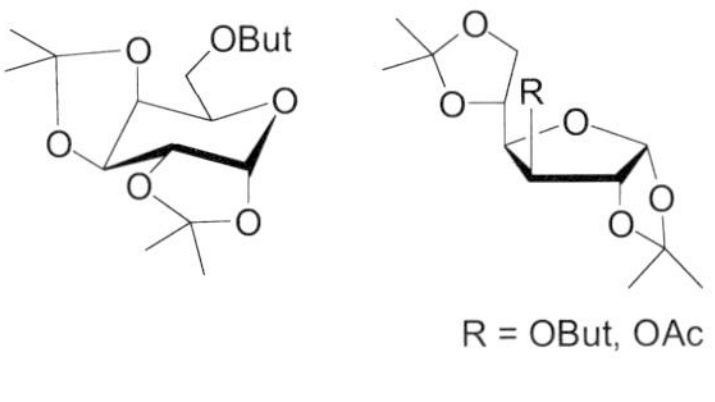

R = OBut, OAc

Fig. 117. PPL-catalyzed hydrolysis of the single ester group in protected sugars.

The most useful reactions are those which selectively protect or deprotect one hydroxyl in the presence of several others. The selectivity of lipases usually parallels the chemical reactivity of the hydroxyls, but with increased selectivity. Thus, in hydrolysis reactions of peracylated sugars, the ester at the anomeric carbon (a secondary hydroxyl) reacts first, followed by the ester at the primary hydroxyl. The remaining esters at the secondary hydroxyls react next. In acylation reactions, the primary hydroxyl group reacts first, followed by the secondary hydroxyls. The relative reactivity among the secondary hydroxyls in either acylation or hydrolysis of the esters remains difficult to predict because it varies with the lipase, reaction conditions and structure of the sugar. Not all reactions follow generalizations. For example, lipases sometimes acylate a secondary hydroxyl group in the presence of a primary hydroxyl.

Primary Hydroxyl Groups in Sugars

Hydrolysis of Esters of Primary Hydroxyl Groups. Sweers and Wong (1986) found that CRL selectively hydrolyzed the ester of the primary alcohol of methyl-2,3,4,6-tetra-*O*-pentanoyl-D-glycopyranosides of glucose, galactose and mannose yielding the corresponding tri-*O*-pentanoates. A later paper included methyl-2-acetamido-2-deoxy-3,4,6-tri-*O*-pentanoyl-D-mannoside (Hennen et al., 1988). The solvent was water containing 9% acetone and the isolated yields were good for the glucoside, but moderate for the galactoside and the two mannosides. The corresponding acetyl esters did not react under these conditions and the octanoyl esters formed emulsions which made isolation difficult (Fig. 118).

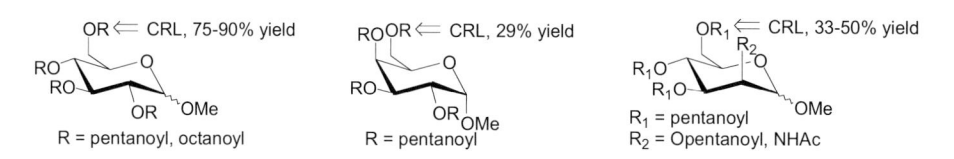

Fig. 118. Selective hydrolysis of esters at the primary position.

Hennen et al. (1988) extended this work to the furanosides shown in Fig. 119 using 10% DMF in buffer. In most cases the ester at the primary hydroxyl reacted selectively. In methyl α-D-2-deoxyriboside, both the primary ester at C-5 and the secondary ester at C-3 reacted at similar rates, while in methyl-β-D-xyloside, the secondary ester at C-3 reacted faster than the primary ester at C-5. Kloosterman et al. (1987) used PPL to selectively hydrolyze the ester from the primary alcohol in the protected D-riboside (Fig. 119).

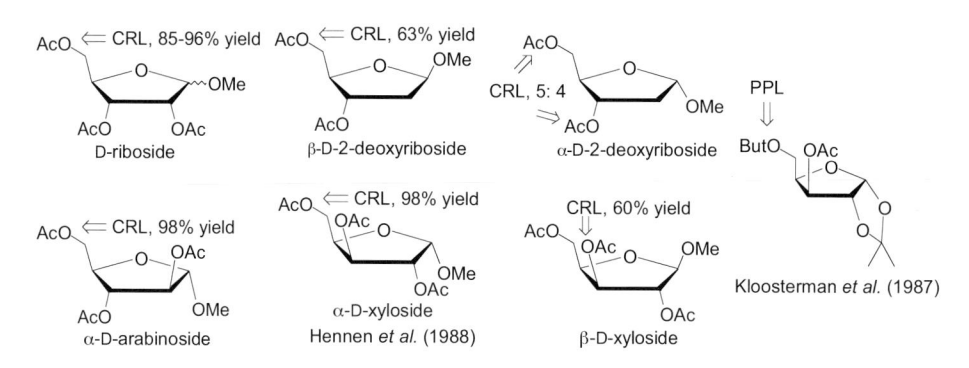

Fig. 119. Hydrolysis usually favors the primary position.

Acylation of Primary Alcohols in Unmodified Sugars. For the reverse reaction, acylation, the biggest problem is finding an organic solvent that dissolves the polar sugar, but does not inactivate the lipase. Therisod and Klibanov (1986) were the first to find that warm pyridine dissolved sugars, yet did not denature crude PPL. They used PPL to selectively acylate glucose at C-6 with 2,2,2-trichloroethyl laurate giving 40% conversion after two days with 95% regioselectivity (Eq. 23). Similarly, PPL selectively acylated the primary alcohol in mannose and galactose, but in fructose, which has primary alcohols at C-1 and C-6, both reacted at similar rates.

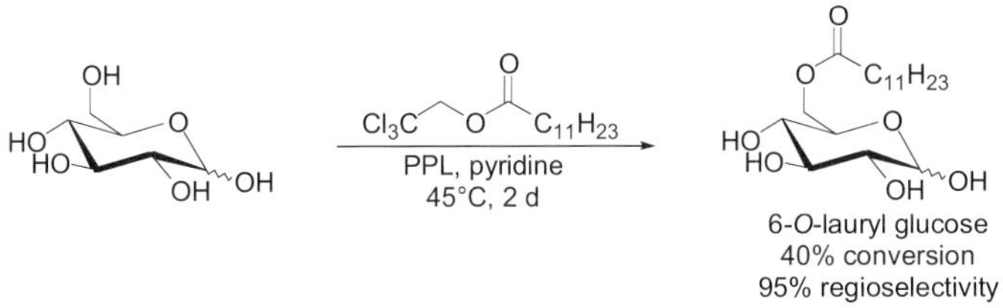

$$\text{glucose} \xrightarrow[\text{45°C, 2 d}]{\text{Cl}_3\text{C-O-CO-C}_{11}\text{H}_{23}, \text{ PPL, pyridine}} \text{6-O-lauryl glucose}$$

6-O-lauryl glucose
40% conversion
95% regioselectivity

(23)

In 2:1 benzene/pyridine, Wang et al. (1988) found that CRL also retained activity (Tab. 15). They acylated mannose and N-acetylmannosamine with the more active acyl donor, vinyl acetate. Using oxime esters as acyl donors, Gotor and Pulido (1991) found that PCL was active in pyridine or 3-methyl-3-pentanol and acylated glucose, L-arabinose, galactose, mannose and sorbose. All acylations favored the primary hydroxyl groups, but the oxime esters were more selective since Gotor and Pulido detected no diacylation. With the thermostable CAL-B, Pulido and Gotor (1993) raised the temperature to 60°C and used the more convenient solvent dioxane and alkoxycarbonyl oximes as acyl donors. This reagent introduced the carbobenzyloxy (Cbz) protective group among others (Eq. 24). More active acyl donors, such as acid anhydrides and even vinyl esters in pyridine, gave nonselective background reactions. None of the conditions above are suitable for the acylation of disaccharides, presumably because they are too insoluble.

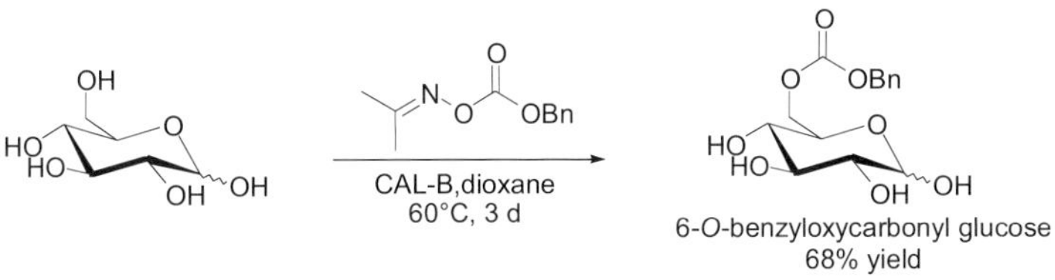

$$\text{glucose} \xrightarrow[\text{60°C, 3 d}]{\text{oxime-OBn, CAL-B, dioxane}} \text{6-O-benzyloxycarbonyl glucose}$$

6-O-benzyloxycarbonyl glucose
68% yield

(24)

Another application of sugar esters is as nonionic surfactants for the food and cosmetic industries (Fig. 120). The advantage of an enzymic route over chemical processes, besides milder reaction conditions and fewer side reactions, would be the ability to label the surfactant 'natural' (Sarney and Vulfson, 1995). In most countries, products produced from natural starting materials using enzymic catalysts are still considered natural. The acylation reactions above all use expensive acylating agents, toxic solvents, and far too much lipase (sometimes four times the weight of sugar). Although they are convenient on a lab scale, they are not practical for surfactant production. For these applications, researchers directly esterified sugars with fatty acids and optimized the reactor configuration to increase yields and reaction rate (Tab. 15).

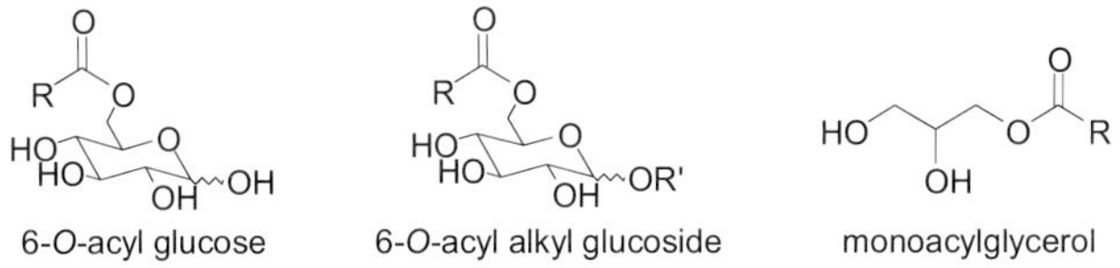

6-O-acyl glucose   6-O-acyl alkyl glucoside   monoacylglycerol

Fig. 120. Examples of surfactants prepared by lipase-mediated reactions. $R = C_7-C_{17}$ chain.

Although Sieno et al. (1984) reported esterification of sugars and fatty acids in aqueous solution, Janssen et al. (1990) found only small amounts of ester, which they extracted using a membrane reactor. Others used polar organic solvents such as 2-pyrrolidone and hindered tertiary alcohols and vacuum or drying agents to remove the water released in the esterification. This removal increased the reaction rate and the yield. Cao et al. (1996; 1997) crystallized the product ester to shift the equilibrium. A suspension of glucose, stearic acid, immobilized CAL-B, and molecular sieves in acetone yielded solid 6-O-stearoyl-D-glucose in 92% conversion after 72 h at 60°C. The acetone created a small catalytic phase, while allowing the product to precipitate (solubility: glucose in acetone = 0.04 mg/mL, glucose stearate = 3.3 mg/mL). No sugar esters formed in reverse micelles (Hayes and Gulari, 1992; 1994). Direct esterification usually works better with longer fatty acids (C_{14}–C_{18}) than with medium chain fatty acids (C_8–C_{12}). This reaction system could also be applied to the CAL-B-catalyzed synthesis of sugar esters based on arylaliphatic carboxylic acids (Fig. 121). CAL-B selectively esterified the primary hydroxyl at C-6 of the glucose moiety of D-(–)-salicin (2-(hydroxymethyl)-phenyl-β-D-glucopyranoside) (Otto et al., 1998a).

Fig. 121. CAL-B-catalyzed synthesis of sugar esters based on arylaliphatic carboxylic acids.

Acylation of Primary Alcohols in Alkyl Glycosides and Other Modified Sugars. Since the poor solubility of sugars in organic solvents is a major limitation of lipase-catalyzed acylations of sugars, many researchers modified the sugars to increase their solubility (Tab. 16). Holla (1989) used glycals (sugar precursors) which are more soluble because they lack two hydroxyl groups. Acetalisation of sugars with acetone increased the solubility so much that researchers eliminated the solvent and dissolved the sugar acetal in the cosubstrate fatty acid (Fregapane et al., 1991). Ikeda and Klibanov (1993) complexed glucose with phenylboronic acid (Fig. 122).

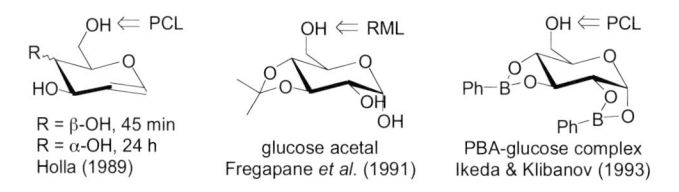

Fig. 122. Increased solubility of modified sugars in organic solvents simplifies acylation reactions.

Tab. 15. Lipase-Catalyzed Acylation of Unmodified Sugars.

Sugar	Acyl donor	Solvent	Lipase	Rate[a]	References
Glucose, mannose, galactose, fructose[b]	trichloro ethylester	pyridine	PPL	0.02	Therisod and Klibanov (1986)
Mannose	vinyl acetate	benzene-pyridine	CRL	0.004	Wang et al. (1988)
Glucose, L-arabinose, galactose, mannose, sorbose	oxime ester	pyridine, 3-methyl-3-pentanol	PCL	0.02β0.03	Gotor and Pulido (1991); Pulido et al. (1992)
Glucose, galactose, mannose, ribose, arabinose	O-(alkoxylcarbo-nyl) oxime	Dioxane	CAL-B	0.08	Pulido and Gotor (1993)
Glucose	fatty acid	water, EMR[c]	CRL	0.02[d]	Janssen et al. (1990)
Glucose	fatty acid	Water	CRL	0.04	Sieno et al. (1984)[e]
Sorbitol	fatty acid	2-pyrrolidone, EMR[c]	CVL	1.4[d]	Janssen et al. (1991)
Glucose	fatty acid	reverse micelles	ROL, CRL	no rxn	Hayes and Gulari (1992)
Glucose, fructose	fatty acid	t-butanol	CAL-A & B	0.3	Oguntimein et al. (1993)
Glucose	fatty acid	Acetonitrile	CAL-B	0.04	Ljunger et al. (1994)
Sorbitol	fatty acid	2-methyl-2-butanol	CAL-B	0.2	Ducret et al. (1995)
Gructose	fatty acid	2-methyl-2-butanol	RML, CAL-B	0.05	Scheckermann et al. (1995)
Glucose	fatty acid	acetone/crystallization	CAL-B	0.2β0.4	Cao et al. (1996)
Fructose	vinyl laurate	THF	PFL	0.05	Sin et al. (1996)
Ribose, 2-deoxyribose	propanic anhydride	THF	CAL-B	3.3	Prasad et al. (1995)
Glucose	vinyl laurate	t-butanol / IL[f]	CAL-B	n.d.	Ganske and Bornscheuer (2005)

[a] mmol sugar ester produced per gram enzyme and hour calculated from literature data. [b] Both primary alcohol groups in fructose reacted at similar rates. [c] EMR = enzyme membrane reactor. [d] Initial rate [e] Later workers found little ester formation under these conditions. [f] IL = ionic liquids.

The complex dissolved in *t*-butanol and PCL efficiently catalyzed the acylation of the primary hydroxyl group with vinyl or trifluoroethyl butyrate. Solubilization of fructose in hexane with phenylboronic acid allowed selective acylation of the C-1 primary hydroxyl, with no reaction at the C-6 primary hydroxyl (Scheckermann et al., 1995; Schlotterbeck et al., 1993). However, the reaction was one hundred times slower than the glucose reaction reported by Ikeda and Klibanov. Complexation with boronic acids or acetalisation also allow acylation of disaccharides (Oguntimein et al., 1993; Sarney et al., 1994).

Alkyl glucosides are more soluble in organic solvents, hence, lipase-catalyzed acylations of these sugar derivatives is simpler than unmodified sugars. In addition, the rate of acylation increases as the size of the acyl group increases. The Wong group acylated methyl-β-D-glucoside using CRL and vinyl acetate in a mixture of benzene and pyridine and several methylfuranosides (D-ribose, D-arabinose or D-xylose) using PPL and 2,2,2-trifluoroethyl acetate in THF (Hennen et al., 1988; Wang et al., 1988). Theil and Schick (1991) significantly improved the rate of acylation using ethyl glycosides, crude PPL and vinyl acetate in a mixture of THF and triethylamine. In all cases, acylation was selective for the primary alcohol group.

For surfactant applications, Novozymes catalyzed the direct esterification of alkyl glucosides in molten fatty acid using immobilized CAL-B (Adelhorst et al., 1990; Björkling et al., 1989). Typical reactions showed excellent yield and good regioselectivity (Eq. 25). Upon scale-up, Unichema International (a subsidiary of Unilever) encountered difficulties with the viscous and heterogenous reaction mixture. Adding 25 vol% *t*-butanol reduced the viscosity and adding 5 mol% product (e.g., 6-*O*-laurylglucopyranoside) emulsified the reactants. In a packed bed reactor with a separate pervaporation compartment to remove water, they achieved 90% conversion in 25 h for 40 batch reactions (Macrae, personal communication, 1996).

Unichema International also produces esters such as decyl oleate, octyl palmitate, isopropyl myristate, isopropyl palmitate, and PEG400 monostearate for skin care products using CAL-B catalyzed esterification (Bosley, 1997). Water is removed by vacuum. Unichema calls these 'bioesters' because products made by lipase-catalyzed process starting from natural materials retain their 'natural' designation. Other companies may produce flavor esters such as isoamyl acetate or geranyl acetate by lipases.

$$\text{(25)}$$

6-*O*-dodecyl ethyl glucoside, 94% yield

2,6-*O*-didodecyl ethyl glucoside, 2.4% yield

RML was less regioselective (14% of the diester), but adding hexane improved the regioselectivity (Fabre et al., 1993a). Pelenc et al. (1993) further combined this process with an α-transglucosidase-catalyzed synthesis of α-butyl glucoside from maltose and butanol (Eq. 26).

maltose $\xrightarrow[\text{n-butanol}]{\alpha\text{-transglucosidase}}$ [α-butyl glucoside, 82% yield] $\xrightarrow[\substack{65°C \\ \text{vacuum}}]{\substack{\text{RML} \\ \text{RCOOH}}}$ [6-O-acyl-α-butyl glucoside, 80% yield] (26)

RML also catalyzed the 6-O-selective acylation of α-butyl glucoside with more complex acids, for example, the protected amino acid $BocNH(CH_2)_3COOCH_2CCl_3$ (Fabre et al., 1994) and bis(2,2,2-trichloroethyl)adipate (Fabre et al., 1993b).

CAL-B selectively acylates the primary alcohol in a wide variety of nucleosides. Gotor and Morís (1992) found that oxime esters of simple acids or protected amino acids selectively acylated the primary hydroxyl, while oxime carbonates gave the 5'-O-carbonates. Interestingly, PCL selectively acylated the secondary 3'-hydroxyl even when the primary alcohol was unprotected (Fig. 123).

PCL
R = H, base = uracil, 5-fluorouracil, 5-trifluoromethyluracil
hexanoic anhydride
Uemura et al. (1989a), Nozaki et al. (1990)

acylation with oxime esters

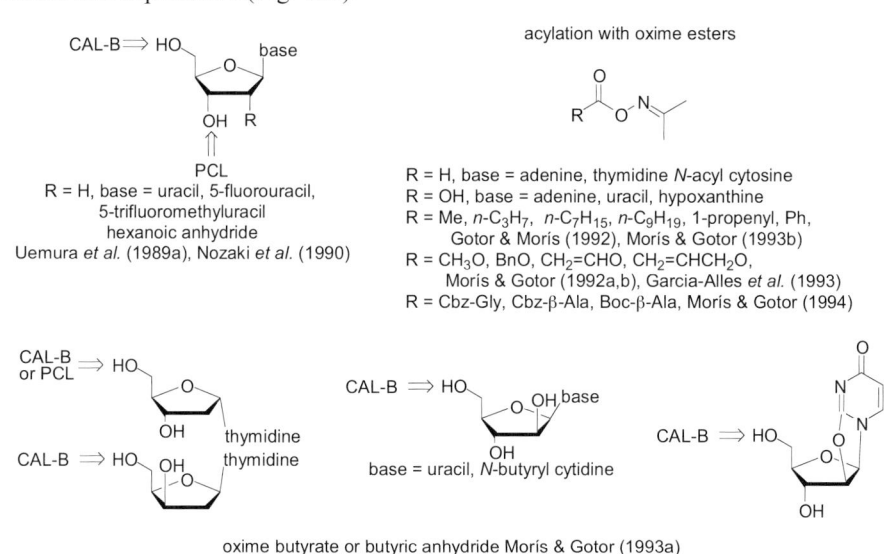

R = H, base = adenine, thymidine N-acyl cytosine
R = OH, base = adenine, uracil, hypoxanthine
R = Me, $n\text{-}C_3H_7$, $n\text{-}C_7H_{15}$, $n\text{-}C_9H_{19}$, 1-propenyl, Ph,
Gotor & Morís (1992), Morís & Gotor (1993b)
R = CH_3O, BnO, CH_2=CHO, CH_2=CHCH$_2$O,
Morís & Gotor (1992a,b), Garcia-Alles et al. (1993)
R = Cbz-Gly, Cbz-β-Ala, Boc-β-Ala, Morís & Gotor (1994)

oxime butyrate or butyric anhydride Morís & Gotor (1993a)

Fig. 123. CAL-B selectively acylates the primary position, while PCL favors the secondary position.

Subtilisin, a protease, also catalyzes the selective acylation of carbohydrates (Carrea et al., 1989; Riva, 1996; Riva et al., 1988). Subtilisin is better suited than lipases for the acylation of disaccharides, and often shows complementary selectivity to lipases (Sect. 6.4.1.1) (Kazlauskas and Weissfloch, 1997).

Danieli et al. (1995) selectively acylated one of the two primary hydroxyl groups in the triterpene oligoglycoside ginsenoside Rg_1 using CAL-B and vinyl acetate or bis(2,2,2-trichloroethyl)malonate (Fig. 124).

Eberling et al. (1996) used lipase from wheat germ to hydrolyze all the acetates from the sugar portion of O-glycosyl amino acid (methoxyethoxy)ethyl (MEE) esters. The MEE esters were removed later using RJL (see Sect. 5.3.1.3). For example, a glucosyl serine derivative is shown in Fig. 125; Eberling et al. (1996) also deprotected a number of similar compounds (galactose, galactosamine and xylose) for the sugar portion and threonine for the amino acid portion.

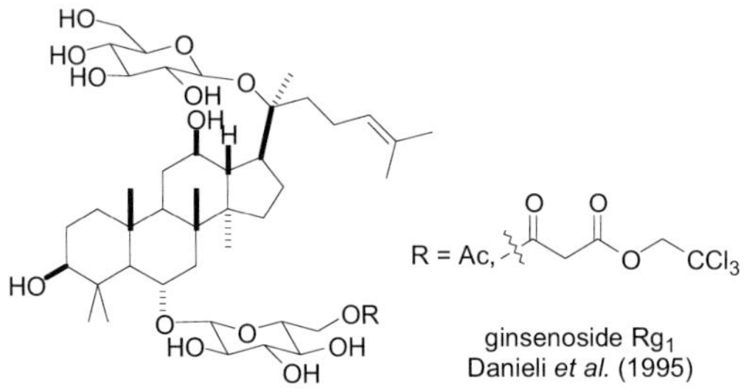

148   5.3   Chemo- and Regioselective Lipase-Catalyzed Reactions

Fig. 124. CAL-B selectively acylates one of the two primary hydroxyls.

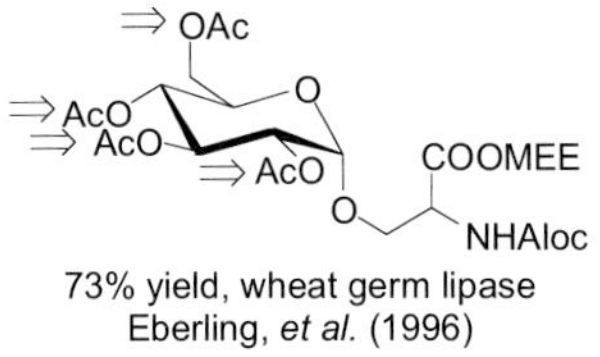

Fig. 125. Wheat germ lipase hydrolyzed the acetyl groups at both primary and secondary positions.

Riva et al. (1996) selectively acylated the only primary hydroxyl group in the flavonoid glycosides isoquercitin and naringin using CAL-B (Fig. 126).

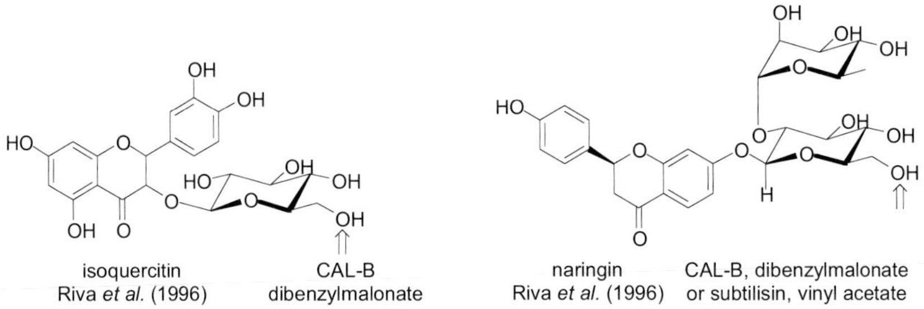

Fig. 126. CAL-B selectively acylates the primary position.

Achromobacter sp. lipase regioselectively acylated the hydroxyl group on *sn*-1 carbon among the two primary hydroxyl groups of 3-*O*-β-D-galactopyranosyl-*sn*-glycerol (Fig. 127).

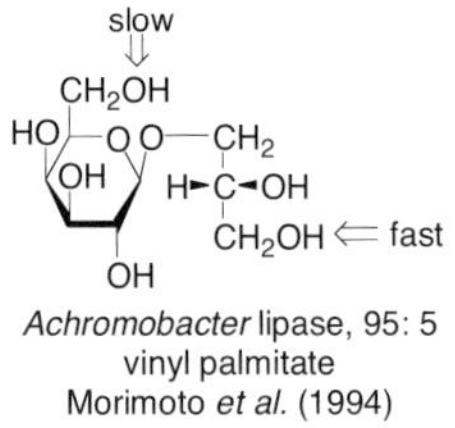

Fig. 127. *Achromobacter* sp. lipase regioselectively acylated one of two primary hydroxyl groups.

Tab. 16. Lipase-Catalyzed Reactions of Modified Sugars.[a]

Sugar-derivative	Acyl donor	Solvent	Lipase	Rate[b]	References
IPG-sugar	fatty acid	none	RML	4.540	Fregapane et al. (1991; 1994)
PBA-β-D- & β-D-glucose and others[c]	vinyl ester	t-BuOH	PCL	4.166	Ikeda and Klibanov (1993)
PBA-fructose	fatty acid	hexane	RML CAL-B	0.068	Scheckermann et al. (1995); Schlotterbeck et al. (1993)
Methyl xylose	TFEA	THF	PPL	0.007	Hennen et al. (1988)
Methyl glucose	vinyl ester	pyridine/benzene	CRL	0.006	Wang et al. (1988)
Ethyl sugar	vinyl ester	THF/Et$_3$N	PPL	1.880	Theil and Schick (1991)
Alkyl-sugar	fatty acid	none	CAL-B	n.a.[d]	Adelhorst et al. (1990); Björkling et al. (1989)
Alkyl-sugar	fatty acid	hexane	RML	n.a.[d]	Fabre et al. (1993)

[a] IPG-sugar: isopropylidene glucose, galactose, or xylose; TFEA: 2,2,2-trifluoroethyl acetate; PBA: phenylboronic acid complex; THF: tetrahydrofuran; DMF: dimethylformamide. [b] mmol sugar ester produced per gram enzyme and hour calculated from literature data. [c] D-galactose, D-fructose, sucrose, lactose maltose, D-mannitol, D-glucosamine, D-gluconic acid. [d] n.a. = data not available.

Secondary Hydroxyl Groups in Sugars

Hydrolysis of Acylated Secondary Hydroxyl Groups. In peracylated sugars, the most chemically reactive ester is the one at the anomeric position. Although it is more hindered than the primary alcohol ester, the anomeric hydroxyl is the best leaving group, because it is a hemiacetal. The ester at the anomeric position is also the most reactive in lipase-catalyzed deacylations. Hennen et al. (1988) found that PPL or ANL in buffer containing 9% DMF catalyzed selective hydrolysis of the acetate at the anomeric position for pyranoses and furanoses (Fig. 128). In most cases, the yields were above 70%. One exception was peracetyl β-D-glucopyranose, where CRL catalyzed selective hydrolysis of the esters at positions 4 and 6 leaving the triacetyl derivative.

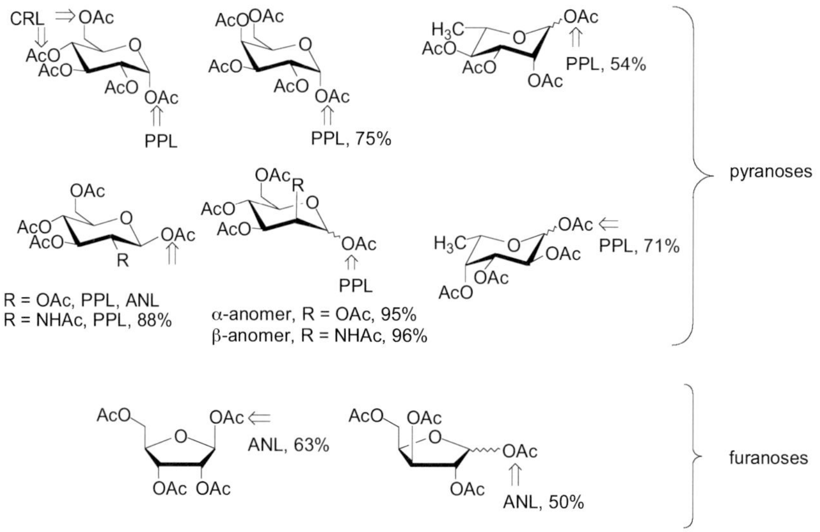

Fig. 128. Selective hydrolysis of the acetate at the anomeric position.

If the anomeric position lacks an ester group, then the most reactive ester is the one at the primary hydroxyl. When the sugar lacks esters both at the anomeric and at the primary hydroxyls, it is not easy to predict which secondary hydroxyl ester will react most rapidly, even for the same lipase. In the series of anhydropyranoses in Fig. 129, CRL catalyzed hydrolysis of both the butyrates at the 2- and 4-positions in the glucose derivative, while several other lipases catalyzed hydrolysis of only the butyrate at the 4-position (Kloosterman et al., 1989). However in the 3-azido-3-deoxy-glucose derivative, CRL selectively hydrolyzed the acetate at the 4-position (Holla et al., 1992), while subtilisin favored the acetate at the 2-position. In the galactose derivative, CRL favored the butyrate at the 2-position (Ballesteros et al., 1989) (Fig. 129). Subtilisin and lipases often show an opposite regioselectivity in reactions involving secondary alcohols (Kazlauskas and Weissfloch, 1997).

5 Lipases and Esterases   151

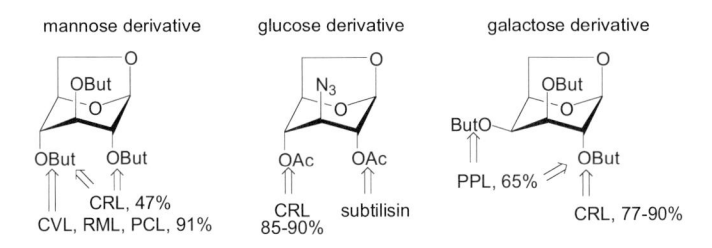

Fig. 129. Selectivity among secondary hydroxyl groups is hard to predict.

PCL selectively hydrolyzed the acetate at the 3-position or 4-position in the structures of Fig. 130 (Holla, 1989; López et al., 1994). The yields were 90% in both cases. Note that this selectivity follows the secondary alcohol rule in Fig. 53.

Fig. 130. Regioselectivity sometimes follows the secondary alcohol rule.

CRL and wheat germ lipase selective hydrolyzed an acetyl group from a secondary hydroxyl in octa-O-acetyl sucrose (Fig. 131). CRL selectively removed the acetyl group at the 4'-position, while wheat germ lipase selectively removed acetyl groups at the 4'- and 6'-positions. Kloosterman et al., 1989 attributed this surprising selectivity to selective hydrolysis of the acetyl group at the 6'-primary hydroxyl followed by acetyl migration from the 4'- to the 6'-position.

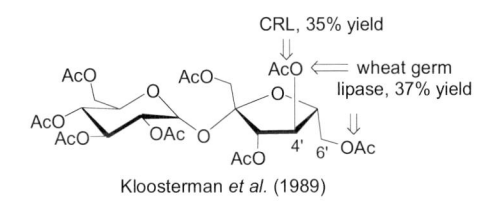

Fig. 131. Lipases selectively deacetylated octa-O-acetyl sucrose at a secondary position.

Several researchers reported selective reaction at the secondary alcohol position in the presence of a chemically more reactive primary alcohol position. For example, PCL selectively cleaved the hexanoate ester of secondary alcohol in several 2-deoxyribonucleosides (Uemura et al., 1989b). Subtilisin selectively cleaved the ester at the primary position. In the reverse reaction, PCL also catalyzed the selective acylation of this secondary hydroxyl (see above) (Garcia-Alles et al., 1993; Uemura et al., 1989a). Protecting the primary alcohol as a hindered ester also allowed selective hydrolysis of the ester at the secondary position in the protected arabinose in Fig. 132 (Kloosterman et al., 1987).

152   5.3   Chemo- and Regioselective Lipase-Catalyzed Reactions

Fig. 132. Lipases sometimes favor hydrolysis of esters at secondary positions over primary positions. Lipases also show selectivity among secondary positions.

To form the monoacetate of 1,4:3,6-dianhydro-D-glucitol, Seemayer et al. (1992) started with the diacetate and selectively hydrolyzed the acetate at the (R)-stereocenter. This selectivity fits the secondary alcohol rule discussed in Sect. 5.2.1.1, but in this case the starting material was derived from a sugar and thus was already enantiomerically pure.

Acylation of Secondary Hydroxyl Groups. Therisod and Klibanov (1987) found that lipases catalyzed the regioselective acylation of the C-2 or C-3 hydroxyl group in C-6 protected glucose. The regioselectivity depended on the lipase. For example, CVL catalyzed the butyrylation of the C-3 hydroxyl of 6-O-butanoyl glucopyranose with trichloroethyl butyrate in THF, while PPL catalyzed butyrylation of the C-2 hydroxyl. Chemical or enzymic methods removed the protecting groups at the 6-position leaving a C-2 or C-3 hydroxyl protected glucose. No lipase acylated at the C-4 hydroxyl. For 6-O-butyryl mannose and galactose the selectivity was low (5:1 at best, typically 2:1).

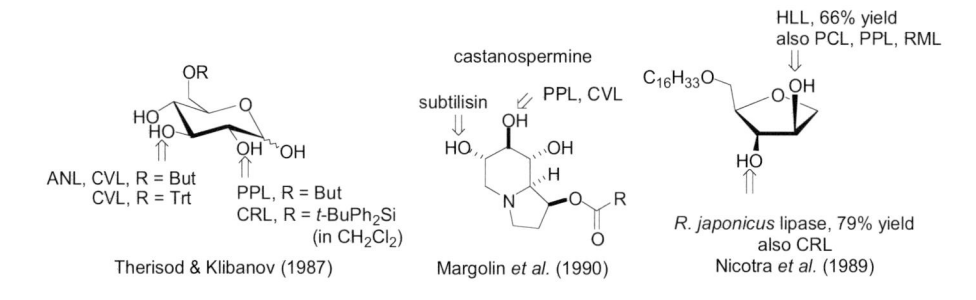

Fig. 133. Selectivity among secondary positions when the primary position is protected.

Another example of lipase-dependent regioselectivity is the butyrylation of 1,4-anhydro-5-O-hexadecyl-D-arabinitol with trichloroethyl butyrate in benzene (Nicotra et al., 1989). HLL catalyzed butyrylation of the C-2 hydroxyl, while *R. japonicus* lipase favored the hydroxyl at C-3 (Fig. 133). In castanospermine, lipases PPL and CVL favored acylation of one hydroxyl group, while subtilisin favored another (Fig. 133) (Margolin et al., 1990).

For methyl glycosides, several lipases (PCL, PPL, CRL) selectively acylated the C-2 hydroxyl in 6-*O*-butyryl methyl α-D-galactopyranoside, but the regioselectivity for the corresponding mannoside was still low (Fig. 134, right column). In a series of methyl pyranosides, Ciuffreda et al. (1990) found that PCL acylated the C-2 hydroxyl in the D-series of sugars, but the C-4 hydroxyl in the L-series. Acylation was much slower when the reacting hydroxyl group was axial (D-rhamnose and L-fucose derivatives). They suggested that efficient acylation requires an axial-equatorial-equatorial arrangement of hydroxyls with acylation occuring at the last equatorial hydroxyl.

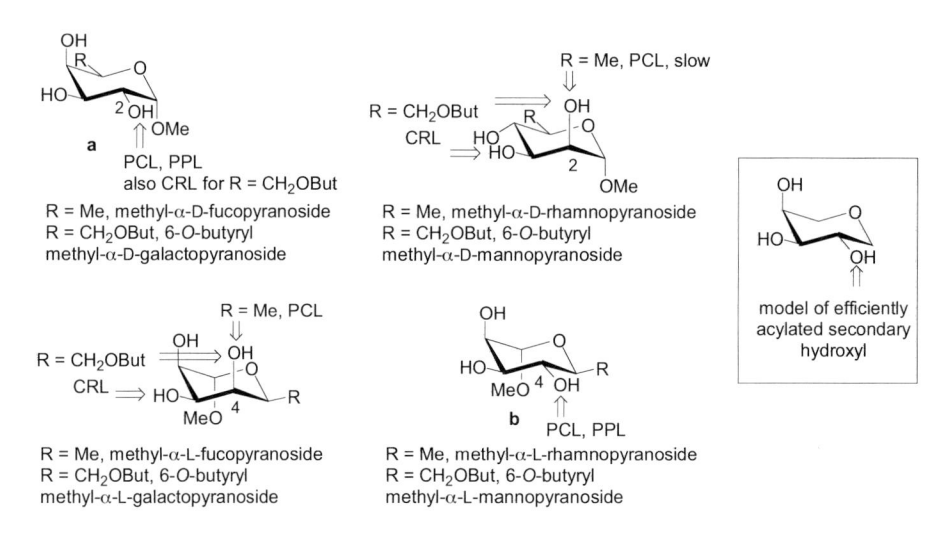

Fig. 134. In methyl α-D- and α-L-glycopyranosides, PCL regioselectively acylated the C-2 hydroxyl group in the D-series (top two structures), but the C-4 hydroxyl group in the L-series (bottom two structures) using trifluoroethyl butyrate in THF (Ciuffreda et al., 1990). Only the sugars **a** and **b** reacted quickly, thus Ciuffreda et al. (1990) suggested that an efficiently acylated sugar contains an axial-equatorial-equatorial arrangement of hydroxyls as shown in the model.

The regioselectivity of the PCL- and PFL-catalyzed acylation of methyl 4,6-*O*-benzylidene glycopyranosides depended on the configuration at the anomeric carbons (Fig. 135) (Chinn et al., 1992; Iacazio and Roberts, 1993; Panza et al., 1993a; b). The α-anomers yielded the C-2 monoester with typical reaction times of 7 h, while the β-anomers reacted in about 1 h and yielded the C-3 monoester. The Roberts group used vinyl acetate as the solvent and acylating agent, while Panza et al. (1993a; b) used a variety of vinyl esters and trifluoroethyl esters in THF. The galactose derivatives reacted significantly more slowly, possibly due to steric hindrance.

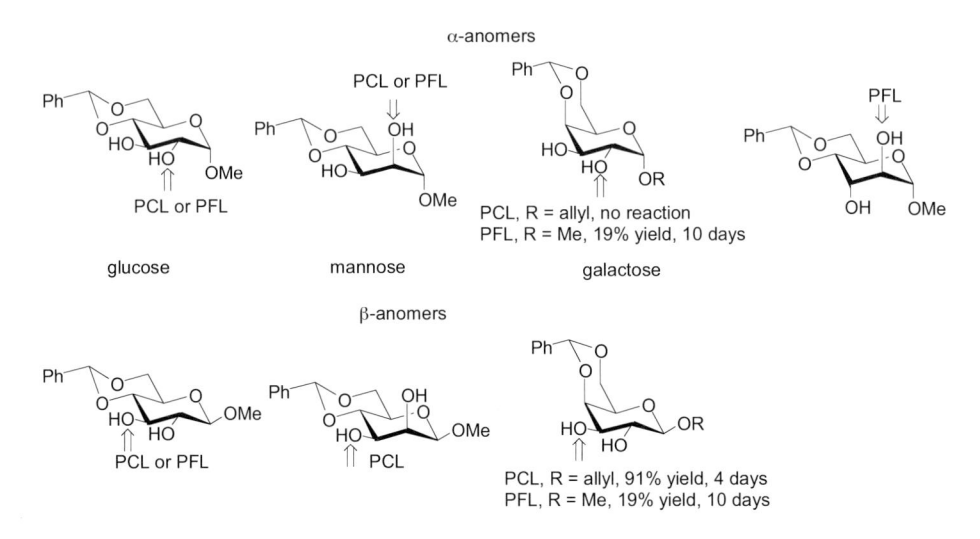

5.3 Chemo- and Regioselective Lipase-Catalyzed Reactions

Fig. 135. The configuration at the anomeric carbon determines the regioselectivity of the acylation of methyl 4,6-O-benzylidene glycopyranosides. PCL and PFL acylate the C-2 hydroxyl in the α-anomers and the C-3 hydroxyl in the β-anomers, regardless of the orientation of the reacting hydroxyl.

PCL also catalyzed acylation of the C-3 hydroxyl in 6-O-acetyl D-glucal and 6-O-acetyl D-galactal with vinyl esters (Holla, 1989) (Fig. 136).

Fig. 136. Regioselectivity among secondary hydroxyl groups sometimes follows the secondary alcohol rule.

López et al. (1994) found that the regioselectivity also varies with the nature of the substituent at the anomeric position. PCL catalyzed formation of the 3,4-diacetate of methyl-β-D-xylopyranoside using vinyl acetate in acetonitrile, whereas the octyl derivative in acetonitrile or hexane gave a mixture of the 2,4- and 3,4-diacetates. At short reaction times, the 2-monoacetate predominated. The choice of solvent and reaction conditions is less critical than for sugars because these sugar derivatives are more soluble (Fig. 137).

Fig. 137. Regioselectivity varies with the nature of the substituent at the anomeric position.

In summary, lipases can react selectively at the different secondary hydroxyls. The selectivity varies with lipase and substrate structure (anomeric orientation, anomeric substituent, orientation of hydroxyl) and one cannot make broad generalizations yet.

Hydroxyl Groups in Non-Sugars

Phenolic hydroxyls. Several lipases, especially PCL and PPL, catalyze the deacetylation of peracetylated polyphenols by transesterification with *n*-butanol in organic solvents (Figs. 138–140). Researchers deacetylated by transesterification instead of hydrolysis because the substrates do not dissolve in water. The regioselectivity of lipases toward phenolic hydroxyls usually paralleled their chemical reactivity – less hindered positions reacted more quickly. For flavone acetates and related compounds, a generalization in Fig. 139 summarizes some of the observed regioselectivity. In addition, PCL catalyzed the regioselective acetylation of polyphenols with vinyl acetate (Figs. 138 and 140). Both acetylation and deacetylation favor the less hindered positions, thus the two reactions yield complementary products. Nicolosi et al. (1993) used the deacetylation reaction in the synthesis of a rare *O*-methyl flavonoid, ombuin.

In a number of symmetrical acylated catechols, PPL selectively removed only one acyl group (Parmar et al., 1996; 1997). Lipases CRL and PPL also showed excellent chemoselectivity. They cleaved the phenolic ester, while leaving the benzoate ester intact.

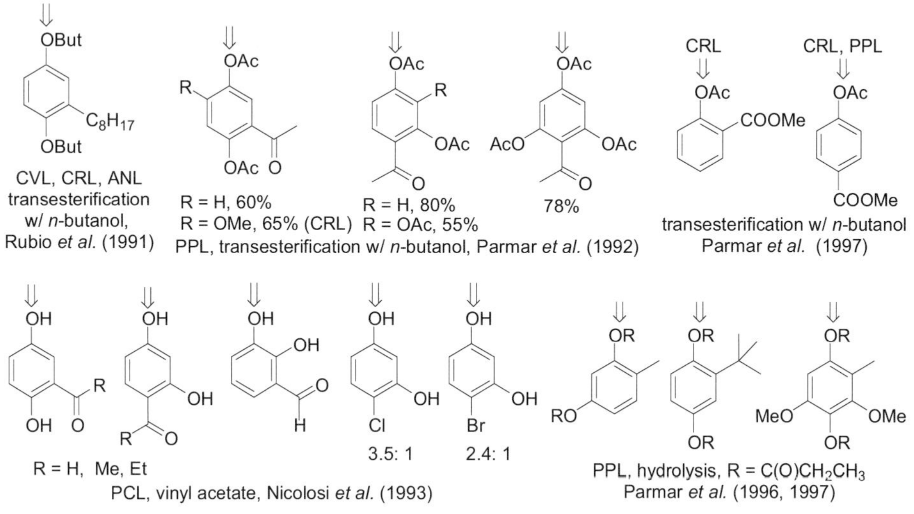

Fig. 138. Regioselectivity of lipases toward polyhydroxylated benzenes. Lipases favored the less hindered position in both deacetylation of peracetylated phenols by transesterification with *n*-butanol and in acetylation of phenols with vinyl acetate. Note that the first two examples of Parmar et al. (1996; 1997) show deacylation at the more hindered ester.

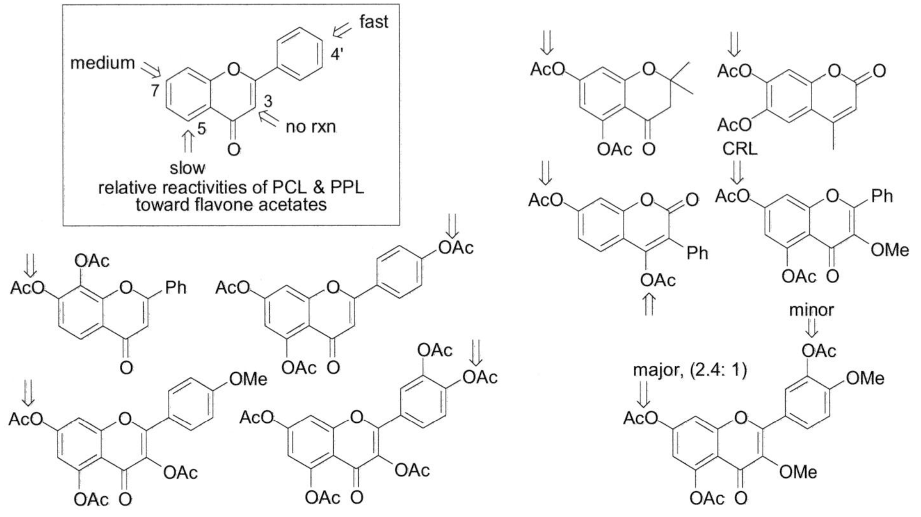

5.3 Chemo- and Regioselective Lipase-Catalyzed Reactions

Fig. 139. Regioselectivity of lipases toward flavone acetates and related compounds. Lipases catalyzed the deacetylation by transesterification with n-butanol. Less hindered acetates react more quickly; a generalization for the observed regioselectivity is suggested above.

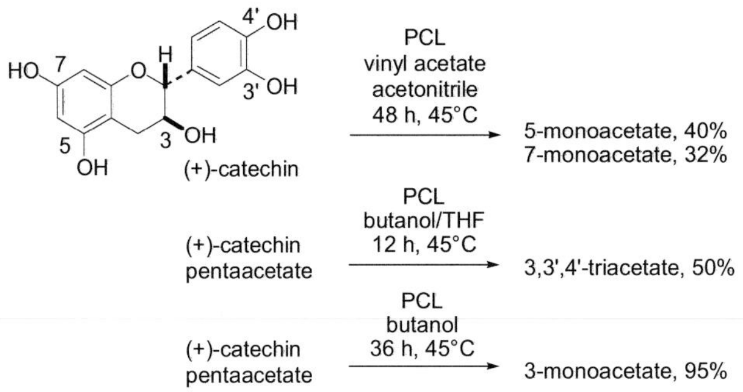

Fig. 140. Regioselective acetylation and deacetylation of catechin. Hydroxyls or acetates at positions 5 and 7 react most quickly, while those at position 3 do not react. Acetylation and deacetylation yield complementary acetates (Lambusta et al., 1993).

Aliphatic Hydroxyls. PPL in acetone selectively acylated the primary hydroxyl group in several diols using trifluoroethyl butyrate (Parmar et al., 1993b). Deacylation of the corresponding diesters showed apparent selectivity for the secondary hydroxyl, but later work showed that deacylation occurred at the primary position, followed by acyl migration to the secondary position (Bisht et al., 1996) Several lipases selectively hydrolyzed one primary hydroxyl in a diacetate (Itoh et al., 1996c) (Fig. 141).

In some cases, the configuration of nearby stereocenters changed the selectivity. For example, PCL showed a low selectivity for the less hindered primary hydroxyl in the (*R*)-enantiomer in Fig. 142, but a moderate selectivity for the more hindered primary hydroxyl in the (*S*)-enantiomer. In another case, CRL acetylated the hydroxyl at the (*S*)-stereocenter only in the (*S,S*)-stereoisomer, not in the (*S,R*)-stereoisomer (Fig. 142).

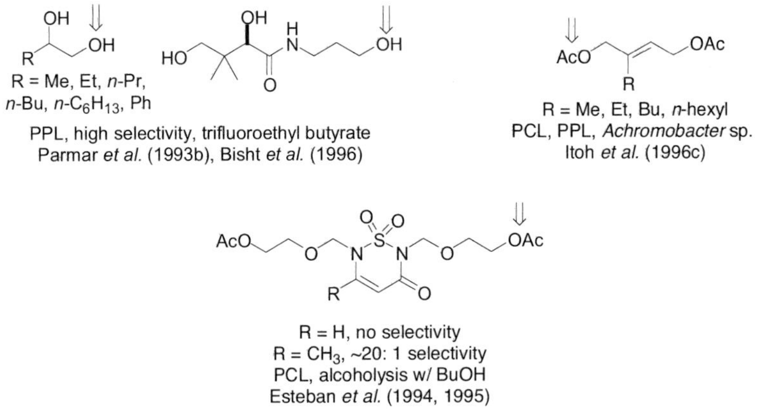

Fig. 141. Selective acylation and deacylation of primary alcohols.

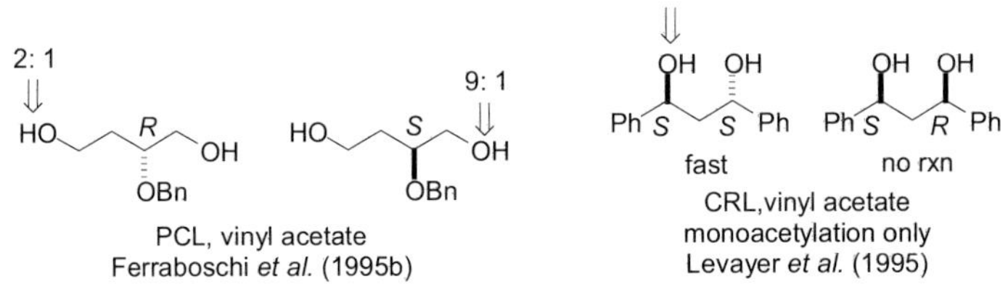

Fig. 142. Selectivity varies with the configuration of nearby stereocenters.

Sattler and Haufe (1995) selectively acylated the primary over the secondary alcohol in a mixture of diastereomers. This regioselective reaction is a more convenient way of separating the isomers than the orginal method of flash chromatography (Fig. 143).

Garcia-Granados et al. (1998) separated a mixture of three diastereomeric sesquiterpene lactones (formed by hydrogenation of a natural product) by selective acetylation. First, either PPL or RML selectively acetylated the diastereomer having the (*R*)-carbinol. After separation of the unreacted diastereomers, CAL-B selectively acetylated the *trans* diastereomer (Fig. 143).

158   5.3   Chemo- and Regioselective Lipase-Catalyzed Reactions

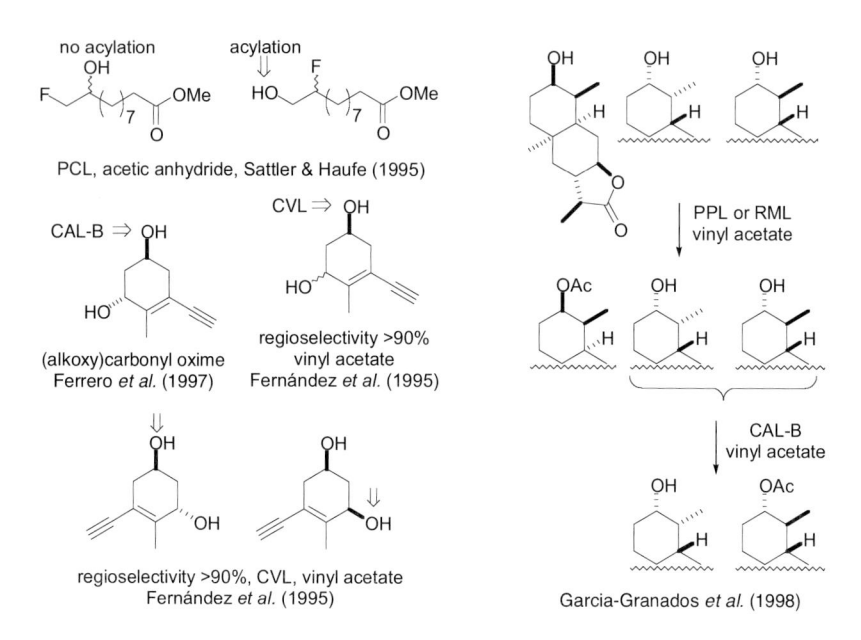

Fig. 143. Selective acylation of one diastereomer simplified separation.

Several lipases (PCL, ANL, CRL, GCL) also resolved hydrophobic amino acids (Ala, Leu, Met, Phe, Val) by hydrolysis of methyl ester in *N*-Cbz protected amino acids (Fig. 144). The enantioselectivity was usually moderate and favored the L-enantiomer (Chiou et al., 1992).

Fig. 144. Resolution of *N*-Cbz protected amino acids by lipase-catalyzed hydrolysis.

Lipases catalyze the chemoselective acylation of 2-mercaptoethanol and similar compounds. Acylation occured on the oxygen yielding *O*-acyl esters, not *S*-acyl thioesters (Baldessari et al., 1994; Iglesias et al., 1996). This chemoselectivity may be in part due to the choice of an unactivated acyl donor – simple carboxylic ethyl esters – which lack the thermodynamic driving force to form a thioester. In steroids, lipases favor the less-hindered and equatorially oriented hydroxyls (Fig. 145).

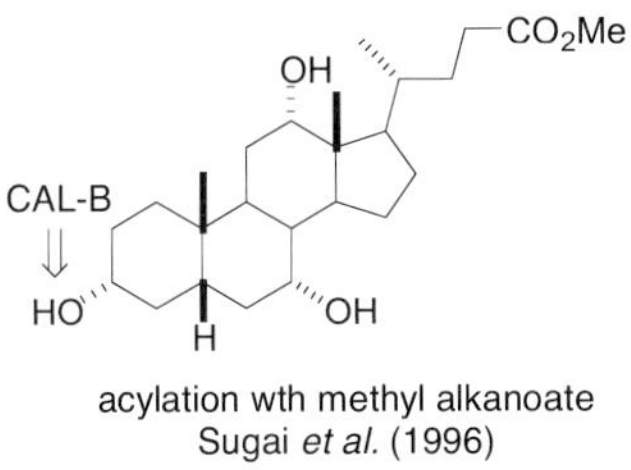

[structure diagram]

Fig. 145. CAL-B favored the less-hindered and equatorially oriented hydroxyls.

5.3.1.2 Amino Groups

Amines react spontaneously with most acylating agents so few lipase-catalyzed reactions have been reported. Gardossi et al., 1991 used dilute solutions and a large amount of lipase to selectively acetylate the ε-amino group in L-Phe-α-L-Lys-O-*t*-Bu and L-Ala-α-L-Lys-O-*t*-Bu with trifluoroethyl acetate. Pozo et al. (1992) used the less reactive vinyl carbonate and CAL-B to form a carbamate, one of the more common amino protective groups (Eq. 27).

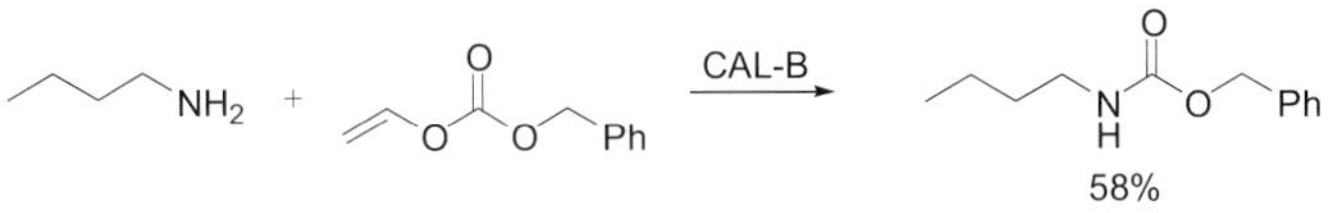

[reaction scheme, 58%] (27)

Adamczyk and Grote (1996) protected amines by PCL-catalyzed acylation using benzyl esters.

Lipases are not used to deprotect amines because lipases rarely cleave amides or carbamates, the most common amino protective groups. Proteases such as penicillin G acylase are normally used for deprotection (reviewed by Waldmann and Sebastian, 1994). However, Waldmann and Naegele (1995) reported an indirect removal of carbamate protective group with an esterase. Upon cleavage of the acetyl group from a *p*-acetoxybenzyloxycarbonyl-protected peptide with acetylesterase, the carbamate link cleaved spontaneously. Lipases should also catalyze this reaction.

PCL catalyzed the hydrazidolysis of α,β-unsaturated esters such as methyl acrylate (Eq. 28) (Astorga et al., 1991; 1993; Gotor et al., 1990). These reactions occur at room temperature with simple esters, while chemical methods require higher temperatures, activated esters or acid chlorides, and suffer from competing Michael additions.

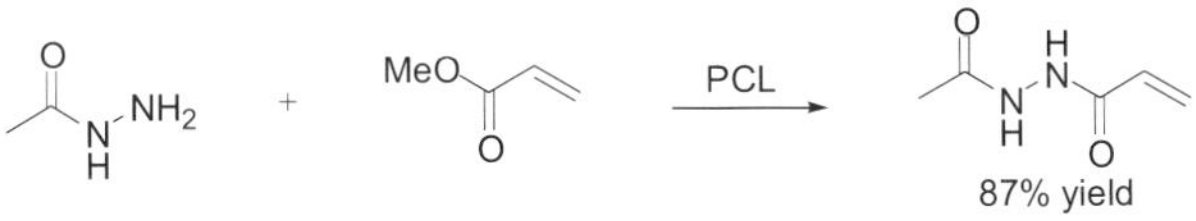

[reaction scheme, 87% yield] (28)

ANL was effectively used for the mild removal of an acetyl protective group in a precursor of *N*-glycolylneuraminic acid – a member of the sialic acid family – to afford *N*-glycosylmannosamine (Fig. 146) (Kuboki et al., 1997). Khmelnitsky et al. (1997) regioselectively acylated paclitaxel to introduce water-solubilizing groups. The same group also used regioselective acylation of a flavonoid to generate a combinatorial library of derivatives (Fig. 146) (Mozhaev et al., (1998).

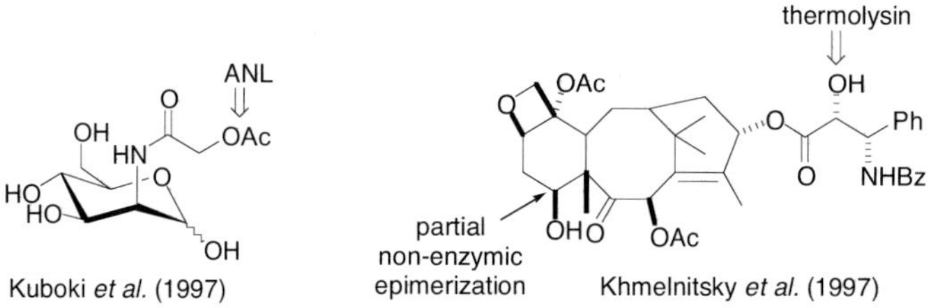

Fig. 146. ANL catalyzed mild removal of an acetyl protective group. Thermolysin regioselectively acylated only one hydroxyl in paclitaxel.

Penicillin G acylase (PGA, penicillin amidase) is highly selective for phenylacetyl groups. Researchers exploited this chemoselectivity to selectively remove N-protective groups from peptides (Waldmann et al., 1991). In addition, Pohl and Waldmann (1996) made a carbamate protective group that can be removed with PGA.

5.3.1.3 Carboxyl Groups

Although many chemical methods exist to protect and deprotect carboxyl groups in amino acids for peptide synthesis, many of these are incompatible with sensitive functional groups such as thioesters, phosphate esters, and polyenes (farnesyl groups). The mild reaction conditions for enzymic reactions makes them ideal for reactions involving sensitive substrates. Chemists have developed a variety of methods, most of which involve proteases, esterases or lipases (Waldmann and Sebastian, 1994). Selected examples are reviewed below.

The advantages of lipases over proteases are that they tolerate water-insoluble substrates, do not cleave peptide bonds (a potential side reaction in protease-catalyzed reactions of peptides), and tolerate both L- and D-amino acids.

tert-Butyl esters are usually inert to lipases and proteases, but thermitase (see Sects. 6.2.3 and 6.4.2.1) is an exception. Thermitase deprotected the carboxyl group in several peptides and glycopeptides by hydrolysis of the *t*-butyl esters (Schultz et al., 1992). Thermitase-catalyzed hydrolysis of ester links is much faster than of peptide links, so the peptide remained intact. In addition, an esterase from *Bacillus subtilis* (BsubpNBE) and lipase A from *Candida antarctica* (CAL-A) were used to hydrolyze a range of *tert*-butyl-esters of protected amino acids (e.g., Boc-Tyr-O^{t}Bu, Z-GABA-O^{t}Bu, Fmoc-GABA-O^{t}Bu) in good to high yields and left Boc, Z and Fmoc-protecting groups intact (Schmidt et al., 2005). A lipase from *Burkholderia* sp. YY62 was able to hydrolyze *tert*-butyl-octanoate (Yeo et al., 1998).

Many groups have used lipases to deprotect carboxyl groups in peptides. Braun et al. (1990; 1991) cleaved the heptyl ester carboxyl protective group from a wide range of dipeptides using ROL (Amano N). This lipase did not cleave the amino protective groups Cbz, Boc, Alloc, or Fmoc and the heptyl protective group survived conditions used to remove these amino protective groups (hydrogenation, HCl/ether, or Pd(0)/C-nucleophile). Although the crude lipase (Amano N) also hydrolyzed the peptide link, pretreatment with PMSF, a serine protease inhibitor, eliminated this side reaction. Hydrolysis of the heptyl group slowed and sometimes did not proceed when the peptide was hindered and/or hydrophobic. In some of these cases, replacing the heptyl ester with the more reactive 2-bromoethyl ester or the more water-soluble 2-(*N*-morpholino)ethyl ester or MEE ester increased reaction rate (Braun et al., 1993; Eberling et al., 1996; Kunz et al., 1994; Waldmann et al., 1991) (Fig. 147).

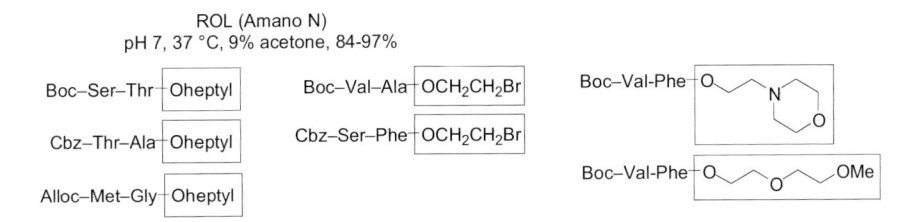

Fig. 147. Deprotection of carboxyl groups in more hydrophobic peptides requires more reactive or more soluble esters.

In other cases where hydrolysis catalyzed by ROL (Amano N) was slow, researchers substituted another lipase. During a glycopeptide synthesis researchers used RJL to cleave the heptyl protective group (Braun et al., 1992; 1993). In a similar glycopeptide, RJL did not cleave the heptyl ester, so Eberling et al. (1996) used the more reactive MEE ester. To cleave *C*-terminal proline MEE esters, researchers used HLL (Kunz et al., 1994) (Fig. 148).

Sutherland and Willis (1997) used CRL to deprotect several α-keto acids by hydrolysis of their methyl esters (Fig. 148). The mild conditions prevented decarboxylation and allowed the researchers to perform the next step, a dehydrogenase-catalyzed reduction, in the same pot.

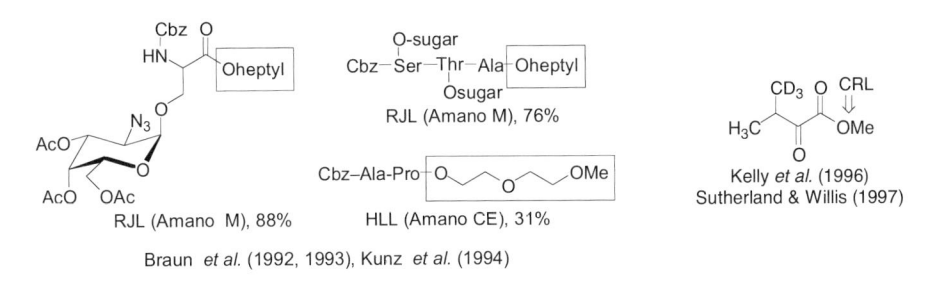

Fig. 148. Deprotection of carboxyl groups using other lipases.

Lipases can also selectively deprotect the two carboxylates in glutamic acid. For the uncommon enantiomer glutamic acid diesters (D-), both CRL and CAL-B favored reaction at the less hindered ester. CRL catalyzed the selective hydrolysis of the dicyclopentylester (Wu et al., 1991), while CAL-B catalyzed the selective amidation of the diethyl ester with pentylamine (Chamorro et al., 1995). On the other hand, in the more common enantiomer L-glutamate diethyl ester, CAL-B selectively amidates the more hindered ester (Chamorro et al., 1995) (Fig. 149). Proteases subtilisin and α-CT showed an opposite regioselectivity toward the cyclohexyl diester of L-aspartic acid. While α-CT favored the 'normal' carboxylate at the α-position, subtilisin favored the less hindered carboxylate at the β-position (Fig. 149).

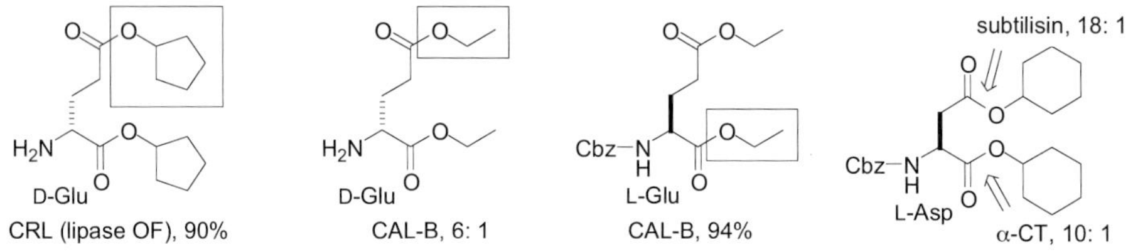

Fig. 149. Regioselective deprotection of glutamic and aspartic acid esters.

To develop immunoassays for drugs, researchers must immunize animals with the drug linked to a protein. In several cases, lipases have simplified this linking process (Adamczyk et al., 1994; 1995). PCL selectively hydrolyzed one of the ester groups in the diacid linker molecule for both the rapamycin and the digoxigenin derivatives. For the digoxigenin derivatives, ester groups on shorter linkers did not react (Fig. 150).

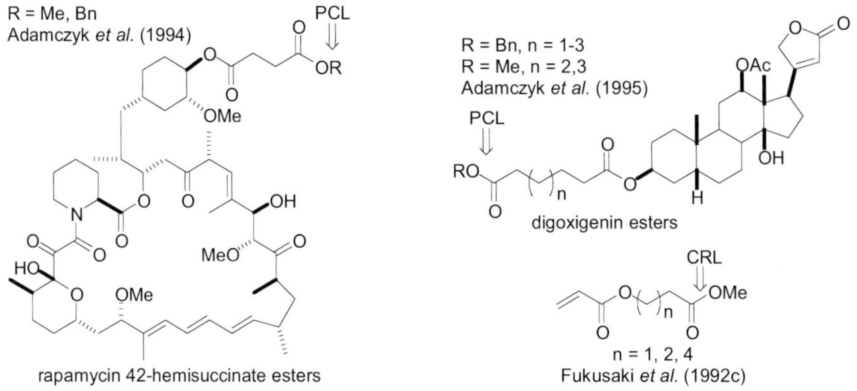

Fig. 150. Regioselective deprotection of carboxyl groups.

Sharma et al. (1995) used PPL to selectively monoesterify aliphatic dicarboxylic acids with butanol. Acid groups containing a carbon-carbon double bond at the α,β-position reacted more slowly than acid groups adjacent to saturated chains. Some CRL-catalyzed chemoselective reactions are summarized in Fig. 151.

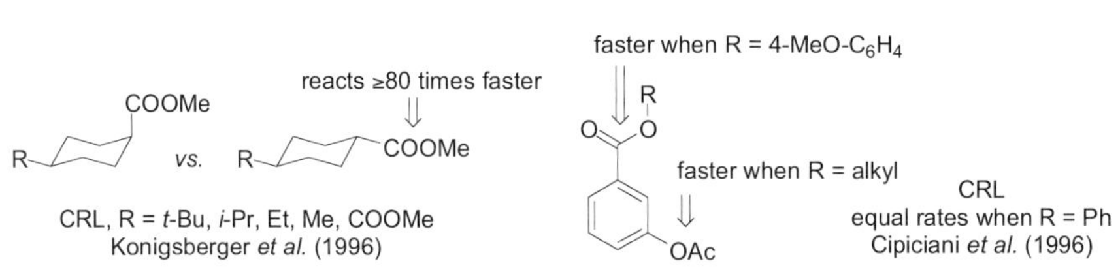

Fig. 151. Chemoselective reactions catalyzed by CRL.

5.3.2 Lipid Modifications

Of the 120 million metric tons of fats and oils produced each year world-wide, most are used directly in food, but about 15 million tons undergo chemical processing such as hydrolysis, glycerolysis and alcoholysis. Current chemical processes require high temperature and pressure which degrade the fats and introduces impurities. For example, sodium methoxide-catalyzed interesterification of triglycerides also catalyzes Claisen condensations and imparts a brown color. Lipases require milder conditions – lower temperatures, near neutral pH – and are also regioselective for the primary vs. secondary positions in glycerol and chemoselective for different fatty acids. Researchers have developed several lipase-catalyzed processes for specialty fats. These processes exploit either the regioselectivity of lipases, the fatty acid selectivity or the mild reaction conditions.

Lipases have been used to produce structured triglycerides, i.e. cocoa-butter equivalents, which are obtained by interesterification of palm oil (predominantly palmitic acid at *sn*1- and *sn*3-positions) with stearic acid changing the melting point. Another lipid of this type is Betapol (1,3-oleoyl-2-palmitoyl-glyceride), which is used in infant nutrition. More recently, the production of zero-*trans* margarine by lipase-catalyzed interesterification was developed. Several examples for the application of lipases in lipid modificaton can be found in a book (Bornscheuer, 2000) and reviews (Adamczak et al., 2002; Adlercreutz, 1994; Haas and Joerger, 1995; Vulfson, 1994).

5.3.3 Oligomerization and Polymerizations

Lipases can catalyze polymerizations by formation of ester links (reviews: Gutman, 1990; Kobayashi et al., 1994; Linko and Seppälä, 1996; Akkara et al., 1999; Gross et al., 2001). The three main approaches are (1) simple condensation of diacids with diols or hydroxy acids with themselves, (2) transesterification of either hydroxy esters or diesters with diols and (3) ring-opening polymerization of lactones. The second two approaches are more common. The most useful lipases are CRL, PCL, PPL, and RML. The potential advantages of lipase-catalyzed polymerizations are their stereoselectivity, the more narrow molecular weight range of the polyesters and the ability to make polyesters that are inaccessible by chemical methods. For example, polymerization of divinyl adipate with glycerol occurs selectively at the primary alcohol positions of glycerol leaving a pendant secondary hydroxyl. Early polymerizations yielded only low molecular weights (usually 1 000–7 000), but later report include examples of high molecular weight (40 000).

Only a few groups reported simple condensation reactions. Either PCL (Ajima et al., 1985) or CRL (O'Hagan and Zaidi, 1993; 1994a) catalyzed the condensation of 10-hydroxydecanoic acid to the corresponding polyester. At 55°C in the presence of 3 Å molecular sieves, O'Hagan and Zaidi (1994a) obtained a polyester with a molecular weight of ~9 000 (50 repeat units) (Eq. 29).

In the simple condensation of a diacid with a diol, Okumura et al. (1984) reported ANL-catalyzed formation of pentamers and heptamers. Binns et al. (1993) used a two-stage reaction to condense adipic acid with butane-1,4-diol. The first condensation formed oligomers, after isolation these oligomers were coupled in the second condensation to form a polyester with an average of 20 repeat units.

Most researchers used transesterification reactions to form the ester link. Early reports showed only oligomers such as pentamers or hexamers (Chaudhary et al., 1995; Geresh and Gilboa, 1990; 1991; Gutman et al., 1987; Knani and Kohn, 1993; Margolin et al., 1987; Park et al., 1994) even when using activated esters such as 2,2,2-trichloroethyl. However, Wallace and Morrow (1989a; b) obtained polyesters with degrees of polymerization up to 25 by using highly purified monomers. In addition, they used exactly two moles of diester for each mole of diol because only one enantiomer of the diester reacted (Fig. 152).

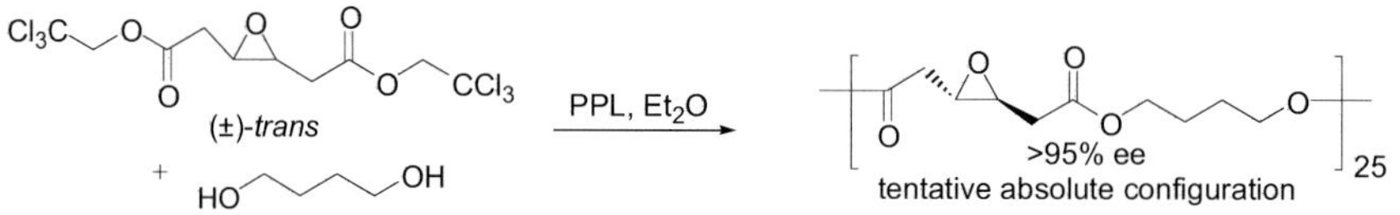

Fig. 152. PPL-catalyzed polymerization of a diester and a diol.

In later work Morrow suggested that the release of the alcohol, even a poorly nucleophilic alcohol like 2,2,2-trichloroethanol, limits the molecular weight of the polymer. Release of alcohol may also promote desorption of water from the enzyme which also limits the molecular weight by permitting hydrolysis of either the starting diester or the product polyester (Brazwell et al., 1995). To minimize this problem, researchers carried out polymerizations under vacuum in high boiling solvent to remove the released alcohol (Brazwell et al., 1995; Linko et al., 1995a; b). For example, a PPL-catalyzed transesterification of bis(2,2,2-trifluoroethyl)glutarate with 1,4-butanediol reached a molecular weight of 39 000 (> 200 repeat units) under vacuum, while only ~2 900 without vacuum (Fig. 153).

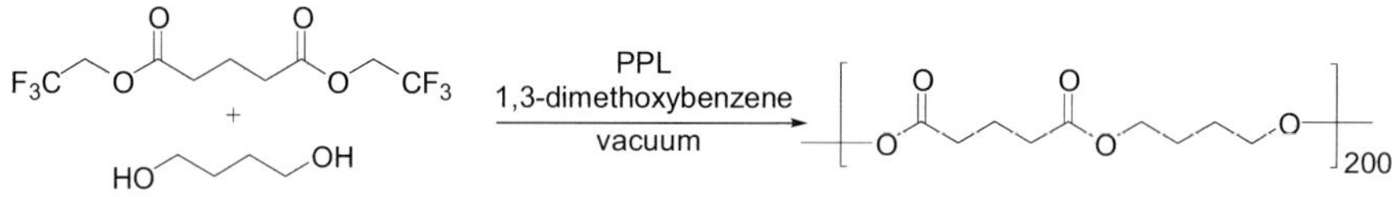

Fig. 153. PPL-catalyzed polymerization under vacuum gives high molecular weight polyester.

Polymerizations starting with vinyl esters (Uyama and Kobayashi, 1994) or oxime esters (Athawale and Gaonkar, 1994) gave molecular weights up to 7 000 (~35 repeat units). For the CAL-B catalyzed polymerization of divinyl adipate and 1,4-butanediol, Chaudhary et al. (1998) found that competing hydrolysis, even under 'dry' conditions, limited the molecular weight of the polymer to ~6000. Upon increasing the substrate concentration to minimize this hydrolysis and decreasing the size of the immobilized enzyme pellets (to minimize diffusional limitations), the molecular weight increased to >30 000 and even >200 000.

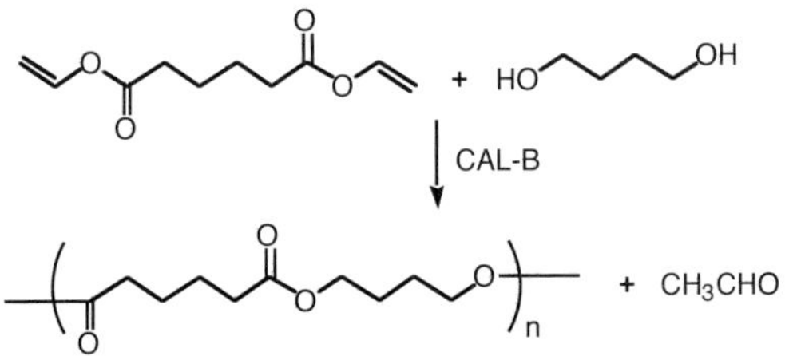

Fig. 154. Polymerization of divinyl adipate with 1,4-butanediol. Trace amounts of water can limit the molecular weight of the polyester.

Ring-opening polymerization is a special case of transesterification polymerization that does not release a molecule of alcohol. Lipase-catalyzed polymerization of ε-caprolactone with either PCL or PPL yields a polyester with a molecular weight up to 7 700 (67 repeat units) (Knani et al., 1993; MacDonald et al., 1995; Uyama and Kobayashi, 1993; Uyama et al., 1993). Researchers added a small amount of alcohol such as butanol to initiate the polymerization. MacDonald et al. (1995) suggested that water bound to the enzyme limits the molecular weight of the polymer by reacting with the oligomers (Fig. 155). A way to remove this water is azeotropic distillation with toluene under reduced pressure at 70° C. This approach gave polycaprolactone with an M_n of 81 kDa (Ebata et al., 2001).

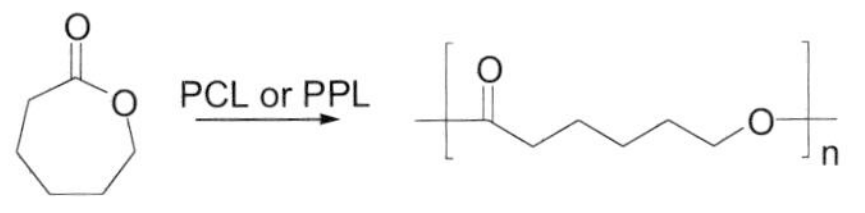

Fig. 155. Ring-opening polymerization of ε-caprolactone.

Similar polymers form upon ring-opening polymerization of the 12-membered 11-undecanolide and the 16-membered 15-pentadecanolide (Uyama et al., 1995) and also the four-membered β-propiolactones, including substituted β-propiolactones (Nobes et al., 1996; Svirkin et al., 1996). Ring-opening polymerization of succinic acid anhydride with diol gave polymers with degrees of polymerization up to 14 (Kobayashi and Uyama, 1993).

Lipases also catalyze the degradation of polyesters (for examples see: Koyama and Doi, 1996; Nagata, 1996; Tokiwa et al., 1979).

5.3.4 Other Lipase-Catalyzed Reactions

In addition to various hydrolysis and transesterification reactions, CAL-B also catalyzed the "esterification" of carboxylic acids and hydrogen peroxide to peroxycarboxylic acids (Björkling et al., 1992; Cuperus et al., 1994; Kirk et al., 1994). Peroxycarboxylic acids are more reactive than hydrogen peroxide and reacted *in situ* with olefins to give epoxides (Eq. 30). Similarly, added ketones underwent Baeyer-Villiger-oxidation (Lemoult et al., 1995).

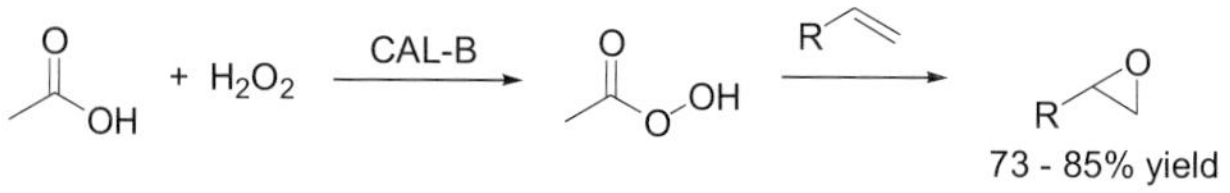

$$\text{AcOH} + H_2O_2 \xrightarrow{\text{CAL-B}} \text{AcOOH} \xrightarrow{R\diagup\!\!\!\diagup} R\text{-epoxide} \quad\quad (30)$$

73 - 85% yield

This method can also be used for the 'self'-epoxidation of unsaturated fatty acids (Fig. 156) such as oleic or linoleic acid and plant oils in high yields (e.g., rapeseed oil 91%, sunflower oil 88%) (Rüsch gen. Klaas and Warwel, 1996). The same group also described a three-step-one-pot synthesis of epoxyalkanolacylates, in which perhydrolysis, epoxidation and interesterification are integrated in a single process (Rüsch gen. Klaas and Warwel, 1998). For instance, 9-octadecen-1-ol was converted to the corresponding epoxyalkanol butyrate in 89% yield after 22 h reaction time.

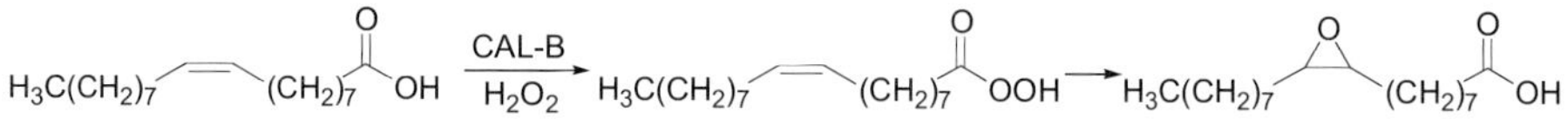

$$H_3C(CH_2)_7\text{-CH=CH-}(CH_2)_7\text{-C(O)OH} \xrightarrow[H_2O_2]{\text{CAL-B}} H_3C(CH_2)_7\text{-CH=CH-}(CH_2)_7\text{-C(O)OOH} \longrightarrow H_3C(CH_2)_7\text{-epoxide-}(CH_2)_7\text{-C(O)OH}$$

Fig. 156. CAL-B-catalyzed access to epoxidized fatty acids via peroxycarboxylic acids.

5.4 Reactions Catalyzed by Esterases

5.4.1 Pig Liver Esterase

5.4.1.1 Biochemical Properties

Pig liver esterase (EC 3.1.1.1) is a serine-type esterase and consists of three subunits. Its physiological role is the hydrolysis of various esters occurring in the pig diet, which might explain its wide substrate tolerance. According to Heymann and Junge (1979) 'most laboratories working on the characterization of highly purified carboxylesterases (especially pig liver esterase) agree that these are highly frustrating enzymes'. Nevertheless, they were able to perform detailed studies of these subunits by isolectric focussing and revealed that they are not identical (Heymann and Junge, 1979). They found up to 16 bands showing esterase activity and identified six major components with isoelectric points ranging from 4.8 to 5.8. Similar observations were reported by Farb and Jencks (1980), who separated seven isozymes. Heymann and Junge reported different substrate specificities of these fractions towards methyl butyrate or butanilicain (N-butylglycyl-2-chloro-6-methylanilide hydrochloride).

The three major fractions of PLE had apparent molecular weights of 58.2 kDa (α-subunit, C-terminus leucine), 59.7 kDa (β-subunit, C-terminus glycine) and 61.4 kDa (γ-subunit, C-terminus alanine) as determined by SDS-PAGE. Amino acid analysis revealed a lower content of aspartic acid and a higher content of arginine in the α-subunit compared to the γ-subunit (Heymann and Junge, 1979).

Fraction I (γ-subunit) has cholinesterase-like properties (Junge and Heymann, 1979), because it hydrolyzes butyrylcholine and is sensitive to physostigmine and fluoride ions. Fraction V (α-subunit) does not hydrolyze butyrylcholine and acts on short-chain aliphatic esters. Main components of PLE are the trimers γγγ, αγγ, ααγ, and ααα. Fractions I and V differ considerably in their substrate spectra and solvents like acetone, methanol or acetonitrile had different effects on fractions I or V.

A trimeric structure of three identical subunits obtained by electron microscopy was reported for a pig intestinal proline-β-naphthylamidase. At pH 4.5 and 4°C, this trimer dissociates into active subunits, which show a 2.5-fold lower activity compared to the trimer (Takahashi et al., 1989; 1991).

Later, the same group determined the nucleotide and deduced amino acid sequence and suggested that this proline-β-naphthylamidase is identical with pig liver esterase (Matsushima et al., 1991). This was further supported by Heymann and Peter (1993), who found that fraction I (γ-subunit) of PLE is capable to hydrolyze proline-β-naphthylamide. However, they claim that the sequence reported by Matsushima et al. is the γ-subunit. Functional expression of the γ-subunit (γ-isoenzyme) of PLE confirmed this (see below).

In contrast to the observations made by Heymann's group and by Öhrner et al. (1990), Jones and co-workers (Lam et al., 1988) claimed that isozymes of commercially available pig liver esterase show essentially the same stereospecificities toward representative monocyclic and acyclic diester substrates. Only differences in activity were observed. However, it cannot be excluded that the stereoselectivity of PLE may vary from batch to batch, because the pig diet probably induces different PLE isozymes (Seebach and Eberle, 1986).

PLE has the major advantage of easy preparation in form of an acetone powder (Adachi et al., 1986; Seebach and Eberle, 1986). A procedure reported by Adachi et al. (1986) is as follows: "Fresh pig liver (about 4 kg) is homogenized in 18 L cold acetone by using a kitchen juicer. After confirming that the homogenized parts became in a well-powdered state, they were collected by filtration. The residue was further washed with cold acetone (18 L) to remove the fatty material as cleanly as possible. The 'acetone powder' thus obtained was dried at room temperature to afford about 1 kg of the crude pig liver esterase. The fibrous material was removed and about 800 g of fine powder was finally obtained after sieving."

However, it must be emphasized that the industrial use of pig liver esterase is nowadays less desirable, as biocatalysts from animal sources tend to be prohibited due to the risk of viral diseases, which also can result in problems with enzyme supply. In addition, products derived therefrom are not kosher or halal further restricting their use.

5.4.1.2 Recombinant PLE

The reported functional overexpression of the γ-isoenzyme of PLE in the yeast *Pichia pastoris* (Lange et al., 2001) allowed for the first time production of a recombinant PLE isoenzyme (γPLE) as a stable product without the interfering influence of other isoenzymes and hydrolases. More importantly, γPLE shows substantially higher E-values in the resolution of some acetates of secondary alcohols. For instance, the resolution of (R,S)-1-phenyl-2-butyl acetate proceeded with E = 1–4 using commercial PLE, but with E>100 using the recombinant enzyme, which contains only the single isoenzyme (Fig.

5.4 Reactions Catalyzed by Esterases

157) (Musidlowska et al., 2001). Interestingly, also changes in enantiopreference were found, i.e. PLE from Fluka preferentially hydrolyzed the (*R*)-enantiomer of 1-phenyl-3-butyl acetate, whereas a preparation from Roche and γPLE showed (*S*)-selectivity. For other substrates all PLEs exhibited the same enantiopreference. Thus, the aspect of varying PLE selectivities controversely debated in a number of publications can now be attributed to changes in isoenzyme ratios depending on the PLE source.

Fig. 157. A recombinant isoenzyme of pig liver esterase (γPLE) shows significantly higher enantioselectivity towards 1-phenyl-2-butyl acetate than crude preparation from commercial suppliers.

The gene encoding γPLE shares 97% identity to the published nucleotide sequence of porcine intestinal carboxyl esterase (PICE). By site-directed mutagenesis, 22 nucleotides encoding 17 amino acids were exchanged step-wise from the PLE gene yielding the recombinant PICE sequence and eight intermediate mutants. All esterases were successfully produced in *Pichia pastoris* as extracellular proteins (Musidlowska-Persson and Bornscheuer, 2003a). Again, significant differences were found for the enantioselectivity of these new esterases in the hydrolysis of a range of acetates of secondary alcohols (Musidlowska-Persson and Bornscheuer, 2003b), Fig. 158, Fig. 159.

1: $n = 0$, $R = CH_3$
2: $n = 0$, $R = CH_2CH_3$
3: $n = 2$, $R = CH_3$
4: $n = 1$, $R = CH_3$
5: $n = 1$, $R = CH_2CH_2CH_3$
6: $n = 1$, $R = CH_2CH_3$

(*R,S*)-**1-6**

Fig. 158. Acetates **1–6** studied as model substrates for γPLE and variants derived therefrom.

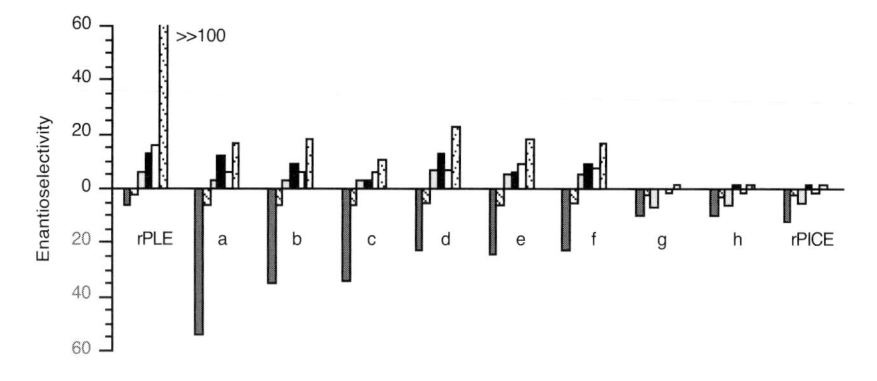

Fig. 159. Changes in enantioselectivity observed for the recombinant isoenzyme of pig liver esterase (rPLE), mutants (a-h) and recombinant porcine intestinal carboxyl esterase (rPICE). Enantioselectivity was studied towards the six acetates of secondary alcohols (Fig. 158) and is shown in the corresponding columns (acetates **1-6** from left to right bars). Note that not only enantioselectivity values changed, but also enantiopreference is altered by amino acid substitutions.

An up to six-fold increase in enantioselectivity (E = 46) compared to γPLE (E = 8) was observed in the hydrolysis of (R,S)-1-phenyl-1-ethyl acetate using a variant containing only one mutation, E77G (a in Fig. 159). For other substrates, a switch in enantiopreference was observed with the introduction of certain mutations.

5.4.1.3 Overview of PLE Substrate Specificity and Models

In contrast to lipases, PLE usually does not accept highly hydrophobic substrates. PLE shows highest activity in aqueous buffered systems (best: pH 7.0), which might be supplemented up to 20% with polar protic water-miscible solvents like methanol, *t*-butanol, DMSO, acetone or acetonitrile. pH as well as solvent should be carefully choosen, because both can strongly influence the stereoselectivity and rate of reaction. For instance, addition of cosolvents to a PLE-catalyzed hydrolysis of *meso*-cyclohexene diacetate increased the enantioselectivity and decreased the rate (Tab. 17) (Guanti et al., 1986). Others reported similar increases in enantioselectivity (Björkling et al., 1987; Lam et al., 1986; Polla and Frejd, 1991), but in some cases addition of cosolvent decreased the enantioselectivity (for a summary see: Faber et al., 1993).

Although PLE converts a huge variety of substrates, only hydrolysis of methylesters of carboxylic acids or acetates of alcohols is recommended. PLE is especially useful in the preparation of optically pure compounds starting from prostereogenic or *meso*-substrates. PLE-catalyzed acylations in organic solvents are extremely slow and therefore have no practical importance. However, activity of PLE in organic solvents could be substantially enhanced after colyophilization of the esterase with methoxypolyethylene glycol (Ruppert and Gais, 1997).

Tab. 17. Effect of Organic Solvents on PLE-Catalyzed Asymmetric Hydrolysis (Guanti et al., 1986).

Organic cosolvent	Relative rate [%]	Optical purity [% ee]
None	100	55
20% DMSO	70	59
40% DMSO	28	72
20% DMF	35	84
5% *t*-BuOH	70	94
10% *t*-BuOH	44	96

Because noone succeeded so far in the crystallization of PLE, no structural data based on x-ray diffraction or NMR measurements is available. However, the recently solved structures of a rabbit (75% identity to γ-PLE, Bencharit et al., 2002) and a human carboxylesterase (76% identity to γ-PLE, Bencharit et al., 2003) can be used to create structural models, which were already used to explain the substrate specificity found for rPLE

5.4 Reactions Catalyzed by Esterases

variants and PICE (Musidlowska-Persson and Bornscheuer, 2003a). Before, a three dimensional cubic model of the active site of PLE (Fig. 160) developed by Jones and coworkers was used, which allowed a prediction of the stereoselectivity and reactivity of PLE towards a given substrate. The model was based on the evaluation of over 100 substrates. The best fit of a substrate is determined by locating the ester group to be hydrolyzed within the locus of the serine residue and then arranging the remaining moieties in the H and P pockets (Provencher et al., 1993; Toone et al., 1990).

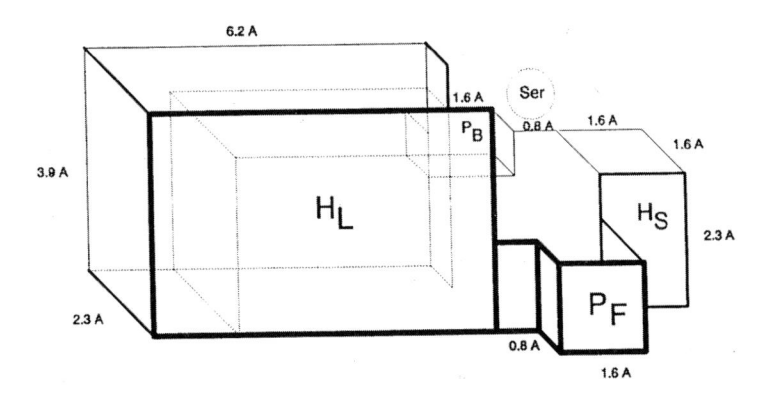

Fig. 160. Active site model for PLE. H_L is the large and H_S the small hydrophobic binding pocket. P_F is the polar, hydrophilic front pocket and P_B the rear polar pocket. Ser stands for the serine residue of the catalytic triad (Provencher et al., 1993; Toone et al., 1990).

In addition, a substrate model for PLE was proposed (Fig. 161) (Mohr et al., 1983). Optimal selectivity with PLE can be ensured by assigning the α- and β-substituents of methyl carboxylates according to their size (S = small, M = medium, L = large) with the preferably accepted enantiomer oriented as shown in Fig. 161.

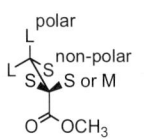

Fig. 161. Substrate model of PLE (Mohr et al., 1983).

In the following survey, various applications of PLE are summarized. Pre-1990 publications dealing with the use of PLE in organic synthesis have been extensively reviewed (Jones, 1993; Ohno and Otsuka, 1990; Phytian, 1998; Tamm, 1992; Zhu and Tedford, 1990) and only a few remarkable examples are included here.

5.4.1.4 Asymmetrization of Carboxylic Acids with a Stereocenter at the α-Position

The examples summarized in Fig. 162 show that the asymmetrization of carboxylic acids with a stereocenter at the α-position varies considerably. Selectivity was largely affected and even inversed by ring size or presence of a double bond. This changes in enantiopreference could be explained by the active site model of PLE shown in Fig. 160 (Toone and Jones, 1991). In addition, enantiomeric excess differed substantially for hydrolysis of either methyl- (E = 9) or ethyl esters (E > 100) of the acetonide of a bicyclic *meso*-diester (Fig. 162, top row, first structure (Adachi et al., 1986; Arita et al., 1983).

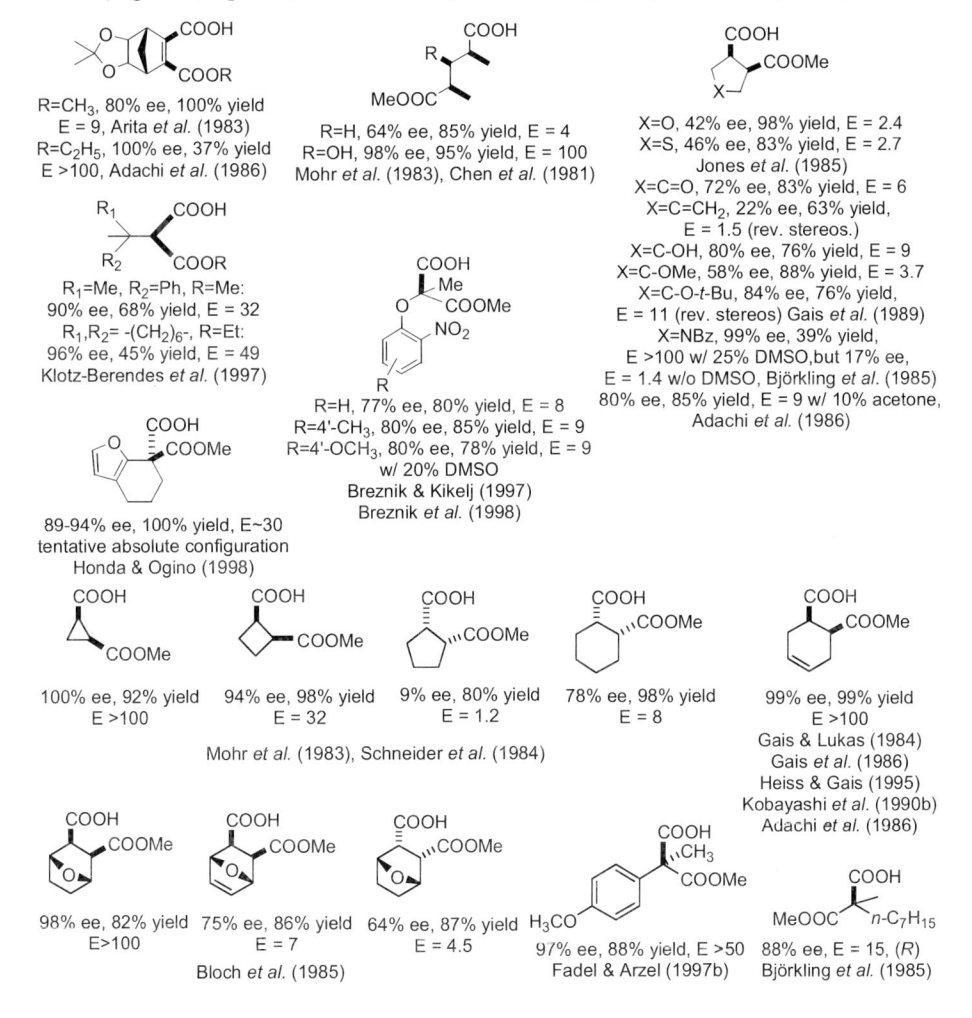

Fig. 162. PLE-catalyzed asymmetrization of *meso*- or prochiral dicarboxylic acids with a stereocenter at the α-position.

5.4.1.5 Asymmetrization of Carboxylic Acids with Other Stereocenters

Similar tendencies but less examples can be found for the asymmetrization of carboxylic acids with other stereocenters (Fig. 163).

Fig. 163. PLE-catalyzed asymmetrization of *meso-* or prostereogenic dicarboxylic acids with other stereocenters.

5.4.1.6 Asymmetrization of Primary and Secondary *meso*-Diols

Asymmetrization of primary or secondary *meso*-diols with PLE often proceeded with very high enantioselectivities (Fig. 164).

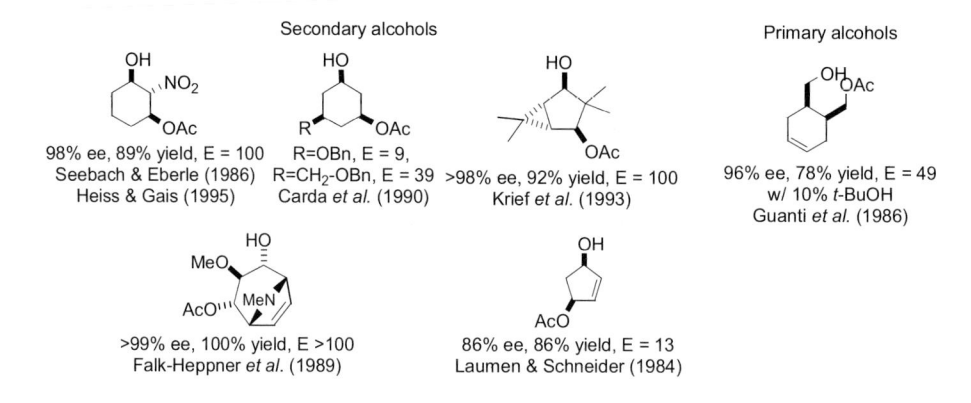

Fig. 164. PLE-catalyzed asymmetrization of *meso*-diols.

5.4.1.7 Kinetic Resolution of Alcohols or Lactones

PLE also catalyzed the kinetic resolution of alcohols. Selected examples are summarized in Fig. 165. Note that enantioselectivities are usually high and that in contrast to most lipases (see Sect. 5.2.1.3) even tertiary alcohols reacted. Selectivity towards lactones parallels that of most lipases (see Sect. 5.2.3).

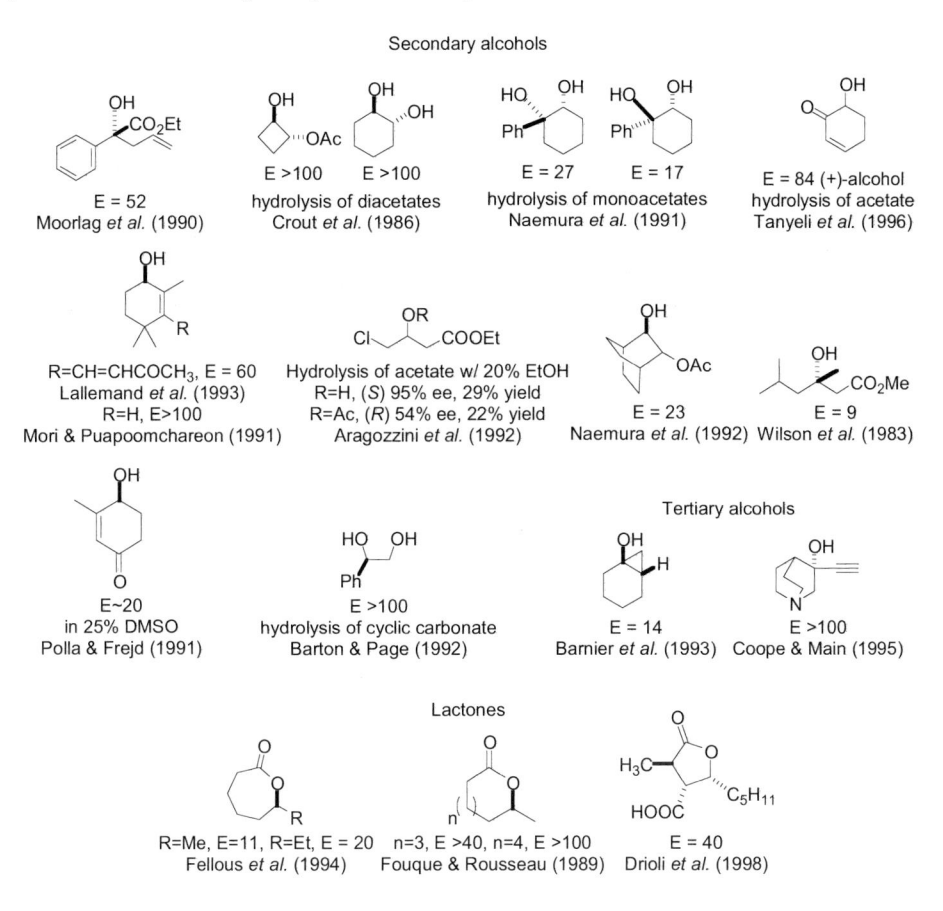

Fig. 165. PLE-catalyzed kinetic resolution of racemic secondary or tertiary alcohols and lactones.

5.4.1.8 Kinetic Resolution of Carboxylic Acids

Kinetic resolution of racemic carboxylic acids with PLE usually proceeded with high enantioselectivities (Fig. 166) and E-values are in the same range as found for *Candida rugosa* lipase, which represents the most enantioselective lipase for these kind of substrates (see Sect. 5.2.2).

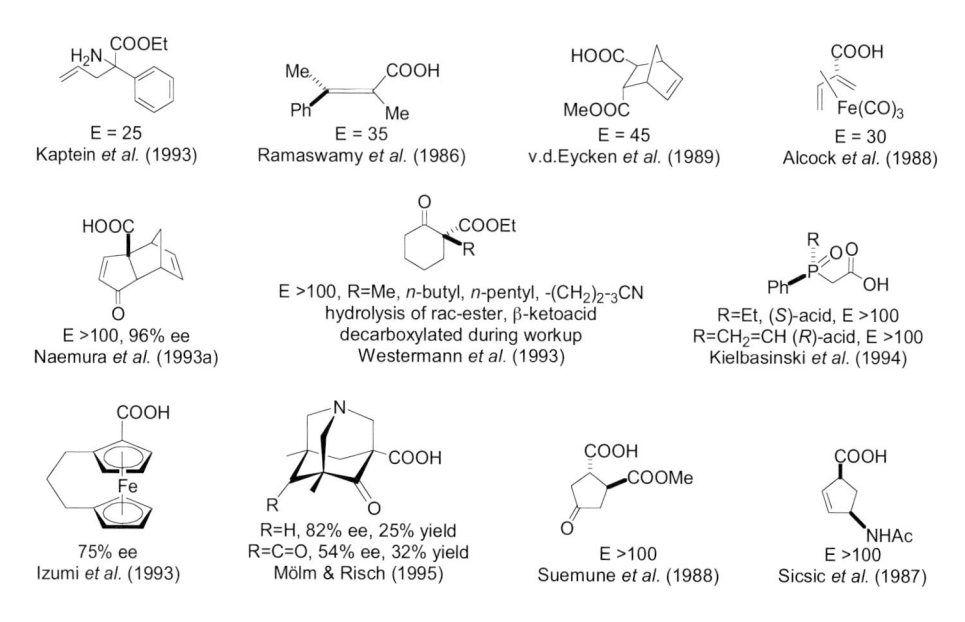

Fig. 166. PLE-catalyzed kinetic resolution of racemic carboxylic acids.

5.4.1.9 Reactions Involving Miscellaneous Substrates

Selected examples exploiting the regioselectivity of PLE are summarized in Fig. 167.

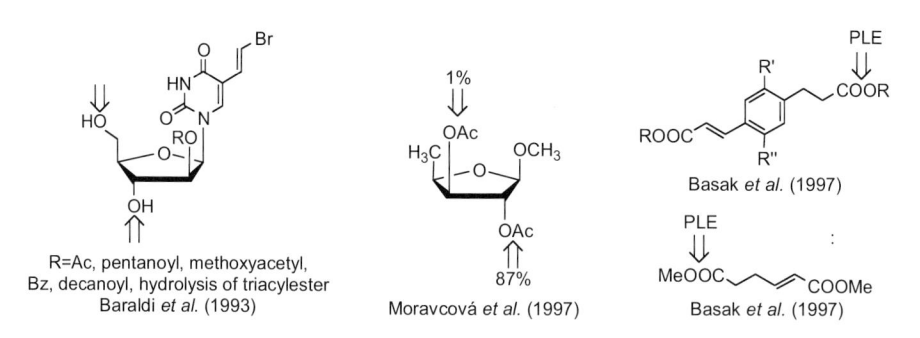

Fig. 167. Other substrates used in PLE-catalyzed biotransformations.

5.4.2 Acetylcholine Esterase

5.4.2.1 Biochemical Properties and Applications in Organic Syntheses

Acetylcholine esterase (AChE, EC 3.1.1.7) plays a key role in cholinergic neurotransmission. By rapid hydrolysis (25 000 mols per second) of the neurotransmitter acetylcholine, the enzyme effectively terminates the chemical impulse, thereby setting the basis for rapid, repetitive responses and enabling the re-uptake (and recycling) of choline (Maelicke, 1991). Historically, first hints that an esterase is involved in the hydrolysis of acetylcholine, date back to 1914 and the first crude extract of AChE was prepared in 1932 by Stedman's group. Since then, numerous papers on the enzymes' mechanism of action appeared. Cholinesterase from different species differ considerably in their substrate specificity. Currently, the major distinction is based on the substrate specificity towards acetylcholine or butyrylcholine (naming the latter enzyme butyrylcholine esterase, BChE).

AChE from the electric eel *Electrophorus electricus* consists of 12 subunits (75 kDa each), which are assemmbled via disulfide bridges as three tetramers. These subunits are further classified into globulary (G_1, G_2, disulfide-bridged dimers, identical subunits, G_4, tetramer) and asymmetric (A_4, A_8, A_{12}, with 4, 8 or 12 catalytic subunits, resp.) forms.

The structure of AChE from *Torpedo californica* (68 kDa, tetramer) is known (resolution 2.8 Å) (Sussman et al., 1991). It consists of a 12-stranded mixed β-sheet surrounded by 14 α-helices and bears a striking resemblance to several hydrolase structures including dienelactone hydrolase, serine carboxypeptidase-II, three neutral lipases (especially lipases from *Geotrichum candidum* and *Candida rugosa*), and a haloalkane dehalogenase. The active site is unusual because it contains Glu_{327}, not Asp, in the Ser_{200}–His_{440}–acid catalytic triad and because the relation of the triad to the rest of the protein approximates a mirror image of that seen in the serine proteases. Furthermore, the active site lies near the bottom of a 20 Å deep and narrow gorge that reaches halfway into the protein. Modeling of acetylcholine binding to the enzyme suggests that the quaternary ammonium ion is bound not to a negatively charged "anionic" site, but rather to some of the 14 aromatic residues that line the gorge.

AChE is inhibited through formation of (in most cases) irreversible acyl-enzyme complexes. The longest known inhibitor is the natural alkaloid physostigmine, which carbamoylates the active site serine residue. Organophosphorous compounds such as Paraoxon, E 605 and diisopropyl fluorophosphate, which are used as agricultural insecticides or as nerve gases in chemical warfare, react with the active site serine forming a very stable covalent phosphoryl-enzyme complex. This leads to breath paralysis, cardiac arrest and death. Some oximes such as pralidoxim iodine and obidoxime chloride can be used as antidots. A connection between Alzheimer's disease and AChE has been proposed.

BChE cleaves butyryl- and propionylcholine faster than acetylcholine and was found in sera, liver and mammalian pancreas as well as in snake venom. Tetraisopropyldiphosphoramide (*iso*-OMPA) is a specific inhibitor of BChE. More information about AChE and BChE is collected in databases (e.g., *http://www.ensam.inra.fr/cholinesterase/*).

Activity of AChE is usually determined with acetylthiocholine as substrate and the released thiocholine is determined spectrophotometrically at 410 nm after derivatization with Ellmanns-reagent (5,5-dithio-bis-(2-nitrobenzoic acid), DTNB).

Only a few biocatalytic applications of AChE can be found (Fig. 168), because the enzyme is only commercially available from electric eel, bovine brain, and human or bovine erythrocytes at high prices. Although heterologous expression of recombinant AChE was achieved (Heim et al., 1998), the expression levels are still far to low for the production of AChE on a practical scale.

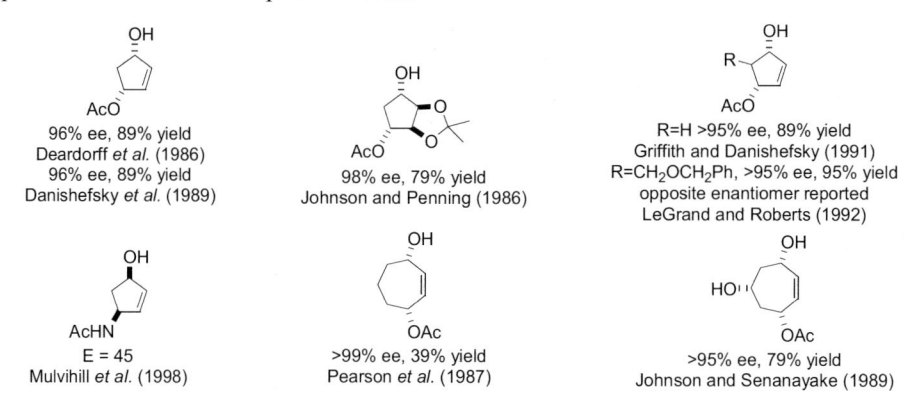

Fig. 168. Products obtained by electric eel AChE-catalyzed hydrolysis of *meso*-diacetates.

5.4.3 Other Mammalian Esterases

Selected examples for other mammalian esterases are summarized in Fig. 169.

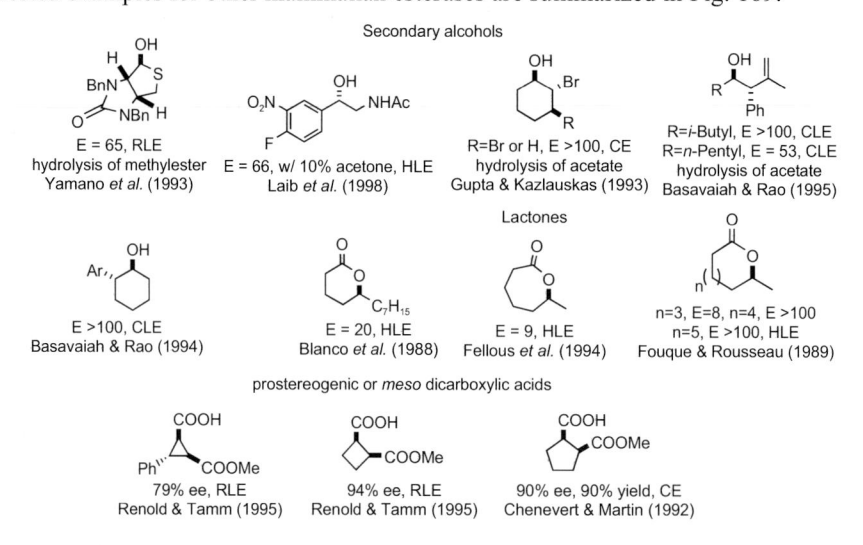

Fig. 169. Application of rabbit liver esterase (RLE), horse liver esterase (HLE), cholesterol esterase (CE) and chicken liver esterase (CLE) in the kinetic resolution of secondary alcohols, lactones or asymmetrization of prostereogenic or *meso*-dicarboxylic acids.

5.4.4 Microbial Esterases

Although a large number of microbial esterases have been described in literature, only a few of them have been used on larger scale. This is most likely due to a narrower substrate range and lower enantioselectivity compared to lipases. In addition, only a limited number of esterases is commercially available. Properties of them have been reviewed in detail (Bornscheuer, 2002) and have been also related structurally to lipases (Arpigny and Jäger, 1999). The x-ray structure of esterase from *Pseudomonas fluorescens* shows the same fold as lipases (Cheeseman et al. 2004). Microbial esterases have been already improved by directed evolution, examples are given in Sect. 4.2.1.3.

Currently, classification of esterases is mostly based on sequence identity and the presence of consensus sequences. A thorough comparison of 53 amino acid sequences of lipases and esterase revealed that beside the G-X-S-X-G consensus sequence also other motifs exist (Arpigny and Jäger, 1999). For instance, some lipases and an esterase from *Streptomyces scabies* contain a GDSL (Gly-Asp-Ser-Leu) consensus sequence. Moreover, structure elucidation of this esterase revealed that it contains a catalytic Ser-His dyad instead of the common Ser-Asp-His triad (Wei et al., 1995). The acidic side chain, which usually stabilizes the positive charge of the active site histidine residue, is replaced by the backbone carbonyl of Trp_{315} located three positions upstream of the His itself. The enzyme also has an α/β tertiary fold, which differs substantially from the α/β-hydrolase fold. Other esterases in the GDSL group includes those from *Pseudomonas aeruginosa* (accession code: AF005091), *Salmonella typhimurium* (AF047014) and *Photorhabdus luminescens* (X66379), the first two being outer-membrane bound esterases. Other enzymes show high homology to the mammalian hormone-sensitive lipase (HSL) family. Here, conserved sequence blocks were found, which initially have been related to be responsible for activity at low temperature. However, it was found that esterases from psychrophilic (e.g. *Moraxella* sp., X53869; *Psychrobacter immobilis*, X67712) as well as mesophilic (*Escherichia coli*, AE000153) and thermophilic (*Archeoglubus fulgidus*, AAE000985) origins belong to this family (Tab. 18).

Members of family V such as esterases from *Sulfolobus acidocaldarius* (AF071233) and *Acetobacter pasteurianus* (AB013096) share significant homology to non-lipolytic enzymes, e.g. epoxide hydrolase, dehalogenase and haloperoxidase. Rather small (23–26 kDa) enzymes are found in family VI, which includes an esterase from *Pseudomonas fluorescens*, of which the structure is known (Kim et al., 1997). The esterase is active as a dimer, has a typical Ser-Asp-His catalytic triad and hydrolyzes small substrates, but not long-chain triglycerides. Interestingly, ~40% homology to eukaryotic lysophospholipases is found for members of this family.

In contrast, esterases from family VII are rather large (~55 kDa) and share significant homology to eukaroytic acetylcholine esterases and intestine or liver carboxyl esterases (e.g., pig liver esterase). A *p*-nitrobenzyl esterase from *Bacillus subtilis* (Zock et al., 1994; Moore and Arnold, 1996) and an esterase from *Arthrobacter oxydans* (Q01470) active against phenylcarbamate herbicides (Pohlenz et al., 1992) belong to this group. In the last family VIII, high homology to class C β-lactamases is observed. These enzymes contain a –G-X-S-X-G– motif and a –S-X-X-L– motif, but it could recently been demonstrated by site-directed mutagenesis studies of an esterase (EstB) from *Burkholderia gladioli* that the –G-X-S-X-G– motif does not play a significant role in enzyme function (Petersen et al., 2001). The most prominent member is an esterase from *Arthrobacter globiformis* (AAA99492) (Nishizawa et al., 1995), which stereoselectively forms an important precursor of pyrethrin insecticides (see Sect. 5.4.4.2).

Many novel esterases were recently discovered by the metagenome approach, but they have not been characterized in detail yet. One exception is an esterase discovered in a deep sea hypersaline anoxic basin (DHAB, Ferrer et al., 2005; Bornscheuer, 2005). This enzyme (named O.16) exhibits several properties typical of most esterases, but shows two highly unique characteristics: First, it efficiently resolves solketal acetate with very high enantioselectivity (E > 100). The much more striking second feature relates to its unique sequence and structural properties as O.16 contains several consensus sequences (two times –G-X-S-X-G– and once –G-X-X-L– motif, one conferring thioesterase and two mediating carboxyl esterase activity). Moreover, the tertiary and quaternary structure of the isolated enzymes was shown to be pressure dependent. Experiments mimicking the pressure prevailing in the DHAB (around 40 MPa) showed that O.16 was even activated (1.9-fold more active at 20 MPa, 1.5-fold more active at 40 MPa) under these conditions, suggesting that O.16 evolved especially to withstand the high pressure in the habitat. O.16 appears to occur in several forms with differences not only in its multimeric forms, but also in the molecular weights of the monomers. For instance, under standard conditions (atmospheric pressure, no salt), it is a monomer (104 kDa). Addition of the reducing agent DTT generates two polypeptides, a 85 kDa fragment hydrolyzing only propionyl-CoA, and a 21 kDa fragment active only towards p-nitrophenyl butyrate. Under pressure (40 MPa) and at high saline content (i.e, 2-4 M NaCl), it is a homotrimeric protein (325 kDa, the largest esterase reported until now), with up to 700-fold increased activity compared to standard conditions. This multimer can also be reversed to the 104 kDa monomer by either salt removal or at atmospheric pressure. Thus, O.16 has a substantially higher level of structural and functional complexity than other known esterases.

More information about biochemical properties and preferred substrates of selected microbial esterases are summarized in Tab. 18.

Carboxylesterase NP is probably still the best-studied esterase and is treated in detail below.

Tab. 18. Comparison of (recombinant) Microbial Esterases.

Origin[a]	Biochemical properties	Remarks	References
Burkholderia gladioli (EstB)	392 aa, 42 kDa	S-x-x-L-motif, β-lactamase-like	Petersen et al. (2001)
Burkholderia gladioli (EstC)	298 aa, 32 kDa	G-x-S-x-G-motif, homology to plant hydroxynitrile lyases	Reiter et al. (2000)
Pseudomonas fluorescens DSM50106	36 kDa, $T_{opt.}$ 43°C	G-x-S-x-G-motif, homology to a haloperoxidase	Khalameyzer et al. (1999)
Pseudomonas fluorescens SIKW1	27 kDa, homodimer	low haloperoxidase activity, altered substrate specificity and improved enantioselectivity by directed evolution	Park et al. (2005); Baumann et al. (2000); Horsman et al. (2003); Choi et al. (1990); Henke and Bornscheuer (1999); Krebsfänger et al. (1998); Liu et al. (2001); Pelletier and Altenbuchner (1995)
Pseudomonas putida MR2068	29 kDa, dimer, $T_{opt.}$ 70°C	–	Ozaki and Sakimae (1997); Ozaki et al. (1995)
Bacillus acidocaldarius	34 kDa, $T_{opt.}$ 70°C	homology to hormone-sensitive lipase	Manco et al. (1998)
Bacillus subtilis NRRL B8079	489 aa, 54 kDa, $T_{opt.}$ 52°C (66.5°C for best mutant)	evolved by directed evolution for increased stability in DMF and thermostability	Arnold (1998); Giver et al. (1998); Moore and Arnold (1996); Zock et al. (1994)
Bacillus stearothermophilus	–	thermostable mutants	Amaki et al. (1992; 1993)
Bacillus subtilis (Thail-8, identical to Carboxylesterase NP)[e]	32 kDa, $T_{opt.}$ 35–55°C	structure known, more stable mutants by SDM[b]	Quax and Broekhuizen (1994)
Thermoanaerobacterium sp.	320 aa, 36 kDa	G-x-S-x-G-motif	Lorenz and Wiegel (1997)
Clostridium thermocellum	31 kDa	esterase activity within cellulosome	Blum et al. (2000)
Pyrococcus furiosus	$T_{opt.}$ 100°C, $t_{1/2}$ 50 min at 126°C	–	Ikeda and Clark (1998)

Tab. 18. Comparison of (recombinant) Microbial Esterases (continued).

Origin[a]	Biochemical properties	Remarks	References
Lactococcus lactis[d]	258 aa, 30 kDa	G-x-S-x-G-motif, function unclear	Fernandez et al. (2000)
Acinetobacter sp.	37 kDa	G-x-S-x-G-motif, involved in catechol branch of β-ketoadipate pathway	Jones et al. (1999)
Rhodococcus ruber DSM 43338[c]	–	two esterases with opposite enantiopreference	Pogorevc et al. (2000)
Rhodococcus sp. H1	34 kDa, tetramer	G-x-S-x-G-motif, conserved His$_{86}$	Rathbone et al. (1997)
Rhodococcus sp. MB1	574 aa, 65 kDa, monomer	G-x-S-Y-x-G-motif, homology to X-prolyl-dipeptidyl aminopeptidases	Bresler et al. (2000)
Streptomyces chrysomallus	42 kDa	G-x-S-x-G-motif, high homology to esterase from *Arthrobacter globiformis*	Berger et al. (1998)
Streptomyces diastatochromogenes	326 aa, 31 kDa	–	Khalameyzer and Bornscheuer (1999); Tesch et al. (1996)
Orpinomyces sp. PC-2	313 aa, 35 kDa, T$_{opt.}$ 30°C	high homology to other acetyl xylan esterases	Blum et al. (1999)
Aspergillus awamori IFO4033[c]	275 aa, 31 kDa	homology to lipases from *Geotrichum candidum* and *Candida cylindracea*	Koseki et al. (1997)
Saccharomyces cerevisiae IFO2347	28 kDa, homodimer, T$_{opt.}$ 25°C	–	Fukuda et al. (2000)

[a] overexpressed in *E. coli*, if not stated otherwise; [b] pNP, *p*-nitrophenyl; MU, 4-methylumbelliferyl; SDM, site-directed mutagenesis; [c] non-recombinant purified enzymes; [d] overexpressed in *L. lactis*; [e] overexpressed in *B. subtilis*.

5.4.4.1 Carboxylesterase NP

Carboxylesterase NP was originally isolated from *Bacillus subtilis* (strain ThaiI-8) and was cloned and expressed in *Bacillus subtilis* (Quax and Broekhuizen, 1994). The esterase shows very high activity and stereoselectivity towards 2-arylpropionic acids, which are e.g., used in the synthesis of (*S*)-naproxen – (+)-(*S*)-2-(6-methoxy-2-naphthyl) propionic acid – a non-steroidal anti-inflammatory drug (Fig. 170). (*S*)-naproxen is ca. 150 times more effective than (*R*)-naproxen, the latter also might promote unwanted gastrointestinal disorders. Therefore, this esterase is usually abbreviated in most references as carboxylesterase NP. It has a molecular weight of 32 kDa, a pH optimum between pH 8.5–10.5 and a temperature optimum between 35–55°C. Carboxylesterase NP is produced as intracellular protein; its structure is unknown.

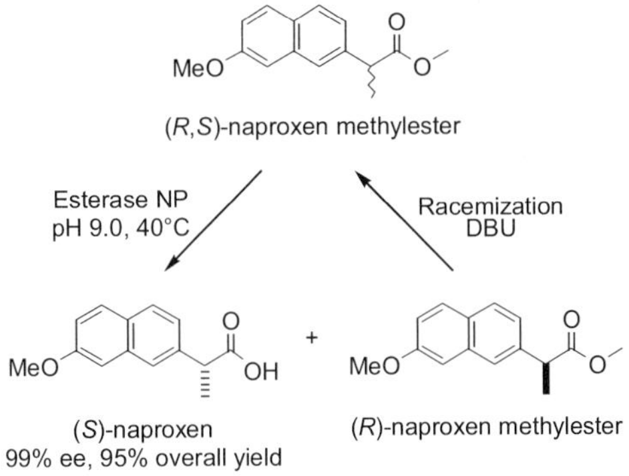

Fig. 170. Synthesis of (*S*)-naproxen by kinetic resolution of the (*R*,*S*)-methylester with carboxylesterase NP followed by chemical racemization of the (*R*)-naproxen methylester (Quax and Broekhuizen, 1994).

In a pilot-scale process, the (*R*,*S*)-naproxen methylester is hydrolyzed in the presence of Tween 80 to increase substrate solubility at pH 9.0. The (*S*)-acid is separated from the remaining (*R*)-methylester by filtration and the latter is racemized with DBU. This reaction yields (*S*)-naproxen with excellent optical purity (99% ee) at an overall yield of 95%. At concentrations > 20 g/L, an irreversible inactivation of carboxylesterase NP by the acid released was observed. Although this could be circumvented by addition of formaldehyde, the activity dropped by 50% . A better solution was based on site-directed mutagenesis by replacement of lysine residues with glutamine, thus eliminating the positively charged target prone to the formation of a Schiff base. 11 Lys residues were all successively replaced by Glu and mutant K34Q turned out to be the best choice (Quax and Broekhuizen, 1994). Carboxylesterase NP was also used in the resolution of (*R*,*S*)-ibuprofen methylester and showed higher selectivity compared to lipase from *Candida rugosa* (Mustranta, 1992).

A detailed investigation of the substrate spectra of carboxylesterase NP towards other chiral acids revealed high enantioselectivity in most cases (Fig. 171) and the (*S*)-enantiomer was usually preferred. All reactions shown have been performed at 20°C in a phosphate buffer:acetone mixture (9:1) at pH 7.2. Carboxylesterase NP was superior compared to α-chymotrypsin and PCL, which had (*R*)-enantiopreference (Azzolina et al., 1995a; b). Variation of the substituents as well as pH had a strong influence on enantioselectivity and reaction rate (Azzolina et al., 1995a).

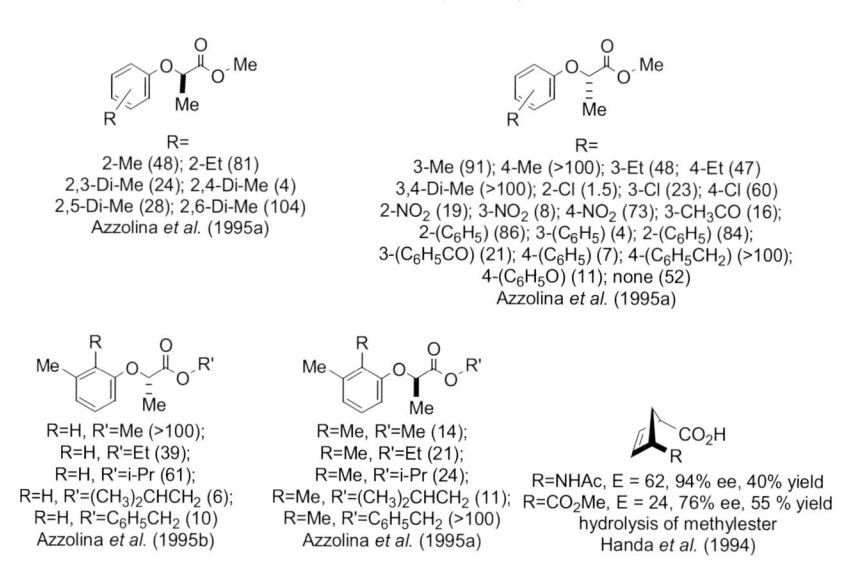

R=
2-Me (48); 2-Et (81)
2,3-Di-Me (24); 2,4-Di-Me (4)
2,5-Di-Me (28); 2,6-Di-Me (104)
Azzolina et al. (1995a)

R=
3-Me (91); 4-Me (>100); 3-Et (48); 4-Et (47)
3,4-Di-Me (>100); 2-Cl (1.5); 3-Cl (23); 4-Cl (60)
2-NO_2 (19); 3-NO_2 (8); 4-NO_2 (73); 3-CH_3CO (16);
2-(C_6H_5) (86); 3-(C_6H_5) (4); 2-(C_6H_5) (84);
3-(C_6H_5CO) (21); 4-(C_6H_5) (7); 4-($C_6H_5CH_2$) (>100);
4-(C_6H_5O) (11); none (52)
Azzolina et al. (1995a)

R=H, R'=Me (>100);
R=H, R'=Et (39);
R=H, R'=i-Pr (61);
R=H, R'=(CH_3)$_2$CHCH$_2$ (6);
R=H, R'=$C_6H_5CH_2$ (10)
Azzolina et al. (1995b)

R=Me, R'=Me (14);
R=Me, R'=Et (21);
R=Me, R'=i-Pr (24);
R=Me, R'=(CH_3)$_2$CHCH$_2$ (11);
R=Me, R'=$C_6H_5CH_2$ (>100)
Azzolina et al. (1995a)

R=NHAc, E = 62, 94% ee, 40% yield
R=CO_2Me, E = 24, 76% ee, 55 % yield
hydrolysis of methylester
Handa et al. (1994)

Fig. 171. Other examples for carboxylesterase NP-catalyzed resolutions of chiral carboxylic acids. Values in brackets refer to enantioselectivity E.

Carboxylesterase NP was also used in the kinetic resolution of α-chloropropionic acid methylester (Wolff et al., 1994), however the enantioselectivity was rather low (E = 4.7). Performing the reaction in combination with a dehalogenase as a sequential kinetic resolution increased E to 15 (Rakels et al., 1994b) (Fig. 172).

Fig. 172. Sequential kinetic resolution of α-chloropropionic acid methylester with carboxylesterase NP and a dehalogenase (Rakels et al., 1994b).

5.4.4.2 Other Microbial Esterases

Selected examples for the application of other microbial esterases are summarized in Fig. 173. (+)-*trans*-(1*R*,3*R*) chrysanthemic acid (Fig. 173, second row, right structure) is an important precursor of pyrethrin insecticides. An efficient kinetic resolution starting from the (±)-*cis-trans* ethylester was achieved using an esterase from *Arthrobacter globiformis* resulting in the sole formation of the desired enantiomer (> 99% ee, at 77% conversion). The enzyme was purified and the gene was cloned in *E. coli* (Nishizawa et al., 1993). In a 160 g scale process, hydrolysis is performed at pH 9.5 at 50°C. Acid produced is separated through a hollow-fiber membrane module and the esterase was very stable over four cycles of 48 h (Nishizawa et al., 1995).

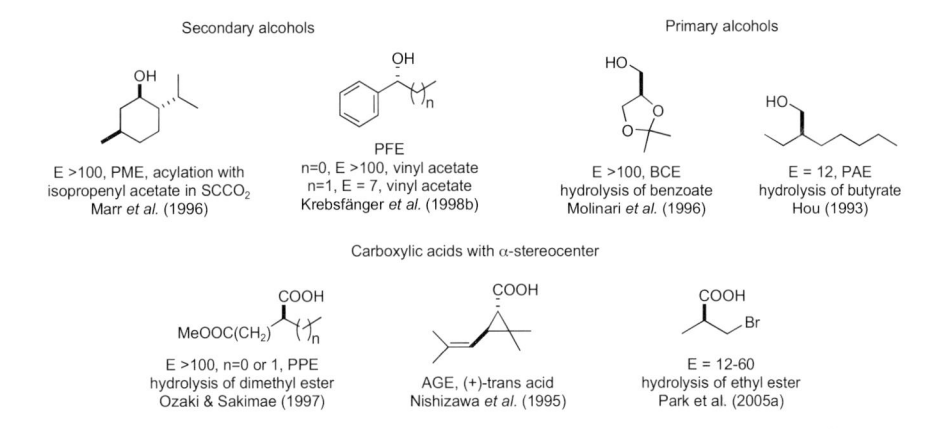

Fig. 173. Application of esterases from *Pseudomonas marginata* (PME), *Pseudomonas fluorescens* (PFE), *Pseudomonas aeruginosa* (PAE), *Pseudomonas putida* (PPE), *Bacillus coagulans* (BCE) and *Arthrobacter globiformis* (AGE) in the kinetic resolution of secondary and primary alcohols or carboxylic acids.

Shimizu et al. (1992b) found a novel lactone-hydrolyzing enzyme in the fungus *Fusarium oxysporum*. This lactonase catalyzes the enantioselective ring opening of several aldonate lactones (e.g., D-galactono-γ-lactone, L-mannono-γ-lactone, D-gulono-γ-lactone and D-glucono-δ-lactone) and also D-pantoyl lactone (Eq. 31). A commercial resolution uses fungal mycelia and yields D-pantoyl lactone, a precursor for the synthesis of pantothenic acid (Kataoka et al., 1995).

$$\text{(31)}$$

6 Proteases and Amidases

6.1 Occurrence and Availability of Proteases and Amidases

By far the most important commercial proteases are the subtilisins. Most laundry detergents contain subtilisins to help remove protein-based stains. Subtilisins are produced on a multi ton scale and are thus very inexpensive. Because of their commercial importance, industrial researchers devoted a lot of effort in protein engineering of subtilisins, especially to improve stability and activity at high pH and at high temperatures. The first recombinant subtilisin, added to detergents starting in 1988, was a subtilisin BPN' variant developed by Genencor with Procter & Gamble. Current subtilisins for laundry applications tolerate high pH, high temperatures, surfactants, oxidants and organic cosolvents.

Proteases and amidases are grouped into four families based on their catalytic mechanism: serine, cysteine, aspartic, or metallo proteases. Subtilisin, subtilisin relatives, chymotrypsin, and penicillin amidase are serine proteases. Papain is a cysteine protease, while acylase, thermolysin and aminopeptidase are metallo proteases. Researchers rarely use the aspartic proteases in organic synthesis.

Tab. 19. Some Commercially-Available Proteases and Amidases[a]

Enzyme	Biological Source	Synonyms
Subtilisins		
Subtilisin BL	*Bacillus lentus*	Savinase
Subtilisin BPN'	*Bacillus amyloliquefaciens*	subtilisin Novozymes
Subtilisin Carlsberg	*Bacillus licheniformis*	Alcalase, subtilisin A, Optimase
Proteases structurally related to subtilisins		
Thermitase	*Thermoactinomyces vulgaris*	
Proteinase K	*Tritirachium album* Limber	
Other proteases		
Chymotrypsin	bovine pancreas	
Thermolysin	*Bacillus thermoproteolyticus*	
Papain	papaya	
Amidases		
Amino acid acylase	porcine kidney, *Aspergillus melleus*	acylase I
Penicillin amidase	*Escherichia coli*	penicillin acylase
Leucine aminopeptidase		

[a] Most proteases are available in research quantities from common suppliers such as Sigma, Fluka, Amresco and others. Industrial quantities are available from Genencor, Novozymes, and Biocatalysts. Thermitase is available from Life Technologies as PreTaq Thermophilic protease.

Unlike lipases, proteases and amidases act only on soluble substrates. Many substrates of interest to synthetic organic chemists dissolve only slightly in water, so researchers often add cosolvents, such as DMF, DMSO or acetone, to help dissolve the substrate. Both subtilisins and chymotrypsin tolerate low levels of organic cosolvents (usually < 10 vol%). Some solvents, e.g., dioxane, sharply reduce catalytic activities even at low levels (Bonneau et al., 1993).

The two main applications of proteases and amidases in organic synthesis are: enantioselective hydrolysis of natural and unnatural α-amino acid esters and other carboxylic acid esters (Sects. 6.4.1 and 6.4.2) and synthesis of di- and oligopeptides by coupling of N-protected amino acids and peptides esters (Sect. 6.2.2). To a lesser extent organic chemists also use proteases for enantioselective hydrolyis of esters of secondary alcohols and for regioselective reactions of sugars.

6.2 General Features of Subtilisin, Chymotrypsin, and Other Proteases and Amidases

6.2.1 Substrate Binding Nomenclature in Proteases and Amidases

Proteases usually contain a channel on their surface which binds the polypeptide substrate. Schechter and Berger (1967) suggested numbering the different regions within this channel according to the amino acid residues that bind there and the distance of these amino acids from the amide link to be cleaved (Fig. 174). Thus, the acyl group of the amino acid undergoing cleavage binds at the S_1 site and this amino acid is the P_1 residue. The amino group to be released belongs to the P_1' residue, which binds at the S_1' site.

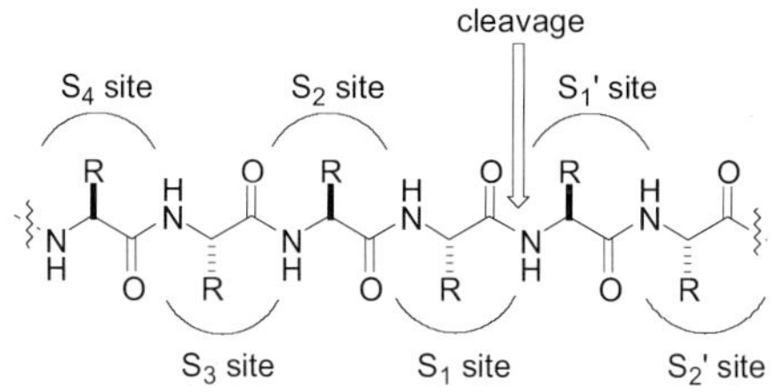

Fig. 174. Naming of the binding site of proteases according to Schechter and Berger (1967). The acyl part of the amide link to be cleaved lies in the S_1, S_2, S_3, etc. binding sites, while the amino part of the amide link to be cleaved lies in the S_1', S_2', etc binding sites. The substrate residues are called P_1, P_2, P_3, etc, and P_1', P_2', etc. according to their location relative to the amide link being cleaved.

6.2.2 Synthesis of Amide Bonds Using Proteases and Amidases

Proteases and amidases catalyze both formation and hydrolysis of amide links. Although their natural role is hydrolysis, researchers also use proteases and amidases to form amide links. They use two different strategies – thermodynamic control or kinetic control (Fig. 175).

In thermodynamically-controlled syntheses, researchers change reaction conditions to shift the equilibrium toward synthesis instead of hydrolysis. Hydrolysis of peptides is favored by ~2.2 kcal/mol and is driven mainly by the favorable solvation of the carboxylate and ammonium ions. One common way to shift the equilibrium toward synthesis is to replace water with an organic solvent. The organic solvent suppresses the ionization of the starting materials and also reduces the concentration of water. Other common ways to shift the equilibrium are to increase the concentrations of the starting materials or to choose protective groups that promote precipitation of the product.

In kinetically controlled syntheses, researchers start with an activated carboxyl component, usually an ester. The ester reacts with the enzyme to form an acyl enzyme intermediate, which then reacts either with an amine to form the desired amide, or with water to form a carboxylic acid. Because the starting material is an activated carboxyl component, reactions are faster in the kinetically-controlled approach than in the thermodynamically-controlled approach. Because the kinetically-controlled approach requires an acyl enzyme intermediate, only serine hydrolases (e.g., subtilisin, lipases) are suitable. Metallo proteases such as thermolysin work only in thermodynamically-controlled syntheses. Kinetically-controlled syntheses are more common.

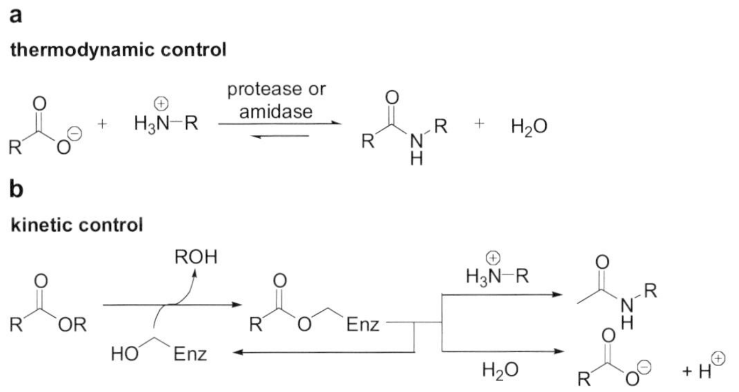

Fig. 175. Synthesis of amide bonds using proteases and amidases. **a** Thermodynamic control shifts the equilibrium toward synthesis by changing the reaction conditions. For example, researchers add organic solvents to reduce the concentration of water and suppress ionization of the starting materials. **b** Kinetic control starts with an activated carboxyl component (e.g., an ester) and forms an acyl enzyme intermediate. The acyl enzyme intermediate then reacts with an amine to form the amide. In a competing side reaction, water may react with the acyl enzyme intermediate. The kinetic control approach requires the formation of an acyl enzyme intermediate; thus, serine hydrolases and cysteine hydrolases are suitable, but not metallo proteases.

Protease-catalyzed peptide synthesis was first reported in 1901 (Savjalov, 1901). Until the end of the 1930's researchers believed that biosynthesis of proteins involved the reverse action of proteases. In the late 1970's synthetic chemists began to use proteases to simplify peptide synthesis and it continues to be an active area of research. Several ex-

6.2 General Features of Proteases and Amidases

cellent reviews on the synthesis of amide links using proteases and amidases are available (Kullmann, 1987; Schellenberger and Jakubke, 1991; Wong and Whitesides, 1994; Drauz and Waldmann, 1995).

The advantages of a hydrolase-catalyzed peptide synthesis over a chemical synthesis are mild conditions, no racemization, minimal need for protective groups, and high regio- and enantioselectivity.

The largest scale application (hundreds to thousands of tons) of protease-catalyzed peptide synthesis is the thermolysin catalyzed synthesis of aspartame, a low calorie sweetener (Fig. 176) (Isowa et al., 1979; Oyama, 1992). Precipitation of the product drives this thermodynamically-controlled synthesis. The high regioselectivity of thermolysin ensures that only the α-carboxyl group in aspartate reacts. Thus, there is no need to protect the β-carboxylate. The high enantioselectivity allows Tosoh (Japan) to use racemic amino acids; only the L-enantiomer reacts.

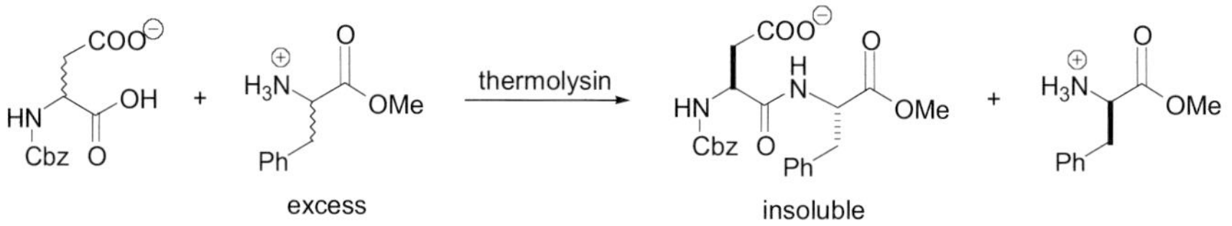

Fig. 176. Commercial process for the production of aspartame (α-L-aspartyl-L-phenylalanine methyl ester) by Tosoh Corporation (Japan). Thermolysin catalyzes the coupling of an N-Cbz protected aspartic acid with phenylalanine methyl ester. The product forms an insoluble salt with excess phenylalanine methyl ester. This precipitation drives this thermodynamically-controlled peptide synthesis. The high regioselectivity of thermolysin for the α-carboxylate allows Tosoh to leave the β-carboxylate in aspartate unprotected. The enantioselectivity of thermolysin allows Tosoh to use racemic starting materials.

An example of a kinetically-controlled peptide synthesis is the α-CT-catalyzed production of kyotorphin (Tyr-Arg) (Fig. 177) (Fischer et al., 1994). To minimize the hydrolysis of the acyl enzyme intermediate, Fischer et al. (1994) used high concentrations of the nucleophile. The charged maleyl protective group increased the solubility of the carboxyl component.

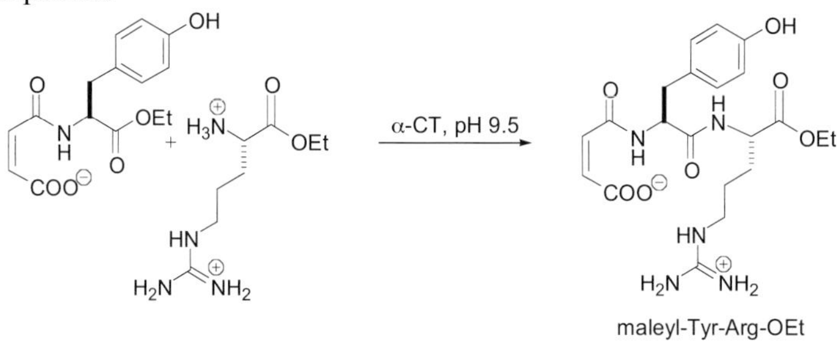

Fig. 177. Large scale synthesis of a dipeptide, kyotorphin. α-CT catalyzes the coupling of the two amino acid derivatives in a concentrated aqueous solution (1.5 M in each). The N-maleyl protective group on the tyrosine moiety increases its solubility. The high concentration of nucleophile minimized the competing hydrolysis of the acyl enzyme intermediate and increased the yield. The remaining N-maleyl and –OEt protective groups were removed by a subsequent treatment with acid.

Subtilisin accepts a broader range of substrates than other proteases, so researchers usually use subtilisin for amide couplings involving unnatural substrates (Moree et al., 1997). When coupling a D-amino acid, it is best to use it as the nucleophile, not the carboxyl donor because subtilisin is more tolerant of changes in the nucleophile than in the carboxyl group.

Researchers also coupled larger peptides using proteases. For example, subtilisin cleaved a protein such as lysozyme and RNase into several peptides (Vogel et al., 1996; Witte et al., 1997). Next, addition of an organic solvent shifted the equilibrium toward peptide synthesis and the same subtilisin now reassembled the protein. In the future, this condensation may create proteins containing unnatural amino acids or sugars. Manufacturers convert porcine insulin to human insulin by a similar process (Morihara and Oka, 1983).

Acylases and amidases also catalyze the formation of peptide bonds. For example, Justiz et al. (1997) coupled a side chain to 7-aminocephalosporanic acid (7-ACA) in the key step of an antibiotic synthesis (Fig. 178). They used the kinetically-controlled approach and obtained a 98% yield in aqueous solution. Amino acid acylases are metallo proteases, so only the thermodynamically-controlled approach can be used.

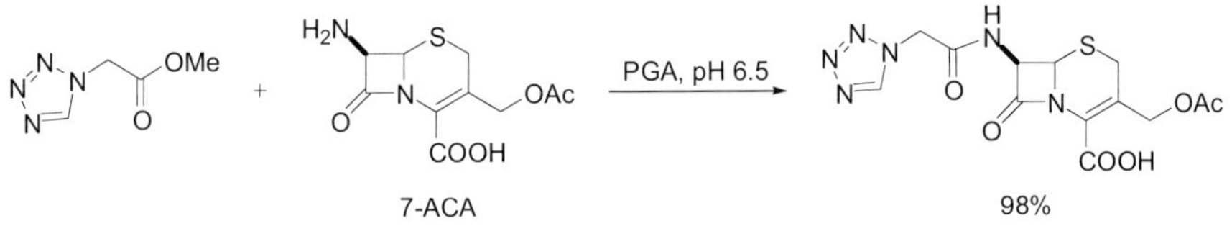

Fig. 178. The penicillin acylase (PGA)-catalyzed coupling of a side chain to a cephalosporin nucleus.

6.2.3 Subtilisin and Related Proteases

Subtilisins are a family of bacterial serine proteases secreted by various *Bacillus* species (Tab. 19). Siezen et al. (1991) reviewed the structures and sequence alignments of amino acid sequences of subtilisins (see also: Siezen and Leunissen, 1997). The mature subtilisin contains approximately 270 amino acids. Subtilisins are endopeptidases with a broad specificity and contain a structural calcium ion. Like chymotrypsins, they favor a large uncharged residue at the P_1 position, for example, phenylalanine. An approximate order of preference is Tyr, Phe > Leu, Met, Lys > His, Ala, Gln, Ser >> Glu, Gly (Estell et al., 1986; Wells et al., 1987). Subtilisins show little preference for amino acids at the P_2 and P_3 positions, but favor hydrophobic residues at the P_4 position (Perona and Craik, 1995). The most important commercial subtilisins are subtilisin BL (subtilisin from *Bacillus lentus*), subtilisin Carlsberg (subtilisin from *B. licheniformis*) and subtilisin BPN' (subtilisin from *B. amyloliquefaciens*). In spite of differences in amino acid sequence (84 of 275 amino acids differ between subtilisins Carlsberg and BPN'), the structures and substrate specificities are very similar.

Subtilisins are alkaline serine proteases, that is, proteases that show maximum activity at alkaline pH. There is no optimum pH. The reaction rate increases in the pH range 6–9 and then remains constant. Above pH 11, both subtilisin Carlsberg and subtilisin BPN' denature, but subtilisin BL remains stable to at least pH 12. This difference likely stems from the fact that *Bacillus lentus* is an alkalophilic bacteria that can grow at higher pH than other *Bacillus* species.

While the amino acid sequences of subtilisin BPN' and subtilisin Carlsberg are 75% identical (for the 194 core residues (Siezen et al., 1991)), the amino acid sequences of thermitase (secreted by a *Thermoactinomyces* bacteria) and proteinase K (secreted by a *Tritirachium* fungus) show respectively only 52% and 44% sequence identity to subtilisin BPN'.

6.2.4 Chymotrypsin

The most widely studied form is α-chymotrypsin (241 amino acids, 25 kDa) from bovine pancreas (α-CT) (review by Jones and Beck, 1976). The pancreas secretes the catalytically-inactive chymotrypsinogen A. Proteases, including trypsin, remove two peptides yielding active α-CT, which consists of three polypeptide chains linked by five disulfide bonds. α-CT is an endopeptidase favoring peptide links with Phe, Tyr, or Trp as the acyl group.

Although crystalline α-CT is stable indefinitely, α-CT digests itself in solution near neutral pH (pH 6–9), so solutions of α-CT should not be stored. The pH optimum for amide hydrolysis is pH 7.8 and most researchers carry out reactions near this pH. For more reactive substrates like esters, reactions also proceed at pH 5, where α-CT is more stable.

Like other proteases, chymotrypsin only acts on dissolved substrates. To dissolve nonpolar organic substrates in aqueous solutions, researchers often add organic cosolvents. Organic cosolvents decrease the rate of α-CT-catalyzed reactions usually by increasing the K_m for the substrate. Some aromatic compounds also inhibit α-CT, presumably by binding to the hydrophobic binding pocket.

Above a concentration of ~0.25 mg/mL, the catalytic efficiency of α-CT decreases because it associates into inactive dimers, similar to that found in the crystal structure. High ionic strength, such as 0.1 M NaCl typically used in synthetic reactions, decreases this association thereby increasing the rate of reaction. For reactions in organic solvent using suspended lyophilized α-CT, Khmelnitsky et al. (1994) dramatically increased the rate of reactions by lyophilizing a salty solution of α-CT. Presumably the salts minimized the association of chymotrypsin in the lyophilized form.

6.2.5 Thermolysin

Thermolysin is a thermostable metalloprotease secreted by *Bacillus thermoproteolyticus*. Thermolysin contains a zinc ion at the active site and and four calcium ions that stabilize the structure. Like subtilisin and chymotrypsin, thermolysin also favors hydrophobic amino acids at the P_1 residue.

6.2.6 Penicillin G Acylase

Penicillin G acylase (PGA, penicillin amidase, for a review see: Baldaro et al., 1992) catalyzes hydrolysis of the phenylacetyl group in penicillin G (benzylpenicillin) to give 6-aminopenicillanic acid (6-APA) (Eq. 32).

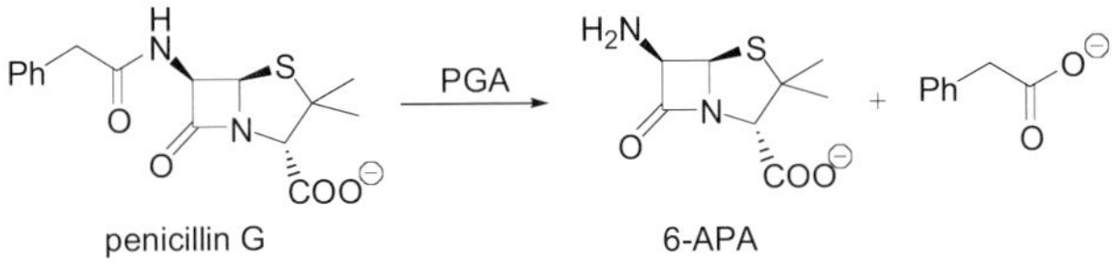

$$\text{penicillin G} \xrightarrow{\text{PGA}} \text{6-APA} + \text{PhCH}_2\text{COO}^- \tag{32}$$

Penicillin G acylase also cleaves the side chain in penicillin V, where the phenylacetyl group is replaced by a phenoxyacetyl group. The commercially available enzymes are derived from *E. coli* strains. Both penicillin G and V are readily available from fermentation, so penicillin manufacturers carry out a PGA-catalyzed hydrolysis to make 6-APA on a scale of approximately 5 000 metric tons per year (Matsumoto, 1992). They use 6-APA to prepare semisynthetic penicillin such as ampicillin, where a D-phenylglycine is linked to the free amino group of 6-APA, or amoxicillin, where a D-4-hydroxyphenyl glycine is linked.

PGA favors hydrolysis of phenylacetyl esters and amides, but accepts acyl groups that are structurally similar to phenylacetyl, such as 4-pyridylacetyl and phenoxyacetyl. On the other hand, PGA accepts a wide range of structures as the leaving group (alcohol part of an ester or amine part of an amide) (Sect. 6.4.2.2).

6.2.7 Amino Acid Acylases

Acylases catalyze the hydrolysis of *N*-acetyl-L-amino acids to the L-amino acid. Derivatives of D-amino acids do not react. Two amino acid acylases are commercially available – from porcine kidney and from *Aspergillus melleus*. Both are metallo proteases containing a zinc ion in the active site. On a lab scale, Chenault et al. (1989) recommended using the porcine kidney enzyme because it accepts a slightly wider range of substrates. On an industrial scale, researchers usually use the *Aspergillus* acylase because it is more stable. Researchers have found several other interesting acylases in microorganisms. Acylases from a *Pseudomonas* strain or a *Comamonas* strain accept a wider range of amino acids than the two commercial acylases, including derivatives of cyclic amino acids such as proline and *N*-alkyl amino acids (Groeger et al., 1990; 1992; Kikuchi et al., 1983).

6.2.8 Protease Assays

The determination of protease activity is usually performed using *p*-nitrophenyl derivatives of amino acids. Representative chromogenic substrates are shown in Fig. 179 for trypsin, elastase, chymotrypsin and subtilisin. A detailed procedure for assaying chymotrypsin or subtilisin activity is:

100 µl N-Succinyl-L-Ala-Ala-Pro-Phe-*p*-nitrophenyl anilide (0.5 mM) dissolved in HEPES buffer (5 mM, pH 8.0) containing 9% DMF and 900 µl protease solution are

mixed in a cuvette and the amount of *p*-nitrophenyl anilide released is determined by recording the absorbance at 410 nm at 25°C using a spectrophotometer. One unit is defined as the amount of protease releasing 1 µmol *p*-nitrophenyl anilide per min under assay conditions (Graham et al., 1993).

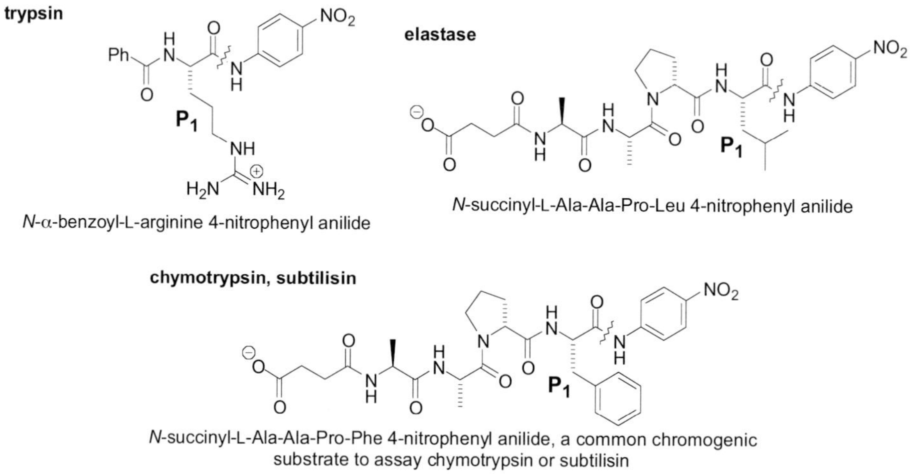

trypsin

N-α-benzoyl-L-arginine 4-nitrophenyl anilide

elastase

N-succinyl-L-Ala-Ala-Pro-Leu 4-nitrophenyl anilide

chymotrypsin, subtilisin

N-succinyl-L-Ala-Ala-Pro-Phe 4-nitrophenyl anilide, a common chromogenic substrate to assay chymotrypsin or subtilisin

Fig. 179. A chromogenic substrates often used to assay subtilisin or chymotrypsin. A hydrophobic residue, Phe, lies at the P_1 position. The cleaved position is indicated by a wavy bond.

6.3 Structures of Proteases and Amidases

Crystallographers have solved the structures of most of the proteases important for organic synthesis (Tab. 20). In most cases, several structures are available, including structures of mutants and of proteases with bound inhibitors. One notable absence from this list is amino acid acylase.

Tab. 20. X-Ray Crystal Structures of Selected Proteases and Amidase.

Proteinase	Number of structures	pdb code (example)	References
Subtilisins and relatives			
Subtilisin BL	3	1jea	Bott et al. (1996)
Subtilisin BPN'	23	1sbt	Drenth et al. (1972); Wright et al. (1969)
Subtilisin Carlsberg	19	1sec	McPhalen et al. (1985)
Thermitase	5	2tec	Gros et al. (1989)
Proteinase K	6	2prk	Betzel et al. (1992)
Other proteinases			
α-Chymotrypsin	9	2cha	Matthews et al. (1967)
Penicillin amidase	10	1pnk, 3pva	Duggleby et al. (1995); Suresh et al. (1999)
Thermolysin	20	4tmn	review: Matthews (1988)

6.3.1 Serine Proteases – Subtilisin and Chymotrypsin

Although the catalytic residues in both subtilisins and chymotrypsins have a similar three-dimensional arrangement, their protein folds are not related. Chymotrypsin has a β/β fold – two antiparallel β-barrel domains. On the other hand, subtilisin has an α/β fold – a core of parallel β-sheets surrounded by four α-helices. The similar three dimensional arrangement of catalytic residues in these two proteases is an example of convergent evolution where completely different loop regions, attached to different framework structures, form similar active sites. Note that this fold is not the same as the α/β-fold of lipases (see Sect. 5.1.3) (Branden and Tooze, 1991).

Subtilisin belongs to the family of subtilases, a superfamily of subtilisin-like serine proteases. The main members of this family are the subtilisins, thermitase and proteinase K. Many subtilisin mutants and variations are available. For examples, see the recent sequence alignment of subtilases (Siezen and Leunissen, 1997).

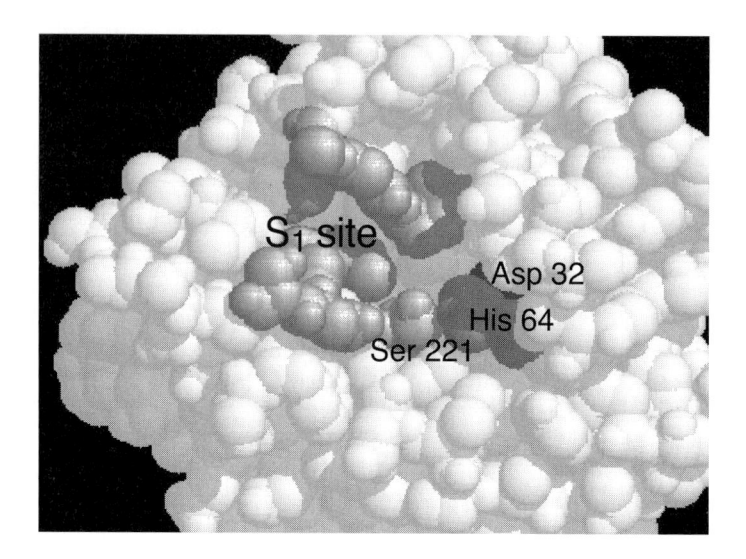

Fig. 180. The crystal structure of subtilisin Carlsberg shown in a space-filling representation. The labels and coloring show the amino acid residues of the catalytic triad and the residues forming the S_1 binding site. This binding site is a shallow groove lined with nonpolar amino acid residues. In contrast, the S_1 binding site of chymotrypsin (Fig. 195, Sect. 6.4.2.1) is a well defined hydrophobic pocket. For this reason subtilisin accepts a wider range of substrates than chymotrypsin. Coordinates are from Brookhaven protein data bank file 1sbc (Neidhart and Petsko, 1988) and the figure was created using RasMac v 2.6 (Sayle and Milner-White, 1995).

The substrate preference in subtilisin is dominated by the S_1 and S_4 sites, which both favor aromatic or large nonpolar residues such as phenylalanine (see Sect. 6.2.1 for explanation of the active site nomenclature). A typical substrate for colorimetric assay of subtilisin activity (see Sect. 6.2.8), has a Phe residue at P_1 and an Ala residue at P_4 (Graham et al., 1993). For organic synthesis applications, the substrates are usually to small to reach the S_4 site, so the S_1 site is the most important one. Since most synthetic organic intermediates are hydrophobic, it is not surprising that the most useful proteases are those that favor hydrophobic substrates. Fig. 180 shows a structure of subtilisin Carlsberg with the S_1 site highlighted. This site is a shallow groove lined with nonpolar amino acids.

Subtilisin's substrate specificity stems mainly from the acyl binding site, not the amide binding site. For this reason, subtilisin shows higher stereoselectivity toward acids and a lower, broader selectivity toward the amine (or alcohol in the case of an ester). Like subtilisin, chymotrypsin also favors hydrophobic residues at the P_1 position. However, the S_1 site in chymotrypsin is a well defined hydrophobic pocket (Blow, 1976).

Penicillin amidase (penicillin G acylase, PGA) is also a serine protease, but its active site structure is unusual – instead of a triad, it contains only a *N*-terminal serine residue (Duggleby et al., 1995). Penicillin V amidase (PVA) from *Bacillus sphaericus* shows, in spite of no detectable similarities in sequence, a structural core similar to that of PGA (Suresh et al., 1999). Both are Ntn-hydrolases (*N*-terminal nucleophile hydrolase), where the *N*-terminal amino acid is the catalytic nucleophile. In PGA this residue is serine, while in PVA it is cysteine. The N-terminal amino group may act as the base during catalysis.

6.4 Survey of Enantioselective Protease- and Amidase-Catalyzed Reactions

6.4.1 Alcohols and Amines

Proteases and amidases usually show their highest selectivities toward the acyl part of an ester or amide, and lower selectivity toward the alcohol or amine part.

6.4.1.1 Secondary Alcohols and Primary Amines

Overview and Models

Like lipases, subtilisin is enantioselective toward secondary alcohols (HOCHRR') and the isosteric primary amines (H$_2$NCHRR'). However, unlike lipases, subtilisin has a shallow binding site for these groups and leaves part of them in the solvent. For this reason, the solvent strongly affects the enantiopreference of subtilisin. A general model (Fig. 181) shows that one substituent ($R_{S1'}$) binds in the hydrophobic S_1'-pocket, which can accommodate a para-substituted phenyl group, while the other substituent (R_{SOLV}) remains in the solvent (Savile and Kazlauskas, 2005).

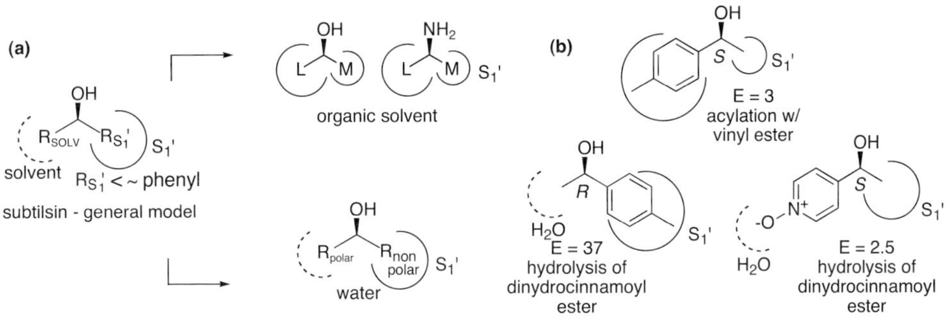

Fig. 181. Enantiopreference of subtilisins toward secondary alcohols and primary amines. (a) A general model shows that one substituent ($R_{S1'}$) binds in the hydrophobic S_1'-pocket, which can accommodate a para-substituted phenyl group, while the other substituent (R_{SOLV}) remains in the solvent. In organic solvents, the size of the subsituents determines the enantiopreference, while in water the polarity of substituents determines the enantiopreference. See text for details. (b) Examples showing the favored enantiomer in organic solvent (top) and in water (lower two examples). Subtilisin favors the (S)-enantiomer of 1-(p-tolyl)ethanol in organic solvents (p-tolyl is larger than methyl), but the (R)-enantiomer in water (methyl is more polar than p-tolyl). Replacing the p-tolyl substituent with the more polar 4-pyridine N-oxide reverses the enantiopreference.

In organic solvents, both the S_1'-pocket and the solvent have similar polarity, so steric effects determine the enantioselectivity. The general model simplifies to a rule based on size where L is the large substituent and M is the medium substituent. The larger substituent avoids the S_1' pocket to minimize steric hindrance (Fitzpatrick and Klibanov, 1991; Kazlauskas and Weissfloch, 1997).

6.4 Survey of Enantioselective Protease- and Amidase-Catalyzed Reactions

In water, the polarity of the substitutents determined the enantiopreference (Savile and Kazlauskas, 2005). Placing a nonpolar substituent in water is unfavorable. Reactions in water involving methyl and nonpolar aryl substituents will favor the nonpolar aryl substituent in the S_1' pocket, opposite to that predicted based on size alone. Thus, subtilisin favors the (S)-enantiomer of 1-(p-tolyl)ethanol in organic solvents (p-tolyl is larger than methyl), but the (R)-enantiomer in water (methyl is more polar than p-tolyl). On the other hand, with a polar aryl group such as (4-pyridine N-oxide), the (S)-enantiomer is favored both in water, where solvation of the pyridine N-oxide is favorable, and in organic solvent, where placing the pyridine N-oxide in the solvent avoids steric interactions in the S_1' pocket.

The enantiopreference of subtilisin in organic solvent is opposite to that for lipases, likely due to differences in protein structure (Kazlauskas and Weissfloch, 1997). Like lipases, subtilisin contains a catalytic triad, an oxyanion hole and follows an acyl enzyme mechanism. However, the three-dimensional arrangement of the catalytic triad in subtilisin is the mirror image of that in lipases (Ollis et al., 1992). Folding the protein to assemble the catalytic machinery creates a pocket with restricted size, the 'M' or stereoselectivity pocket for lipases and the S_1' pocket for subtilisin. The mirror image relationship places the catalytic histidine on opposite sides of this pocket in lipases and subtilisin. This difference likely accounts for the opposite enantiopreference. In addition, lipases and subtilisin often show opposite regioselectivity (Sect. 5.3.1.1). Chymotrypsin, which has a catalytic triad similar to that in subtilisin, but a different protein fold, shows little enantioselectivity toward alcohols (see Sect. 6.4.1.1).

Subtilisin

Sixteen of the seventeen examples of secondary alcohols in Fig. 182 follow the general model above. In organic solvents, relative substituent size determines enantiopreference, while in water, relative substituent polarity determines enantiopreference.

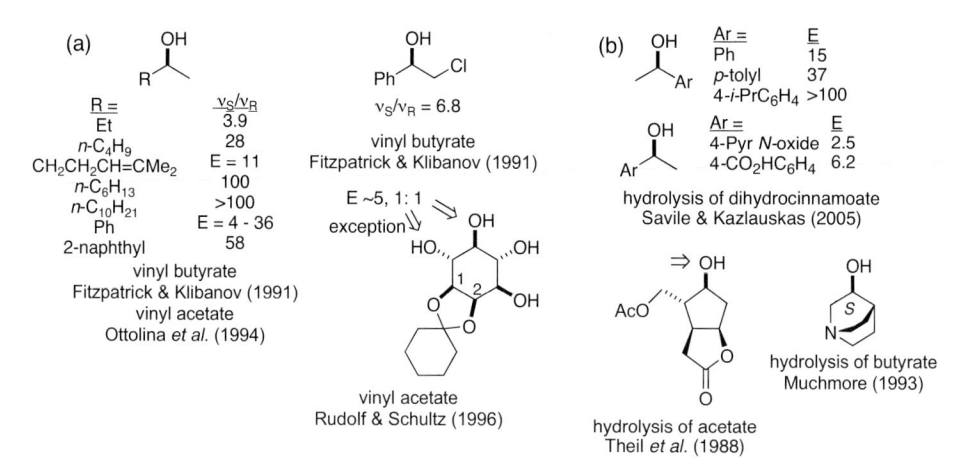

Fig. 182. Examples of secondary alcohols resolved by subtilisin. (a) For reactions in organic solvents, size of the substituents determines the enantiopreference – larger group is on the left. (b) For reactions in water, polarity of the substituents determines the enantiopreference – more polar group is on the left. Where no enantioselectivities are given, it was difficult to calculate from the reported data.

Subtilisin also catalyzes the enantioselective acylation of various amines in organic solvents (Fig. 183). All examples follow the empirical rule above and the enantioselectivity is usually good to excellent. Cleavage of the resulting amides requires vigorous reaction conditions. For this reason, Orsat et al. (1996) introduced diallyl carbonate as an acylating agent (see Sect. 5.1.6.1) The resulting allyl carbamate cleaves readily in the presence of palladium.

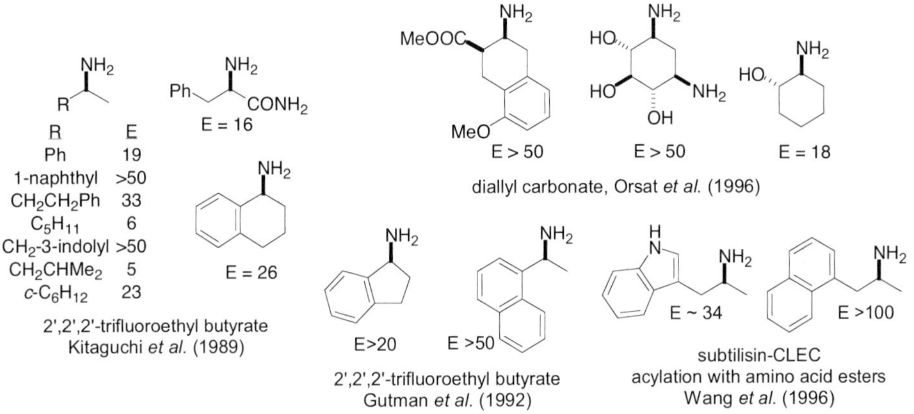

Fig. 183. Examples of primary amines of the type H_2NCHRR' resolved by subtilisin-catalyzed acylation in organic solvents.

Subtilisin-catalyzed hydrolysis of *N*-acyl sulfinamide (an analog of amines) in water is also highly enantioselective and is a good route to sulfinamide chiral auxiliaries (Fig. 184). The high enantioselectivity likely stems from the large difference in polarity of the substituents at sulfur (oxygen vs. aryl).

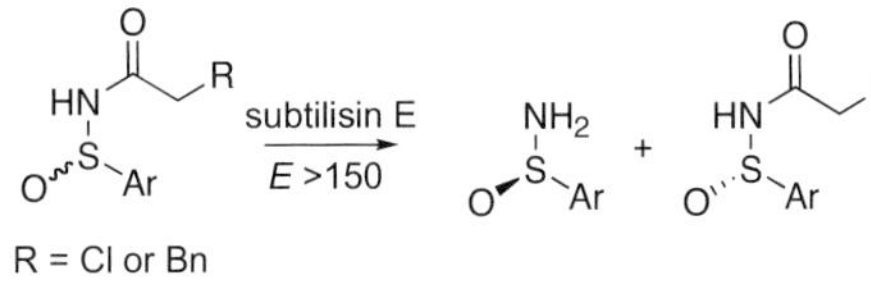

Fig. 184. Subtilisin-catalyzed kinetic resolution of sulfinamides.

Other Proteases and Amidases

Chymotrypsin shows low enantioselectivity toward secondary alcohols (E < 6 for esters of 2-butanol, 2-octanol or α-phenylethanol (Lin et al., 1974). Chymotrypsin also showed low enantioselectivity (E < 3) in the acetylation of the prochiral 1,3-propanediol shown below in most organic solvents, but in diisopropyl ether the enantioselectivity was E = 16 (Fig. 185) (Ke and Klibanov, 1998).

Penicillin amidase (penicillin G acylase, PGA) resolves alcohols and amines, usually by hydrolysis of their phenylacetyl derivatives (Fig. 186). The enantioselectivity of PGA toward alcohols is often lower than toward amines. To increase the enantioselectivity Pohl and Waldmann (1995) replaced the phenylacetyl ester with a 4-pyridylacetyl ester. For example, PGA catalyzed hydrolysis of the phenylacetyl ester of 1-phenyl propanol

198   6.4 Survey of Enantioselective Protease- and Amidase-Catalyzed Reactions

with an enantioselectivity of E = 31–36, but catalyzed hydrolysis of the 4-pyridylacetyl ester with an enantioselectivity of E >100.

γ-CT, E = 16
diisopropyl ether/vinyl acetate
Ke & Klibanov (1998)

Fig. 185. Chymotrypsin shows moderate enantioselectivity in diisopropyl ether.

secondary alcohols

PGA, E = 6-8
R = Me, CN, COOMe
Waldmann (1989)

n = 0,1, 2
E = 60-80
Baldaro et al. (1993)

X = H, E = 36
X = Cl, E >100

primary alcohols

PGA, E >30
Fuganti et al. (1988)

amines and amino acids

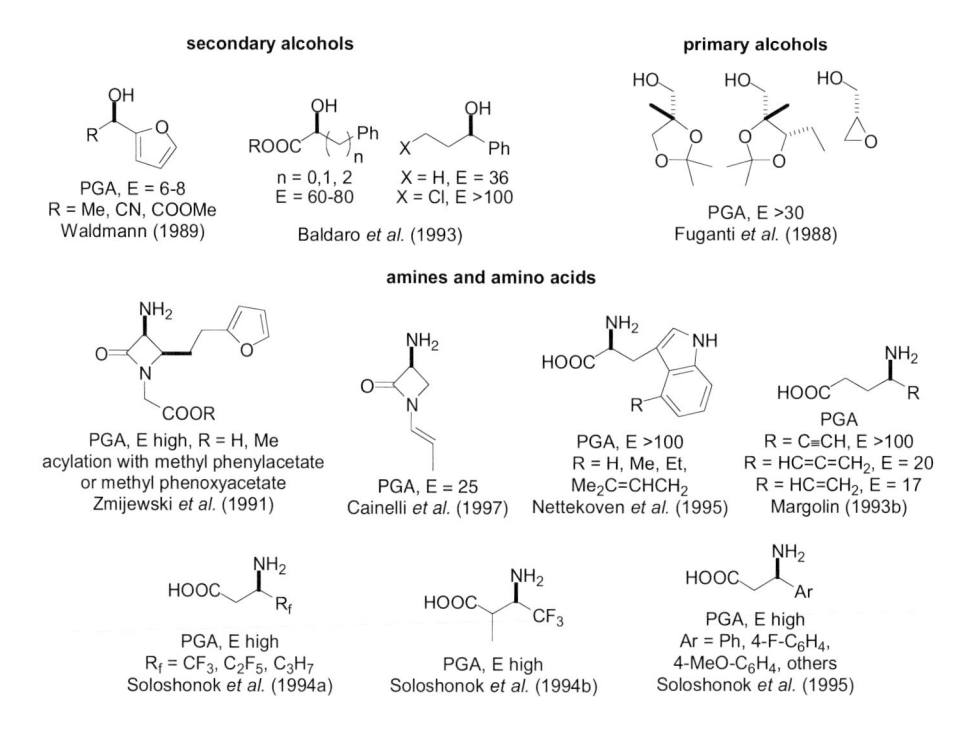

PGA, E high, R = H, Me
acylation with methyl phenylacetate
or methyl phenoxyacetate
Zmijewski et al. (1991)

PGA, E = 25
Cainelli et al. (1997)

PGA, E >100
R = H, Me, Et,
Me$_2$C=CHCH$_2$
Nettekoven et al. (1995)

PGA
R = C≡CH, E >100
R = HC=C=CH$_2$, E = 20
R = HC=CH$_2$, E = 17
Margolin (1993b)

PGA, E high
R$_f$ = CF$_3$, C$_2$F$_5$, C$_3$H$_7$
Soloshonok et al. (1994a)

PGA, E high
Soloshonok et al. (1994b)

PGA, E high
Ar = Ph, 4-F-C$_6$H$_4$,
4-MeO-C$_6$H$_4$, others
Soloshonok et al. (1995)

Fig. 186. Examples of alcohols and amines resolved by penicillin amidase (PGA). PGA resolves secondary and primary alcohols as well as amines. The amines include an α-amino acid, a γ-amino acid, and several β-amino acids. The enantioselectivity of PGA toward amines appears higher than toward alcohols. Unless otherwise noted, resolutions involve hydrolysis of the phenylacetyl derivative.

6.4.1.2 Secondary amines

Simple acyl amides are poor substrates for hydrolases presumably because they both chemically unreactive and sterically hindered. Switching to an oxalamic acid ester solved both of these problems (Fig. 187) allowing the protease-catalyzed resolution of secondary amines (Hu et al., 2005). The oxalamic acid ester changed the reacting group from an amide to an ester thereby increasing reactivity. The oxalamic ester also shifts the position of reacting group away from the sterically hindered secondary amide to the less hindered ester of ethanol, a primary alcohol.

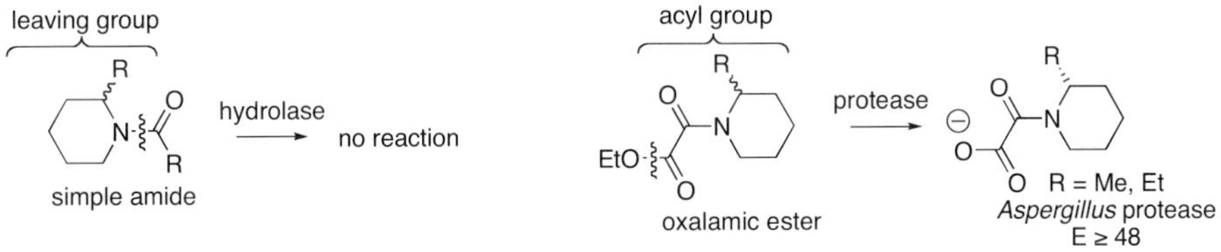

Fig. 187. Simple amides of secondary alcohols are poor substrates for hydrolases, but the oxalamic esters are good substrates for several proteases.

The change to an oxalamic ester also shifts the location of the stereocenter in the active site. The secondary amide moiety is in the leaving group pocket when it is a simple amide, but the secondary amide moiety is in the acyl group pocket when it is the oxalamic ester. Proteases have better defined acyl-binding pockets and often show higher stereoselectivity toward stereocenter in the acyl part as compared to the leaving group part. The oxalic ester approach has also been used to resolve sterically hindered alcohols, see Sect. 5.2.1.3.

6.4.1.3 α-Amino Acids via Reactions at the Amino Group

Amino Acid Acylases
Amino acid acylases catalyze the hydrolysis of the *N*-acyl group from *N*-acyl-amino acids and show high enantioselectivity for the natural L-enantiomer (Eq. 33) (for a review see Greenstein and Winitz (1961).

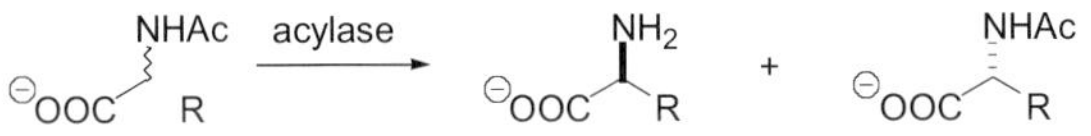

(33)

Tanabe commercialized an acylase-catalyzed resolution of natural amino acids in 1954 (Chibata et al., 1992) and Degussa introduced a continuous process in 1981. Both acylase from porcine kidney and from *Aspergillus* sp. are available, but commercial processes (Bommarius et al., 1992; Chibata et al., 1992) use the *Aspergillus* enzyme because it is more stable. Beside natural amino acids, acylases accept non-proteinogenic amino acids (Bommarius et al., 1997; Chenault et al., 1989). A few examples are shown in Fig. 188. The carboxyl group must be free. The *N*-acetyl group may be replaced by chloroacetyl to increase the reaction rate. Dowpharma Chirotech Technology Ltd. (Cambridge, UK) sells many unnatural amino acids resolved by acylases.

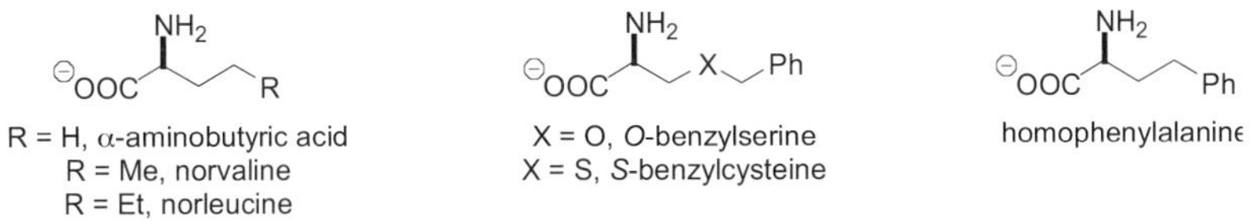

Fig. 188. Examples of unnatural amino acids resolved by acylase.

6.4.2 Carboxylic Acids

6.4.2.1 α-Amino Acids via Reactions at the Carboxyl Group

Amino acids, especially unnatural α-amino acids (or nonproteinogenic α-amino acids), are the type of carboxylic acids most commonly resolved by proteases and amidases (for reviews see: Kamphuis et al., 1992; Williams, 1989). Several derivatives are suitable starting materials (Fig. 189). Proteases or lipases can resolve esters of amino acids, where the amino group may be protected. Amidases can resolve amino acid amides, while nitrilases resolve α-amino nitriles. Both of these substrates come directly from a Strecker synthesis of amino acids. Hydantoinases resolve hydantoin derivatives of amino acids, while lipase also resolve 2-phenyloxazolin-5-one derivatives of amino acids.

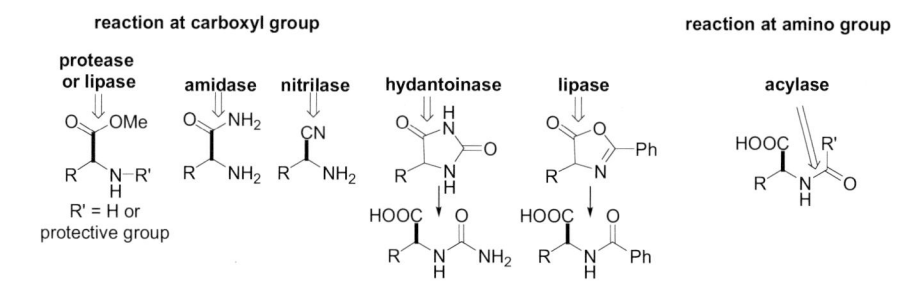

Fig. 189. Hydrolases can resolve several different derivatives of α-amino acids. Most resolutions involve reactions at the carboxyl group, but acylase-catalyzed resolutions (Sect. 6.4.1.3) involve reactions at the amino group.

All these resolutions involve reaction at the carboxyl group of the amino acid. Acylases resolve N-acyl amino acids by reaction at the amino group of the amino acid. None of these routes is clearly superior; which route is best depends on the cost of starting materials, ability to recycle the unwanted enantiomer, ease of isolation, as well as on other details for each amino acid.

The sections below cover resolutions involving proteases, amidases, and hydantoinases. The less common resolutions with nitrilases are covered in Sect. 9 . Lipase-catalyzed resolutions are covered in Sect. 5.2.1 (hydrolysis of esters) and Sect. 2.1.2 (hydrolysis of 2-phenyloxazolin-5-ones), while acylase-catalyzed resolutions are covered in Sect. 6.4.1.3.

Like proteases and lipases, carbonic anhydrase also catalyzes the hydrolysis of esters of amino acids. However, proteases and lipases favor the natural L-enantiomer, while carbonic anhydrases favor the unnatural D-enantiomer. The enantioselectivity is high for N-acetyl esters of Phe and N-acetyl diesters of Asp and Glu. For Glu, it is the δ-ester that reacts (Chênevert et al., 1993). Researchers also used other proteases to resolve natural and unnatural amino acids. For example, Lankiewicz et al. (1989) resolved several N-Boc protected analogs of phenylalanine by hydrolysis of their methyl esters catalyzed by thermitase.

Subtilisin

Subtilisin accepts a wide range of amino acid esters, but favors esters of hydrophobic amino acids. The approximate preference is Tyr, Phe > Leu, Met, Lys > His, Ala, Gln, Ser >> Glu, Gly (Estell et al., 1986; Wells et al., 1987). Subtilisin favors the natural L-enantiomer, but this preference is smaller than one might expect. For example, the enantiomeric ratio toward methyl esters of Tyr or Phe is only E ~ 100 and only E = 45 for Ala methyl ester (Fig. 190). Kawashiro et al. (1996) even reported a reversal of the enantioselectivity for the 2'2'2'-trifluoroethyl ester of *N*-trifluoroacetyl phenylalanine. Transesterification of this ester in *tert*-amyl alcohol favored the unnatural D-enantiomer by ~2. In comparison, hydrolysis of racemic amino acid esters yielded the L-amino acid and the unreacted D-amino acid ester (Chen et al., 1986).

natural amino acids

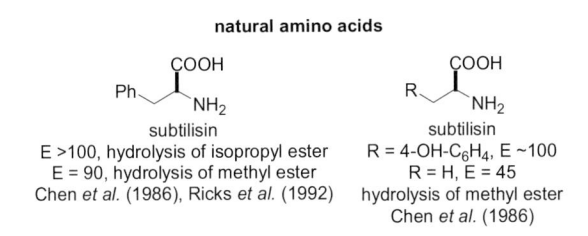

Ph—CH(NH$_2$)—COOH
subtilisin
E >100, hydrolysis of isopropyl ester
E = 90, hydrolysis of methyl ester
Chen *et al.* (1986), Ricks *et al.* (1992)

R—CH(NH$_2$)—COOH
subtilisin
R = 4-OH-C$_6$H$_4$, E ~100
R = H, E = 45
hydrolysis of methyl ester
Chen *et al.* (1986)

Fig. 190. Subtilisin resolves the hydrophobic natural amino acids with good to excellent enantioselectivity.

Subtilisin also accepts a wide range of unnatural amino acids (Fig. 191). Although chymotrypsin also resolves hydrophobic amino acids, often with higher enantioselectivity, most researchers favor subtilisin for large scale applications because of its low cost and stability. Cross-linked crystals of subtilisin are available from Altus Biologics (for a review see: Wang et al., 1997). The cross-linking simplifies recovery and also stabilizes subtilisin against autolysis.

6.4 Survey of Enantioselective Protease- and Amidase-Catalyzed Reactions

unnatural amino acids

R = H, 2-Cl, 3-Cl, 4-Cl
4-SMe, 4-OH
subtilisin, E high, hydrolysis of Me ester
Schutt et al. (1985)

X = O, S

X = CH, N

subtilisin, E >100
2,2'-bipyridin-4-yl, 5-yl or 6-yl isomers
hydrolysis of Me esters
Imperiali et al. (1993)

subtilisin, E ~70
R = H or OMe
hydrolysis of Et ester
Morgan et al. (1997c)

subtilisin, E = 34
hydrolysis of Et ester
Morgan et al. (1997c)

subtilisin, E high
X = SO₃Na, COOMe, COOBn, CONHOBn
hydrolysis of Et ester
Garbay-Jaureguiberry et al. (1992)

subtilisin, E high
hydrolysis of Et ester
Solladie-Cavallo et al. (1994)

subtilisin, E high
hydrolysis of Et ester
Baczko et al. (1996)

subtilisin, E>100
hydrolysis of Me ester
Chênevert & Thiboutot (1989)

subtilisin, E >100
hydrolysis of Et ester
Leanna & Morton (1993)

subtilisin, E >100
hydrolysis of Me ester
stereocenter '∗' set by synthesis
Bogenstätter & Steglich (1997)

subtilisin, 81% ee, 97% yield
hydrolysis of dimethyl ester
Wang et al. (1997b)

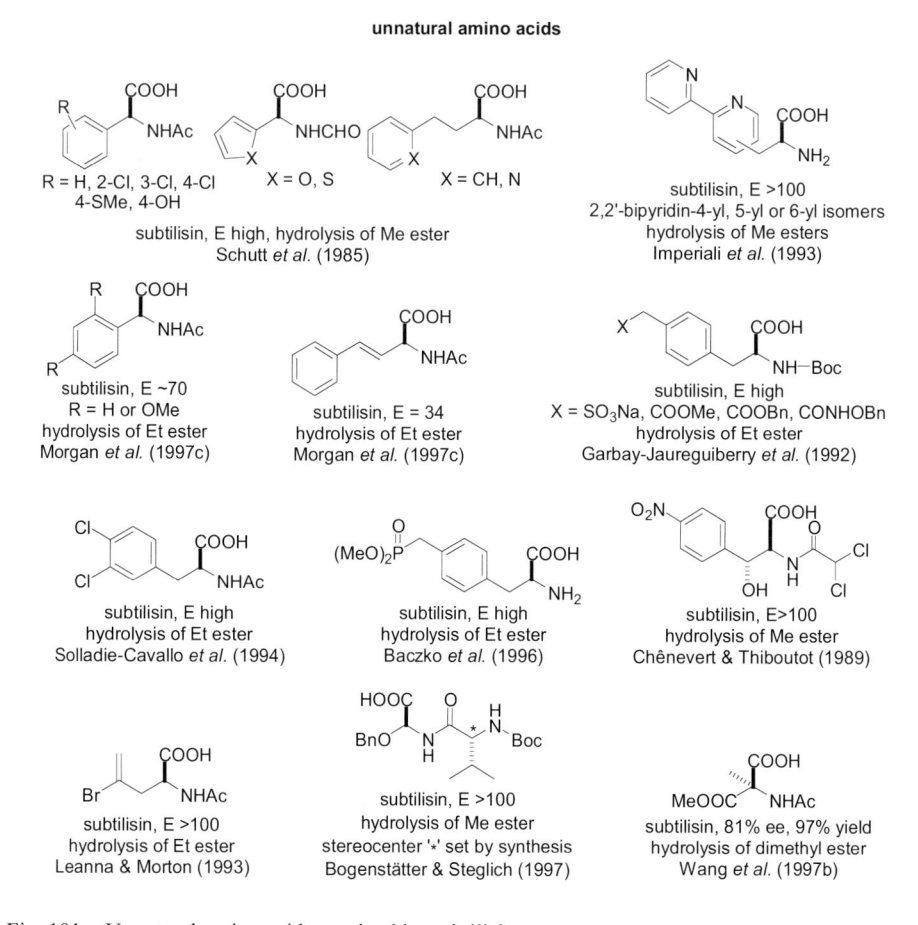

Fig. 191. Unnatural amino acids resolved by subtilisin.

Chymotrypsin

Chymotrypsin (α-CT), like other proteases, shows high enantioselectivity for the natural enantiomer of amino acids. Esters and amides containing aromatic amino acids – Phe, Tyr, and Trp – as the acyl group react most readily. Esters hydrolyze more rapidly than amides, so most preparative reactions use esters. For example, α-chymotrypsin resolves the ethyl esters of N-acetyl of Phe, Trp, and Trp with very high enantioselectivity. One exception to the high enantioselectivity rule is the moderate enantioselectivity (E = 8) of α-CT toward the methyl ester of N-benzoyl alanine (Fig. 192). The lower enantioselectivity is likely due to the binding of the benzoyl group in the hydrophobic pocket in the 'wrong' enantiomer.

Chymotrypsin also resolves unnatural amino acids (Fig. 193). For example, α-CT resolves phenylglycine (Cohen et al., 1966) and ring substituted phenylalanine esters (3-OH, 3,4-OH₂, 4-Cl, 4-F, (Tong et al., 1971), 2-Me (Berger et al., 1973)). O-substitution of tyrosine by alkyl, aryl, or acyl groups completely abolishes activity (Kundu et al., 1972), presumably because the aromatic moiety is too large to fit in the hydrophobic pocket. Chymotrypsin also resolves analogs of tryptophan where the indoyl group is replaced by 2-naphthyl or 6-quinolyl (Berger et al., 1973) and unnatural amino acids containing an alkenyl side chain (Schricker et al., 1992). Esters of Glu and Asp, which contain a –COO– in the side chain, are not hydrolyzed, but upon esterification of the side chain, the diesters can be resolved (Cohen and Crossley, 1964; Cohen et al., 1963). Note that α-CT removes the ester group from the more hindered carbonyl, the one at the α-position.

Fig. 192. α-Chymotrypsin shows high enantioselectivity toward *N*-acetyl amino acids, but low enantioselectivity toward the methyl ester of *N*-benzoyl alanine.

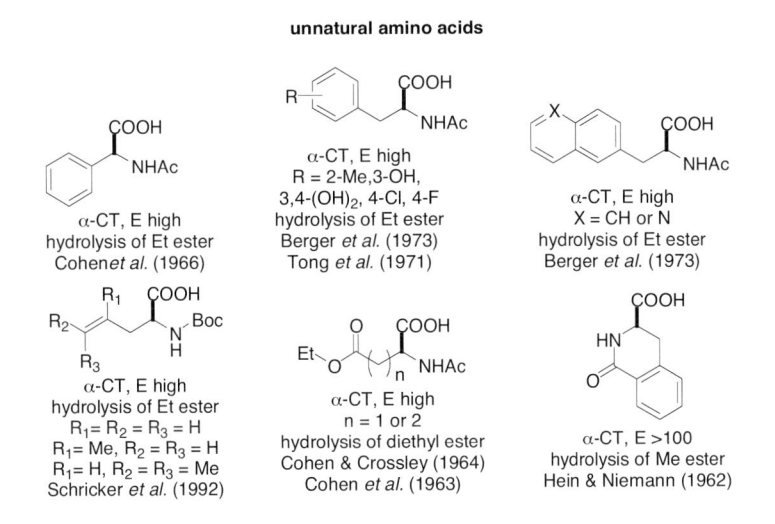

Fig. 193. α-Chymotrypsin also resolves unnatural amino acids.

Chymotrypsin also resolves hindered α-methyl amino acids (Anantharamaiah and Roeske, 1982) and the related α-methyl α-nitro acids (Lalonde et al., 1988) (Fig. 194). The α-methyl α-nitro acids spontaneously decarboxylated, but the unreacted esters were recovered and reduced to α-methyl α-amino acid esters. Interestingly, in both cases chymotrypsin favors the D-enantiomer, suggesting that, for the favored enantiomer, the amino or nitro groups bind in the *h*-site. Consistent with this notion, addition of an *N*-acetyl group slows the reaction rate dramatically. Chymotrypsin also resolves cyclic, unnatural α-amino acids (Dirlam et al., 1987; Hein and Niemann, 1962a; Hein and Niemann, 1962b; Matta and Rohde, 1972).

6.4 Survey of Enantioselective Protease- and Amidase-Catalyzed Reactions

hindered α-methyl acids

α-CT, E high, R = H, F
hydrolysis of Me ester
Anantharamaiah & Roeske (1982)

α-CT, E high
R = H, Me
Dirlam et al. (1987)

α-CT, E = 22
hydrolysis of
p-NO$_2$Ph ester
Matta & Rohde (1972)

R = CH$_2$CH=CH$_2$, Ph, 2-indoyl
Lalonde et al. (1988)

>95% ee

Fig. 194. α-Chymotrypsin also resolves hindered α-methyl amino acids.

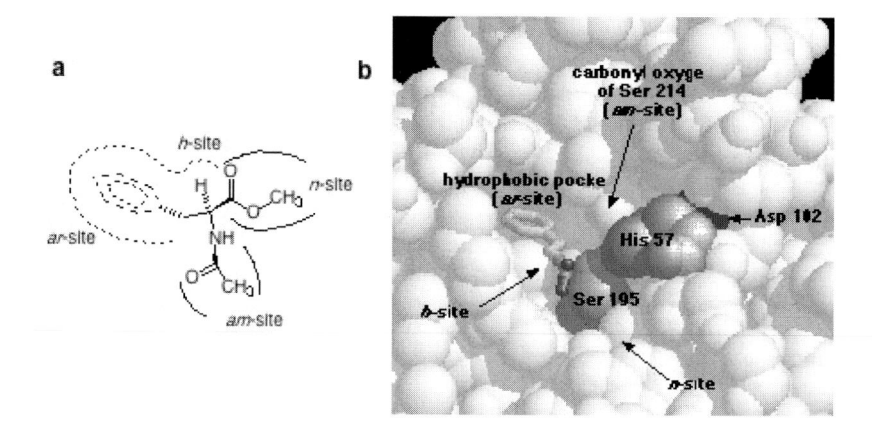

Fig. 195. The active site of α-chymotrypsin. **a** Cohen's model of α-CT (Cohen et al., 1969) containing a good substrate, *N*-acetyl-L-phenylalanine methyl ester. The *ar*-site (*aromatic*-site) is a hydrophobic pocket, the *am*-site (*amide*-site) is the amide carbonyl of Ser-124, which hydrogen bonds to the amide N-H of the substrate. An OH in the substrate can also form a hydrogen bond here. Other groups such as CH$_3$, Cl, AcO, also fit in the *am*-site, but do not form a hydrogen bond. The *am*-site is open to the solvent and can accommodate long groups. The *h*-site (*hydrogen*-site) is a small region in the active site. It is large enough to fit an H, Cl, or OH, but not a CH$_3$. The *n*-site (*nucleophile*-site) bind the leaving group, an ester or amide. Like the *am*-site, the *n*-site is open-ended, that is, it is open to the solvent and can accommodate large groups. In contrast, both the *ar*- and the *h*-sites are closed, that is, they point toward the center of the protein. **b** X-ray structure of α-chymotrypsin (spheres) contained transition state analog bound to the active site serine (Tulinsky and Blevins, 1987). The transition state analog is a 2-phenylethyl boronic acid (stick representation). The amino acid residues of the catalytic triad are labeled, as are the regions that correspond to the four sites of Cohen's model.

Chymotrypsin also catalyzes the hydrolysis of 2-phenyloxazolin-5-ones, but the enantioselectivity was low, e.g., E = 8 for the 2-phenyloxazolin-5-one derived from phenylalanine (Daffe and Fastrez, 1980). Lipases show much higher enantioselectivity toward these compounds (Sect. 2.1.2).

Researchers usually use Cohen's active site model (Cohen et al., 1969) to rationalize the selectivity of α-CT toward acids (Fig. 195). This model defines four sites that bind each of the four substituents at the stereocenter of a good substrate like N-acetyl-L-phenylalanine methyl ester. The strongest binding comes from the *ar*-site, a hydrophobic pocket, which binds aromatic or ether hydrophobic groups. When the substrate has the incorrect configuration, for example, N-acetyl-D-phenylalanine methyl ester, it still binds to the *ar*-site, but the remaining substituents cannot orient in a productive manner. One observes nonproductive binding. Jones and Beck (1976) reviewed the substrate specificity of α-CT in detail.

Computer modeling can also help explain the enantioselectivity of chymotrypsin toward natural substrates (DeTar, 1981; Wipff et al., 1983), unnatural substrates (Norin et al., 1993) and explain more subtle phenomenon such as changes in enantioselectivity in different solvents (Ke and Klibanov, 1998).

Hydantoinases

Hydantoinases (for reviews see: Ogawa and Shimizu, 1997; Syldatk et al., 1999) catalyze the hydrolysis of 5-monosubstituted hydantoins to N-carbamoyl α-amino acids (Fig. 196). Some hydantoinases are identical to dihydropyrimidase, an enzyme in pyrimidine degradation, but others are different enzymes. The most common hydantoinases favor the unnatural D-enantiomer, but L-selective hydantoinases are also known. 5-Substituted hydantoins, especially 5-aryl hydantoins racemize readily, thus permitting dynamic kinetic resolution (see Sect. 2.1.2). Synthetic applications use either isolated enzymes or whole microorganisms.

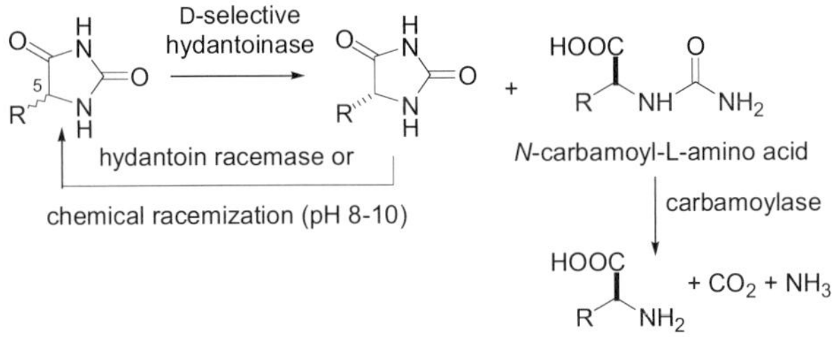

Fig. 196. Hydantoinases catalyze the hydrolysis of 5-monosubstituted hydantoins to N-carbamoyl α-amino acids. The equation shows a D-selective hydanoinase; L-selective hydanoinases are less common. 5-Substituted hydantoins racemize readily either enzymatically or chemically under basic reaction conditions (pH 8–10). An *in situ* racemization of the hydantoin permits a dynamic kinetic resolution (see Sect. 2.1.2). Many hydantoinase-catalyzed resolutions also include a carbamoylase that removes the N-carbamoyl group to give the free α-amino acid (Olivieri et al., 1979).

Both Kanegafuchi Industries (Japan) and DEBI Recordati (Italy) produce unnatural D-amino acids using D-selective hydantoinase. Most of the production focuses on D-phenylglycine and D-4-hydroxyphenyl glycine for the semi-synthetic penicillins ampicillin and amoxilicillin (for reviews see: Syldatk et al., 1992a; b).

Hydantoinases tolerate a wide range of substituents at the 5-position, but 5,5-disubstituted react very slowly (Garcia and Azerad, 1997; Keil et al., 1995; Olivieri et al., 1981; Takahashi et al., 1979).

Amidases

DSM and Lonza resolve several unnatural amino acids using amidases (Eichhorn et al., 1997; Kamphuis et al., 1992) (Fig. 197). Both groups use whole cells of microorganisms that produce amidases. Unlike proteases, which always favor the natural L-enantiomer, amidases occasionally favor the D-enantiomer; for example, the *Burkholderia* amidase catalyzes hydrolysis of piperazine-2-carboxamide (Eichhorn et al., 1997).

Fig. 197. Examples of unnatural α-amino acids resolved by amidases.

The DSM group also found an amidase that catalyzes hydrolysis of α,α-disubstituted α-amino acid amides (Kamphuis et al., 1992; Kaptein et al., 1993; Kruizinga et al., 1988) and proposed an active site model for its substrate specificity (Fig. 198).

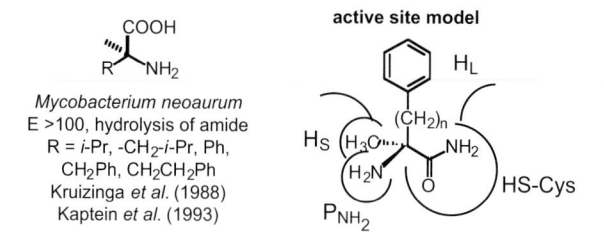

Fig. 198. Typical α,α-disubstituted α-amino acids resolved by amidase(s) from *Mycobacterium neoaurum*. The active site model summarizes the known substrate specificity. The P_{NH_2} site represents a polar site for the NH_2 group. The amidase does not tolerate substituents on the nitrogen. The H_S site is a small hydrophobic site that tolerates methyl, ethyl, and allyl groups. The H_L site is a large hydrophobic site that tolerated a range of groups.

6.4.2.2 Other Carboxylic Acids

Subtilisin

Subtilisin also catalyzes the enantioselective hydrolysis of esters of carboxylic acids that are not amino acids (Fig. 199). In the case of the 2-phenoxypropanoic acids chymotrypsin and subtilisin favored opposite enantiomers (Chênevert and D'Astous, 1988).

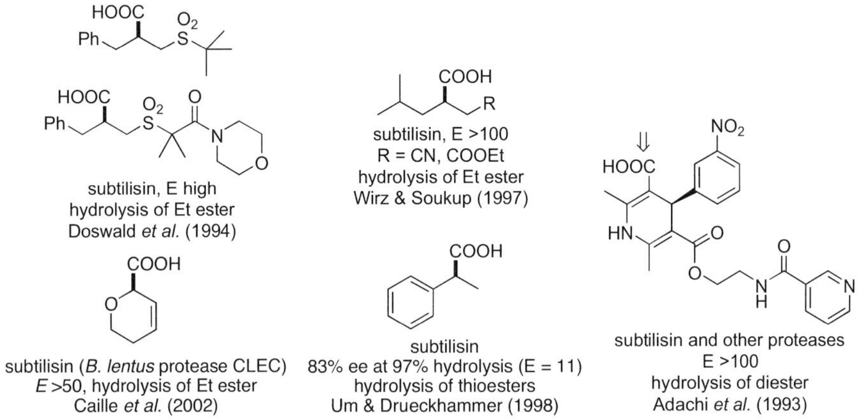

Fig. 199. Other carboxylic acids resolved by subtilisin.

Chymotrypsin

α-CT often shows lowered stereoselectivity upon replacing the amino group of α-amino acids with an α-hydroxy or α-*O*-acyl group. For example, α-CT showed no stereoselectivity toward mandelic acid, but high stereoselectivity toward the corresponding amino acid, phenylglycine (Fig. 199). On the other hand, α-CT efficiently resolves β-phenyllactate (analog of Phe) by hydrolysis of its ethyl ester (Cohen and Weinstein, 1964). However, hydrolysis of the same substrate in dichloromethane containing 0.2% water showed no stereoselectivity (Ricca and Crout, 1993). Hydrolysis of *O*-benzoyl lactate ethyl ester favors the D-enantiomer (analog of the unnatural enantiomer of Ala). Binding of the benzoyl group in the *ar*-site presumably accounts for this reversal. Like the corresponding amino compound, the resolution of α-methyl-β-phenyllactate is efficient, but the stereoselectivity is reversed (Cohen and Lo, 1970). Presumably the OH group, but not the NH$_2$ group, favors the *h*-site. Replacement of the amino group in the hindered cyclic acid below with either hydrogen or *O*-acetyl abolished stereoselectivity (Matta and Rohde, 1972). α-CT efficiently resolved the dihydrobenzofuran derivative in Fig. 200 (Lawson, 1967). α-CT can also resolve acids without a polar substituent, for example the 3-cyclohexenyl acids in Fig. 200 (Cohen et al., 1969; Jones and Marr, 1973).

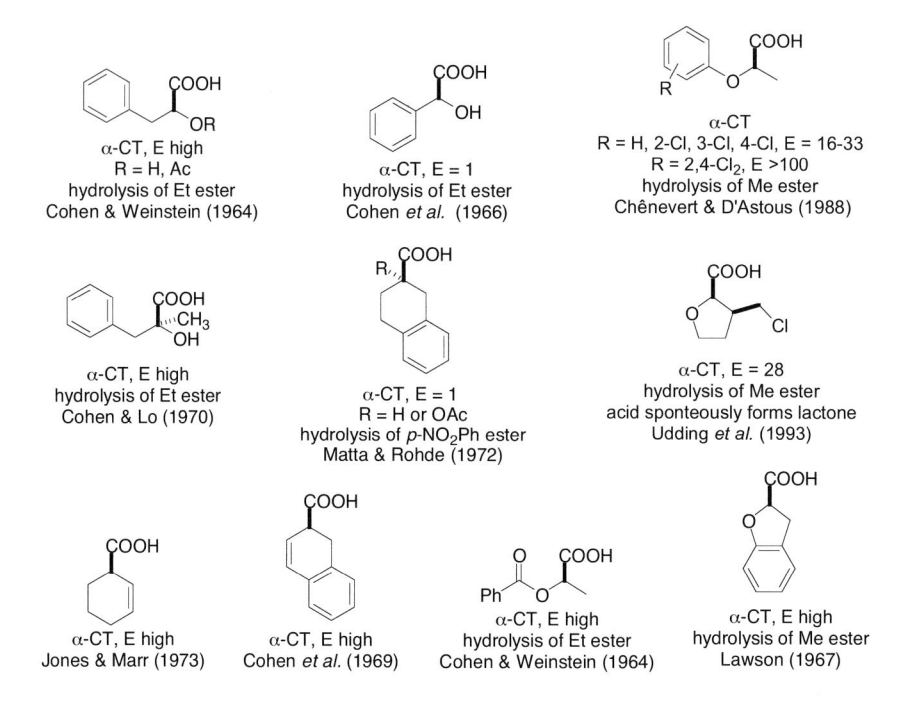

6.4 Survey of Enantioselective Protease- and Amidase-Catalyzed Reactions

Fig. 200. Other carboxylic acids resolved by chymotrypsin.

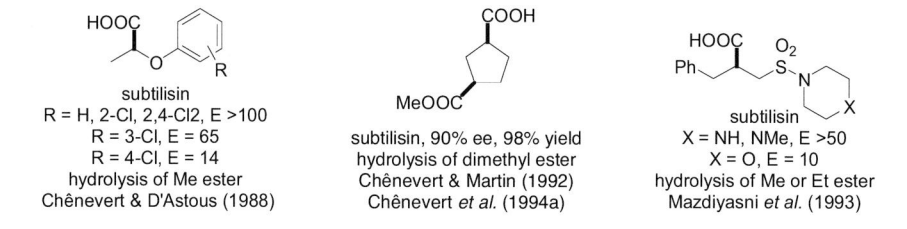

Fig. 200. Other carboxylic acids resolved by chymotrypsin (continued).

Other Proteases and Amidases

Other proteases and amidases can also resolve carboxylic acids that are not amino acids (Fig. 201). Penicillin acylase from *E. coli* strongly favors phenylacetyl esters or amides, but it also accepts close analogs. The 2-substituted phenylacetic acids are chiral and PGA showed high enantioselectivity toward several of these. The Lonza group isolated and cloned a *Klebsiella* amidase that shows high enantioselectivity toward hindered propionic acids.

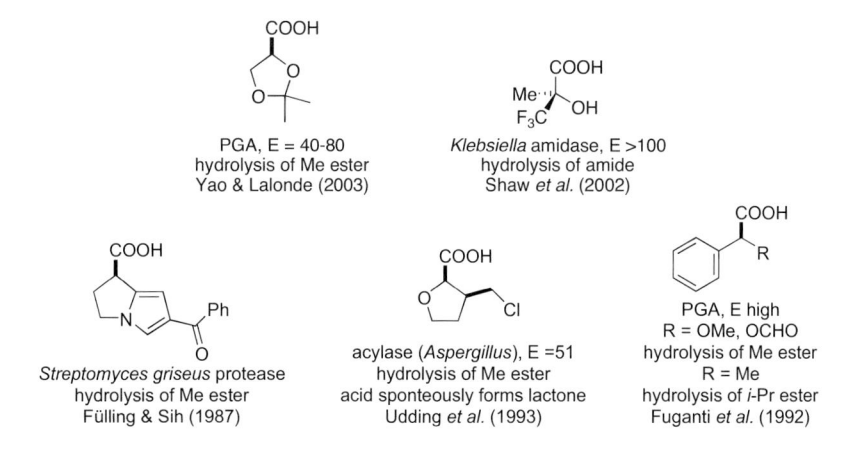

Fig. 201. Non-amino acids resolved by proteases and amidases.

6.4.2.3 Commercial Enantioselective Reactions

Unnatural Amino Acids

Unnatural amino acids are produced commercially via acylase-, hydantoinase- and amidase-catalyzed routes (for reviews see Bommarius et al., 1992; Kamphuis et al., 1992). A few examples are shown in Fig. 202. Acylase-catalyzed routes are best for the L-amino acids because acylases favor the L-enantiomer. Racemization of unreacted N-acyl derivative of the D-amino acid avoids wasting half of the starting material. Hydantoinase-catalyzed routes are best for producing D-amino acids because the most common hydantoinases favor the D-enantiomer. Amidase-catalyzed resolution are best for unusual amino acids whose derivatives are not substrates for acylases or hydantoinases. For example, α-alkyl-α-amino acids are resolved by amidases as are the cyclic amino acids 2-piperidine carboxylic acid (pipecolic acid) and piperazine-2-carboxylic acid. Amidases usually favor the natural L-enantiomer.

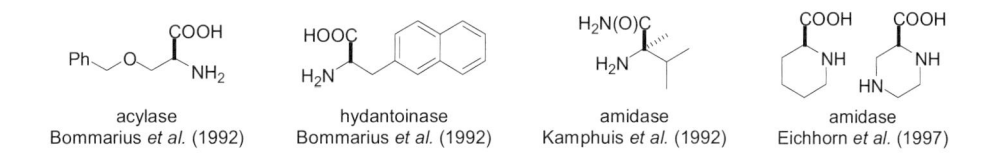

Fig. 202. Examples of unnatural amino acids produced via acylase, hydantoinase or amidase.

6.4 Survey of Enantioselective Protease- and Amidase-Catalyzed Reactions

Other Carboxylic Acids

Other carboxylic acids are also important intermediates in the synthesis of pharmaceuticals. Researchers have published kilogram-scale resolutions by using proteases (Fig. 203).

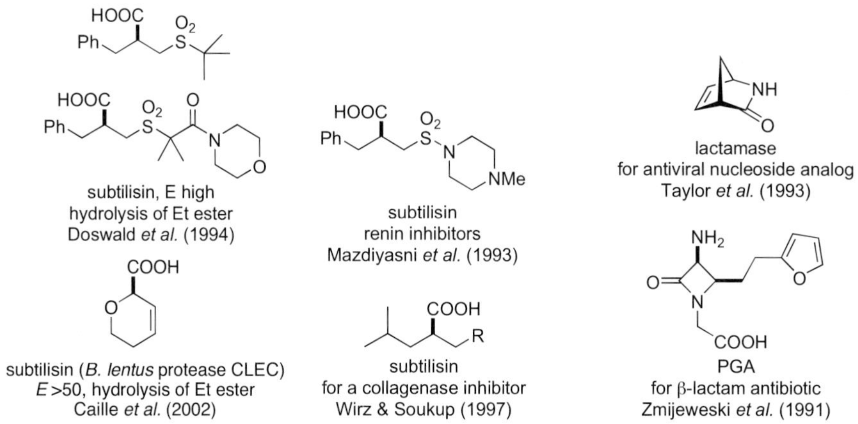

Fig. 203. Kilogram-scale routes to pharmaceutical precursors involving proteases or amidases.

Regioselective and Chemoselective Reactions of Proteases

The most common route to 6-aminopenicillanic acid (6-APA) is hydrolysis of penicillin G catalyzed by penicillin G acylase (PGA), but a small amount (10-15%) is produced by penicillin V acylase catalyzed hydrolysis of penicillin V (PVA). Although both acylases catalyze hydrolysis of both penicillins G and V, PGA favors penicillin G and PVA favors penicillin V (Fig. 204, left). The main disadvantage of PVA is the higher cost of penicillin V, while the main advantage is the ability to operate at lower pH where the product, 6-APA, is more stable (review: Shewale and Suhakaran, 1997; see also Buchholz et al., 2005).

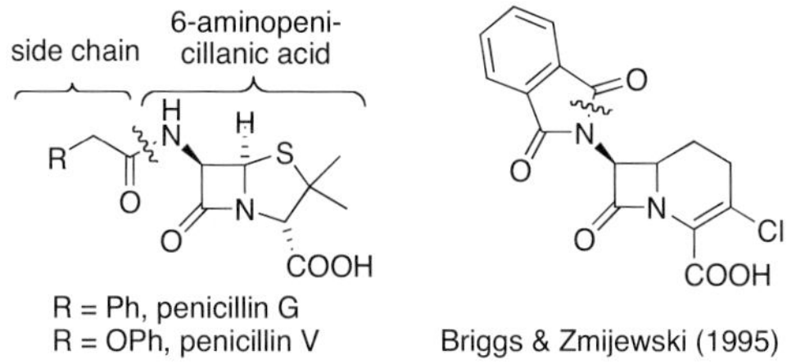

Fig. 204. 6-APA is produced from penicillin G or V using PGA (left). A phthalyl amidase cleaves the phthalyl protecting group from a β-lactam (right).

Briggs and Zmijewski (1995) cloned and overexpressed a phthalyl amidase that selectively cleaves the phthaloyl protective group under mild conditions. This enzyme removed a phthalyl group from a β-lactam (Fig. 204, right) and from phthaloyl aspartame. In both cases the amidase cleaved only the phthaloyl group and not the β-lactam or peptide links.

7 Phospholipases

Phospholipids are amphiphilic molecules that are ubiquitous in nature, where they are basic components of natural membranes and cell walls. Phospholipases catalyze the hydrolysis of these phospholipids. Four types of enzymes with different regioselectivities have been identified and their cleavage sites are given in Fig. 205 for a phosphatidyl choline. The physiological role of phospholipases is believed to be the degradation of phospholipid components of cell membranes and the digestion of phospholipid-containing fats in food.

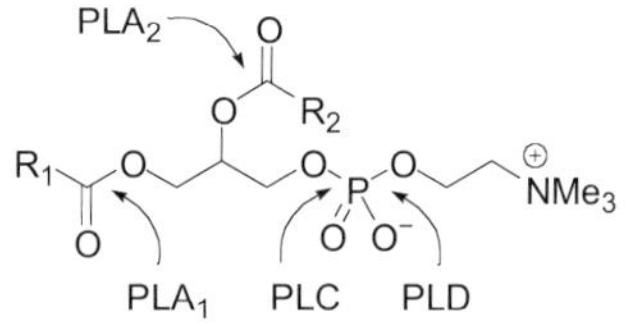

Fig. 205. Regioselectivity of different phospholipases.

The interest in new phospholipids and phospholipid analogs stems from their potential use as biodegradable surfactants, carriers of drugs or genes or as biologically active compounds in medicine and agriculture (New, 1993). However, synthesis of these substances is difficult by chemical means since control of regio- and stereoselectivity must be ensured.

The survey below focused on the application of phospholipases in organic synthesis. Further applications and examples can be found in reviews (D'Arrigo et al., 1996) including detailed procedures (Kötting and Eibl, 1994).

7.1 Phospholipase A_1

PLA_1 (EC 3.1.1.32) is the only phospholipase, which is rarely used. This is mainly because the resulting phospholipid is unstable due to facile acyl migration of residues at the *sn*-2 position to the *sn*-1 position. In addition the enzyme is not commercially available and similar reactions are catalyzed by a number of lipases, e.g., from *Rhizopus* sp.

7.2 Phospholipase A_2

The best-characterized enzyme is PLA_2 (EC 3.1.1.4) from cobra venom. Other sources are bovine or porcine pancreas or microorganisms (mainly *Streptomyces* sp.). Structural analysis of PLA_2 revealed that catalysis occurs at the interface of aggregated substrates (Scott et al., 1990). PLA_2 cleaves fatty acids from lecithin or cephalin forming lysolecithin or lysocephalin resulting in disruption of the cell wall structure and haemolysis of erythrocytes. Arachidonic acid ($C_{20:4}$) released from phospholipids by the action of PLA_2 is an important precursor of prostaglandins (Kudo et al., 1993).

PLA_2 usually shows higher efficiency in hydrolysis than in transesterification and the enzyme is specific for the natural absolute configuration at the *sn*-2 position. Thus, resolution of racemic phospholipids as well as the determination of absolute configuration are feasible. Sahai and Vishwakarma (1997) synthesized optically pure radiolabeled precursors of cell surface conjugates of the parasite *Leishmania donovani* by kinetic resolution of phospholipids using PLA_2 from the snake *Naja mocambique mocambique*. Hosokawa et al. (1995) and Na et al. (1990) quantitatively incorporated polyunsaturated fatty acids into the *sn*-2 position of (lyso)phosphatidylcholine using PLA_2 from porcine pancreas.

7.3 Phospholipase C

Most microbial phospholipase C (EC 3.1.4.3) hydrolyse L-α-phosphatidylcholine to diglyceride and choline phosphate. Other microbial PLC are specific for phosphatidylinositol (Griffith et al., 1991; Iwasaki et al., 1994) or sphyngomyelin. Direct transphosphatidylation with PLC seems to be more difficult compared to PLD. Structural analogs of inositol phosphate diesters have been obtained by PLC-catalyzed hydrolysis of phosphatidylinositol, isolation of inositol-1,2-cyclic-phosphate and subsequent transesterification using the same enzyme to afford a broad range of *O*-alkyl inositol-1-phosphates (Fig. 206).

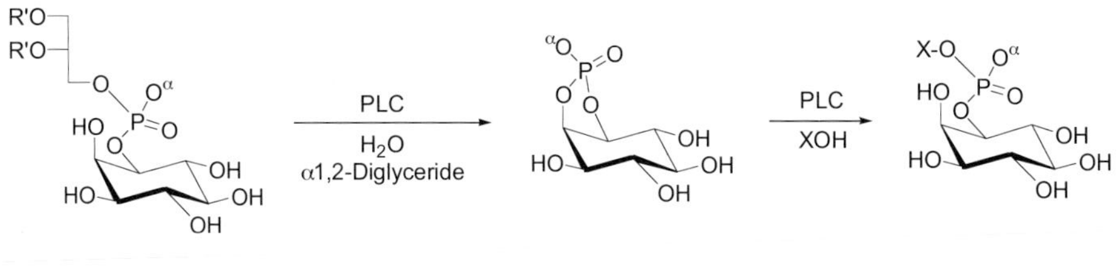

Fig. 206. Access to *O*-alkyl inositol-1-phosphates using PLC (Bruzik et al., 1996).

7.4 Phospholipase D

PLD (EC 3.1.4.4) is the most commonly used phospholipase in organic syntheses. Hydrolysis with PLD leads to phosphatidic acids, but in the presence of suitable alcohols, e.g., ethanolamine, glycerol, serine, PLD also catalyzes an efficient head group exchange. These reactions are usually performed in a biphasic mixture in order to ensure high enzymatic activity. A disadvantage of the biphasic system is the competing hydrolysis to phosphatidic acid. PLD from cabbage, peanut and several microbial sources, such as *Streptomyces* sp., are commercially available.

Although crystallographers have not solved the x-ray crystal structure of phospholipase D, Stuckey and Dixon (1999) solved the structure of a close relative, a bacterial endonuclease. This structure suggests a dimeric active site containing two histidines. One is the nucleophile that displaces the leaving group and forms a phosphohistidine intermdiate. The second histidine acts as an acid/base catalyst.

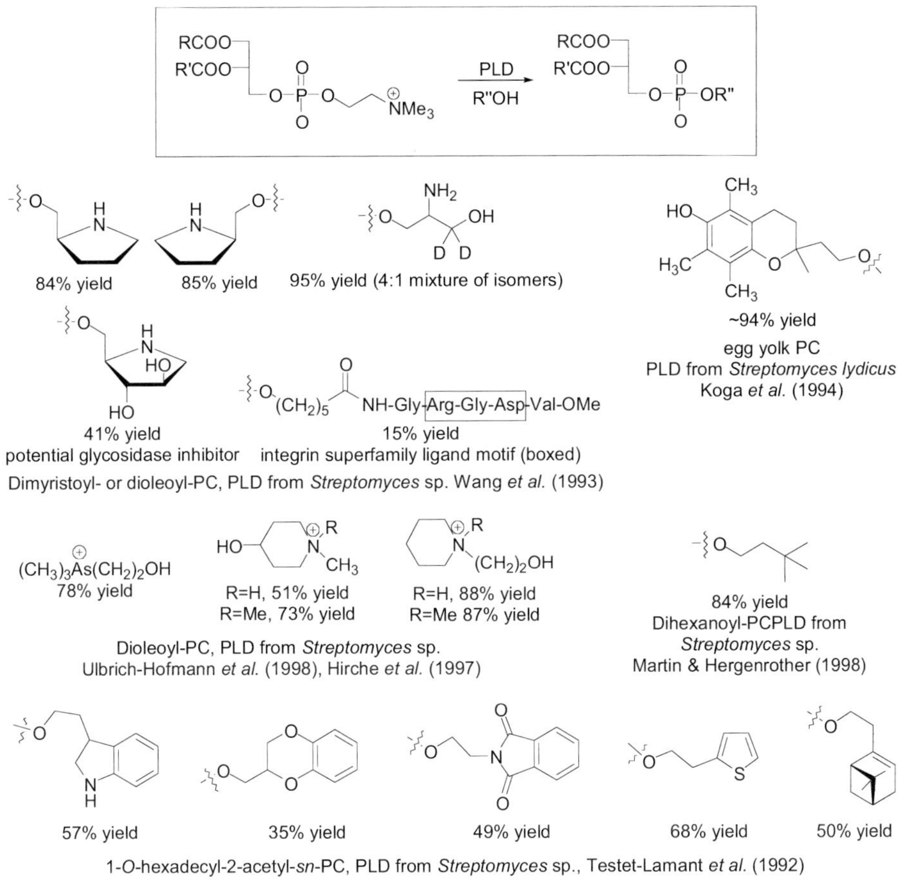

Fig. 207. Examples for PLD-catalyzed head group exchange starting from phosphatidylcholine.

Pisch et al. (1997) synthesized a range of phospholipids bearing acetylenic fatty acids (4- or 14-octadecynoic acid) by head group exchange from the corresponding phosphatidylcholine (PC) using a recombinant PLD from *Streptomyces antibioticus* (Iwasaki et al., 1995). Ethanolamine, glycerol or L-serine were quantitatively converted into the corresponding phospholipids within 1 h using a biphasic system (CHCl$_3$:buffer, 1:1.5) and isolated in 74–87% yield. PLD also accepts a wide range of nucleophiles, e.g., *N*-heterocyclic and As-containing compounds (Hirche et al., 1997; Ulbrich-Hofmann et al., 1998), however amine or thiols do not react. Martin and Hergenrother, 1998 synthesized a phospholipid bearing a *t*-butyl group by PLD-catalyzed head group exchange. The activity of PLC from *Bacillus cereus* – often used as a probe for signal transduction pathways in mammalian systems – toward the resulting phospholipid was reduced by a factor > 1000 (Fig. 207).

214   7.4   Phospholipase D

Fig. 207. Examples for PLD-catalyzed head group exchange starting from phosphatidylcholine (continued).

Only a few examples can be found, where secondary alcohols or phenols were accepted by PLD (D'Arrigo et al., 1996; Takami et al., 1994). However, competing hydrolysis of PC to phosphatidic acid substantially affected transphosphatidylation (Fig. 208). Sugars containing solely secondary hydroxyls like L-Rhamnose and *myo*-inositol were not converted.

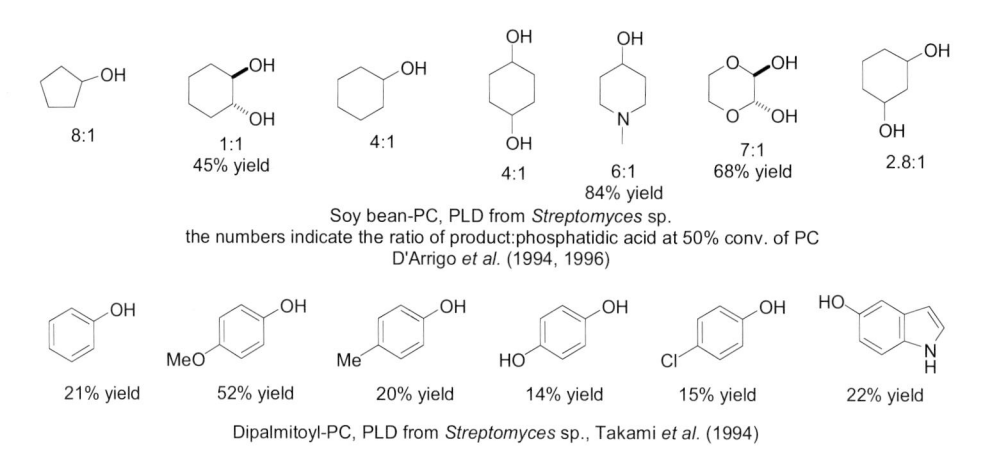

Fig. 208. Secondary alcohols and phenols accepted by PLD from *Streptomyces* sp.

8 Epoxide Hydrolases

8.1 Introduction

Optically pure epoxides are versatile building blocks in organic synthesis. Several chemical ways for the preparation of optically pure epoxides have been developed, such as the methods described by Sharpless or Jacobsen-Katsuki. However, the Sharpless epoxidation (Katsuki and Martin, 1996) is limited to allylic alcohols and the Jacobsen-Katsuki method (Hosoya et al., 1994; Linker, 1997) gives poor enantiomeric excess (≤ 60% ee) with *gem*- or *trans*-disubstituted olefins.

Direct stereospecific epoxidation of alkenes by isolated monooxygenases (cytochrome P450) and bacterial monooxygenases have been reported (for reviews see: Besse and Veschambre, 1994; de Bont, 1993; Onumonu et al., 1994; Pedragosa-Moreau et al., 1995). However, although enantiomeric excess are often high, yields are typically low, most enzymes require cofactors and these biocatalysts accept only a narrow range of substrates making them less attractive for organic synthesis.

Alternatively, epoxides can be resolved by using epoxide hydrolases (EC 3.3.2.3), which catalyze the hydrolysis of an epoxide to furnish the corresponding vicinal diol. The reaction proceeds via an S_N2-specific opening of the epoxide leading to the formation of the corresponding *trans*-configured 1,2-diols.

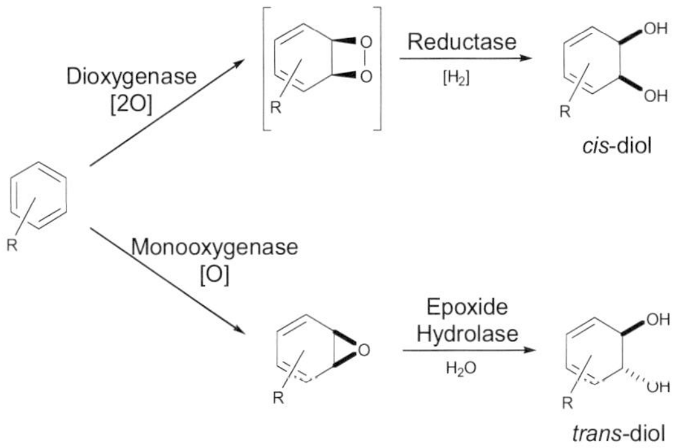

Fig. 209. Degradation of aromatics can proceed via dioxygenase-catalyzed formation of a dioxetane leading after reduction to *cis*-diols or via a monooygenase-catalyzed epoxidation followed by epoxide hydrolase-catalyzed ring-opening to *trans*-diols.

Epoxide hydrolases (EH's) do not require cofactors, and occur in mammals (Bellucci et al., 1991), plants (Blée and Schuber, 1995), insects (Linderman et al., 1995), yeasts (Weijers, 1997), filamentous fungi (Grogan et al., 1996; Pedragosa-Moreau et al., 1996b) and bacteria (Mischitz et al., 1995a; Osprian et al., 1997). They are catalytically active in the presence of organic solvents and often show high regio- and enantioselectivity. In eukaryotes, EH's play a key role in the metabolism of xenobiotics, in particular of aromatic systems. In prokaryotes, EH's hydrolyze epoxides formed by the action of

8.1 Introduction

P450-monooxygenases from olefinic or aromatic compounds giving the organisms access to these carbon sources (Fig. 209).

In mammals, microsomal (mEH's) and soluble (sEH's, earlier referred to as cytosolic EH's) epoxide hydrolases have been identified and the most widely studied enzymes are mammalian liver mEH's (review: Seidegard and de Pierre, 1983). In addition some EH's act on cholesterol, leukotriene A_4 or hepoxilin (Pinot et al., 1995). Several mammalian EH's have been purified and characterized and the rat mEH was cloned into *E. coli* (Bell and Kasper, 1993). However, it is unlikely that these enzymes will be widely used as biocatalysts in organic synthesis, since they are not available in reasonable amounts for large-scale biotransformations.

Mechanism

One proposed mechanism of epoxide ring-opening involves the attack of a nucleophilic carboxylate residue at one end of the epoxide which again has been activated by protonation. This leads to an α-hydroxyester intermediate covalently bound to the active site of the enzyme. This intermediate is hydrolyzed by the nucleophilic attack of a water molecule which is activated by a histidine, followed by the release of the diol product and regeneration of the enzyme (Lacourciere and Armstrong, 1993) (Fig. 210). Evidence for the likelihood of this mechanism came from experiments involving ^{18}O-labeled enzyme in unlabeled water and *vice versa*. For soluble rat epoxide hydrolase, Asp_{333}, Asp_{495}, His_{523} (Arand et al., 1996) and for microsomal epoxide hydrolase, Asp_{226}, Asp_{352}, His_{431} were identified to form the catalytic triad (Laughlin et al., 1998).

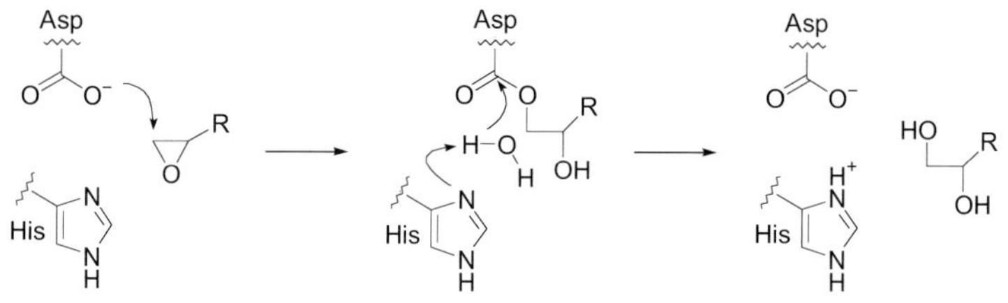

Fig. 210. Mechanism proposed for mammalian epoxide hydrolases.

However, new insights into the mechanism were obtained from the X-ray structure of the EH from *Agrobacterium radiobacter* AD1 (Nardini et al., 1999), which together with mutation studies showed that two tyrosine residues (Tyr_{152} and Tyr_{215}) are responsible for substrate activation and also for transition state stabilization (Rink et al., 1999). The water molecule required for the second catalytic step is activated by a His-Asp pair before the diol product is formed (Fig. 211). The EH from *Aspergillus niger* has a similar structure and mechanism (Zou et al., 2000).

Some epoxide hydrolases may follow a different mechanism where a water molecule directly attacks the epoxide. Mammalian leukotriene A4 epoxide hydrolase contains a zinc ion in the active site (Tholander et al., 2005). Limonene epoxide hydrolase from *Rhodococcus erythropolis* has a novel protein fold with a hydrophobic pocket containing a cluster of polar groups, including an Asp-Arg-Asp triad, at its deepest point. Other experiments suggest a one-step mechanism involving the protonation of the epoxide ring by Asp_{101} with concomitant direct attack of a water molecule on a carbon atom of the epoxide ring (Arand et al., 2003). This mechanism would allow – in contrast to epoxide hydrolases that use an aspartate nucleophile – the acceptance of nucleophiles other than water.

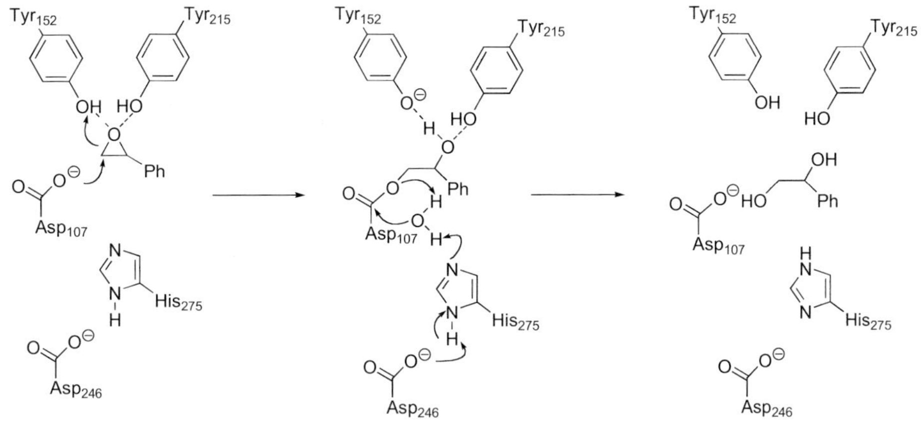

Fig. 211. Mechanism proposed for epoxide hydrolase from *Agrobacterium radiobacter* AD1.

Stereochemistry

The enzymatic hydrolysis of terminal epoxides may proceed via attacking either the less hindered oxirane carbon leading to retention of configuration (most common) or at the stereogenic center resulting in inversion of configuration (Faber et al., 1996) (Fig. 212).

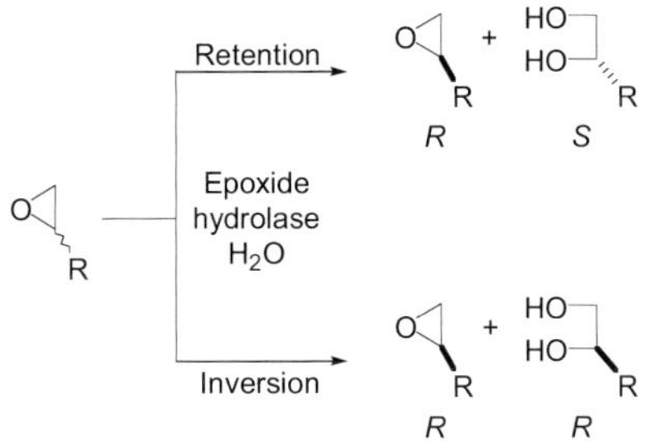

Fig. 212. Hydrolysis of epoxides can proceed with retention or inversion of configuration.

A detailed study of the regioselectivity of EH from *Rhodococcus* sp. NCIMB11216 using ^{18}O-labeled water revealed that also the position of substituents influenced the stereospecifity of the reaction (Fig. 213) (Mischitz et al., 1996). Mossou et al. (1998a; b) developed equations to determine the regio- and enantioselectivity of epoxide hydrolase-catalyzed ring opening of epoxides without the need for ^{18}O-labeling. By using experimentally obtained data for conversion,% ee diol and% ee epoxide, one can distinguish between (1) total regioselectivity on the same carbon, (2) partial regioselectivity between the two carbon atoms of the oxirane ring, and (3) total but opposite regioselectivity (enantioconvergence) (Moussou et al., 1998a; b).

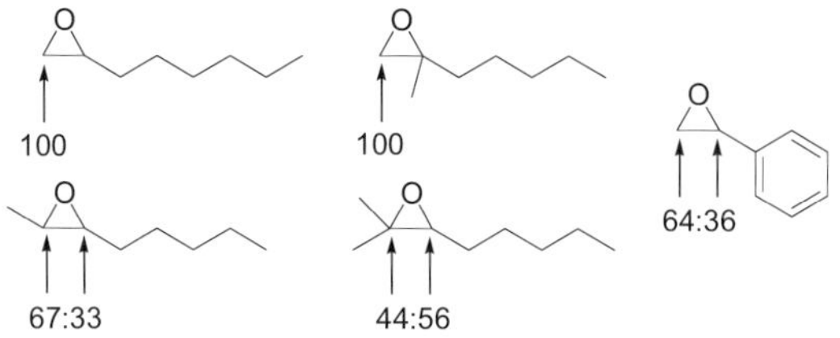

Fig. 213. Regioselectivity of epoxide hydrolase from *Rhodococcus* sp. NCIMB11216 studied using ^{18}O-labeled water revealed that insertion of oxygen usually occurs at the less hindered carbon. Note that in the case of styrene oxide non-enzymatic incorporation took place at the more hindered carbon (Mischitz et al., 1996).

Assays

Discovery of epoxide hydrolase activity in strain collections is usually performed by incubation of an epoxide followed by GC or HPLC analysis to analyze epoxide depletion and/or diol formation. Although these methods also allow the determination of enantiomeric excess, they are rather time-consuming and not useful for large strain collections or directed evolution experiments. Thus, several methods for a rapid prescreening for epoxide hydrolase activity have been developed. One assay monitors amount of starting epoxide by reaction with 4-*p*-nitrobenzyl pyridine forming a blue dye. The method works in microtiter plates and as a filter-based assay (Zocher et al., 1999). Reetz et al. (2004) used this assay to increase the enantioselectivity of an epoxide hydrolase by directed evolution (see Sect. 8.3.2). Doderer et al. (2003) developed an assay that measures the product diol formation. Periodate cleaves the diol yielding a ketone or aldehyde and NaIO$_3$, Excess periodiate and iodate are removed with sulfurous acid. The ketone or aldehyde react with fuchsin and Na$_2$SO$_3$ (Schiff's reagent) to a magenta dye quantified at 560 nm. Based on an assay initially developed for dehydrogenases and lipases/esterases, Badalassi et al. (2004), developed a fluorescence screening method for epoxide hydrolases using umbelliferyl derivatives (see Fig. 31).

The following sections review selected examples for the application of mammalian and microbial epoxide hydrolases in organic synthesis proceeding with good to excellent enantioselectivities. A number of reviews contain further examples (Archelas, 1998; Archelas and Furstoss, 1998; Faber et al., 1996; Orru et al., 1998a; Swaving and de Bont, 1998; Moussou et al., 1998c; Weijers and de Bont, 1999; Archer, 1997; de Vries and Janssen, 2003; Smit, 2004; Steinreiber and Faber, 2001).

8.2 Mammalian Epoxide Hydrolases

Although the biological role and mechanism of mammalian epoxide hydrolases has been studied in detail, only a few groups studied their application in organic syntheses. Generally, mEH's show higher enantioselectivity than sEH's. For example, Bellucci et al. (1994a) observed remarkable differences in enantio- and regioselectivity between soluble and microsomal EH from rabbit liver in the resolution of phenyl substituted epoxides. In case of alkyl oxiranes, sEH showed only high enantioselectivity for a *t*-butyl derivative. Here attack of the nucleophile occured at the least hindered carbon (Bellucci et al., 1991) (Fig. 214). Stereoconvergent hydrolyses were reported for 3,4-epoxytetrahydropyran (Bellucci et al., 1981), a series of *cis*-disubstituted aliphatic epoxides (Chiappe et al., 1998) and β-*n*-alkyl-substituted styrene oxides (Bellucci et al., 1996) yielding the corresponding (*R*,*R*)-diols.

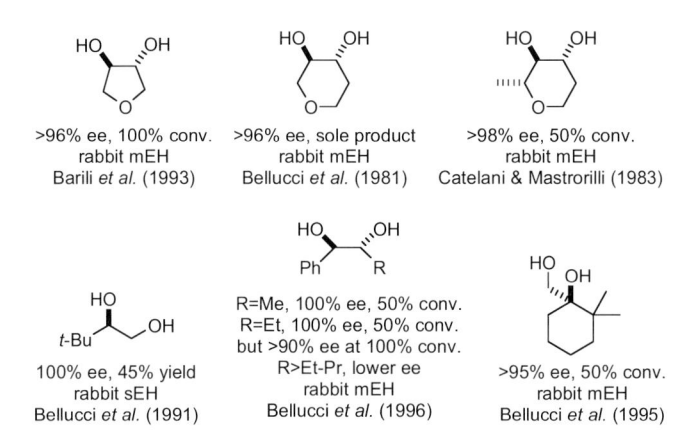

Fig. 214. Examples for the resolution of epoxides by mammalian EH's. Note that in the case of 3,4-epoxytetrahydropyran (top row, middle structure) hydrolysis occured with complete enantioconvergence.

In general, lipophilic aryl- or alkyl chains located close to the oxirane ring enhance the reaction rate in mEH-catalyzed reactions, whereas polar groups decrease the reaction rate, which explains why most (apolar) xenobiotics are converted to more hydrophilic ones prior to their elimination (Jerina and Daly, 1974). In addition, steric hindrance is an important factor, because terminal monosubstituted epoxides are hydrolyzed more rapidly than their *cis*-1,2-disubstituted analogs. *Trans*-1,2-di-, tri- or tetra-substituted compounds are usually not accepted as substrates (Oesch, 1973; 1974).

Mammalian EH's also efficiently resolved several *meso*-epoxides, but the optical purity of the resulting diols varied with the substitution pattern, ring size of the substrates or whether microsomal or soluble EH's were used (Fig. 215). More examples on the application of mammalian epoxide hydrolases can be found in an excellent review (Archer, 1997).

Fig. 215. Asymmetrization of *meso*-epoxides by mammalian EH's. Note that substitution pattern, ring size or origin of EH had a strong influence on enantioselectivity.

8.3 Microbial Epoxide Hydrolases

8.3.1 Bacterial Epoxide Hydrolases

In 1969, Allen reported the first application of an epoxide hydrolase isolated from a *Pseudomonas putida* strain catalyzing the synthesis of L- and *meso*-tartaric acid from an epoxide precursor (Allen and Jacoby, 1969). More than 20 years passed until researchers started to thoroughly investigate the application of epoxide hydrolases of microbial origin. Especially the groups of K. Faber (Austria) and R. Furstoss (France) have shown that a wide variety of microorganisms produce EH's, which often exhibit good to excellent enantioselectivity. In contrast to mammalian epoxide hydrolases, conventional fermentation techniques enable the production of EH's from bacteria or fungi in sufficient quantities for large-scale biotransformations. In addition, several epoxide hydrolases have been cloned and functionally expressed in common cultivation hosts raising access to sufficient quantities of these biocatalysts for preparative scale applications.

Initially, Faber's group discovered that the biocatalyst preparation SP409 produced by Novozymes also exhibited epoxide hydrolase activity. SP409 (from *Rhodococcus* sp.) was already widely used for the hydrolysis of nitriles (see Sect. 9), but was also able to hydrolyze a wide variety of mono- and 1,1-disubstituted epoxides with low to moderate enantioselectivity. In addition, also nucleophiles other than water (e.g., azide) were accepted (Fig. 216) (Mischitz and Faber, 1994).

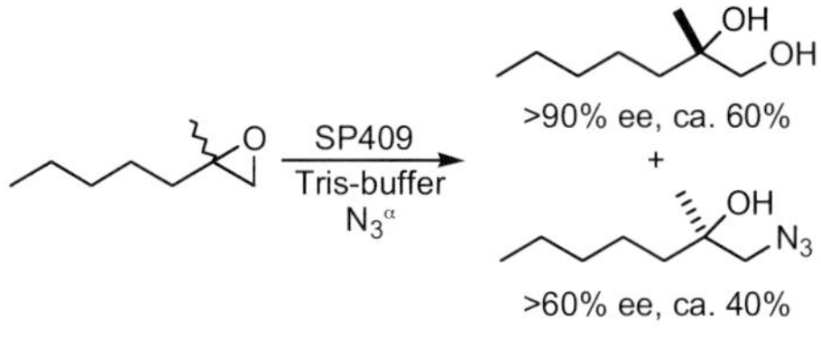

Fig. 216. *Rhodococcus* sp. SP409 also catalyzed epoxide ring-opening and accepts an azide as nucleophile (Mischitz and Faber, 1994).

Stimulated by this discovery, the Faber group screened a wide variety of organisms. Of 43 strains investigated, four bacterial and three fungal strains were found to show epoxide hydrolase activity (Mischitz et al., 1995b). *Rhodococcus* sp. NCIMB11216 and *Corynebacterium* sp. UPT9 preferentially hydrolyzed the (*R*)-epoxide of 1,2-epoxyoctane yielding the (*R*)-1,2-diol, however enantioselectivity was only E = 2.8 and E = 2.6 resp. Similar enantioselectivities were found for an epoxide hydrolase produced by *Corynebacterium* sp. C12 (Carter and Leak, 1995). Only the resolution of 1-methyl-1,2-epoxycyclohexane proceeded with good enantioselectivity, Fig. 217 (Archer et al., 1996). The purified enzyme is a multimer (probably tetrameric) with a subunit size of 32 kDa. The gene encoding *Corynebacterium* EH was isolated and sequenced. Sequence comparison revealed high similarity to mammalian and plant soluble EH's and to the EH cloned from *Agrobacterium radiobacter* AD1 (see below) (Misawa et al., 1998).

>99% ee, 30% yield
Corynebacterium sp. C12
Archer *et al.* (1996)

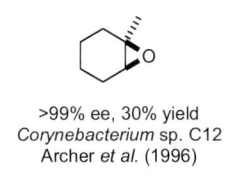

Fig. 217. Resolution of 1-methyl-1,2-epoxycyclohexane using *Corynebacterium* sp. C12 proceeded with good enantioselectivity.

Significantly enhanced enantioselectivities were achieved by hydrolyzing 1,1-disubstituted epoxides using the *Rhodococcus* sp. strain, e.g., for 2-methyl-1,2-epoxyheptane enantioselectivity was E > 100 and simultaneously the enantiopreference was inversed. Later work showed that *Rhodococcus* sp. epoxide hydrolases accept a wide range of substrates which are often converted with high enantioselectivity.

Selected examples for the resolution of racemic epoxides are summarized in Fig. 218. Note that in most cases only optical purity and yield of remaining epoxide are given in literature, which is probably due to difficult isolation of the more hydrophilic diol. In most cases epoxide hydrolases were not isolated. Instead, substrates were added to the culture medium, to centrifuged and washed whole cells or to lyophilized whole cells.

R=n-C_5H_{11}, E >100
Rhodococcus ruber DSM43338
R=CH_2-Ph, E >100
Rhodococcus sp. NCIMB11216
Osprian *et al.* (1997)
R=n-C_5H_{11}, E >100; R=n-C_7H_{15}, E >100;
R=n-C_9H_{19}, E>100
Rhodococcus sp. NCIMB11216
(lyophilized cells or purified enzyme)
Mischitz *et al.* (1995a)

E = 39, 20% ee (*R*)-epoxide,
94% ee (*S*)-diol at 18% conv.
Rhodococcus equi IFO3730
Kroutil *et al.* (1997d)

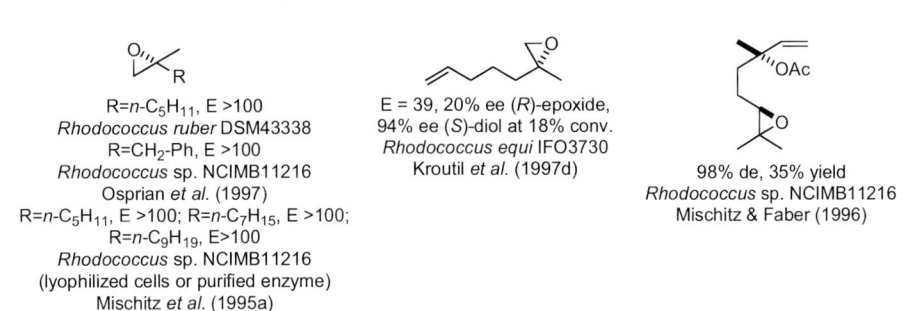

98% de, 35% yield
Rhodococcus sp. NCIMB11216
Mischitz & Faber (1996)

Fig. 218. Examples for the resolution of racemic epoxides using epoxide hydrolases from *Rhodococcus* sp. Here and in the following schemes, only the slow reaction enantiomer is shown.

In general, EH from *Rhodococcus* sp. NCIMB11216 shows high enantioselectivity for methyl-alkyl substituted oxiranes (Mischitz *et al.*, 1995a; Osprian *et al.*, 1997) but low E for ethyl-alkyl substituted oxiranes (Wandel *et al.*, 1995). The strain was also used in the synthesis of optically pure linalool oxide (Fig. 218, right structure) leading to a product with high diastereomeric excess (98% de) (Mischitz and Faber, 1996). Epoxide hydrolase activity present in lyophilized cells of *Rhodococcus equi* IFO3730 allowed the synthesis of a precursor of (*S*)-frontalin, a sex pheromone (94% ee for the produced diol), however conversion was only 18% (Kroutil *et al.*, 1997d).

Another very enantioselective EH is produced by the strain *Mycobacterium paraffinicium* NCIMB10420 (Faber *et al.*, 1996). Botes *et al.* (1998a) described an EH produced by the bacteria *Chryseomonas luteola* which converted straight-chain aliphatic epoxides with moderate to excellent enantioselectivity. Highest selectivity was found for 1,2-epoxyoctane, where > 98% ee for the remaining (*S*)-epoxide and 86% ee for the (*R*)-diol were determined. No enantioselectivity was observed for 2,2-disubstituted epoxides, benzyl glycidyl ether and 2-methyl-1,2-epoxyheptane (Botes *et al.*, 1998a).

8.3 Microbial Epoxide Hydrolases

Several strains from *Nocardia* sp. (H8, EH1, TB1) produce epoxide hydrolases. Especially *Nocardia* sp. EH1 shows high enantioselectivity at 50% conversion in the resolution of 2-methyl-1,2-epoxyheptane (Fig. 219, middle structure). However, introduction of a phenyl group into the side chain almost destroyed enantioselectivity (E = 5.6). The enzyme was purified to homogeneity via a four-step procedure. It is a monomer with a molecular weight of 34 kDa, a pH optimum of 8–9 and a temperature optimum of 35–40°C. The pure enzyme is much less stable than a whole cell preparation, but addition of Tween 80 or Triton-X-100 stabilizes it (Kroutil et al., 1998a). Immobilization on DEAE-cellulose doubled specific activity and allowed five repeated batch reactions, however enantioselectivity was slightly lowered (Kroutil et al., 1998b). Using *Nocardia* sp. EH1, the synthesis of naturally occuring (R)-(–)-mevalonolactone was achieved by deracemization of 10 g 2-benzyl-2-methyloxirane. The enzymatic reaction gave the corresponding (S)-diol, addition of catalytic amounts of sulfuric acid hydrolyzed the remaining (R)-epoxide under inversion of configuration, thus allowing the isolation of (S)-diol in an overall yield of 94% at 94% ee. Subsequent chemical steps afforded (R)-(–)-mevalonolactone in a total yield of 55% (Orru et al., 1997; 1998c).

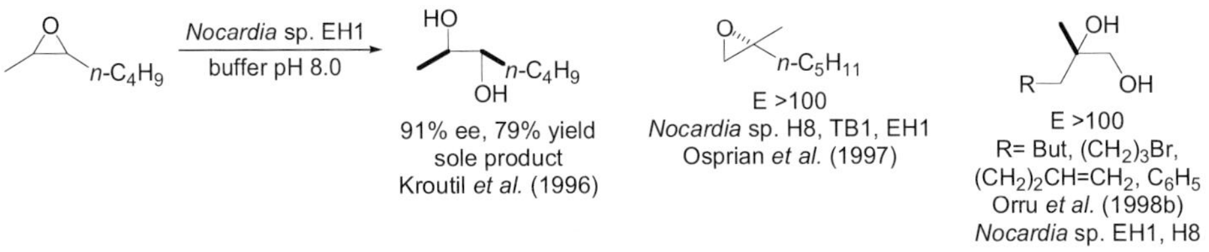

Fig. 219. Examples for the resolution of epoxides using epoxide hydrolases from *Nocardia* sp.

Interestingly, hydrolysis of (±)-*cis*-2,3-epoxyheptane with rehydrated lyophilized cells of *Nocardia* sp. EH1 proceeded in an enantioconvergent fashion and only (2R,3R)-heptane-2,3-diol was obtained as the sole product (Fig. 219) (Kroutil et al., 1996). Further examples for enantioconvergent reactions can be found in Kroutil et al. (1997c).

A range of epoxide hydrolases can now be recombinantly produced. The first example was the enzyme from *Agrobacterium radiobacter* AD1 overexpressed in *E. coli* (Rink et al., 1997) and accepts a broad range of styrene oxide derivatives and phenyl glycidyl ether which are converted with excellent enantioselectivity (Fig. 220) (Spelberg et al., 1998).

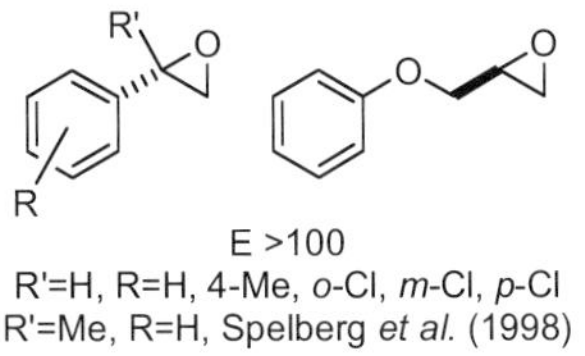

Fig. 220. Examples for the resolution of epoxides using recombinant EH from *Agrobacterium radiobacter* AD1.

Thorough modeling to avoid mass transfer limitation allowed the resolution of a 39 g/L solution of racemic styrene oxide in octane and (S)-styrene oxide (>95% ee) was isolated at 30% yield (Baldascini et al., 2001).

8.3.2 Fungal and Yeast Epoxide Hydrolases

The first report on the use of a fungal epoxide hydrolase appeared in 1972 by Suzuki and Marumo, who found that *Helminthosporum sativum* catalyzed the enantioselective hydrolysis of 10,11-epoxyfarnesol yielding the corresponding (*S*)-diol in 73% ee.

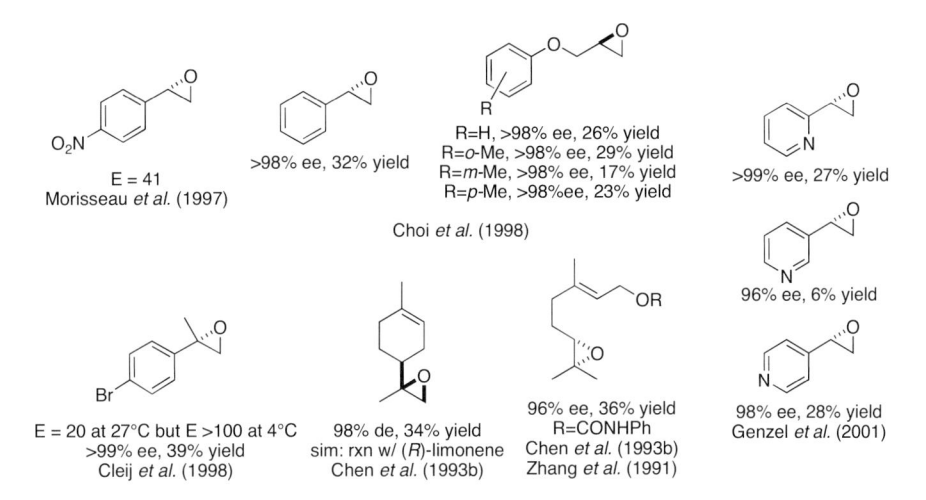

Fig. 221. Examples for the resolution of epoxides using epoxide hydrolase from *Aspergillus niger*.

The first preparative-scale epoxide hydrolysis was reported by the group of Furstoss, who discovered that the fungus *Aspergillus niger* enantioselectively converts geraniol-N-phenylcarbamate yielding the (*S*)-epoxide in high optical purity (96% ee, Fig. 221, bottom row, 3rd structure), which was further converted to Bower's compound, an analog of insect juvenile hormone (Zhang et al., 1991). Furstoss' group first used the mycelium, but due to several problems, they later used a lyophilized enzyme preparation obtained after concentration and desalting, resulting in a 7-fold higher specific activity. This preparation also showed increased tolerance to higher concentrations of substrate in the resolution of *p*-nitrostyrene oxide. The reaction proceeded with acceptable enantioselectivity (E = 41) and up to 20% DMSO could be added without significant loss of activity (Morisseau et al., 1997; Nellaiah et al., 1996). Lowering the temperature from 27°C to 4°C increased the enantioselectivity 13-fold to E = 260 in the resolution of *p*-bromo-α-methyl styrene oxide (Cleij et al., 1998). A few more substrates hydrolyzed with high enantioselectivity by *Aspergillus niger* are shown in Fig. 221.

Fig. 222. Kinetic resolution of an epoxide yields enantiomers, which both can be used in the synthesis of an azole antifungal compound.

More recently, Monfort et al. (2002) described a high yield synthesis of an azole antifungal compound using the recombinant version of EH from *Aspergillus niger* (expressed in *E. coli*). Resolution of the racemic substrate afforded not only the remaining chloro-substituted epoxide in excellent optical purity. The diol produced was then converted via the corresponding alcohol-substituted epoxide into the chloro derivative. Thus, both enantiomers can be used for the synthesis of the target compound overcoming the 50% yield limitation in standard kinetic resolutions (Fig. 222).

Directed evolution improved slightly the enantioselectivity of recombinant EH from *Aspergillus niger* towards glycidyl phenyl ether (from E = 4.6 to E = 10.8). Libraries obtained by error-prone PCR were first prescreened using an assay with 4-p-nitrobenzylpyridine. Active mutants were then assayed for improved enantioselectivity using a mixture of *pseudo*-enantiomers of phenyl glycidyl ether (one enantiomer deuterated at the aromatic ring) followed by an ESI-MS assay. The best variant contained three mutations, two of them were spatially far from the catalytically active center (Reetz et al., 2004).

Epoxide hydrolase activity was also discovered in *Beauveria sulfurescens* ATCC7159, which converts styrene oxide (Fig. 223) as well as two-membered ring epoxides with high enantioselectivity (Pedragosa-Moreau et al., 1996a).

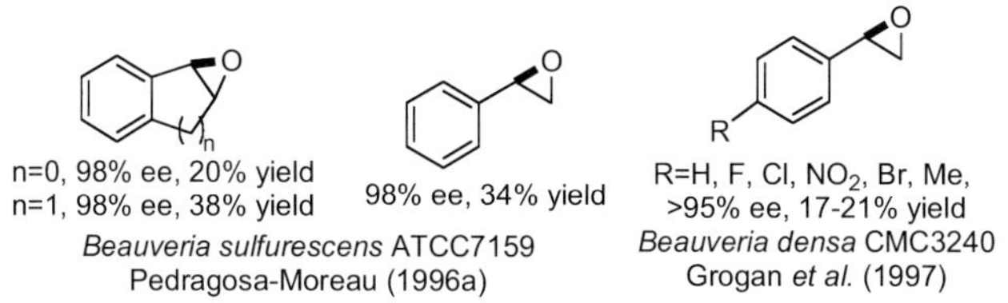

n=0, 98% ee, 20% yield
n=1, 98% ee, 38% yield   98% ee, 34% yield

R=H, F, Cl, NO_2, Br, Me,
>95% ee, 17-21% yield

Beauveria sulfurescens ATCC7159
Pedragosa-Moreau (1996a)

Beauveria densa CMC3240
Grogan et al. (1997)

Fig. 223. Examples for the resolution of epoxides using EH from *Beauveria* sp.

Interestingly, *Aspergillus niger* and *Beauveria sulfurescens* produced the (*R*)-diol in the hydrolysis of styrene oxide. Thus, the reaction catalyzed by *A. niger* proceeded with retention of configuration (via attack at C-2), whereas the hydrolysis with *B. sulfurescens* occurred with inversion of configuration (via attack at C-1, benzylic position).

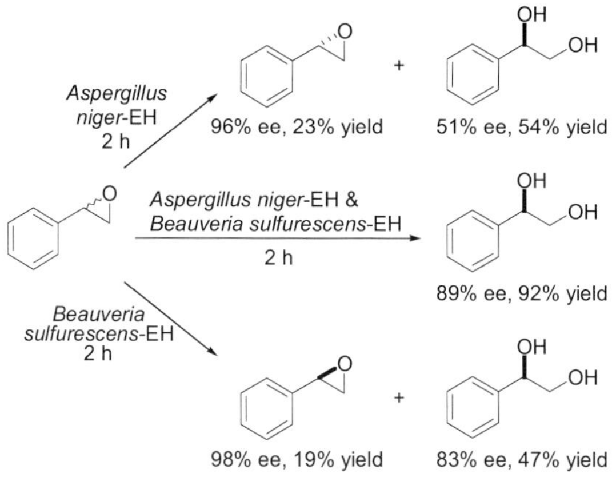

Aspergillus niger-EH 2 h
96% ee, 23% yield   51% ee, 54% yield

Aspergillus niger-EH & *Beauveria sulfurescens*-EH
2 h
89% ee, 92% yield

Beauveria sulfurescens-EH 2 h
98% ee, 19% yield   83% ee, 47% yield

Fig. 224. Resolution of styrene oxide using fungal epoxide hydrolases from *Aspergillus niger* or *Beauveria sulfurescens* or a mixture of both for an enantioconvergent synthesis (Pedragosa-Moreau et al., 1993; 1996c).

Employing a mixture of both organisms permitted the enantioconvergent synthesis of (*R*)-1-phenyl-1,2-dihydroxyethane in 92% yield and 89% ee (Fig. 224).

In search for an enantioselective EH capable to resolve indene oxide, a precursor to the side chain of HIV protease inhibitor MK639, researchers at Merck found that out of 80 fungal strains investigated, *Diploida gossipina* ATCC16391 and *Lasiodiploida theobromae* MF5215 showed excellent enantioselectivity yielding exclusively the desired (1*S*,2*R*)-enantiomer. Two other strains from *Gilmaniella humicola* MF5363 and from *Altenaria enius* MF4352 showed opposite enantiopreference. Preparative biotransformation using whole cells of *Diploida gossipina* ATCC16391 allowed isolation of optically pure (1*S*,2*R*)-indene oxide in 14% yield after 4 h reaction time (Fig. 225) (Zhang et al., 1995).

Fig. 225. Resolution of indene oxide catalyzed by EH from *Diploida gossipina* yields a HIV protease inhibitor precursor (Zhang et al., 1995).

Another useful and simple-to-grow strain is the yeast *Rhodotorula glutinis* which converts a wide range of aryl-, alkyl- and alicyclic epoxides with very high enantioselectivity. Best substrates are phenyl substituted epoxides (Weijers, 1997; Weijers et al., 1998) (Fig. 226). The same group discovered eight other yeast strains – out of 187 strains investigated – showing good to excellent enantioselectivity (E > 100) in the resolution of 1,2-epoxyoctane yielding the (*R*)-diol. Most stereoselective ones were from *Rhodotorula araucariae* CBS6031 and *Rhodosporidium toruloides* CBS0349 (Botes et al., 1998b).

An EH from *Rhodotorula glutinis* was recombinantly produced in the yeast *Pichia pastoris*. Kinetic resolution of styrene oxide (526 mM) in an epoxide–water two-liquid phase gave (*S*)-styrene oxide with 98% ee and 36% yield (Lee et al 2004). As an advantage, this approach does not require the use of organic solvents to dissolve the substrate.

rac-epoxides

>98% ee, 22% yield (1*R*,2*S*)   >98% ee, 14% yield (*R*)   R=H, >98% ee, 18% yield (*S*)
R=Me, >98% ee, 45% yield (1*R*,2*S*)

R	E
Me	<10
Et	20
n-Pr	>80
n-Bu	30
n-Pe	15

R	R'	R"	% ee	% yield	conf.
Me	H	H	>98	15	S
ClCH$_2$	H	H	>98	10	R
HOCH$_2$	H	H	no rxn	no rxn	no rxn
Et	H	H	>98	16	S
Me	Me	H	>98	47	2*R*,3*R*
Me	H	Me	90 (diol)	n.d.	2*R*,3*R*
Et	Me	H	>98	48	2*R*,3*R*
Et	H	Me	>98	48	2*R*,3*R*

>98% ee, 48% yield (1*S*,2*R*,4*S*)
>98% ee, 28% yield (1*S*,2*R*,4*R*)

Rhodotorula glutinis, Weijers (1997), Weijers *et al.* (1998)

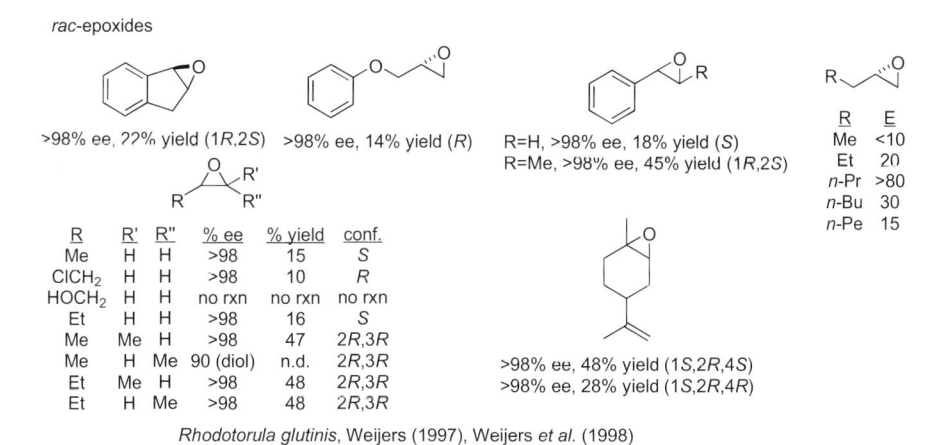

Fig. 226. Examples for the resolution of racemic epoxides using EH from *Rhodotorula glutinis*.

Meso epoxides

In contrast to the kinetic resolution of racemic epoxides, examples for the desymmetrization of *meso*-epoxides are scarce. Weijers (1997) resolved alicyclic epoxides using EH from *Rhodotorula glutinis* with high enantioselectivity furnishing the (R,R)-diols. Researchers at Diversa recently discovered more than 50 novel and diverse epoxide hydrolases in DNA libraries made from environmental libraries and demonstrated their synthetic utility for the desymmetrization of *cis*-stilbene and its derivatives (Zhao et al., 2004). For instance, BD8877 yielded (R,R)-1,2-diphenyl-1,2-ethanediol in 83% yield and 99% ee. Furthermore, they identified epoxide hydrolases (BD9126, BD10090), which also produced the corresponding (S,S)-diol with 99% ee and also accepted several other substrates (Fig. 227).

R	% ee
C_6H_5	>99
2-Cl-C_6H_5	98
2-F-C_6H_5	80
3-Cl-C_6H_5	98.5
4-Cl-C_6H_5	>99.5
2-pyridyl	99
3-pyridyl	97
4-pyridyl	98

to R,R-diol >98% ee

to R,R-diol 90% ee

Rhodotorula glutinis Weijers (1997)

to R,R-diol BD8877

to S,S-diol BD9126, 99% ee

to S,S-diol BD10090, 76% ee

Zhao et al. (2004)

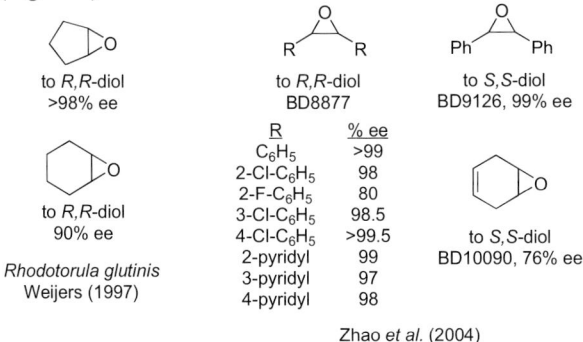

Fig. 227. Examples for the resolution of *meso*-epoxides.

9 Hydrolysis of Nitriles

9.1 Introduction

Although nitriles can also be hydrolyzed to the corresponding carboxylic acid by strong acid or base at high temperatures, nitrile hydrolyzing enzymes have the advantages that they require mild conditions (Sect. 9.2) and do not produce by-products. During hydrolysis of dinitriles, they are often regioselective for one nitrile group (Sect. 9.3) The nitrile hydrolyzing enzymes are often enantioselective and can be used for the synthesis of optically active substances (Sect. 9.4).

The enzymatic hydrolysis of nitriles follows two different pathways (Fig. 228). Nitrilases (EC 3.5.5.1) directly catalyze the conversion of a nitrile into the corresponding acid plus ammonia. (The amide may be a side product in a nitrilase-catalyzed reaction, but it is not an intermediate, see below.) In the other pathway, a nitrile hydratase (NHase, EC 4.2.1.84; a lyase) catalyzes the hydration of a nitrile to the amide, which may be converted to the carboxylic acid and ammonia by an amidase (EC 3.5.1.4).

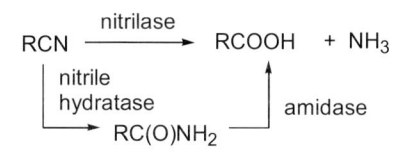

$$RCN \xrightarrow{\text{nitrilase}} RCOOH + NH_3$$
$$RCN \xrightarrow{\text{nitrile hydratase}} RC(O)NH_2 \xrightarrow{\text{amidase}} RCOOH + NH_3$$

Fig. 228. Hydrolysis of nitriles follows two different pathways.

The biosynthesis of the phytohormone indole-3-acetic acid from indole-3-acetonitrile can follow both pathways. *Alcaligenes faecalis* JM3 (Kobayashi et al., 1993) utilizes a nitrilase and in *Agrobacterium tumefaciens* and *Rhizobium* sp. a nitrile hydratase/amidase system was found for this reaction (Kobayashi et al., 1995).

Pure nitrilases and nitrile hydratases are usually unstable so most researchers use them in whole cell preparations. Furthermore, the nitrile hydrolyzing activity must be induced first. Common inducers are benzonitrile, isovaleronitrile, crotononitrile, acetonitrile but the inexpensive inducer urea also works. In addition, inducing with ibuprofen or ketoprofen nitriles yields enantioselective enzymes (Layh et al., 1997). After induction, preparative conversion is usually performed by adding the nitriles either during cultivation or to resting cells. Most commonly used strains are from *Rhodococcus* sp. and the most important ones are subspecies of *R. rhodochrous*. Probably due to the low stability of isolated enzymes, only one biocatalyst was produced commercially by Novozymes under the trade names SP361 or SP409. Both contained nitrile hydratase and amidase activity derived from *Rhodococcus* sp. CH5 and were immobilized on an ion-exchange resin. However, these materials are no longer available.

Nitrilases are cysteine hydrolases that act via an enzyme-bound imine intermediate, Fig. 229. Thiol reagents (e.g., 5,5'-dithiobis(2-nitrobenzoic acid)) inactivate nitrilases and replacement of a conserved cysteine residue by alanine in a nitrilase destroys its activity (Kobayashi and Shimizu, 1994). The nitrilase-catalyzed reaction of nitriles may

228   9.1   Introduction

form amides as a side product or even as a major product (for example, Osswald et al., 2002). This amide is not an intermediate nor a substrate for the nitrilase, but forms via an alternative breakdown of the tetrahedral intermediate, Fig. 229. The amount of side product amide varies with the structure of the nitrile and with the nitrilase.

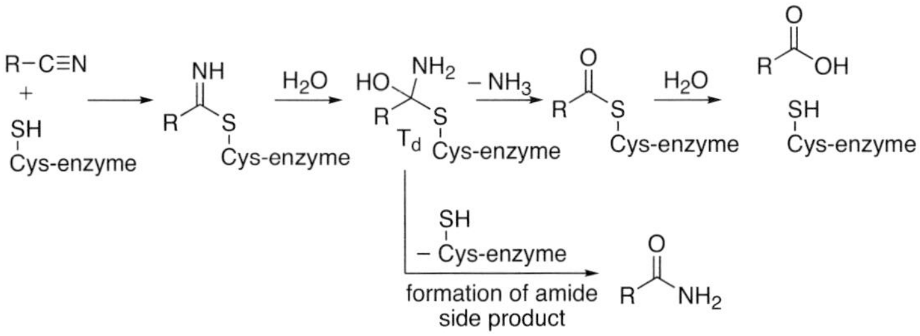

Fig. 229. Proposed mechanism for hydrolysis of nitriles by nitrilases and formation of the amide side product. The breakdown of the tetrahedral intermediate Td normally proceeds by loss of ammonia to form an acyl enzyme and then the carboxylic acid. An alternate breakdown of the tetrahedral intermediate by loss of the Cys-SH leads to the amide side product. Note that the amide is not an intermediate in the normal reaction path and only forms as a side product.

Nitrile hydratases (NHases) contain either Fe(III)- or Co(III)-ions in the active site (Nagasawa et al., 1991; Sugiura et al., 1987). NHase from *Pseudomonas chlororaphis* B23, *Brevibacterium* R312 and *Rhodococcus* sp. N-771 contain iron, the enzymes from *R. rhodochrous* J1 (Nagasawa et al., 1991) and from *Pseudomonas putida* NRRL 18668 (Payne et al., 1997) are cobalt-dependent (Shimizu et al., 1997). A NHase from *Agrobacterium tumefaciens* requires both metal ions (Kobayashi et al., 1995). In addition to their role in catalysis, the metal ions may enhance the folding or stabilize subunit association.

The biochemical properties and substrate specificities of some nitrilases are summarized in Tab. 21.

Ferric NHases

The NHases from *Rhodococcus* R312 (formerly known as *Brevibacterium* R312) and *P. chlororaphis* B23 are the first examples of non-heme iron enzymes containing a low spin Fe(III)-ion (Sugiura et al., 1987). Recent studies of the crystal structures and photoreactivity of a non-heme ferric nitrile hydrates from *Rhodococcus* sp. N-771 of significantly improved the understanding at the molecular level of the catalytic mechanism of the NHase.

In the active site, the iron ion of NHase is ligated by an apical cysteinate sulfur (*trans* to the inhibitor/substrate binding site), two peptide amides, and two cysteinate sulfurs in the equatorial plane (Fig. 230).

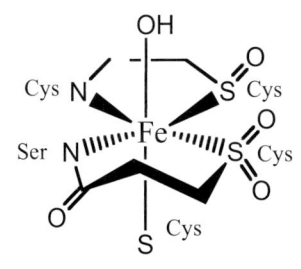

Fig. 230. Active site of cysteinate-ligated nitrile hydratase.

Tab. 21. Characteristics of some nitrilases from different microorganisms (Banerjee et al., 2002).

Microorganism	Properties of nitrilase					
	Mw [kDa]	No. sub-units (Mw [kDa])	Optimum		Substrate specificity	References
			pH	T[°C]		
Nocardia sp. NCIB11216	560	– (45)	8.0	-	aromatic nitriles	Harper (1977)
Nocardia sp. NCIB11215	560	12 (46)	8.0	40	aromatic and heterocyclic	Harper (1985)
R. rhodochrous J1	78	2 (41.5)	7.5	45	aliphatic and aromatic nitriles	Kobayashi et al. (1989)
R. rhodochrous K22	650	15-16 (41)	5.5	50	aliphatic nitriles	Kobayashi et al. (1990b)
Rhodococcus sp. ATCC39484	560	– (40)	7.5	30	aromatic nitriles	Stevenson et al. (1992)
Alcaligens faecalis JM3	260	6	7.5	45	arylacetonitriles	Nagasawa et al. (1990a)
Acinetobacter sp. AK226	580	–	8.0	50	aliphatic and aromatic nitriles	Yamamoto and Komatsu (1991)
Alcaligenes faecalis ATCC8750	32	–	7.5	40-45	arylacetonitriles	Yamamoto et al. (1992)
Comamonas testoteroni	–	Oligomer (38)	7.0	25	adiponitrile	Levy-Schil et al. (1995)
Psedomonas fluorescens DSM 7155	–	2 (40,38)	9.0	55	arylacetonitriles	Layh et al. (1998)
Bacillus pallidus Dac 521	600	– (41)	7.6	65	aromatic nitriles	Almatawah et al. (1999)
Klebsiella ozaenae	37	2 (–)	9.2	35	bromoxynil	Stalker et al. (1988)
Fusurium oxysporum	550	Oligomer (27)	-	4	aliphatic and aromatic nitriles	Goldhust and Bohak, (1989)

9.1 Introduction

The two cysteinate sulfurs in the equatorial plane appear to be oxygenated (*i.e.*, post-translationally modified) (Murakami et al., 2000; Nagashima et al., 1998; Nojiri et al., 1999; Tsujimura et al., 1997), however, one to a sulfenate (114Cys-SOH) and the other to a sulfinate (112Cys-SO$_2$H). It is not clear whether both of these oxidized sulfurs are required for catalytic activity (Nojiri et al., 1999).

Nitric oxide appears to regulate NHase activity in whole cells by binding to the Fe(III) ion (Endo et al., 1999). Storing whole-cell samples of NHase in the dark at low temperatures inactivates the NHase. Light exposure reactivates the enzyme (Honda et al., 1994; Nagamune et al., 1990; Noguchi et al., 1995; Odaka et al., 1997; Tsujimura et al., 1996). On the other hand, purified, isolated samples of the NHase in the dark do not inactivate it, implying something in the whole cells causes the inactivation. Several groups found that nitric oxide (NO•), produced by NO synthase, caused the dark inactivation (Noguchi et al., 1995; Odaka et al., 1997; Bonnet et al., 1997). Nitric oxide (NO) may bind to the metal and prevent binding of nitriles, or may replace the catalytically active Fe-OH species (vide infra). Light photolytically cleaves the Fe-NO, releasing NO• and reactivating the enzymes.

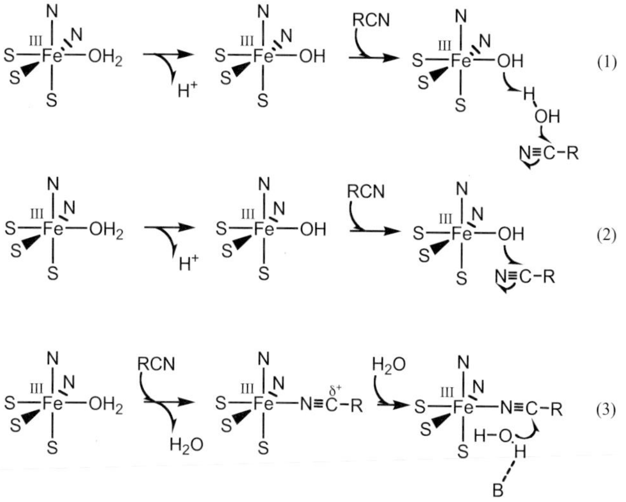

Fig. 231. Proposed reaction mechanisms for nitrile hydrolysis by the metalloenzymes nitrile hydratase (Huang et al., 1997; Kovacs, 2004).

Although the structure of NHase has solved, the mechanism of nitrile hydrolysis remains unsolved. Nitriles are extremely resistant to hydrolysis. The nitrile carbon atom is only slightly electrophilic, making it susceptible to attack by water (or hydroxide) only at elevated temperatures. Nitrile hydratase hydrolyses nitriles under mild conditions (pH 7.5, ambient temperature). Three distinct mechanisms of nitrile hydrolysis by NHase have been proposed (Fig. 231).

Mechanisms 1 and 2 (Fig. 231) involve a metal-bound hydroxide nucleophile that attacks free nitrile. Mechanism 1 also involves an intervening water molecule. Mechanism 3 coordinates the nitrile to the metal increasing the nitrile carbon's electrophilicity. The last two mechanisms involve metal-bound intermediates, while the first one does not (Kovacs, 2004).

Assays for nitrile hydratase activity frequently use either propionitrile or methacrylonitrile (MAN) as substrates. The Michaelis-Menten constant K_M and V_{max} values for hydrolysis of MAN vary from $K_M = 1.95$ (*Rhodococcus sp.* N-771) (Piersma et al., 2000) to 0.282 mM (*Rhodococcus sp.* YH3-3) (Kato et al., 1999) and from Vmax $= 1600$ (*Rhodococcus sp.* N-771) to 287 μmol-product·min^{-1}·mg-protein^{-1} (*Rhodococcus sp.* YH3-3), respectively. Reported K_M and V_{max} values for hydrolysis of propionitrile are 77.8 mM and 1280 μmol-product·min^{-1}·mg-protein^{-1}, respectively (Kato et al., 1999).

Cobalt NHase

Cobalt(III) replaces Fe(III) in NHases obtained from *Pseudomonas putida, Rhodococcus rhodochrous* J1, and *Pseudonocardia thermophila* (Brennan et al., 1996; Nagasawa et al., 1991; Nojiri et al., 2000; Payne et al., 1997). Like in the iron enzyme, the cobalt enzyme contains low-spin Co(III). In the presence of cobalt ions, the actinomycete *R. rhodochrous* J1 produces two NHases, depending on the inducer. Culturing in medium containing urea and cyclohexane carboxamide induces both, high and low molecular weight NHases (H- and L-NHases) (Yamada and Kobayashi, 1996). H-NHases (Nagasawa et al., 1991) favor aliphatic nitriles, whereas L-NHases favor aromatic nitriles.

Sequence homology in the active site regions of Fe-NHase and Co-NHase suggested similar active sites, which was confirmed by X-ray crystallography (Miyanaga et al., 2001). Replacing the Fe^{3+} ion of Fe-NHase from *Rhodococcus* sp. N-771 with Co^{3+} yields an enzyme with properties identical to those of Co-NHase from *Pseudomonas putida* NRRL 18668 (Nojiri et al., 2000). The difference in the metal cofactors may be due to a threonine in the VC(**T**/S)LCSC sequence in the active site of Co-NHases (Payne et al., 1997), as compared to a serine (VC(T/**S**)LCSC) for the Fe-NHases.

Co-NHases are more stable than Fe-NHases and favor aromatic nitriles, in contrast to the Fe-NHases, which favor aliphatic nitriles. This difference in substrate preference may be due to differences in conserved residues in the α-subunit. The Co(III)-containing enzyme has a tryptophan residue (Trp72) near the substrate binding site, while the Fe(III)-containing enzyme has tyrosine. This difference may account for the preference of cobalt NHases for aromatic rather than aliphatic nitriles (Miyanaga et al., 2001).

Nitrile hydratase activity is also possible without a metal ion. A single mutation (Gln19Glu) converted papain, a cysteine protease, into a nitrile hydratase. It converted MeOCO-PheAla-CN to the corresponding amide, which was then hydrolyzed to the carboxylic acid by the amidase activity of papain (Dufour et al., 1995). The mutation placed a proton donor in the oxyanion hole to protonate the nitrogen. This change increased the k_{cat}/K_M value for nitrile hydratase activity by at least 4×10^5 at pH 5, but the reaction is still too slow for practical applications. Although the wild-type shows some NHase activity, it requires 2-mercaptoethanol as an external nucleophile. The stereoselectivity of nitrile hydratase activity (WT or mutant) is lower than for ester hydrolysis.

Various NHase genes have been cloned and characterized (for reviews see: Banerjee et al., 2002; Kobayashi et al., 1992; Shimizu et al., 1997) and it was found that the amidase gene is closely located to the NHase gene supporting the theory that both enzymes are involved in the two-step degradation of nitriles to carboxylic acids. The properties and applications of nitrile-hydrolyzing enzymes are further summarized in a number of reviews (Banerjee et al., 2002; Kobayashi and Shimizu, 1994; 2000; Ohta, 1996; Sugai et al., 1997) or book chapters (Bunch, 1998; Drauz and Waldmann, 1995; Faber, 1997; Ingvorsen et al., 1988; Shimizu et al., 1997; Wieser and Nagasawa, 1999; Wyatt and Linton, 1988).

9.1 Introduction

The biochemical properties and substrate specificities of some nitrile hydratases are summarized in Tab. 22 (Banerjee et al., 2002).

Tab. 22. Characteristics of some nitrile hydratases from different microorganisms.

Microorganism	Properties of nitrile hydratase						
	Metal and PQQ	Mw [kDa]	No. sub-units (Mw [kDa])	Optimum		Substrate specificity	References
				pH	T[°C]		
Rhodococcus rhodochrous J1 L-NHase	Co (+)	101	18–20 α: 26 β: 29	8.8	40	aromatic nitriles	Kobayashi et al. (1991)
Rhodococcus rhodochrous J1 H-NHase	Co (+)	505	4–5 α: 26 β: 29	6.5	35-40	aliphatic nitriles	Komeda et al. (1996)
Rhodococcus sp. N-774	Fe (+)	70	2 α: 28.5 β: 29	7.7	35	aliphatic nitriles	Endo and Watanabe (1989)
Rhodococcus sp. N-771	Fe (+)	70	2 α: 27.5 β: 28	7.8	30	aliphatic nitriles	Yamada and Kobayashi (1996)
Rhodococcus equi A4	– (–)	60	2 α: 25 β: 25	–	–	aliphatic nitriles	Prepechalova et al. (2001)
Rhodococcus erythropolis	–	–	α: 23.5 β: 23	–	–	heterocyclic / aromatic nitriles	Duran et al. (1993)
Rhodococcus sp. YH3-3	Co (–)	130	2 α: 27.1 β: 34.5	–	–	aliphatic & aromatic nitriles	Kato et al. (1999)
Pseudonocardia thermophila JCM3095	Co (–)	–	– α: 29 β: 25		60	acrylonitrile	Yamaki et al. (1997)
Brevibacterium sp. R312	Fe (+)	85	3–4 α: 26 β:27.5	7.8	25	aliphatic nitriles	Nagasawa et al. (1986)
Pseudomonas chlororaphis B23	Fe (+)	100	4 α:25 β:25	7.5	20	aliphatic nitriles	Nagasawa et al. (1987)
Pseudomonas putida NRRL-18668	Co (+)	54, 95	– α: 23 β: 25	–	–	aliphatic nitriles	Payne et al. (1997)
Bacillus sp. RAPc8	Fe (–)	–	4 α: 28 β: 29	7.0	60	alkylnitrile	Pereira et al. (1998)
Agrobacterium tumefaciens d3	Fe (–)	69	4 (27)	7.0	40	2-arylpropionitrile	Bauer et al. (1994)
Myrothecium verrucaria	Zn (–)	170	6 (27.7)	7.7	55	cyanamide	Marier-Greiner et al. (1991)

9.2 Mild Conditions

Two applications based on nitrile hydratase in *Rhodococcus rhodochrous* originally isolated by Yamada's group (Nagasawa and Yamada, 1995) have been commercialized (Fig. 232). Nitto Chemical (Yokohama, Japan) produces the commodity chemical acrylamide from acrylonitrile on a > 30 000 metric tons per year scale. Initial production used strains from *Rhodococcus* sp. N-774 or *Pseudomonas chlororaphis* B23, but the current process uses the 10-fold more productive strain *Rhodococcus rhodochrous* J1. The productivity is > 7 000 g acrylamide per g cells at 99.97% conversion of acrylonitrile. Acrylic acid is barely detectable at the reaction temperature of 2–4°C. Lab-scale experiments with resting cells achieved up to 656 g acrylamide per liter, which solidifies the reaction mixture (Kobayashi et al., 1992).

Very recently, a NHase from *Comamonas testosteroni* 5-MGAM-4D was described for the kg-scale production of acrylamide. The encoding gene was expressed recombinantly in *E. coli* immobilized with alginate. This preparation showed good activity up to 35°C at 2 M acrylonitrile, but the best productivity (1035 g acrylamide / g dry cell weight) was still achieved at low temperatures (5°C) during 206 consecutive batch reactions (Mersinger et al., 2005).

One of the most important characteristics of the *R. rhodochrous* strain is its tolerance toward up to 50% acrylamide. Induction of NHase activity with urea gives more than 50% of the total soluble protein as NHase. Added cobalt ions are essential for active NHase. Besides acrylamide, also a wide range of other amides can be produced, e.g., acetamide (150 g/L), isobutyramide (100 g/L), methacrylamide (200 g/L), propionamide (560 g/L) and crotonamide (200 g/L) (Kobayashi et al., 1992).

Switzerland's Lonza commercialized the conversion of 3-cyanopyridine to nicotinamide (a vitamin in animal feed supplementation) by *R. rhodochrous* J1 (Nagasawa et al., 1988). Production is 3 000 metric tons per year scale in Lonza's plant in Guangzhou, China. In contrast to the chemical process, the nitrile hydratase process does not form the by-product nicotinic acid.

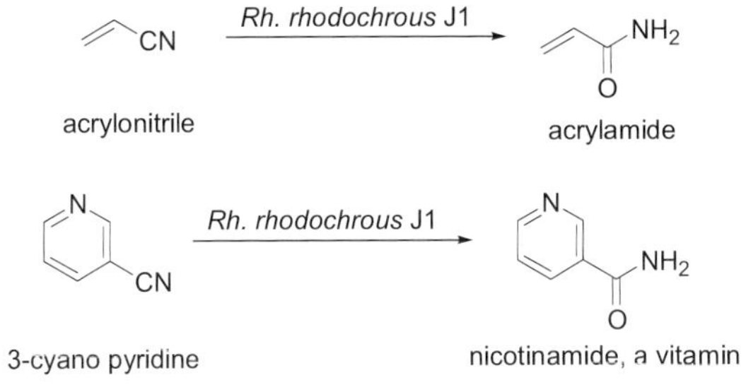

Fig. 232. Commercial production of acrylamide and nicotinamide using resting cells of *Rhodococcus rhodochrous* J1.

(*S*)-Pipecolic acid is a building block for a number of pharmaceuticals, such as Incel from Vertex for the treatment of cancer multi-drug resistance and the local anaesthetic Naropin (ropivacaine) from Astra-Zeneca. The Lonza process for the (*S*)-pipecolic acid production efficiently combines both chemistry and biotechnology (Fig. 233).

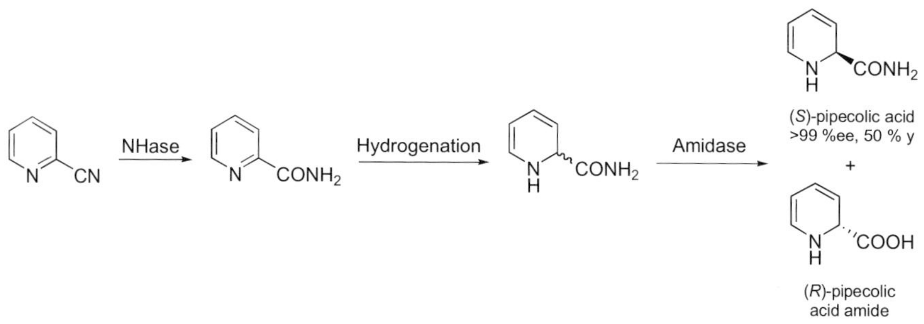

234   9.2 Mild Conditions

Fig. 233. Production of (S)-pipecolic acid by Lonza AG, Switzerland (Shaw et al., 2003).

In the first step whole cells containing nitrile hydratase convert 2-cyanopyridine to pyridine-2-carboxamide, which is then chemically hydrogenated to (R,S)-piperidine-2-carboxamide. This compound is the substrate for an amidase-catalysed enantiomer resolution with whole cells of *Pseudomonas fluorescens* DSM9924. Finally, the product is isolated by precipitation at acid pH.

One of the most attractive features of whole-cell nitrile and amide converting enzymes is their ability to carry out versatile stereo-, regio- and chemoselective reactions. However, examples of undesirable enzyme activities in whole-cell biocatalysts, *e.g.* in chemoselective hydrolysis of labile nitriles such as methyl-3-cyanobenzoate, methyl-4-cyanobenzoate, methoxy derivatives of benzonitrile, (R,S)-3-benzyloxy-4-cyanobutanoate and methyl (R,S)-3-benzyloxy-4-cyanobutanoate were reported (Klempier et al., 1996; Martinkova et al., 1998; 2001). Polarity of some compounds was important to their susceptibility to esterase attack. Cyanobenzoates, but not the corresponding amidobenzoates or hemiesters, were substrates of an esterase from *Rhodococcus equi* A4. Therefore, high activity of the nitrile hydratase, which transformed cyanobenzoates into amidobenzoates determined high chemoselectivity. The purified nitrile hydratase from this strain was the biocatalyst for the chemoselective hydration of cyanobenzoates (Martinkova et al., 1998) and cyanobutanoates (Martinkova et al., 2001).

Rhodococcus butanica (~*R. rhodochrous*) ATCC 21197 also catalyzed the mild hydrolysis of a variety of substituted benzonitriles, arylacetonitriles and α- or β-naphthylnitriles. Selected examples are shown in Fig. 234 (Kakeya et al., 1991b).

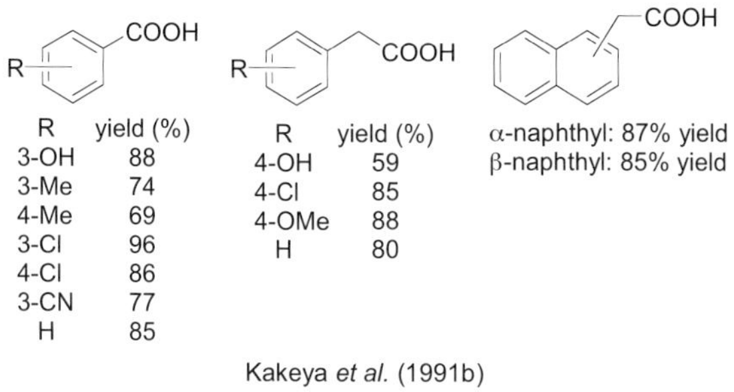

R	yield (%)	R	yield (%)	
3-OH	88	4-OH	59	α-naphthyl: 87% yield
3-Me	74	4-Cl	85	β-naphthyl: 85% yield
4-Me	69	4-OMe	88	
3-Cl	96	H	80	
4-Cl	86			
3-CN	77			
H	85			

Kakeya et al. (1991b)

Fig. 234. Acids obtained by mild hydrolysis of substituted aromatic nitriles using cells of *Rhodococcus butanica* ATCC 21197.

Griengl's group used the commercial biocatalyst SP409 (Novozymes, no longer available) for the mild hydrolysis of a wide range of aliphatic or alicyclic nitriles either bearing other hydrolyzable groups or functionalities prone to side reactions such as elimination or aldol-type reactions (de Raadt et al., 1992; Klempier et al., 1991). In most cases high yields of acids could be achieved, selected examples are shown in Fig. 235. Note that no enantioselectivity was observed, when racemates were used and in a few cases also the corresponding amides were formed. α-Branched and crystalline substrates did not react.

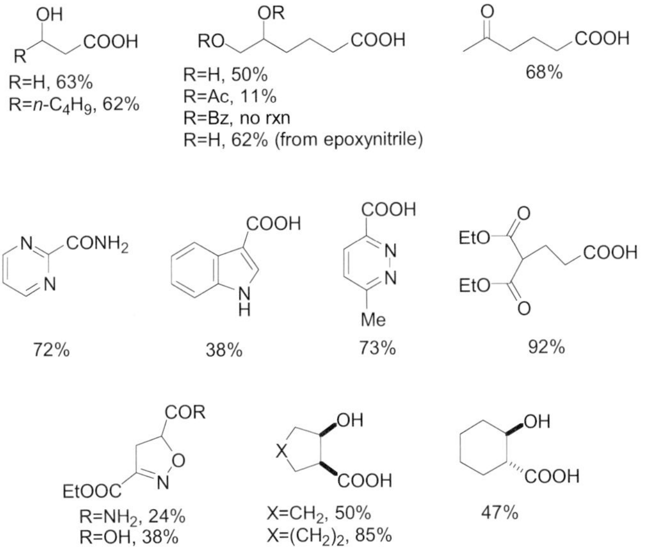

Fig. 235. Selected examples of carboxylic acids obtained by SP409-catalyzed hydrolysis of nitriles. Note that a 1,2-diol-carboxylic acid (R = H) was obtained from the epoxynitrile (top row, middle structure) and in one case the amide could be isolated in substantial amounts (last row, first structure) (de Raadt et al., 1992; Klempier et al., 1991).

3-Hydroxy alkanoic acids such as 3-hydroxyvaleric acid, could be obtained in up to quantitative yields from the corresponding nitriles using a combination of NHase and amidase activity present in two *Comamonas testosteroni* and one *Dietzia* sp. strains (Hann et al., 2003). Best results were achieved with the Ca-alginate immobilized *C. testosteroni* 5-MGAM-4D strain with a productivity of 670 g product per g dry cell weight after 85 consecutive batch reactions. Interestingly, no biocatalyst exhibited any enantioselectivity in this reaction.

Fewer examples can be found for the use of nitrilases, presumably due to insufficient stability. A nitrilase from *R. rhodochrous* J1 produces nicotinic acid (172 g/L), p-aminobenzoic acid (110 g/L), acrylic acid (390 g/L) and metacrylic acid (260 g/L) (Mathew et al., 1988; Nagasawa et al., 1990b).

Treatment of polyacrylonitrile with nitrilase (Gübitz and Cavaco-Paulo (2003) yields surface carboxyl groups while treatment with nitrile hydratase (Battistel et al., 2001) or ntirile hydratase/amidase from *Rhodococcus rhodochrous* (Tauber et al., 2000) yields surface carboxyamide groups. Either of these changes increased the polarity of the surface and increased the absorption of dyes up to two-fold. Chemical treatment of polyacrylonitrile fibers with acid or alkali to hydrolyze the nitrile groups irreversibly yellowed the fibers.

9.3 Regioselective Reactions of Dinitriles

A wide variety of dinitriles can be hydrolyzed with moderate to excellent regioselectivity to the corresponding monocarboxylic acids (Fig. 236). When using NHase from *Rhodococcus* sp. (SP409) monoamides were the major products. Note that either poor or no regioselectivity was observed in the hydrolysis of non-fluorinated analogs of the substrates shown in Fig. 236 (last row) (Crosby et al., 1994).

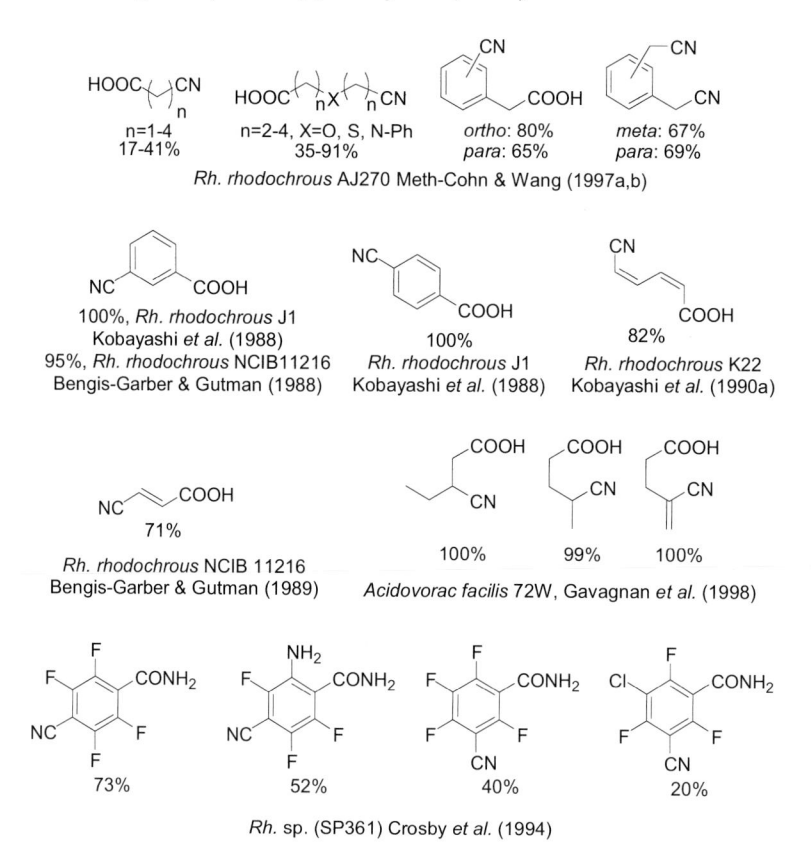

Fig. 236. Examples of the regioselectivity of NHases and nitrilases.

9.4 Enantioselective Reactions

Nitrilases and amidases are often enantioselective, but only recently have researchers showed that nitrile hydratases can also be enantioselective. Resting cells from *Rhodococcus erythropolis* MP50, *Agrobacterium tumefaciens* d3 or *Pseudomonas putida* NRRL18668 formed optically active amides from racemic nitriles, even after addition of an amidase inhibitor such as diethyl phosphoramidate (Bauer et al., 1994; 1998; Layh et al., 1994). In the nitrile hydratase/amidase system, the two enzymes sometimes show opposite stereopreference (Ohta, 1996).

Many kinetic resolutions involve precursors of non-steroidal anti-inflammatory drugs (NSAID) such as ketoprofen, ibuprofen or naproxen (Fig. 237). Enantioselectivities varied with the nitrile-hydrolyzing strain used and whether the substrate was a nitrile or an amide. Both enzymes in the NHase/amidase system contributed to enantioselectivity. For instance, in the resolution of naproxen (Fig. 237, top row, right structure) hydrolysis of the nitrile using *R. butanica* proceeded with E ~ 20-40, but the hydrolysis of the amide with only E ~ 10. Reactions in biphasic systems (phosphate buffer/hexane) or immobilization of cells gave no improvement. Increasing the reaction temperature increased the enantioselectivity moderately (Bauer et al., 1998; Fallon et al., 1997).

Rhodococcus sp. C3II
Hydr. of amide to R=OH: E >100
Hydr. of nitrile to R=NH$_2$: E >100
Rhodococcus erythropolis MP50
Hydr. of amide to R=OH: E >100
Hydr. of nitrile to R=NH$_2$: slow rxn
Layh et al. (1995)

Rhodococcus butanica
Hydr. of nitrile: E ~ 20-40
Hydr. of amide: E ~ 10
Rhodococcus sp. C3II
Hydr. of nitrile: E >100
Hydr. of amide: E >100
Effenberger & Böhme (1994)

Rhodococcus equi A4
R=OMe: E ~ 40; R=Cl: E >50
Martínková et al. (1996)
Acinetobacter sp. AK226
R=*i*-Bu: E ~ 25
Yamamoto et al. (1990a)
Rhodococcus butanica ATCC21197
R=*i*-Bu, Cl, OMe, R'=Me: E >100
Kakeya et al. (1991a,b)

Pseudomonas putida NRRL 18668
Hydrolysis of nitrile: E >50
Fallon et al. (1997)

Agrobacterium tumefaciens d3
(purified NHase)
R=H, R'=Me: E >100
R=H, R'=Et: E ~ 40
R=Cl, R'=Me: E ~ 50
Bauer et al. (1998)

Rhodococcus butanica ATCC21197
R=*i*-Bu, Cl, OMe, R'=Me: E >100
Kakeya et al. (1991a,b)

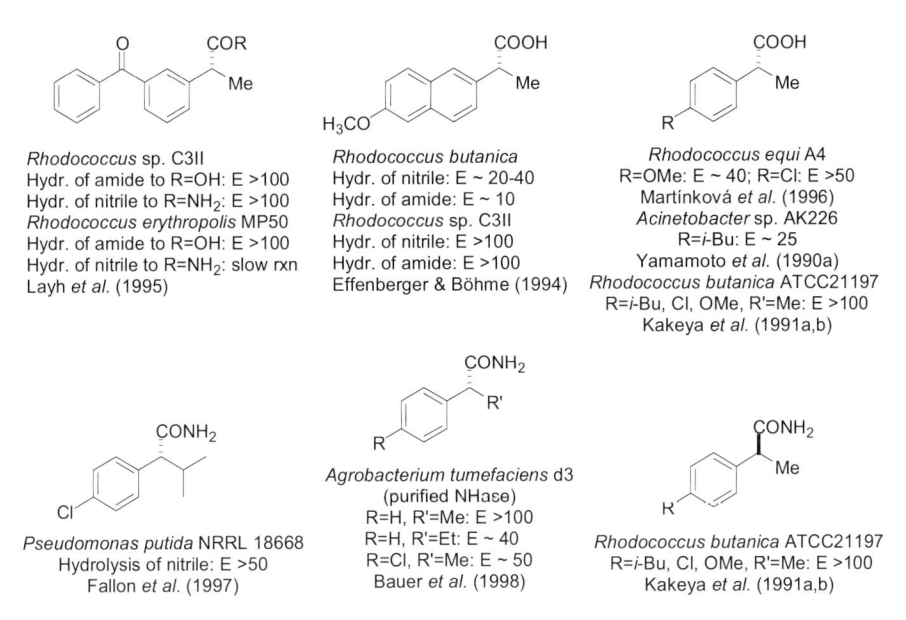

Fig. 237. Examples for the synthesis of non-steroidal anti-inflammatory drugs by hydrolysis of nitrile or amide precursors. Purified NHase from *Agrobacterium tumefaciens* d3 showed opposite enantiopreference compared to *Rhodococcus butanica* ATCC21197 (last row).

The situation becomes more confusing, because some strains produce nitrilases, NHases and amidases, while others produce only nitrilases. In some cases the amidase is not active and the conversion of nitriles stops at the carboxylic amides. For example, *Rhodococcus* sp. C3II (Effenberger and Böhme, 1994) and *Rhodococcus equi* A4 (Martínková et al., 1996) contain no nitrilase, but *R. butanica* ATCC 21197 (renamed *R.*

9.4 Enantioselective Reactions

rhodochrous (Yokoyama et al., 1993)) exhibits nitrilase, nitrile hydratase and amidase activities. Here, the nitrilase preferentially hydrolyzes the (*S*)-nitrile to give the (*S*)-acid. The stepwise conversion of racemic nitriles gave enriched (*R*)-amide by the modest enantioselective nitrile hydratase. Faster hydrolysis of the minor (*S*)-amides by the amidase allowed for the isolation of the (*R*)-amide in high optical purity (Effenberger and Böhme, 1994; Kakeya et al., 1991b). The strain *Acinetobacter* sp. AK226 produces only a nitrilase, because no amide could be detected as intermediate product (Yamamoto and Komatsu, 1991; Yamamoto et al., 1990).

The first nitrile hydratase from a gram-negative organism was found in *Pseudomonas putida* NRRL 18668. The NHase yields (*S*)-amides, but the stereoselectivity primarily resides in the amidase (Fallon et al., 1997). A cobalt-containing nitrile hydratase with a very broad substrate spectrum is produced by *Agrobacterium tumefaciens* d3. To ensure that no amidase activity is present, the NHase was purified (Bauer et al., 1998; Stolz et al., 1998).

Besides NSAID-precursors, only a few other nitriles were subjected to kinetic resolution (Fig. 238). Using the strain *Rhodococcus rhodochrous* NCIMB11216 (for properties see: Hoyle et al., 1998) in the hydrolysis of 2-substituted aliphatic nitriles and arylaliphatic nitriles, high enantioselectivity was found only for 2-methylhexanitrile (Gradley et al., 1994; Gradley and Knowles, 1994).

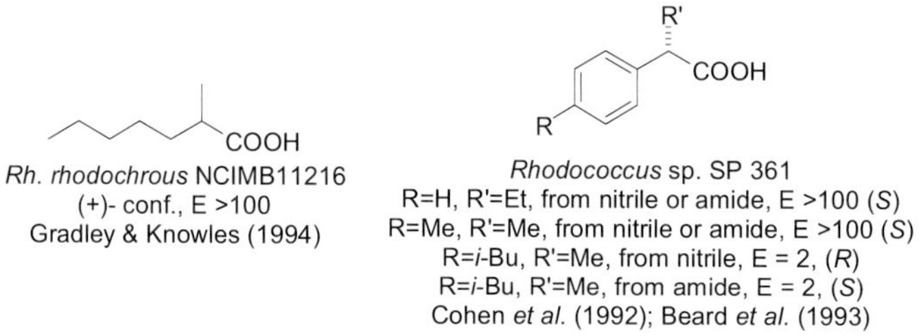

Fig. 238. Other substrates resolved by nitrile-hydrolyzing enzymes.

Hydrolysis of arylacetonitriles using the Novozymes catalyst SP361 was (*R*)-selective in the amide generating step, but the further hydrolysis to the acid is (*S*)-selective. Reactions with the corresponding racemic amide confirmed, that only the amidase contributed to high enantioselectivity. Moreover, hydrolysis of 4-*i*-butyl nitrile gave the (*R*)-acid, but the reaction of the amide produced the (*S*)-acid (but at very low E) (Beard et al., 1993; Cohen et al., 1992). Other recent examples are shown in Fig. 239.

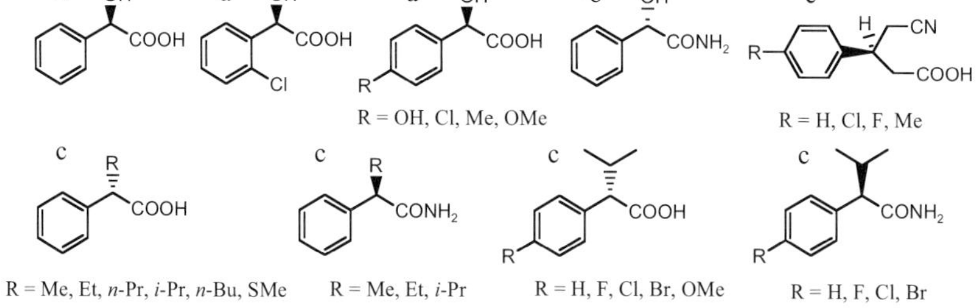

Fig. 239. Other substrates resolved by nitrile-hydrolyzing enzymes (**a** *Gordona terrae* MA-1, **b** *Rhodococcus* sp. HT40-6, **c** *Rhodococcus* sp. AJ270 (Wang et al., 2000; 2001).

The strain *Alcaligenes faecalis* ATCC8750 efficiently produces (*R*)-(−)-mandelic acid (100% ee) from mandelonitrile (Yamamoto et al., 1991). The remaining (*S*)-mandelonitrile racemized creating a dynamic kinetic resolution with a 91% yield of mandelic acid. The rapid racemization is likely to proceed via an equilibrium with benzaldehyde and HCN because in the presence of these substrates also mandelic acid was formed (Fig. 240). After partial purification, nitrilase and amidase were investigated separately and it was found that the nitrilase is highly enantioselective and active, but the amidase is not (24% ee for mandelic acid in the hydration of the amide). *A. faecalis* contains a nitrilase and an amidase, but no NHase, which otherwise could have contributed to the reaction. Hydrolysis of *O*-acetylmandelonitrile by resting cells of different *Pseudomonas* strains proceeded with lower enantioselectivity (Layh et al., 1992).

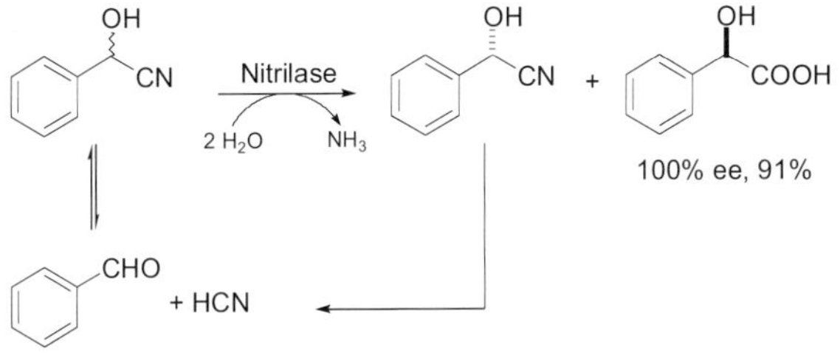

Fig. 240. Dynamic kinetic resolution of (*R*)-(−)-mandelic acid by a nitrilase in *A. faecalis* resting cells. Racemization of mandelonitrile occurs via an equilibrium with benzaldehyde and HCN (Yamamoto et al., 1991; 1992).

An impressive example of biodiversity from the metagenome (see Sect. 4.1) is the discovery of >130 novel nitrilases from more than 600 biotope-specific environmental DNA libraries (Robertson and Steer, 2004), compared to less than 20 nitrilases isolated by classical cultivation methods. Screening these novel nitrilases revealed 27 nitrilase that formed (*S*)-mandelic acid in >90% ee in a dynamic kinetic resolution (see Fig. 240) and one nitrilase that afforded (*R*)-mandelic acid in 86% yield and 98% ee (DeSantis et al., 2002). Also, aryllactic acid derivatives were accepted at high conversion and selectivity (Tab. 23).

Tab. 23. Preparation of Optically Active Aryllactic Acid Derivatives Using a Nitrilase Under Dynamic Kinetic Resolution Conditions (modified from DeSantis et al., 2002).

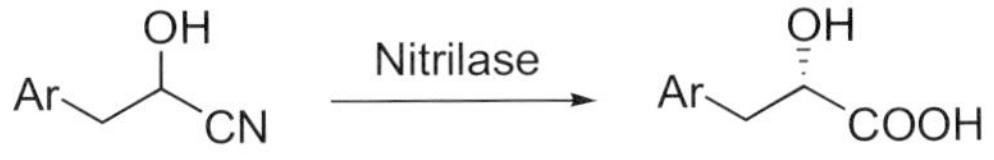

Ar	Relative activity [%]	Optical purity [% ee]
C_6H_5	25	96
2-Me-C_6H_5	160	95
2-Br-C_6H_5	121	95
3-F-C_6H_5	22	99
1-naphthyl	64	96
2-pyridyl	10.5	99

A few groups investigated the hydrolysis of prochiral dinitriles (Fig. 241). Only the use of Bn- or Bz-proctected 3-hydroxyglutaronitriles gave acceptable enantiomeric excess for the monocarboxylic acid. Using SP361, a *pro*-(S)-selective NHase transforms the dinitrile followed by a fast, non-selective amidase-catalyzed hydrolysis to the carboxylic acid (Crosby et al., 1992). In contrast, hydrolysis of a disubstituted malononitrile with R. *butanica* involved a fast non-selective nitrilhydratase forming the prochiral diamide followed by a slow *pro*-(R)-selective hydrolysis catalyzed by an amidase (Yokoyama et al., 1993). The product 2-carbamoyl-2-methylhexanoic acid formed in high purity and yield.

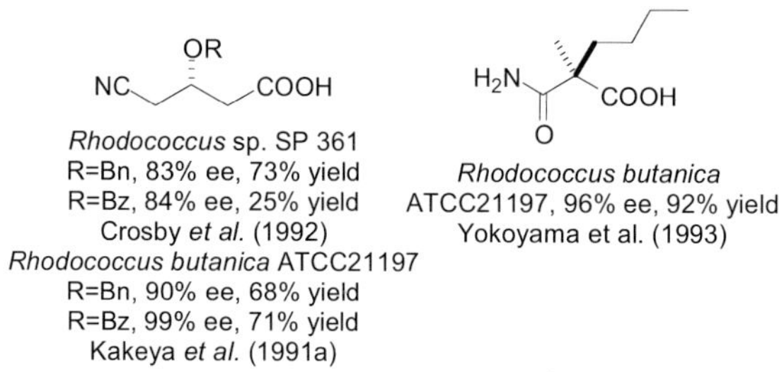

Fig. 241. Resolution of prochiral dinitriles proceeds with good enantioselectivity.

Several of the Diversa nitrilases showed high conversion (>95%) and selectivity (>90% ee) in the desymmetrization of 3-hydroxyglutaronitriles without the need for protecting groups (Fig. 242). The best enzyme gave 98% yield and 95% ee for the (R)-product (DeSantis et al., 2002) and 22 enzymes gave the opposite enantiomer with 90-98% ee. The most effective (R)-nitrilase was later optimized by directed evolution to withstand high substrate concentrations while maintaining high enantioselectivity. The best variant obtained by 'gene-site saturation mutagenesis' technique contained a single mutation (Ala190His) and allowed the production of the (R)-acid at 3 M substrate concentration with 96% y at 98.5% ee (DeSantis et al., 2003). Detailed phylogenetic analysis confirmed that these novel nitrilases identified from the environmental libraries were extremely diverse (Robertson et al., 2004).

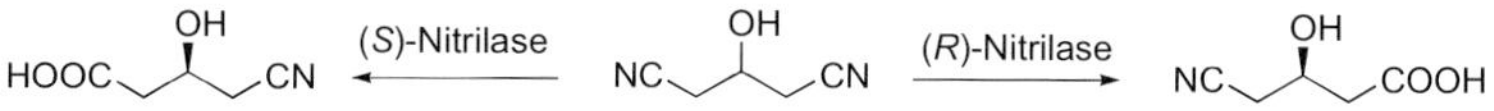

Fig. 242. Desymmetrization of 3-hydroxyglutaronitrile using novel and improved nitrilases (DeSantis et al., 2002; 2003).

10 Other Hydrolases

10.1 Glycosidases

Glycosidases (EC 3.2) are cofactor-independent enzymes catalyzing the hydrolysis of glycosidic bonds. Glycosidases also catalyze the reverse reaction, formation of a glycosidic link, and this ability is their most important synthetic application. Major advantages of glycosidase-catalyzed glycosyl transfers are that there is no – or minimal – need for protection and that the stereochemistry at the newly formed anomeric center can be controlled by the use of either α- or β-glycosidases.

Retaining β-glycosidases normally catalyze hydrolysis of β-glycosidic links, but also catalyze glycoside exchange in low-water conditions. The reaction involves a starting β-glycoside (sugar–OR) reacting with an incoming nucleophile (HOR'), Fig. 243. The incoming nucleophile is water for hydrolysis, or a second glycoside for glycoside exchange. Retaining glycosidases use a double displacement mechanism with a catalytic acid/base and a catalytic nucleophile. The initial step forms an α-linked covalent intermediate by attack of the catalytic nucleophile on the starting β-glycoside. The catalytic acid assists this step by protonating the leaving group. The second step releases this covalent intermediate by the catalytic base-assisted attack of incoming nucleophile - water or a new glycoside.

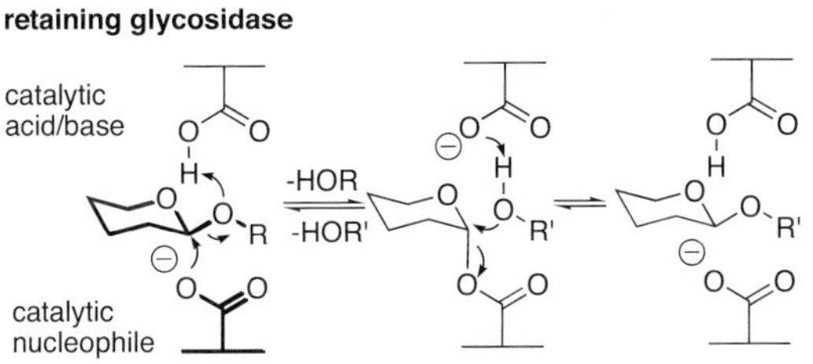

Fig. 243. Glycoside exchange using retaining β-glycosidases involves a double displacement. A glycosyl donor (β-sugar–OR) forms an α-linked glycosyl enzyme, which then reacts with an incoming nucleophile or acceptor (HOR') to make a new β-glycosidic link (sugar–OR').

Most glycosidases are specific for both the glycosyl moiety and the nature of the glycosidic linkage, e.g., α-D-galactosidase only accepts derivatives of α-D-galactose (Fig. 246). Site-directed mutagenesis of glycosyl-binding region can broaden the glycosyl donor specificity of glycosidases (Hancock et al., 2005). On the other hand, most glycosidases show low specificity towards the leaving group. This is usually a sugar, but can also an activated group such as *p*-nitrophenyl or fluoride. The regioselectivity towards the nucleophile is also low. For this reason, a synthesis involving a sugar nucleophile may give a mixture where different hydroxyl groups reacted.

Glycosides can be synthesized by either thermodynamic or kinetic control – similar to the formation of peptide bonds (Sect. 6.2.2) – of the reaction (Fig. 244). For thermodynamic control, researchers shift the reaction equilibrium toward synthesis by increasing the concentrations of donor and nucleophile. Since the free energy of hydrolysis of a glycosyl link (–3.8 kcal/mol) is large, the required concentrations are quite high usually yielding a thick syrup. The concentration of water cannot be reduced below 10% because glycosidases require >10% water to remain active. Yields of glycoside from the thermodynamic approach are usually below 15%. By comparison, the free energy of hydrolysis of a peptide is smaller (–2.2 kcal/mol) so the thermodynamic approach is more successful.

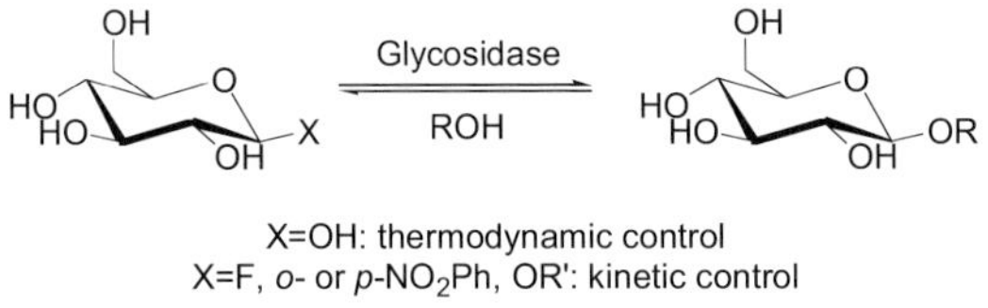

X=OH: thermodynamic control
X=F, o- or p-NO$_2$Ph, OR': kinetic control

Fig. 244. Glycoside synthesis using glycosidases. For thermodynamic control, researchers shift the equilibrium toward synthesis by increasing the concentrations of donor or nucleophile. For kinetic control researchers use an activated donor and stop the reaction before hydrolysis of product takes place.

A combination of ß-glucosidase-catalyzed synthesis of *n*-butyl glucose under thermodynamic control followed by lipase-catalyzed acylation using 4-phenylbutyric acid gave a moderate yield of 21% (Fig. 245). However, with *n*-octanol yields of *n*-octyl glucose were < 10% (Otto et al., 1998a). Yields of *n*-alkyl-glycosides are also affected by water activity (Chahid et al., 1992).

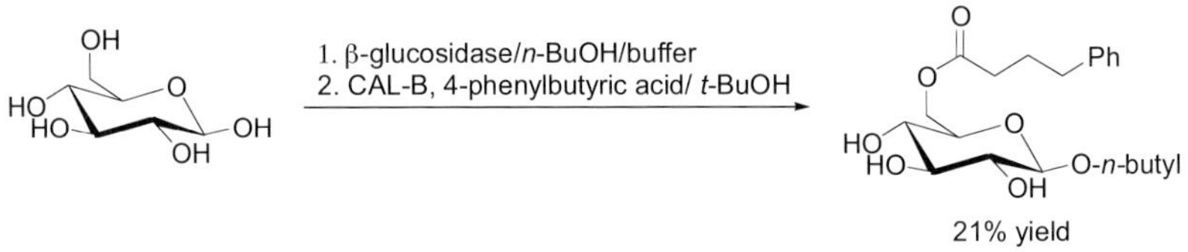

Fig. 245. Synthesis of 1-*n*-butyl-6-(4'-phenylbutyryl)-β-D-glucose in a coupled β-glucosidase and lipase reaction.

The alternative kinetically controlled synthesis starts from an activated glycoside (e.g., fluoride, *o*- or *p*-nitrophenyl derivative, azides), which reacts with the nucleophile (Fig. 246). This transglycosidation gives acceptable yields but requires activated donors. Two competing reactions are hydrolysis of the glycosyl donor and hydrolysis of the product glycoside.

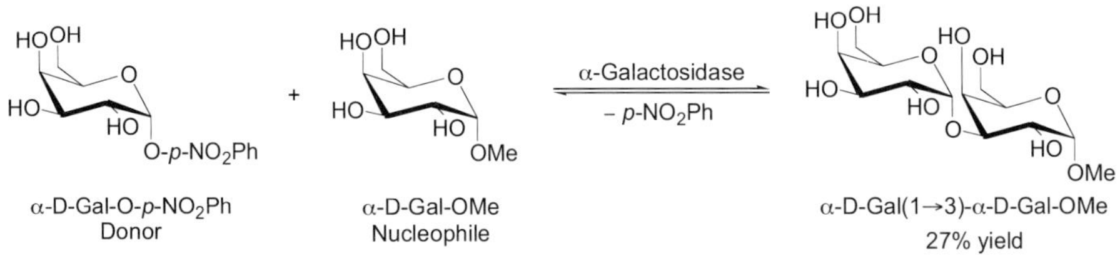

Fig. 246. Kinetically controlled synthesis of α-D-Gal(1→3)-α-D-Gal-OMe using α-D-galactosidase from coffee beans (Nilsson, 1987).

For instance, Gopalan et al. (1992) found that the rate of alcoholysis is only 24-fold greater than the rate of competing hydrolysis in the formation of octyl-β-D-glycoside from octanol and D-glucose using guinea pig liver cytosolic β-glucosidase.

A major advance in the area of glycoside synthesis with glycosidases was the discovery of glycosynthases by Withers and colleagues (Fig. 247). Removing the catalytic nucleophile (e.g., a Glu to Ser mutation) by site-directed mutagenesis prevents formation of the key covalent intermediate, dramatically altering the mechanism (Mackenzie et al., 1998). Normal glycosides no longer react, but α-glycosyl fluorides do react, likely via a direct displacement mechanism. Glycosynthases, like the starting enzyme, form β-glycoside links. However, glycosynthases no longer catalyze hydrolysis of product, which is a non-activated glycoside, and thus give higher yields. More than ten different glycosynthases have been reported with differing glycosyl fluorides specificity and differing regioselectivity (formation of β-1,3 vs. β-1,4 links) (reviews: Williams and Withers, 2002; Jahn and Withers, 2003; Perugino et al., 2004; 2005). The rate of the glycoside synthesis by glycosynthases is slow, but can be increased by directed evolution (Kim et al., 2004; Lin et al., 2004) or by altering the reaction conditions (Perugino et al., 2003).

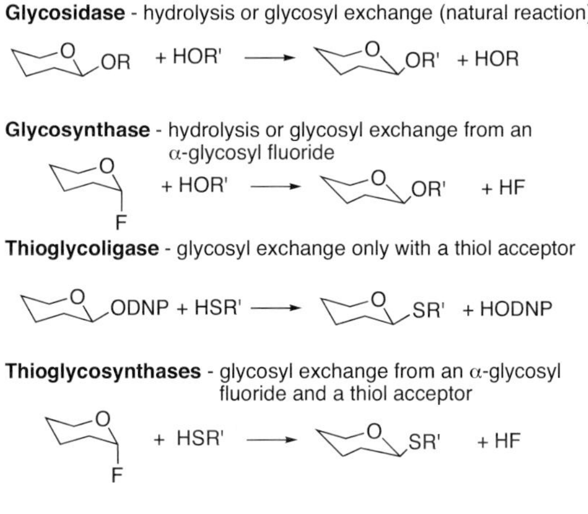

Fig. 247. Disabling key mechanistic steps in a retaining β-glycosidase creates new catalytic activities. Removal of the catalytic nucleophile creates a glycosynthases where only α-glycosyl fluorides react presumably via a single displacement mechanism. Removal of the catalytic acid/base creates a thioglycoligases (only strong incoming nucleophiles such as thiols react), and removal of both catalytic nucleophile and catalytic acid/base creates a thioglycosynthase (only α-glycosyl fluorides and strong incoming nucleophiles react). DNP = 2,4-dinitrophenyl

The second type of new catalytic activity, thioglycoligase, results upon removing the catalytic acid/base (Jahn et al., 2003). One role of this catalytic acid/base is activation of the incoming nucleophile. Absence of this activator precludes reaction with normal incoming nucleophiles and requires strong nucleophile such as thiols. Introducing single amino acid mutation into β-glycosidases from *Agrobacterium* sp. Abg (mutation: E171A) created a variant that formed *S*-glycosidic linkages in high yield (Fig. 248). The wild-type enzyme gave no product, possibly due to sterics caused by the larger sulfur atom.

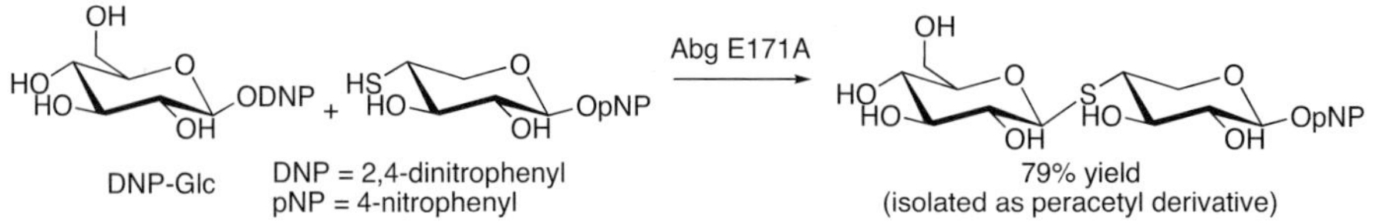

10.1 Glycosidases

Fig. 248. A β-glucosidase lacking the catalytic acid/base no longer catalyzed hydrolysis. However using an activated glycosyl donor overcomes the lack of a proton donor for the leaving group and using a nucleophilic thiol overcome the lack of base to activate the incoming nucleophile. Thus, the mutant now catalyzed formation of an *S*-glycosidic link.

Finally, removing both the catalytic nucleophile and the catalytic acid/base creates a thioglycosynthase, which requires both an α-glycosyl fluoride and a thiol acceptor (Jahn et al., 2004). An alternate route to glycosides uses glycosyl transferases, which require phosphorylated glycosyl donors (e.g., uridine diphosphate sugars) (review: Sears and Wong, 2001).

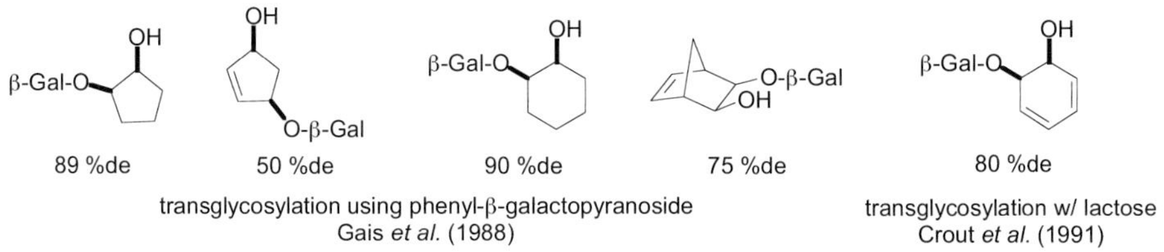

Fig. 249. Examples for the asymmetrization of *meso*-diols using β-galactosidase from *E. coli*.

Glycosidases can also make glycosides of non-sugar alcohols (Huber et al., 1984; Ooi et al., 1985). In some cases, these glycosylations were enantioselective, but this selectivity was too low to yield enantiopure compounds. Fig. 249 contains examples of the desymmetrization of *meso*-diols while Fig. 250 contains example of kinetic resolutions of racemic alcohols. Hydrolysis of β-galactopyranoside derivatives bearing racemic alcohol moieties proceeded with e.g., $E = 7$ for the isopropylidene glycerol derivative (Werschkun et al., 1995).

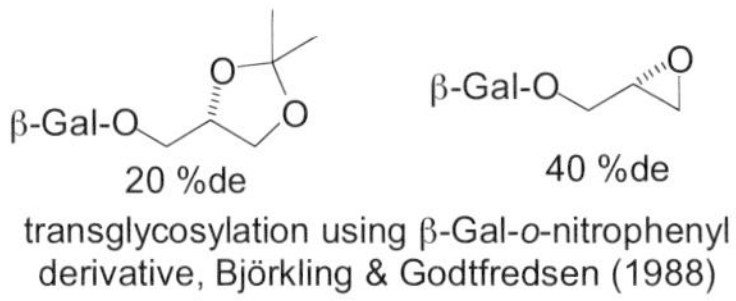

Fig. 250. Examples for the resolution of racemic alcohols using β-galactosidase from *E. coli*.

10.2 Haloalcohol Dehalogenases

Haloalcohol dehalogenase (HAD, EC 3.8.1.5, official name: haloalkane dehalogenase, also named halohydrin dehalogenases or halohydrin hydrogen-halide lyases) belong to the group of dehalogenating enzymes, which also includes atrazine chlorohydrolase (EC 3.8.1.8) but also several lyases (EC 4.5.1.x). HAD catalyze the reversible dehalogenation of vicinal haloalcohols by an intramolecular displacement of a halogen to yield an epoxide and a halide. The mechanism proposed for a HAD from *Agrobacterium radiobacter* – based on the sequence similarity with short-chain dehydrogenase/reductase protein family and the recently solved 3D-structure of the enzyme from *A. radiobacter* (de Jong et al., 2003) – suggests activation of the epoxide by hydrogen bonding to a tyrosine (Fig. 251).

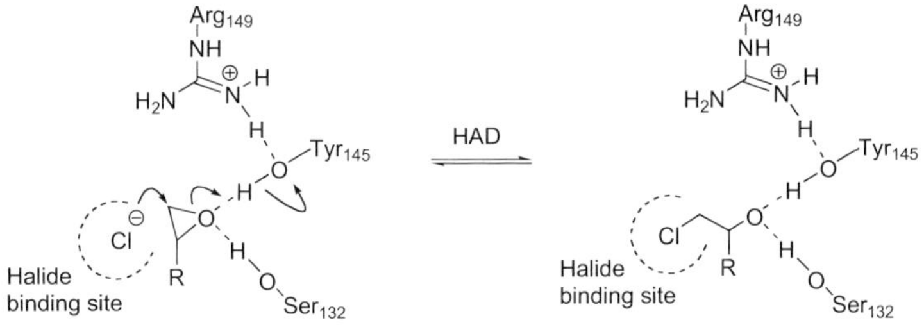

Fig. 251. Proposed mechanism for the reversible epoxide ring-opening catalyzed by *Agrobacterium radiobacter* AD1 haloalcohol dehalogenase. The Arg-Tyr pair is involved in leaving group protonation.

The interest in HAD was initially directed towards the degradation of xenobiotics and soil remediation (Swanson, 1999). More recently, they also became of interest for organic synthesis as the non-covalent pathway depicted in Fig. 251 suggests that these enzymes should in principle also accept nucleophiles other than water (de Vries and Janssen, 2003). Indeed, it could be demonstrated, that not only natural nucleophile halides (Cl^-, Br^-, I^-), but also CN^-, NO_2^- and N_3^- could be used in the ring-opening of epoxides. Consequently, the haloalcohol dehalogenase are synthetically more useful than epoxide hydrolases, as not only 1,2-diols are accessible products (Fig. 252). In addition, the HAD from *Agrobacterium radiobacter* AD1 also gave excellent β-regio- and enantioselectivity in the kinetic resolution of *p*-nitrostyrene oxide (Spelberg et al., 2001; 2002). Enantioselectivity was considerably lower for *p*-Cl-styrene oxide (E = 51) and styrene oxide (E = 15). The reaction was irreversible, as the azide is a very poor leaving group.

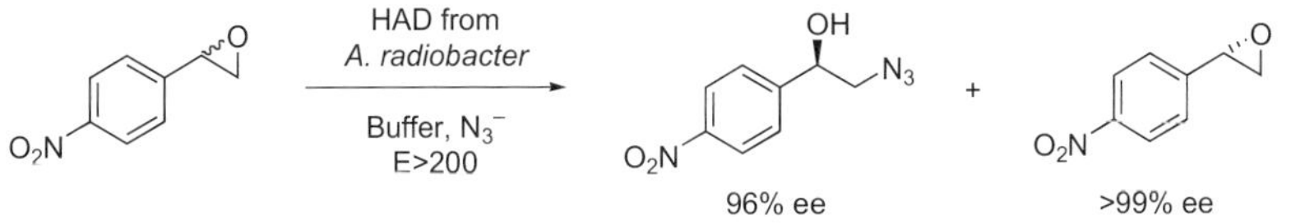

Fig. 252. Haloalcohol dehalogenase-catalyzed ring-opening of *p*-nitrostyrene oxide in the presence of azide proceeds with excellent β-regio- and enantioselectivity. and yields an azido alcohol as product.

10.3 Phosphotriesterases

Phosphotriesterases catalyze the hydrolysis of organophosphorus compounds (Fig. 253). The paroxonases or organophosphorus hydrolases catalyze hydrolysis of a wide range of phosphorus triesters, but favors substrates where a P–O bond is cleaved. In contrast, diisopropyl fluorophosphatases favor substrates where a P–F or P–CN bond is cleaved. Phosphorus triesters do not occur in nature, so the true biochemical substrate of phosphotriesterases may be something else. For example, phosphotriesterases found in human serum (Harel et al., 2004) may be lactonases.

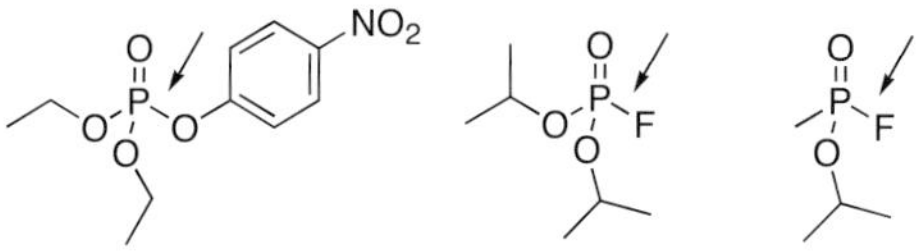

Fig. 253 Structure of paroxon, an insecticide, diisopropyl fluorophosphate, a less toxic analog of soman, and soman, a military nerve agent.

10.3.1 Organophosphorus Hydrolases

Organophosphorus hydrolases (OPH, EC 3.1.8.1) from *Pseudomonas diminuta* or from *Flavobacterium* are identical zinc metalloenzymes (38 kDa) containing two zinc ions bridged by a hydroxide ion in the active site (review: Raushel and Holden, 2000). OPH catalyzes hydrolysis of a wide range of phosphorus triesters and can detoxify organophosphate pesticides such as paraoxon. Although the enzyme is stable for only a few days in aqueous solution, upon immobilization either on polymers or within *E.coli* cells, it is stable for at least a year (Lejeune et al., 1997; Ember, 1997; Mulchandani et al., 1999). This dramatic stabilization may be due to multipoint attachment of OPH to the support which hinders unfolding. Griffiths and Tawfik (2003) used directed evolution to create a variant with 63-fold higher k_{cat} toward paraoxon (1.8-fold increase in k_{cat}/K_M).

The x-ray crystal structure of OPH complexed with a substrate analog reveals why it has such a broad specificity (Vanhooke et al., 1996). A hydrophobic pocket binds part of the substrate analog, while the leaving group lies in a shallow, solvent-exposed pocket. Further analysis revealed high enantioselectivity of OPH toward phosphate triesters (Hong and Raushel, 1999) and that the hydrophobic pocket contains a small and a large subsite that fit small and large substituents in the phosphate triesters, Fig. 254 (Chen-Goodspeed et al., 2001a; Li et al., 2001).

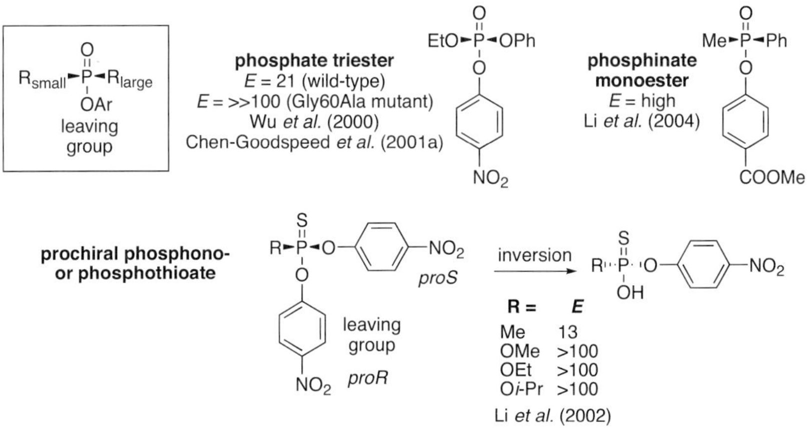

Fig. 254. Organophosphorus hydrolase favors the hydrolysis of phosphate triesters, phosphinate esters and phosphono- and phosphothioates with the shape shown. Examples show the fast reacting enantiomer. For the phosphate triesters and the phosphinate ester, the products are achiral: $(P(O)(OR^1)(OR^2)O^-$ and $P(O)R^1R^2O^-$. For the phosphono- and phosphothioates, hydrolysis removes the *proR* *p*-nitrophenolate group and inverts the configuration at the phosphorus stereocenter as shown.

Rational design within these small and large subsites dramatically altered the enantioselectivity. First, decreasing the size of small subsite by a Gly-to-Ala mutation, dramatically increased the enantioselectivity. The wild-type enzyme favored the S_P enantiomer of ethyl phenyl 4-nitrophenyl phosphate with an enantioselectivity of 21, while the mutant favored this enantiomer by 11 000 (Chen-Goodspeed et al., 2001a). Presumably the slow enantiomer must place its large group in the small subsite to react. Decreasing the size of the small subsite further slows the reaction of the slow-reacting enantiomer and thus increased the enantioselectivity.

Conversely, increasing the size of the small subsite allowed the large group to fit, thus increasing the reaction rate of the slow-reacting enantiomer, which decreased the enantioselectivity. The best mutant was a triple mutant where glycines replaced each of three large residues in the small subsite (Phe132, Ser308, Ile106). The enantioselectivity dropped from 21 to 0.77.

Further mutations even reversed the enantioselectivity. Starting with the nonenantioselective mutant above that had an enlarged small subsite, Chen-Goodspeed et al. (2001b) reduced His-to-Tyr mutation. This mutant now favored the R_P enantiomer of ethyl phenyl 4-nitrophenyl phosphate with an enantioselectivity of 460. A similar mutant shows a reversed enantiopreference to methyl phenyl phosphinate esters (Li et al., 2004).

A different approach to increasing the enantioselectivity of organophosphorus hydrolase was altering the leaving group (Li et al., 2003). Kinetic analysis of the hydrolysis of the phosphonate Ph–P(O)OMe(O-4-nitrophenyl) showed that the rate determining step was substrate binding for the fast-reacting enantiomer, but P–O bond cleavage for the slow-reacting enantiomer. Changing the substituent on the 4-nitrophenoxide leaving group to make it a poorer leaving group (to a 4-carboxymethyl substituent) had little effect on the binding step and thus little change in the overall reaction rate of the fast reacting enantiomer. However, this change dramatically slowed the P–O bond cleavage step for the slow enantiomer and thus, the overall enantioselectivity increased ~50-fold from ~100 to ~5000. For similar reasons, changing the substituent to make the phenoxide a better leaving group decreased the enantioselectivity to as low as ~1. The overall range of enantioselectivity was from ~1 to ~5000 by changing only the leaving group.

Most organophosphorus nerve agents (chemical weapons) contain P–F, P–CN, or P–S bonds and are poor substrates for organophosphorus hydrolase (review: Raushel, 2002). Replacing the zinc ion with cobalt (II) ion (Kolakowski et al., 1997) or enlarging the active site using site-directed mutagenesis (diSioudi et al., 1999) increased the activity toward these substrates five to twenty-fold. The most dramatic increase – ~500-fold – came from saturation mutagenesis in the active site (Hill et al., 2003). This increase was for a 4-nitrophenolate analog of a nerve agent, which simplified screening of the mutants. Mutagenesis of a pair of residues gave 400 possible mutants, from which Hill et al. identified mutant His254Gly/His257W with ~100-fold increased activity. A subsequent single mutagenesis created 19 additional mutants from which they identified a triple mutant which showed ~500-fold increased activity over wild type for the nerve-agent analog.

10.3.2 Diisopropyl Fluorophosphatases

Diisopropyl fluorophosphatases (DFP-ases, EC 3.1.8.2) catalyze the hydrolysis of the P–F bond in disopropyl fluorophosphate and related compounds, such as soman, a military nerve agent, see Fig. 253 above. Since this hydrolysis detoxifies nerve agents, there is considerable military interest in this enzyme.

The x-ray structure of diisopropyl fluorophosphatase from squid (Scharff et al., 2001) shows a histidine and calcium ion in the active site. This structure, combined with biochemical studies (Hartleib and Ruterjans, 2001), suggests that catalysis involves direct attack by an activated water molecule on the phosphorus center.

The marine bacterium *Alteromonas* also contains a diisopropyl fluorophosphatase, but it is not related to the squid enzyme. The *Alteronomas* DFP-ase is a metalloproteinase like OPH above, but contain manganese instead of zinc (DeFrank and Cheng, 1991). Sequence comparison (Cheng et al., 1996) yielded a surprise. The DFP-ases were not new hydrolases, but were prolidases, that is, peptidases that catalyze hydrolysis of X-Pro dipeptides (E.C. 3.4.13.9). The prolidase showed no activity against the nerve agent VX, which contains a P–S link. Researchers noted that the three-dimensional shape and electrostatic distribution of both Soman and Leu-Pro are similar and suggested that this similarilty may account for the catalytic promiscuity of prolidases toward Soman and DFP. Jao et al. (2004) found that a prolidase from *E. coli*, which also contains two manganese ions in the active site, catalyzes hydrolysis of phosphate triesters, Fig. 255. They did not check for DFP-ase activity, but the similarity to the *Alteromonas* enzyme suggests that it may be a DFP-ase. Siebert and Raushel (2005) recently reviewed these promiscuous enzymes from a structural point of view.

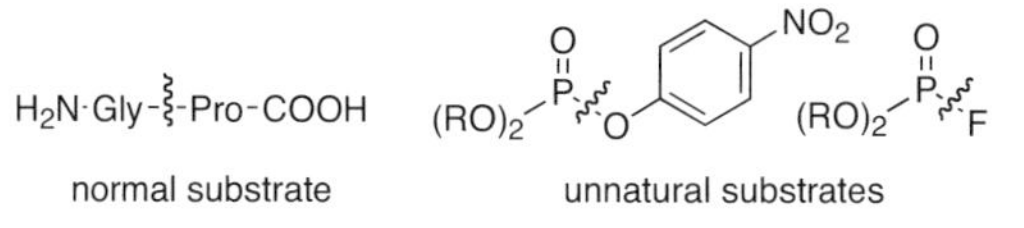

Fig. 255. Prolidases show catalytic promiscuity and also catalyze hydrolysis of organophosphorus compounds.

Abbreviations

7-ACA	7-Aminocephalosporanic acid	DNA	desoxyribonucleic acid
AChE	acetylcholine esterase	DTNB	5,5-dithio-bis-(2-nitrobenzoic acid
AGE	*Arthrobacter globiformis* esterase	E	enantioselectivity / enantiomeric ratio
Alloc	allyloxycarbonyl		
ANL	*Aspergillus niger* lipase	*E.*	*Escherichia*
6-APA	6-aminopenicillanic acid	ee	enantiomeric excess
AOT	bis(2-ethylhexyl)sodium sulfo-succinate	EH	epoxide hydrolase
		Eq.	equation
Ar	aryl	ep	error-prone
ATTC	American Type Culture Collection (Manassas, Virginia)	ESR	electron spin resonance
		Et	ethyl
a_w	water activity	Fmoc	9-fluorenylmethoxycarbonyl
BCE	*Bacillus coagulans* esterase	GC	gas chromatography
BChE	butyrylcholine esterase	GCL	*Geotrichum candidum* lipase
BMIM	1-butyl-3-methyl-imidazolium	HAD	Haloalcohol dehalogenase
Bn	benzyl	HEPES	*N*-(2-hydroxyethyl)-piperazine-*N'*-ethanesulfonic acid
Boc	*t*-butyloxycarbonyl		
bp	base pair(s), boiling point	HLE	horse liver esterase
BSE	*Bacillus stearothermophilus* esterase	HLL	*Humicola lanuginosa* lipase
		HPLC	high performance liquid chromatography
BsubE	*Bacillus subtilis* esterase		
BTL2	*Bacillus thermocatenulatus* lipase	HTS	High-throughput screening
Bu	butyl	Hx	hexyl
But	butyryl	IPG	isopropylidene glycerol, solketal
Bz	benzoyl	kDa	kilo Dalton
c	conversion	Me	methyl
CAL	*Candida antarctica* lipase	MEE	(methoxyethoxy)ethyl
Cbz	*N*-carbobenzyloxy	mEH	microsomal epoxide hydrolase
CE	cholesterol esterase	MOM	methoxymethyl
cEH	cytosolic epoxide hydrolase	MPGM	*trans*-3-(4-methoxy- phenyl)-glycidic acid methylester
cHX	cyclohexyl		
CLE	chicken liver esterase	MTBE	methyl-*t*-butylether
CLEA	Cross-linked enzyme aggregates	MTP	microtiterplate
CLEC	Cross-linked enzyme crystals	NCIMB	National Collections of Industrial and Marine Bacteria (Aberdeen, Scotland)
CLL	*Candida lipolytica* lipase		
CRL	*Candida rugosa* lipase		
CT	Chymotrypsin	NHase	nitrile hydratase
CTAB	cetyltrimethyl ammonium bromide	NRRL	Northern Regional Research Laboratory, now called National Center For Agricultural Utilization Research (Peoria, Illinois)
CVL	*Chromobacterium viscosum* lipase		
		NSAID	non-steroidal anti-inflammatory drugs
DKR	dynamic kinetic resolution		
DMF	*N,N*'-dimethylformamide		
DMSO	dimethylsulfoxid	OPH	Organophosphorus hydrolase

PAE	*Pseudomonas aeruginosa* esterase	PPL	porcine pancreatic lipase
		PQQ	pyrroloquinoline quinone
		Pr	Propyl
PAGE	polyacrylamide gelelectrophoresis	ProqL	*Penicillium roquefortii* lipase
		pTol	*p*-toluyl
PBA	phenylboronic acid	RDL	*Rhizopus delemar* lipase
PC	phosphatidyl choline	RJL	*Rhizopus javanicus* lipase
PcamL	*Penicillium camembertii* lipase	RLE	rabbit liver esterase
PCL	*Pseudomonas cepacia* lipase	RML	*Rhizomucor miehei* lipase
PCMC	protein-coated micro-crystals	RNL	*Rhizopus niveus* lipase
PCR	polymerase chain reaction	ROL	*Rhizopus oryzae* lipase
pdb	brookhaven data base	$SCCO_2$	supercritical carbon dioxide
PEG	polyethylene glycol	SDS	sodium dodecylsulfate
PFE	*Pseudomonas fluorescens* esterase	sEH	soluble epoxide hydrolase
		sn	stereochemical numbering
PFL	*Pseudomonas fluorescens* lipase	StEP	staggered extension process
PfragiL	*Pseudomonas flragi* lipase	*t.*	tertiary
PGA	Penicillin G amidase	TFEA	2,2,2-trifluoroethyl acetate
PL	phospholipase	TG	triglyceride
PLE	pig liver esterase	THF	tetrahydrofuran
PME	*Pseudomonas marginata* esterase	TLC	thin layer chromatography
PMSF	phenylmethylsulfonyl fluoride	Tr	trityl
pNPA	*p*-nitrophenyl acetate	U	Units
pNPP	*p*-nitrophenyl palmitate	Z	*t*-butyloxycarbonyl
PPE	*Pseudomonas putida* esterase		

References

<u>General books about enzymes in organic syntheses</u>

Buchholz, K., Bornscheuer, U. T., Kasche, V. (2005), *Biocatalysts and Enzyme Technology*, Wiley-VCH, Weinheim.

Bommarius, A. S., Riebel, B. R. (2004), *Biocatalysis*, Wiley-VCH, Weinheim.

Faber, K. (2004), *Biotransformations in Organic Chemistry*, 5th ed., Springer, Berlin.

Drauz, K., Waldmann, H., Eds. (2002), *Enzyme Catalysis in Organic Synthesis*, 2nd ed., **Vol. 1-3**, Wiley-VCH, Weinheim.

Patel, R. N., Ed. (2000), *Stereoselective Biocatalysis*, Marcel Dekker, New York.

Roberts S. M. (1999), *Biocatalysts for Fine Chemical Synthesis*, Wiley-VCH, Weinheim.

<u>Specific literature</u>

Aaltonen, O., Rantakylae, M. (1991), Biocatalysis in supercritical carbon dioxide, *CHEMTECH* **21**, 240-248.

Adachi, T., Ishii, M., Ohta, Y., Ota, T., Ogawa, T., Hanada, K. (1993), Chemoenzymatic synthesis of optically active 1,4-dihydropyridine derivatives via enantioselective hydrolysis and transesterification, *Tetrahedron: Asymmetry* **4**, 2061-2068.

Adachi, K., Kobayashi, S., Ohno, M. (1986), Chiral synthons by enantioselective hydrolysis of *meso*-diesters with pig liver esterase: substrate-stereoselectivity relationships, *Chimia* **40**, 311-314.

Adam, W., Groer, P., Humpf, H.-U., Saha-Möller, C. R. (2000), Synthesis of optically active α-methylene β-lactams through lipase-catalyzed kinetic resolution, *J. Org. Chem.* **65**, 4919-4922.

Adam, W., Groer, P., Saha-Moeller, C. R. (1997a), Enzymic preparation of optically active α-methylene β-lactones by lipase-catalyzed kinetic resolution through asymmetric transesterification, *Tetrahedron: Asymmetry* **8**, 833-836.

Adam, W., Mock-Knoblauch, C., Saha-Möller, C. R. (1997b), Kinetic resolution of hydroxy vinylsilanes by lipase-catalyzed enantioselective acetylation, *Tetrahedron: Asymmetry* **8**, 1441-1444.

Adam, W., Diaz, M. T., Fell, R. T., Sahamoller, C. R. (1996), Kinetic resolution of racemic α-hydroxy ketones by lipase-catalyzed irreversible transesterification, *Tetrahedron: Asymmetry* **7**, 2207-2210.

Adamczyk, M., Grote, J. (1996), *Pseudomonas cepacia* lipase mediated amidation of benzyl esters, *Tetrahedron Lett.* **37**, 7913-7916.

Adamczyk, M., Gebler, J. C., Grote, J. (1995), Chemo-enzymic transformations of sensitive systems. Preparation of digoxigenin haptens via regioselective lipase mediated hydrolysis, *Tetrahedron Lett.* **36**, 6987-6990.

Adamczyk, M., Gebler, J. C., Mattingly, P. G. (1994), Lipase mediated hydrolysis of Rapamycin 42-hemisuccinate benzyl and methyl esters, *Tetrahedron Lett.* **35**, 1019-1022.

Adelhorst, K., Björkling, F., Godtfredsen, S. E., Kirk, O. (1990), Enzyme catalysed preparation of 6-*O*-acylglucopyranosides, *Synthesis* 112-115.

Ader, U., Schneider, M. P. (1992), Enzyme assisted preparation of enantiomerically pure β-andrenergic blockers III. Optically active chlorohydrin derivatives and their conversion, *Tetrahedron: Asymmetry* **3**, 521-524.

Adjé, N., Breuilles, P., Uguen, D. (1993), Desymmetrisation of *meso* propargylic diols, *Tetrahedron Lett.* **34**, 4631-4634.

Adlercreutz, P. (1994), Enzyme-catalyzed lipid modification, *Biotechnol. Genet. Eng. Rev.* **12**, 231-254.

Adlercreutz, P. (1991), On the importance of the support material for enzymatic synthesis in organic media. Support effects at controlled water activity, *Eur. J. Biochem.* **199**, 609-614.

Adshiri, T., Akiya, H., Chin, L. C., Arai, K., Fujimoto, K. (1992), Lipase-catalyzed interesterification of triglyceride with supercritical carbon dioxide extraction, *J. Chem. Eng. Jpn.* **25**, 104-105.

Affleck, R., Xu, Z. F., Suzawa, V., Focht, K., Clark, D. S., Dordick, J. S. (1992), Enzymatic catalysis and dynamics in low-water environments, *Proc. Natl. Acad. Sci. USA* **89**, 1100-1104.

Aharoni, A., Griffiths, A. D., Tawfik, D. S. (2005), High-throughput screens and selections of enzyme-encoding genes, *Curr. Opin. Chem. Biol.* **9**, 210-216.

Ahmed, S. N., Kazlauskas, R. J., Morinville, A. H., Grochulski, P., Schrag, J. D., Cygler, M. (1994), Enantioselectivity of *Candida rugosa* lipase toward carboxylic acids: a predictive rule from substrate mapping and X-ray crystallography, *Biocatalysis* **9**, 209-225.

Ajima, A., Yoshimoto, T., K. Takahashi, Tamaura, Y., Saito, Y., Inada, Y. (1985), Polymerization of 10-hydroxydecanoic acid in benzene with polyethylene glycol-modified lipase, *Biotechnol. Lett.* **7**, 303-306.

Akai, S., Naka, T., Takebe, Y., Kita, Y. (1997), Enzyme-catalyzed asymmetrization of 2,2-disubstituted 1,3-propanediols using 1-ethoxyvinyl esters, *Tetrahedron Lett.* **38**, 4243-4246.

Akita, H., Nozawa, M., Futagami, Y., Miyamoto, M., Saotome, C. (1997a), Simple approach to optically active drimane sesquiterpenes based on enzymatic resolution, *Chem. Pharm. Bull.* **45**, 824-831.

Akita, H., Umezawa, I., Matsukura, H. (1997b), Enzymatic hydrolysis in organic solvents for kinetic resolution of water-insoluble α-acyloxy esters with immobilized lipases, *Chem. Pharm. Bull.* **45**, 272-278.

Akita, H. (1996), Recent advances in the use of immobilized lipases directed toward the asymmetric syntheses of complex molecules, *Biocatal. Biotransform.* **13**, 141-156.

Akkara, J. A., Ayyagari, M. S. R., Bruno, F. F. (1999), Enzymatic synthesis and modification of polymers in nonaqueous solvents, *Trends Biotechnol.* **17**, 67-73.

Alberghina, L., Schmid, R. D., Verger, R. (1991), Lipases: Structure, mechanism and genetic engineering, **Vol. 16**, Weinheim: VCH.

Alcock, N. W., Crout, D. H. G., Henderson, C. M., Thomas, S. E. (1988), Enzymatic resolution of a chiral organometallic ester - enantioselective hydrolysis of 2-ethoxycarbonylbuta-1,3-dienetricarbonyliron by pig liver esterase, *J. Chem. Soc., Chem. Commun.* 746-747.

Alfonso, I., Astorga, C., Rebolledo, F., Gotor, V. (1996), Sequential biocatalytic resolution of (±)-*trans*-cyclohexane-1,2- diamine. Chemoenzymic synthesis of an optically active polyamine, *Chem. Commun.* 2471-2472.

Allen, J. V., Williams, J. M. J. (1996), Dynamic kinetic resolution with enzyme and palladium combinations, *Tetrahedron Lett.* **37**, 1859-1862.

Allen, R. H., Jacoby, W. B. (1969), Tartaric acid metabolism. IX. Synthesis with tartrate epoxidase, *J. Biol. Chem.* **244**, 2078-2084.

Allenmark, S. G., Andersson, A. C. (1993), Lipase -catalyzed kinetic resolution of a series of esters having a sulfoxide group as the stereogenic center, *Tetrahedron: Asymmetry* **4**, 2371-2376.

Allenmark, S., Ohlsson, A. (1992a), Enantioselectivity of lipase-catalyzed hydrolysis of some 2-chloroethyl 2-arylpropanoates studied by chiral reversed-phase liqud chromatography, *Chirality* **4**, 98-102.

Allenmark, S., Ohlsson, A. (1992b), Studies of the heterogeneity of a *Candida cylindracea (rugosa)* lipase: monitoring of esterolytic activity and enantioselectivity by chiral liquid chromatography, *Biocatalysis* **6**, 211-221.

Allevi, P., Ciuffreda, P., Tarocco, G., Anastasia, M. (1996), Enzymatic resolution of (R)- and (S)-2-(1-hydroxyalkyl)thiazoles, synthetic equivalents of (R)- and (S)-2-hydroxy aldehydes, *J. Org. Chem.* **61**, 4144-4147.

Allevi, P., Anastasia, M., Cajone, F., Ciuffreda, P., Sanvito, A. M. (1993), Enzymatic resolution of (R)- and (S)-(E)-4-hydroxyalk-2-enals related to lipid peroxidation., *J. Org. Chem.* **58**, 5000-5002.

Almatawah, Q. A., Cramp, R., Cowan, D. A. (1999), Characterization of an inducible nitrilase from a thermophilic *Bacillus*, *Extremophiles* **3**, 283-291.

Amaki, Y., Nakano, H., Yamane, T. (1993), Role of cysteine residues in esterase from *Bacillus stearothermophilus* and increasing its thermostability by the replacement of cysteines, *Appl. Microbiol. Biotechnol.* **40**, 664-668.

Amaki, Y., Tulin, E. E., Ueda, S., Ohmiya, K., Yamane, T. (1992), Purification and properties of a thermostabile esterase of *Bacillus stearophilus* produced by recombinant *Bacillus brevis*, *Biosci. Biotech. Biochem.* **56**, 238-241.

Ampon, K., Basri, M., Salleh, A. B., Yunus, W. M. Z. W., Razak, C. N. A. (1994), Immobilization by adsorption of hydrophobic lipase derivatives to porous polymer beads for use in ester synthesis, *Biocatalysis* **10**, 341-351.

Anantharamaiah, G. M., Roeske, R. W. (1982), Resolution of α-methyl amino esters by chymotrypsin, *Tetrahedron Lett.* **23**, 3335-3336.

Andersch, P., Schneider, M. P. (1993), Enzyme assisted synthesis of enantiomerically pure myo-inositol derivatives. Chiral building blocks for inositol polyphosphates, *Tetrahedron: Asymmetry* **4**, 2135-2138.

Anderson, E. M., Larsson, K. M., Kirk, O. (1998), One biocatalyst - many applications: The use of *Candida antarctica* B-lipase in organic synthesis, *Biocatal. Biotransform.* **16**, 181-204.

Anthonsen, H. W., Hoff, B. H., Anthonsen, T. (1995), A simple method for calculating enantiomer ratio and equilibrium constants in biocatalytic resolutions, *Tetrahedron: Asymmetry* **6**, 3015-3022.

Antikainen, N. M., Hergenrother, P. J., Harris, M. M., Corbett, W., Martin, S. F. (2003), Altering substrate specificity of phosphatidylcholine-preferring phospholipase C of *Bacillus cereus* by random mutagenesis of the headgroup binding site, *Biochemistry* **42**, 1603-1610.

Aouf, N. E., Djerourou, A. H., Blanco, L. (1994), Preparation et essais de dedoublement enzymatique d'acetoxymethyl- et hydroxymethylsilanes chiraux, *Phosphorus, Sulfur, and Silicon* **88**, 207-215.

Aragozzini, F., Valenti, M., Santaniello, E., Ferraboschi, P., Grisenti, P. (1992), Biocatalytic, enantioselective preparations of (R)- and (S)-ethyl 4-chloro-3-hydroxybutanoate, a useful chiral synthon, *Biocatalysis* **5**, 325-332.

Arand, M., Hallberg, B. M., Zou, J., Bergfors, T., Oesch, F., van der Werf, M. J., de Bont, J. A., Jones, T. A., Mowbray, S. L. (2003), Structure of *Rhodococcus erythropolis* limonene-1,2-epoxide hydrolase reveals a novel active site, *Embo. J.* **22**, 2583-2592.

Arand, M., Wagner, H., Oesch, F. (1996), Asp^{333}, Asp^{495}, and His^{523} from the catalytic triad of rat soluble epoxide hydrolase, *J. Biol. Chem.* **271**, 4223-4229.

Archelas, A. (1998), Epoxide hydrolases: new tools for the synthesis of enantiopure epoxides and diols, *J. Mol. Catal. B: Enzym.* **5**, 79-85.

Archelas, A., Furstoss, R. (1998), Epoxide hydrolases: new tools for the synthesis of fine chemicals, *Trends Biotechnol.* **16**, 108-116.

Archer, I. V. J. (1997), Epoxide hydrolases as asymmetric catalysts, *Tetrahedron* **53**, 15617-15662.

Archer, I. V. J., Leak, D. J., Widdowson, D. A. (1996), Chemoenzymic resolution and deracemization of (±)-1-methyl-1,2-epoxycyclohexane: the synthesis of (1-S,2-S)-1-methylcyclohexane-1,2-diol, *Tetrahedron Lett.* **37**, 8819-8822.

Arita, M., Adachi, K., Ito, Y., Sawai, H., Ohno, M. (1983), Enantioselective synthesis of the carbocyclic nucleosides (-)-aristeromycin and (-)-neplanocin A by a chemicoenzymatic approach, *J. Am. Chem. Soc.* **105**, 4049-4055.

Arnold, F. H., Georgiou, G. (Eds.) (2003a), Directed evolution library creation: Methods and protocols, **Vol. 231**, Totawa: Humana Press.

Arnold, F. H., Georgiou, G. (2003b), Directed enzyme evolution: Screening and selection methods, **Vol. 230**: Totawa: Humana Press.

Arnold, F. H. (1998), Design by directed evolution, *Acc. Chem. Res.* **31**, 125-131.

Arnold, F. H., Moore, J. C. (1997), Optimizing industrial enzymes by directed evolution, *Adv. Biochem. Eng./Biotechnol.* **58**, 1-14.

Arpigny, J. L., Jaeger, K. E. (1999), Bacterial lipolytic enzymes: classification and properties, *Biochem. J.* **343**, 177-183.

Arroyo, M., Sinisterra, J. V. (1995), Influence of chiral carvones on selectivity of pure lipase-B from *Candida antarctica*, *Biotechnol. Lett.* **17**, 525-530.

Arroyo, M., Sinisterra, J. V. (1994), High enantioselective esterification of 2-arylpropionic acids catalyzed by immobilized lipase from *Candida antarctica*: A mechanistic approach, *J. Org. Chem.* **59**, 4410-4417.

Asano, Y., Yamaguchi, S. (2005), Dynamic kinetic resolution of amino acid amide catalyzed by D-aminopeptidase and α-amino-ε-caprolactam racemase, *J. Am. Chem. Soc.* **127**, 7696-7697.

Asano, Y. (2002), Overview of screening for new microbial catalysts and their uses in organic synthesis - selection and optimization of biocatalysts, *J. Biotechnol.* **94**, 65-72.

Astorga, C., Rebolledo, F., Gotor, V. (1993), Enzymic hydrazinolysis of diesters and synthesis of N-aminosuccinimide derivatives, *Synthesis* 287-289.

Astorga, C., Rebolledo, F., Gotor, V. (1991), Synthesis of hydrazides through an enzymatic hydrazinolysis reaction, *Synthesis* 350-352.

Azzolina, O., Collina, S., Vercesi, D. (1995a), Stereoselective hydrolysis by esterase: a strategy for resolving 2-(R,R'-phenoxy)propionyl ester racemates, *Farmaco* **50**, 725-733.

Azzolina, O., Vercesi, D., Collina, S., Ghislandi, V. (1995b), Chiral resolution of methyl 2-aryloxypropionates by biocatalytic stereospecific hydrolysis, *Farmaco* **50**, 221-226.

Athawale, V. D., Gaonkar, S. R. (1994), Enzymatic synthesis of polyesters by lipase catalysed polytransesterification, *Biotechnol. Lett.* **16**, 149-154.

Baba, N., Mimura, M., Oda, J., Iwasa, J. (1990a), Lipase-catalyzed stereoselective hydrolysis of thiol acetate, *Bull. Inst. Chem. Res. Kyoto Univ.* **68**, 208-212.

Baba, N., Tateno, K., Iwasa, J., Oda, J. (1990b), Lipase-catalyzed kinetic resolution of racemic methyl 13-hydroperoxy-9Z,11E-octadecadienoate in an organic solvent, *Agric. Biol. Chem.* **54**, 3349-3350.

Baba, N., Yoneda, K., Tahara, S., Iwasa, J., Kaneko, T., Matsuo, M. (1990c), A regioselective, stereoselective synthesis of a diacylglycerophosphocholine hydroperoxide by use of lipoxygenase and lipase, *J. Chem. Soc., Chem. Commun.* 1281-1282.

Baba, N., Mimura, M., Hiratake, J., Uchida, K., Oda, J. (1988), Enzymic resolution of racemic hydroperoxides in organic solvent, *Agric. Biol. Chem.* **52**, 2685-2687.

Babiak, K. A., Ng, J. S., Dygos, J. H., Weyker, C. L., Wang, Y.-F., Wong, C. H. (1990), Lipase-catalyzed irreversible transesterification using enol esters: Resolution of prostaglandin synthons 4-hydroxy-2-alkyl-2-cyclopentenones and inversion of the 4S enantiomer to the 4R enantiomer, *J. Org. Chem.* **55**, 3377-3381.

Backes, A. C., Hotta, K., Hilvert, D. (2003), Promiscuity in antibody catalysis: Esterolytic activity of the decarboxylase 21D8, *Helv. Chim. Acta*, **86**, 1167-1174.

Backlund, S., Eriksson, F., Kanerva, L. T., Rantala, M. (1995), Selective enzymic reactions using microemulsion-based gels, *Colloids Surf., B* **4**, 121-127.

Baczko, K., Liu, W. Q., Roques, B. P., Garbay-Jaureguiberry, C. (1996), New synthesis of D,L-Fmoc protected 4-phosphonomethylphenylalanine derivatives and their enzymatic resolution, *Tetrahedron* **52**, 2021-2030.

Badalassi, F., Klein, G., Crotti, P., Reymond, J.-L. (2004), Flourescence assay and screening of epoxide opening by nucleophiles, *Eur. J. Org. Chem.* 2557-2566.

Badalassi, F., Wahler, D., Klein, G., Crotti, P., Reymond, J.-L. (2000), A versatile periodate-coupled fluorogenic assay for hydrolytic enzymes, *Angew. Chem. Int. Ed.* **40**, 4457-4460.

Bakke, M., Takizawa, M., Sugai, T., Ohta, H. (1998), Lipase-catalyzed enantiomeric resolution of ceramides, *J. Org. Chem.* **63**, 6929-6938.

Bakker, M., van Rantwijk, F., Sheldon R. A. (2002), Metal substitution in thermolysin: Catalytic properties of tungstate thermolysin in sulfoxidation with H_2O_2, *Can. J. Chem.* **80**, 622-625.

Balcao, V. M., Paiva, A. L., Malcata, F. X. (1996), Bioreactors with immobilized lipases: state of the art, *Enzyme Microb. Technol.* **18**, 392-416.

Baldaro, E., D'Arrigo, P., Pedrocchi-Fantoni, G., Rosell, C. M., Servi, S., Tagliani, A., Terreni, M. (1993), Pen G acylase catalyzed resolution of phenyl acetate esters of secondary alcohols, *Tetrahedron: Asymmetry* **4**, 1031-1034.

Baldaro, E., Fuganti, C., Servi, S., Tagliani, A., Terreni, M. (1992), The use of immobilized penicillin G acylase in organic synthesis, in.: *Microbial Reagents in Organic Synthesis* (Servi, S.; Eds.), pp. 175-188. Dordrecht: Kluwer Academic.

Baldascini, H., Ganzeveld, K. J., Janssen, D. B., Beenackers, A. A. (2001), Effect of mass transfer limitations on the enzymatic kinetic resolution of epoxides in a two-liquid-phase system, *Biotechnol. Bioeng.* **73**, 44-54.

Baldessari, A., Iglesias, L. E., Gros, E. G. (1994), An improved procedure for chemospecific acylation of 2-mercaptoethanol by lipase-catalyzed transesterification, *Biotechnol. Lett.* **16**, 479-484.

Balkenhohl, F., Ditrich, K., Hauer, B., Ladner, W. (1997), Optically active amines via lipase-catalyzed methoxyacetylation, *J. Prakt. Chem.* **339**, 381-384.

Balkenhohl, F., Hauer, B., Ladner, W., Schnell, U., Pressler, U., Staudenmaier, H. R. (1993), Lipase-catalyzed acylation of alcohols with diketenes, *German Patent* DE 4 329 293.

Ballesteros, A., Bornscheuer, U., Capewell, A., Combes, D., Condorét, J. S., König, K., Kolisis, F. N., Marty, A., Menge, U., Scheper, T., Stamatis, H., Xenakis, A. (1995), Enzymes in non-conventional phases, *Biocatal. Biotransform.* **13**, 1-42.

Ballesteros, A., Bernabé, M., Cruzado, C., Martin-Lomas, M., Otero, C. (1989), Regioselective deacylation of 1,6-anhydro-β-D-galactopyranose derivatives catalyzed by soluble and immobilized lipases, *Tetrahedron* **45**, 7077-7082.

Banerjee, A., Sharma, R., Banerjee, U. C. (2002), The nitrile-degrading enzymes: current status and future prospects, *Appl. Microbiol. Biotechnol.* **60**, 33-44.

Banfi, L., Guanti, G., Riva, R. (1995), On the optimization of pig pancreatic lipase catalyzed monoacetylation of prochiral diols, *Tetrahedron: Asymmetry* **6**, 1345-1356.

Bänziger, M., Griffiths, G. J., McGarrity, J. F. (1993a), A facile synthesis of (2R,3E)-4-iodobut-3-en-2-ol and (2S,3E)-4-iodobut-3-en-2-yl chloroacetate, *Tetrahedron: Asymmetry* **4**, 723-726.

Bänziger, M., McGarrity, J. F., Meul, T. (1993b), A facile synthesis of N-protected statine and analogues via a lipase-catalyzes kinetic resolution, *J. Org. Chem.* **58**, 4010-4012.

Baraldi, P. G., Bazzanini, R., Manfredini, S., Simoni, D., Robins, M. J. (1993), Facile access to 2'-O-acyl prodrugs of 1-(β-D-arabinofuranosyl)-5(E)-(2-bromovinyl)uracil (BVARAU) via regioselective esterase-catalyzed hydrolysis of 2',3',5'-triester, *Tetrahedron Lett.* **48**, 7731-7734.

Barco, A., Benetti, S., Risi, C. D., Pollini, G. P., Romagnoli, R., Zanirato, V. (1994), A chemoenzymic approach to chiral phenylisoserinates using 4-isopropyl-2-oxazolin-5-one as masked umpoled synthon for hydroxycarbonyl anion, *Tetrahedron Lett.* **35**, 9289-9292.

Barili, P. L., Berti, G., Mastrorilli, E. (1993), Regio- and stereochemistry of the acid catalyzed and of a highly enantioselective enzymatic hydrolysis of some epoxytetrahydrofurans, *Tetrahedron* **49**, 6263-6276.

Barnier, J. P., Blanco, L., Rousseau, G., Guibé-Jampel, E., Fresse, I. (1993), Enzymic resolution of cyclopropanols. An easy access to optically active cyclohexanones possessing an α-quaternary chiral carbon, *J. Org. Chem.* **58**, 1570-1574.

Barnier, J. P., Blanco, L., Guibe-Jampel, E., Rousseau, G. (1989), Preparation of (*R*)-Veratryl- and (*R*)-(3-Methoxybenzyl)succinates, *Tetrahedron* **45**, 5051-5058.

Barton, P., Page, M. I. (1992), The resolution of racemic 1,2-diols by the esterase-catalyzed hydrolysis of the corresponding cyclic carbonate, *Tetrahedron* **48**, 7731-7734.

Barz, M., Herdtweck, E., Thiel, W. R. (1996), Kinetic resolution of *trans*-2-(1-pyrazolyl)cyclohexan-1-ol catalyzed by lipase B from *Candida antarctica*, *Tetrahedron: Asymmetry* **7**, 1717-1722.

Basak, A., Bhattacharya, G., Palit, S. K. (1997), Novel regioselective ester hydrolysis by pig-liver esterase, *Bull. Chem. Soc. Jpn.* **70**, 2509-2513.

Basavaiah, D., Rao, P. D. (1995), Synthesis of enantiomerically enriched *anti*-homoallyl alcohols mediated by crude chicken liver esterase (CCLE), *Tetrahedron: Asymmetry* **6**, 789-800.

Basavaiah, D., Krishna, P. R. (1994), Pig liver acetone powder (PLAP) as biocatalyst: enantioselective synthesis of *trans*-2-alkoxycyclohexan-1-ols, *Tetrahedron* **50**, 10521-10530.

Basavaiah, D., Rao, P. D. (1994), Enzymatic resolution of *trans*-2-arylcyclohexan-1-ols using crude chicken liver esterase (CCLE) as biocatalyst, *Tetrahedron: Asymmetry* **5**, 223-234.

Bashir, N. B., Phythian, S. J., Roberts, S. M., (1995), Enzymatic regioselective acylation and deacylation of carbohydrates, in.: *Preparative Biotransformations* (Roberts, S. M., Ed.), pp. 0:0.43-40:40.74. New York: Wiley.

Basri, M., Ampon, K., Yunus, W. M. Z. W., Razak, C. N. A., Salleh, A. B. (1995), Synthesis of fatty esters by polyethylene glycol-modified lipase, *J. Chem. Technol. Biotechnol.* **64**, 10-16.

Bassindale, A. R., Brandstadt, K. F., Lane, T. H., Taylor, P. G. (2003), Enzyme-catalysed siloxane bond formation, *J. Inorg. Biochem.* **96**, 401-406.

Battistel, E., Morra, M., Marinetti, M. (2001), Enzymatic surface modification of acrylonitrile fibers, *Appl. Surf. Sci.* **177**, 32-41.

Battistel, E., Bianchi, D., Cesti, P., Pina, C. (1991), Enzymatic resolution of (*S*)-(+)-naproxen in a continuous reactor, *Biotechnol. Bioeng.* **38**, 659-664.

Bauer, R., Knackmuss, H.-J., Stolz, A. (1998), Enantioselective hydration of 2-arylpropionitriles by a nitrile hydratase from *Agrobacterium tumefaciens* strain d3, *Appl. Microbiol. Biotechnol.* **49**, 89-95.

Bauer, R., Hirrlinger, B., Layh, N., Stolz, A., Knackmuss, H.-J. (1994), Enantioselective hydrolysis of racemic 2-phenylpropionitrile and other (*R*,*S*)-2-arylpropionitriles by a new bacterial isolate, *Agrobacterium tumefaciens* strain d3, *Appl. Microbiol. Biotechnol.* **42**, 1-7.

Baumann, M., Stürmer, R., Bornscheuer, U. T. (2001), A high-throughput-screening method for the identification of active and enantioselective hydrolases, *Angew. Chem. Int. Ed.* **40**, 4201-4204.

Baumann, M., Hauer, B., Bornscheuer, U. T. (2000), Rapid screening of hydrolases for the enantioselective conversion of "difficult-to-resolve substrates", *Tetrahedron: Asymmetry* **11**, 4781-4790.

Beard, T., Cohen, M. A., Paratt, J. S., Turner, N. J., Crosby, J., Moilliet, J. (1993), Stereoselective hydrolysis of nitriles and amides under mild conditions using a whole cell catalyst, *Tetrahedron: Asymmetry* **4**, 1085-1104.

Beer, H. D., Bornscheuer, U. T., McCarthy, J. E. G., Schmid, R. D. (1998), Cloning, expression, characterization and role of the leader sequence of a lipase from *Rhizopus oryzae*, *Biochim. Biophys. Acta* **1399**, 173-180.

Bell, G., Halling, P. J., Moore, B. D., Partidge, J., Rees, D. G. (1995), Biocatalyst behaviour in low-water systems, *Trends Biotechnol.* **13**, 468-473.

Bell, I. M., Fisher, M. L., Wu, Z. P., Hilvert, D. (1993), Kinetic studies on the peroxidase activity of selenosubtilisin, *Biochemistry*, **32**, 3754-3762.

Bell, P. A., Kasper, C. B. (1993), Expression of rat microsomal epoxide hydrolase in *Escherichia coli*. Identification of a histidyl residue essential for catalysis, *J. Biol. Chem.* **268**, 14011-14017.

Bello, M., Thomas, D., Legoy, M. D. (1987), Interesterification and synthesis by *Candida cylindracea* lipase in microemulsions, *Biochem. Biophys. Res. Commun.* **146**, 361-367.

Bellucci, G., Chiappe, C., Cordoni, A. (1996), Enantioconvergent transformation of racemic *cis*-β-alkyl substituted styrene oxides to (*R,R*) threo diols by microsomal epoxide hydrolase catalysed hydrolysis, *Tetrahedron: Asymmetry* **7**, 197-202.

Bellucci, G., Chiappe, C., Ingrosso, G., Rosini, C. (1995), Kinetic resolution by epoxide hydrolase catalyzed hydrolysis of racemic methyl substituted methylenecyclohexene oxides, *Tetrahedron: Asymmetry* **6**, 1911-1918.

Bellucci, G., Chiappe, C., Cordoni, A., Marioni, F. (1994a), Different enantioselectivity and regioselectivity of the cytosolic and microsomal epoxide hydrolase catalyzed hydrolysis of simple phenyl substituted epoxides, *Tetrahedron Lett.* **35**, 4219-4222.

Bellucci, G., Chiappe, C., Ingrosso, G. (1994b), Kinetics and stereochemistry of the microsomal epoxide hydrolase-catalyzed hydrolysis of *cis*-stilbene oxides, *Chirality* **6**, 577-582.

Bellucci, G., Chiappe, C., Marioni, F., Benetti, M. (1991), Regio- and enantio-selectivity of the cytosolic epoxide hydrolase-catalyzed hydrolysis of racemic monosubstituted alkyloxiranes, *J. Chem. Soc., Perkin Trans. 1* 361-363.

Bellucci, G., Capitani, I., Chiappe, C., Marioni, F. (1989a), Product enantioselectivity of the microsomal and cytosolic epoxide hydrolase catalysed hydrolysis of *meso* epoxides, *J. Chem. Soc., Chem. Commun.* 1170-1171.

Bellucci, G., Chiappe, C., Marioni, F. (1989b), Enantioselectivity of the enzymatic hydrolysis of cyclohexene oxide and (±)-1-methylcyclohexene oxide: A comparison between microsomal and cytosolic epoxide hydrolases, *J. Chem. Soc., Perkin Trans. 1* 2369-2373.

Bellucci, G., Berti, G., Catelani, C., Mastrorilli, E. (1981), Unusual steric course of the epoxide hydrolase catalyzed hydrolysis of (±)-3,4-epoxytetrahydropyran. A case of complete stereoconvergence, *J. Org. Chem.* **46**, 5148-5150.

Bencharit, S., Morton, C. L., Hyatt, J. L., Kuhn, P., Danks, M. K., Potter, P. M., Redinbo, M. R. (2003), Crystal structure of human carboxylesterase 1 complexed with the Alzheimer's drug tacrine: from binding promiscuity to selective inhibition, *Chem. Biol.* **10**, 341-349.

Bencharit, S., Morton, C. L., Howard-Williams, E. L., Danks, M. K., Potter, P. M., Redinbo, M. R. (2002), Structural insights into CPT-11 activation by mammalian carboxylesterases, *Nat. Struct. Biol.* **9**, 337-342.

Bengis-Garber, C., Gutman, A. L. (1989), Selective hydrolysis of dinitriles into cyano-carboxylic acids by *Rhodococcus rhodochrous* N.C.I.B. 11216, *Appl. Microbiol. Biotechnol.* **32**, 11-16.

Bengis-Garber, C., Gutman, A. L. (1988), Bacteria in organic synthesis: Selective conversion of 1,3-dicyanobenzene into 3-cyanobenzoic acid, *Tetrahedron Lett.* **29**, 2589-2590.

Berger, R., Hoffmann, M., Keller, U. (1998), Molecular analysis of a gene encoding a cell-bound esterase from *Streptomyces chrysomallus*, *J. Bacteriol.* **180**, 6396-6399.

Berger, A., Smolarsky, M., Kurn, N., Bosshard, H. R. (1973), A new method for the synthesis of optically active α-amino acids and their Na derivatives via acylamino malonates, *J. Org. Chem* **38**, 457-460.

Berger, B., Faber, K. (1991), 'Immunization' of lipase against acetaldehyde emerging in acyl transfer reactions from vinyl acetate, *J. Chem. Soc., Chem. Commun.* 1198-1200.

Berger, B., Rabiller, C. G., Königsberger, K., Faber, K., Griengl, H. (1990), Enzymatic acylation using acid anhydrides: crucial removal of acid, *Tetrahedron: Asymmetry* **1**, 541-546.

Berglund, P., Voerde, C., Hoegberg, H.-E. (1994), Esterification of 2-methylalkanoic acids catalyzed by lipase from *Candida rugosa*: enantioselectivity as a function of water activity and alcohol chain length, *Biocatalysis* **9**, 123-130.

Berkowitz, D. B., Maeng, J. H. (1996), Enantioselective entry into benzoxabicyclo[2.2.1]heptyl systems via enzymatic desymmetrization - toward chiral building blocks for lignan synthesis, *Tetrahedron: Asymmetry* **7**, 1577-1580.

Berkowitz, D. B., Pumphrey, J. A., Shen, Q. (1994), Enantiomerically enriched α-vinyl amino acids via lipase-mediated reverse transesterification, *Tetrahedron Lett.* **35**, 8743-8746.

Berkowitz, D. B., Danishefsky, S. J., Schulte, G. K. (1992), A route to artificial glycoconjugates and oligosaccharides via enzymatically resolved glycals: Dramatic effects of the handedness of the sugar domain upon the properties of an anthracycline drug, *J. Am. Chem. Soc.* **114**, 4518-4529.

Bernard, P., Barth, D. (1995), Internal mass transfer limitation during enzymic esterification in supercritical carbon dioxide and hexane, *Biocatal. Biotransform.* **12**, 299-308.

Bernard, P., Barth, D., Perrut, M. (1992), The integration of biocatalysis and downstream processing in supercritical carbon dioxide, in.: *High Press. Biotechnol.* (Balny, C., Hayashi, R., Heremans, K., Masson, P.; Eds.), **Vol. 224**, pp. 451-455.

Bernhardt, P., Hult, K., Kazlauskas, R. J. (2005), Molecular basis of perhydrolase activity in serine hydrolases, *Angew. Chem., Int. Ed.* **44**, 2742-2746.

Berry, D. R., Paterson, A. (1990), Enzymes in the food industry, in.: *Enzyme Chemistry* (Suckling, C. J.; Ed.), pp. 306-351. London: Chapman and Hall.

Besse, P., Veschambre, H. (1994), Chemical and biological synthesis of chiral epoxides, *Tetrahedron* **50**, 8885-8927.

Betzel, C., Klupsch, S., Papendorf, G., Hastrup, S., Branner, S., Wilson, K. S. (1992), Crystal structure of the alkaline proteinase Savinase from *Bacillus lentus* at 1.4 A resolution, *J. Mol. Biol.* **223**, 427-445.

Beveridge, A. J., Ollis, D. L. (1995), A theoretical study of substrate-induced activation of dienelactone hydrolase, *Prot. Eng.* **8**, 135-142.

Bevinakatti, H. S., Newadkar, R. V. (1993), Lipase catalysis in organic solvents. In search of practical derivatizing agents for the kinetic resolution of alcohols, *Tetrahedron: Asymmetry* **4**, 773-776.

Bevinakatti, H. S., Banerji, A. A., Newadkar, R. V., Mokashi, A. A. (1992), Enzymic synthesis of optically active amino acids. Effect of solvent on the enantioselectivity of lipase-catalyzed ring-opening of oxazolin-5-ones, *Tetrahedron: Asymmetry* **3**, 1505-1508.

Bevinakatti, H. S., Banerji, A. A. (1991), Practical chemoenzymic synthesis of both enantiomers of propranolol, *J. Org. Chem.* **56**, 5372-5375.

Bevinakatti, H. S., Newadkar, R. V., Banerji, A. A. (1990), Lipase-catalyzed enantioselective ring-opening of oxazol-5(4H)-ones coupled with partial in situ racemization of the less reactive isomer, *J. Chem. Soc., Chem. Commun.* 1091-1092.

Bhalerao, U. T., Dasaradhi, L., Neelankantan, P., Fadnavis, N. W. (1991), Lipase-catalyzed regio- and enantioselective hydrolysis: molecular recognition phenomenon and synthesis of *R*-dimorphecolic acid, *J. Chem. Soc., Chem. Commun.* 1197-1198.

Bhaskar-Rao, A., Rehman, H., Krishnakumari, B., Yadav, J. S. (1994), Lipase catalyzed kinetic resolution of racemic (±)-2,2-dimethyl-3-(2-methyl-1-propenyl)cyclopropane carboxyl esters, *Tetrahedron Lett.* **35**, 2611-2614.

Biadatti, T., Esker, J. L., Johnson, C. R. (1996), Chemoenzymatic synthesis of a versatile cyclopentenone: (+)-(3a*S*,6a*S*)-2,2-dimethyl-3aβ,6aβ-dihydro-4H-cyclopenta-1,3-dioxol-4-one, *Tetrahedron: Asymmetry* **7**, 2313-2320.

Bianchi, D., Bosetti, A., Golini, P., Cesti, P., Pina, C. (1997), Resolution of isopropylideneglycerol benzoate by sequential enzymatic hydrolysis and preferential crystallization, *Tetrahedron: Asymmetry* **8**, 817-819.

Bianchi, D., Cesti, P. (1990), Lipase-catalyzed stereoselective thiotransesterification of mercapto esters, *J. Org. Chem.* **55**, 5657-5659.

Bianchi, D., Cabri, W., Cesti, P., Francalanci, F., Rama, F. (1988a), Enzymatic resolution of 2,3-epoxyalcohols, intermediates in the synthesis of the gypsy moth sex pheromone, *Tetrahedron Lett.* **29**, 2455-2458.

Bianchi, D., Cabri, W., Cesti, P., Francalanci, F., Ricci, M. (1988b), Enzymic hydrolysis of alkyl 3,4-epoxybutyrates. A new route to (R)-(−)-carnitine chloride, *J. Org. Chem.* **53**, 104-107.

Bianchi, D., Cesti, P., Battistel, E. (1988c), Anhydrides as acylating agents in lipase-catalyzed stereoselective esterification of racemic alcohols, *J. Org. Chem.* **53**, 5531-5534.

Binns, F., Roberts, S. M., Taylor, A., Williams, C. F. (1993), Enzymic polymerization of an unactivated diol/diacid system, *J. Chem. Soc., Perkin Trans. I* 899-904.

Bis, S. J., Whitaker, D. T., Johnson, C. R. (1993), Lipase asymmetrization of *cis*-3,7-dihydroxycycloheptene derivatives in organic and aqueous media, *Tetrahedron: Asymmetry* **4**, 875-878.

Bisht, K. S., Kumar, A., Kumar, N., Parmar, V. S. (1996), Preparative and mechanistic aspects of interesterification reactions on diols and peracetylated polyphenolic compounds catalysed by lipases, *Pure Appl. Chem.* **68**, 749-752.

Bisht, K. S., Parmar, V. S. (1993), Diastereo- and enantioselective esterification of butane-2,3-diol catalyzed by the lipase from *Pseudomonas fluorescens*, *Tetrahedron: Asymmetry* **4**, 957-958.

Björkling, F., Frykman, H., Godtfredsen, S. E., Kirk, O. (1992), Lipase-catalyzed synthesis of peroxycarboxylic acids and lipase-mediated oxidations, *Tetrahedron* **48**, 4587-4592.

Björkling, F., Godtfredsen, S. E., Kirk, O. (1989), A highly selective enzyme-catalyzed esterification of simple glucosides, *J. Chem. Soc., Chem. Commun.* 934-935.

Björkling, F., Godtfredsen, S. E. (1988), New enzyme catalyzed synthesis of monoacyl galactoglycerides, *Tetrahedron* **44**, 2957-2962.

Björkling, F., Boutelje, J., Hjalmarsson, M., Hult, K., Norin, T. (1987), Highly enantioselective route to (R)-proline derivates via enzyme catalysed hydrolysis of *cis*-N-benzyl-2,5-bismethoxycarbonylpyrrolidine in an aqueous dimethyl sulphoxide medium, *J. Chem. Soc., Chem. Commun.* 1041-1042.

Björkling, F., Boutelje, J., Gatenbeck, S., Hult, K., Norin, T. (1985), Enzyme catalysed hydrolysis of dialkylated propanedioic acid diesters, synthesis of optically pure (S)-α-methylphenylalanine, (S)-α-methyltyrosine, and (S)-α-methyl-3,4-dihydroxyphenylalanine, *Tetrahedron Lett.* **26**, 4957-4958.

Blackman, R. L., Spence, J. M., Field, L. M., Devonshire, A. L. (1995), Chromosomal location of the amplified esterase genes conferring resistance to insecticides in *Mycus persicae* (Homoptera: Aphidae), *Heredity* **75**, 297-302.

Blackwood, A. D., Curran, L. J., Moore, B. D., Halling, P. J. (1994), Organic phase buffers control biocatalyst activity independent of initial aqueous pH, *Biochim. Biophys. Acta* **1206**, 161-165.

Blanco, L., Rousseau, G., Barnier, J. P., Guibé-Jampel, E. (1993), Enzymic resolution of 3-substituted-4-oxoesters, *Tetrahedron: Asymmetry* **4**, 783-792.

Blanco, L., Guibé-Jampel, E., Rousseau, G. (1988), Enzymatic resolution of racemic lactones, *Tetrahedron Lett.* **29**, 1915-1918.

Blée, E., Schuber, F. (1995), Stereocontrolled hydrolysis of the linoleic acid monoepoxide regioisomers catalyzed by soy-bean epoxide hydrolase, *Eur. J. Biochem.* **230**, 229-234.

Bloch, R., Guibé-Jampel, E., Girard, G. (1985), Stereoselective pig-liver esterase-catalyzed hydrolysis of rigid bicyclic *meso*-diester - preparation of optically pure 4,7-epoxytetrahydrophthalides and hexa-hydrophthalides, *Tetrahedron Lett.* **26**, 4087-4090.

Botta, M., Cernia, E., Corelli, F., Manetti, F., Soro, S. (1997), Probing the substrate specificity for lipases. 2. Kinetic and modeling studies on the molecular recognition of 2-arylpropionic esters by *Candida rugosa* and *Rhizomucor miehei* lipases, *Biochim. Biophys. Acta* **1337**, 302-310.

Bourne, Y., Martinez, C., Kerfelec, B., Lombardo, D., Chapus, C., Cambillau, C. (1994), Horse pancreatic lipase - the crystal structure refined at 2.3 Å resolution, *J. Mol. Biol.* **238**, 709-732.

Bovara, R., Carrea, G., Ottolina, G. Riva, S. (1993) Water activity does not influence the enantioselectivity of lipase PS and lipoprotein lipase in organic solvents, *Biotechnol. Lett.* **15**, 169-174.

Bovara, R., Carrea, G., Ferrara, L., Riva, S. (1991), Resolution of *trans*-Sobrerol by lipase PS-catalyzed transesterification and effects of organic solvents on enantioselectivity, *Tetrahedron: Asymmetry* **2**, 931-938.

Brackenridge, I., McCague, R., Roberts, S. M., Turner, N. J. (1993), Enzymatic resolution of oxalate esters of a tertiary alcohol using porcine pancreatic lipase, *J. Chem. Soc., Perkin Trans. 1* 1093-1094.

Brady, L., Brzozowski, A. M., Derewenda, Z. S., Dodson, E., Dodson, G., Tolley, S., Turkenburg, J. P., Christiansen, L., Huge-Jensen, B., Norskov, L., Thim, L., Menge, U. (1990), A serine protease triad forms the catalytic centre of a triacylglycerol lipase, *Nature* **343**, 767-770.

Brakmann, S., Schwienhorst, A. (Eds.) (2004), *Evolutionary Methods in Biotechnology: Clever Tricks for Directed Evolution*, Weinheim: Wiley-VCH.

Brakmann, S., Johnsson, K. (Eds.) (2002), *Directed Molecular Evolution of Proteins*, Weinheim: Wiley-VCH.

Brand, S., Jones, M. F., Rayner, C. M. (1995), The first examples of dynamic kinetic resolution by enantioselective acetylation of hemithioacetals: an efficient synthesis of homochiral α-acetoxysulfides, *Tetrahedron Lett.* **36**, 8493-8496.

Branden, C., Tooze, J. (1991), An example of enzyme catalysis: serine proteinases, in.: *Introduction to Protein Structure*, pp. 231-246. New York: Garland.

Branneby, C.; Carlqvist, P.; Hult, K.; Brinck, T.; Berglund, P. (2004), Aldol additions with mutant lipase: analysis by experiments and theoretical calculations, *J. Mol. Catal. B: Enzym.* **31**, 123-128.

Branneby, C.; Carlqvist, P.; Magnusson, A.; Hult, K.; Brinck, T.; Berglund, P. (2003), Carbon-carbon bonds by hydrolytic enzymes, *J. Am. Chem. Soc.* **125**, 874-875.

Braun, P., Waldmann, H., Kunz, H. (1993), Chemoenzymic synthesis of *O*-glycopeptides carrying the tumor associated TN-antigen structure, *Bioorg. Med. Chem.* **1**, 197-207.

Braun, P., Waldmann, H., Kunz, H. (1992), Selective enzymatic removal of protecting functions: heptyl esters as carboxy protecting groups in glycopeptide synthesis, *Synlett* 39-40.

Braun, P., Waldmann, H., Vogt, W., Kunz, H. (1991), Selective enzymatic removal of protecting groups: *n*-heptyl esters as carboxy protecting functions in peptide synthesis, *Liebigs Ann. Chem.* 165-170.

Braun, P., Waldmann, H., Vogt, W., Kunz, H. (1990), Selective enzymatic removal of protecting functions: heptyl esters as carboxy protecting groups in peptide synthesis, *Synthesis* 105-107.

Brazwell, E. M., Filos, D. Y., Morrow, C. J. (1995), Biocatalytic synthesis of polymers. III. Formation of a high molecular weight polyester through limitation of hydrolysis by enzyme-bound water and through equilibrium control, *J. Polym. Sci., Part A: Polym. Chem.* **33**, 89-95.

Breitgoff, D., Laumen, K., Schneider, M. P. (1986), Enzymatic differentiation of the enantiotopic hydroxymethyl groups of glycerol; synthesis of chiral building blocks, *J. Chem. Soc., Chem. Commun.* 1523-1524.

Brennan, B. A., Alms, G., Nelson, M. J., Durney, L. T., Scarrow, R. C. (1996), Nitrile hydratase from *Rhodococcus rhodochrous* J1 contains a non-corrin cobalt ion with two sulfur ligands, *J. Am. Chem. Soc.* **118**, 9194-9195.

Bresler, M. M., Rosser, S. J., Basran, A., Bruce, N. C. (2000), Gene cloning and nucleotide sequencing and properties of a cocaine esterase from *Rhodococcus* sp. strain MB1, *Appl. Environ. Microbiol.* **66**, 904-908.

Breuer, M., Ditrich, K., Habicher, T., Hauer, B., Kesseler, M., Stuermer, R., Zelinski, T. (2004), Industrial methods for the production of optically active intermediates, *Angew. Chem. Int. Ed.* **43**, 788-824.

Breznik, M., Mrcina, A., Kikelj, D. (1998), Enantioselective synthesis of (R)- and (S)-2-methyl-3-oxo-3,4-dihydro-2H-1,4-benzoxazine-2-carboxamides, *Tetrahedron:Asymmetry* **9**, 1115-1116.

Breznik, M., Kikelj, D. (1997), Pig liver esterase catalyzed hydrolysis of dimethyl and diethyl 2-methyl-2-(o-nitrophenoxy)malonates, *Tetrahedron:Asymmetry* **8**, 425-434.

Brieva, R., Crich, J. Z., Sih, C. J. (1993), Chemoenzymatic synthesis of the C-13 side chain of taxol: optically-active 3-hydroxy-4-phenyl β-lactam derivatives, *J. Org. Chem.* **58**, 1068-1075.

Brieva, R., Rebolledo, F., Gotor, V. (1990), Enzymatic synthesis of amides with two chiral centres, *J. Chem. Soc., Chem. Commun.* 1386-1387.

Brocca, S., Schmidt-Dannert, C., Lotti, M., Alberghina, L., Schmid, R. D. (1998), Design, total synthesis, and functional overexpression of the *Candida rugosa lip1* gene encoding for a major industrial lipase, *Protein Sci.* **7**, 1415-1422.

Broos, J., Engbersen, J. F. J., Verboom, W., Reinhoudt, D. N. (1995a), Inversion of enantioselectivity of serine proteases, *Recl. Trav. Chim. Pays-Bas* **114**, 255-257.

Broos, J., Visser, A. J. W. G., Engbersen, J. F. J., Verboom, W., van Hoek, A., Reinhoudt, D. N. (1995b), Flexibility of enzymes suspended in organic solvents probed by time-resolved fluorescence anisotropy. Evidence that enzyme activity and enantioselectivity are directly related to enzyme flexibility, *J. Am. Chem. Soc.* **117**, 12657-12663.

Brown, S. A.; Parker, M.-C.; Turner, N. J. (2000), Dynamic kinetic resolution: synthesis of optically active α-amino acid derivatives, *Tetrahedron: Asymmetry* **11**, 1687-1690.

Brown, S. M., Davies, S. G., de Sousa, J. A. A. (1993), Kinetic resolution strategies II: Enhanced enantiomeric excesses and yields for the faster reacting enantiomer in lipase mediated kinetic resolutions, *Tetrahedron: Asymmetry* **4**, 813-822.

Brozozowski, A. M., Savage, H., Verma, C. S., Turkenburg, J. P., Lawson, D. M., Svendsen, A., Patkar, S. (2000) Structural origins of the interfacial activation in *Thermomyces* (*Humicola*) *lanuginosa* lipase, *Biochemistry* **39**, 15071-15082.

Bruce, M. A., St. Laurent, D. R., Poindexter, G. S., Monkovic, I., Huang, S., Balasubramanian, N. (1995), Kinetic resolution of piperazine-2-carboxamide by leucine aminopeptidase. An application in the synthesis of the nucleoside transport blocker (–) draflazine, *Synth. Commun.* **25**, 2673-2684.

Bruzik, K. S., Guan, Z., Riddle, S., Tsai, M.-D. (1996), Synthesis of inositol phosphodiesters by phospholipase C-catalyzed transesterification, *J. Am. Chem. Soc.* **118**, 7679-7688.

Brzozowski, A. M., Derewenda, U., Derewenda, G. G., Dodson, D. M., Lawson, J., Turkenburg, P., Bjorkling, F., Huge-Jensen, B., Patkar, S. A., Thim, L. (1991), A model for interfacial activation in lipases from the structure of a fungal lipase-inhibitor complex, *Nature* **351**, 491-494.

Bucciarelli, M., Formi, A., Moretti, I., Prati, F. (1988), Enzymatic resolution of chiral N-alkyloxaziridine-3,3-dicarboxylic esters, *J. Chem. Soc., Chem. Commun.* 1614-1615.

Bunch, A. W. (1998), Nitriles, in.: *Biotechnology-Series* (Rehm, H. J., Reed, G., Pühler, A., Stadler, P. J. W., Kelly, D. R.; Eds.), Vol. 8a, pp. 277-324. Weinheim: VCH-Wiley.

Burgess, K., Henderson, I., Ho, K.-K. (1992), Biocatalytic resolutions of sulfinylalkanoates: a facile route to optically active sulfoxides, *J. Org. Chem.* **57**, 1290-1295.

Burgess, K., Ho, K.-K. (1992), Asymmetric syntheses of all four stereoisomers of 2,3-methanomethionine, *J. Org. Chem.* **57**, 5931-5936.

Burgess, K., Henderson, I. (1991), Biocatalytic desymmetrizations of pentitol derivatives, *Tetrahedron Lett.* **32**, 5701-5704.

Burgess, K., Jennings, L. D. (1991), Enantioselective esterifications of unsaturated alcohols mediated by a lipase prepared from *Pseudomonas* sp., *J. Am. Chem. Soc.* **113**, 6129-6139.

Burgess, K., Henderson, I. (1990), Lipase-mediated resolution of SPAC reaction products, *Tetrahedron: Asymmetry* **1**, 57-60.

Burgess, K., Henderson, I. (1989), A facile route to homochiral sulfoxides, *Tetrahedron Lett.* **30**, 3633-3636.

Cadwell, R. C., Joyce, G. F. (1992), Randomization of genes by PCR mutagenesis, *PCR Meth. Appl.* **2**, 28-33.

Cai, Y., Yao, S.-P., Wu, Q., Lin, X-F. (2004), Michael addition of imidazole with acrylates catalyzed by alkaline protease from *Bacillus subtilis* in organic media, *Biotechnol. Lett.* **26**, 525-528.

Caille, J. C., Govindan, C. K., Junga, H., Lalonde, J., Yao, Y. (2002), Hetero Diels-Alder-biocatalysis approach for the synthesis of (S)-3-[2-{(methylsulfonyl) oxy}ethoxy]-4-(triphenylmethoxy)-1-butanol methanesulfonate, a key intermediate for the synthesis of the PKC inhibitor LY333531, *Org. Proc. Res. Dev.* **6**, 471-476.

Cainelli, G., Giacomini, D., Galletti, P., DaCol, M. (1997), Penicillin G acylase mediated synthesis of the enantiopure (S)-3-amino- azetidin-2-one, *Tetrahedron: Asymmetry* **8**, 3231-3235.

Cambillau, C., van Tilbeurgh, H. (1993), Structure of hydrolases - lipases and cellulases, *Curr. Opin. Struct. Biol.* **3**, 885-895.

Cambou, B., Klibanov, A. M. (1984), Preparative production of optically-active esters and alcohols using esterase-catalyzed stereospecific transesterification in organic media, *J. Am. Chem. Soc.* **106**, 2687-2692.

Cao, L. (2005), *Carrier-bound Immobilized Enzymes*, Weinheim: Wiley-VCH.

Cao, L., Langen, L., Sheldon, R. A. (2003), Immobilised enzymes: carrier-bound or carrier-free? *Curr. Opin. Biotechnol.* **14**, 387-394.

Cao, L., van Langen, L. M., van Rantwijk, F., Sheldon, R. A. (2001), Cross-linked aggregates of penicillin acylase: robust catalysts for the synthesis of β-lactam antibiotics, *J. Mol. Catal. B: Enzym.* **11**, 665-670.

Cao, L., van Rantwijk, F., Sheldon, R. A. (2000), Cross-linked enzyme aggregates: a simple and effective method for the immobilization of penicillin acylase, *Org. Lett.* **2**, 1361-1364.

Cao, L., Fischer, A., Bornscheuer, U. T., Schmid, R. D. (1997), Lipase-catalyzed solid phase preparation of sugar fatty acid esters, *Biocatal. Biotransform.* **14**, 269-283.

Cao, L., Bornscheuer, U. T., Schmid, R. D. (1996), Lipase-catalyzed solid-phase synthesis of sugar esters, *Fett/Lipid* **98**, 332-335.

Capewell, A., Wendel, V., Bornscheuer, U., Meyer, H. H., Scheper, T. (1996), Lipase-catalyzed kinetic resolution of 3-hydroxy esters in organic solvents and supercritical carbon dioxide, *Enzyme Microb. Technol.* **19**, 181-186.

Carda, M., van der Eycken, J., Vandewalle, M. (1990), Enantiotoposelective PLE-catalyzed hydrolysis of *cis*-5-substituted-1,3-diacyloxycyclohexanes. Preparation of some useful chiral building blocks, *Tetrahedron: Asymmetry* **1**, 17-20.

Cardellicchio, C., Naso, F., Scilimati, A. (1994), An efficient biocatalyzed kinetic resolution of methyl (Z)-3-(arylsulfinyl)propenoates, *Tetrahedron Lett.* **35**, 4635-4638.

Carlqvist, P., Svedendahl, M., Branneby, C., Hult, K., Brinck, T., Berglund, P. (2005), Exploring the active-site of a rationally redesigned lipase for catalysis of Michael-type additions, *ChemBioChem*, **6**, 331-336.

Caron, G., Kazlauskas, R. J. (1994), Isolation of racemic 2,4-pentanediol and 2,5-hexanediol from commercial mixtures of *meso* and racemic isomers by way of cyclic sulfites, *Tetrahedron: Asymmetry* **5**, 657-664.

Caron, G., Kazlauskas, R. J. (1993), Sequential kinetic resolution of (±)-2,3-butanediol using lipase from *Pseudomonas cepacia*, *Tetrahedron: Asymmetry* **4**, 1995-2000.

Caron, G., Kazlauskas, R. J. (1991), An optimized sequential kinetic resolution of trans-1,2- cyclohexanediol, *J. Org. Chem.* **56**, 7251-7256.

Carrea, G., Ottolina, G., Riva, S. (1995), Role of solvents in the control of enzyme selectivity in organic media, *Trends Biotechnol.* **13**, 63-70.

Carrea, G., Danieli, B., Palmisano, G., Riva, S., Santagostino, M. (1992), Lipase mediated resolution of 2-cyclohexen-1-ols as chiral building blocks en route to eburane alkaloids, *Tetrahedron: Asymmetry* **3**, 775-784.

Carrea, G., Riva, S., Secundo, F., Danieli, B. (1989), Enzymatic synthesis of various 1-O-sucrose and 1-O-fructose esters, *J. Chem. Soc., Perkin Trans. I* 1057-1061.

Carrillo-Munoz, J. R., Bouvet, D., Guibé-Jampel, E., Loupy, A., Petit, A. (1996), Microwave-promoted lipase-catalyzed reactions. Resolution of (±)-1-phenylethanol, *J. Org. Chem.* **61**, 7746-7749.

Carter, S. F., Leak, D. J. (1995), The isolation and characterization of a carbocyclic epoxide-degrading *Corynebacterium* sp., *Biocatal. Biotransform.* **13**, 111-129.

Castillo, E., Marty, A., Combes, D., Condoret, J. S. (1994), Polar substrates for enzymatic reactions in supercritical CO_2: How to overcome the solubility limitation, *Biotechnol. Lett.* **16**, 169-174.

Catelani, C., Mastrorilli, E. (1983), Acid-catalyzed and enzymatic hydrolysis of trans- and cis-2-methyl-3,4-epoxytetrahydropyran, *J. Chem. Soc., Perkin Trans. 1* 2717-2721.

Celia, E. C., Cernia, E., D'Acquarica, I., Palocci, C., Soro, S. (1999), High yield and optical purity in biocatalysed acylation of trans-2-phenyl-1-cyclohexanol with *Candida rugosa* lipase in non-conventional media, *J. Mol. Catal. B: Enzym.*, **6**, 495-503.

Cernia, E., Palocci, C., Gasparrini, F., Misiti, D. (1994a), *Pseudomonas* lipase catalytic activity in supercritical carbon dioxide, *Chem. Biochem. Eng. Q.* **8**, 1-4.

Cernia, E., Palocci, C., Gasparrini, F., Misiti, D., Fagnano, N. (1994b), Enantioselectivity and reactivity of immobilized lipase in supercritical carbon dioxide, *J. Mol. Catal.* **89**, L11-L18.

Chadha, A., Manohar, M. (1995), Enzymic resolution of 2-hydroxy-4-phenylbutanoic acid and 2-hydroxy-4-phenylbutenoic acid, *Tetrahedron: Asymmetry* **6**, 651-652.

Chahid, Z., Montet, D., Pina, M., Graille, J. (1992), Effect of water activity on enzymatic synthesis of alkylglycosides, *Biotechnol. Lett.* **14**, 281-284.

Chamorro, C., Gonzalez-Muniz, R., Conde, S. (1995), Regio- and enantioselectivity of the *Candida antarctica* lipase catalyzed amidations of Cbz-L- and Cbz-D-glutamic acid diesters, *Tetrahedron: Asymmetry* **6**, 2343-2352.

Chan, C., Cox, P. B., Roberts, S. M. (1990), Chemo-enzymatic synthesis of 13-(S)-hydroxyoctadecadienoic acid (13-S-HODE), *Biocatalysis* **3**, 111-118.

Chang, P. S., Rhee, J. S., Kim, J. J. (1991), Continuous glycerolysis of olive oil by *Chromobacterium viscosum* lipase immobilized on liposome in reversed micelles, *Biotechnol. Bioeng.* **38**, 1159-1165.

Chang, R. C., Chou, S. J., Shaw, J. F. (1994), Multiple forms and functions of *Candida rugosa* lipase, *Biotechnol. Appl. Biochem.* **19**, 93-97.

Chapman, D. T., Crout, D. H. G., Mahmoudian, M., Scopes, D. I. C., Smith, P. W. (1996), Enantiomerically pure amines by a new method: biotransformation of oxalamic esters using the lipase from *Candida antarctica*, *Chem. Commun.* 2415-2416.

Chartrain, M., Katz, L., Marcin, C., Thien, M., Smith, S., Fisher, E., Goklen, K., Salmon, P., al., T. B. e. (1993), Purification and characterization of a novel bioconverting lipase from *Pseudomonas aeruginosa* MB 5001, *Enzyme Microb. Technol.* **15**, 575-580.

Chattapadhyay, S., Mamdapur, V. R. (1993), Enzymic esterification of 3-hydroxybutyric acid, *Biotechnol. Lett.* **15**, 245-250.

Chaudhary, A. K., Beckman, E. J., Russell, A. J. (1998), Nonequal reactivity model for biocatalytic polytransesterification, *AICHE J.*, **44**, 753-764.

Chaudhary, A., Beckman, E. J., Russell, A. J. (1995), Rational control of polymer molecular weight and dispersity during enzyme-catalyzed polyester synthesis in supercritical fluids, *J. Am. Chem. Soc.* **117**, 3728-3733.

Cheeseman, J. D., Tocilj, A., Park, S., Schrag, J. D., Kazlauskas, R. J. (2004), X-Ray crystal structure of an aryl esterase from *Pseudomonas fluorescens*, Acta Crystallogr. D, **60**, 1237-1243.

Cheetham, P. S. J. (1997), Combining the technical push and the business pull for natural flavors, *Adv. Biochem. Eng./Biotechnol.* **55**, 1-49.

Cheetham, P. S. J. (1993), The use of biotransformations for the production of flavours and fragrances, *Trends Biotechnol.* **11**, 478-488.

Chen, C. S., Liu, Y. C. (1991), Amplification of enantioselectivity in biocatalyzed kinetic resolution of racemic alcohols, *J. Org. Chem.* **56**, 1966-1968.

Chen, C. S., Sih, C. J. (1989), Enantioselective biocatalysis in organic solvents. Lipase catalyzed reactions, *Angew. Chem. Int. Ed.* **28**, 695-707.

Chen, C. S., Wu, S. H., Girdaukas, G., Sih, C. J. (1987), Quantitative analyses of biochemical kinetic resolution of enantiomers. 2. Enzyme-catalyzed esterifications in water-organic solvent biphasic systems, *J. Am. Chem. Soc.* **109**, 2812-2817.

Chen, C. S., Fujimoto, Y., Girdaukas, G., Sih, C. J. (1982), Quantitative analyses of biochemical kinetic resolutions of enantiomers, *J. Am. Chem. Soc.* **104**, 7294-7299.

Chen, C. S., Fujimoto, Y., Sih, C. J. (1981), Bifunctional chiral synthons via microbiological methods. 1. Optically active 2,4-dimethylglutaric acid monomethyl esters, *J. Am. Chem. Soc.* **103**, 3580-3582.

Chen, P.-Y., Wu, S.-H., Wang, K.-T. (1993), Double enantioselective esterification of racemic acids and alcohols by lipase from *Candida cylindracea*, *Biotechnol. Lett.* **15**, 181-184.

Chen, J. C., Miercke, L. J., Krucinski, J., Starr, J. R., Saenz, G., Wang, X., Spilburg, C. A., Lange, L. G., Ellsworth, J. L., Stroud, R. M. (1998) Structure of bovine pancreatic cholesterol esterase at 1.6 Å: novel structural features involved in lipase activation, *Biochemistry* **37**, 5107-5117.

Chen, S. T., Fang, J. M. (1997), Preparation of optically active tertiary alcohols by enzymatic methods. Application to the synthesis of drugs and natural products, *J. Org. Chem.* **62**, 4349-4357.

Chen, S.-T., Wang, K.-T., Wong, C.-H. (1986), Chirally selective hydrolysis of D,L-amino acid esters by alkaline protease, *J. Chem. Soc., Chem. Commun.* 1514-1516.

Chen, X.-J., Archelas, A., Furstoss, R. (1993), Microbiological transformations. 27. The first examples for preparative-scale enantioselective or diastereoselective epoxide hydrolyses using microorganisms. An unequivocal access to all four Bisabolol stereoisomers, *J. Org. Chem.* **58**, 5528-5532.

Chen, X., Siddiqi, S. M., Schneller, S. W. (1992), Chemoenzymatic synthesis of (−)-carbocyclic 7-deazaoxetanocin G, *Tetrahedron Lett.* **33**, 2249-2252.

Chenault, H. K., Dahmer, J., Whitesides, G. M. (1989), Kinetic resolution of unnatural and rarely occurring amino acids: enantioselective hydrolysis of N-acyl amino acids catalyzed by acylase I, *J. Am. Chem. Soc.* **111**, 6354-6364.

Chênevert, R., Dickman, M. (1996), Enzymatic route to chiral, nonracemic cis-2,6- and cis,cis-2,4,6-substituted piperidines. Synthesis of (+)-dihydropinidine and dendrobate alkaloid (+)-241D, *J. Org. Chem.* **61**, 3332-3341.

Chênevert, R., Desjardins, M. (1996), Enzymatic desymmetrization of meso-2,6-dimethyl-1,7-heptanediol. Enantioselective formal synthesis of the vitamin E side chain and the insect pheromone tribolure, *J. Org. Chem.* **61**, 1219-1222.

Chênevert, R., Courchesne, G. (1995), Enzymatic desymmetrization of meso(anti-anti)-2,4-dimethyl-1,3,5-pentanetriol, *Tetrahedron: Asymmetry* **6**, 2093-2096.

Chênevert, R., Lavoie, M., Courchesne, G., Martin, R. (1994a), Chemoenzymatic enantioselective synthesis of amidinomycin, *Chem. Lett.* 93-96.

Chênevert, R., Pouliot, R., Bureau, P. (1994b), Enantioselective hydrolysis of (±)-chloramphenicol palmitate by hydrolases, *Bioorg. Med. Chem. Lett.* **4**, 2941-2944.

Chênevert, R., Gagnon, R. (1993), Lipase-catalyzed enantioselective esterification or hydrolysis of 1-*O*-alkyl-3-*O*-tosylglycerol derivatives. Practical synthesis of (*S*)-(+)-1-*O*-hexadecyl-2,3-di-*O*-hexadecanoylglycerol, a marine natural product, *J. Org. Chem.* **58**, 1054-1057.

Chênevert, R., Rhlid, R. B., Letourneau, M., Gagnon, R., D'Astous, L. (1993), Selectivity in carbonic anhydrase catalyzed hydrolysis of standard N-acetyl-D,L-amino acid methyl esters, *Tetrahedron: Asymmetry* **4**, 1137-1140.

Chênevert, R., Dickman, M. (1992), Enzyme-catalyzed hydrolysis of *N*-benzyloxycarbonyl-*cis*-2,6-(acetoxymethyl)piperidine. A facile route to optically active piperidines, *Tetrahedron: Asymmetry* **3**, 1021-1024.

Chênevert, R., Martin, R. (1992), Enantioselective synthesis of (+) and (-)-*cis*-3-aminocyclopentane carboxylic acids by enzymatic asymmetrization, *Tetrahedron: Asymmetry* **3**, 199-200.

Chênevert, R., Desjardins, M., Gagnon, R. (1990), Enzyme catalyzed asymmetric hydrolysis of chloral acetyl methyl acetal, *Chem. Lett.* 33-34.

Chênevert, R., Thiboutot, S. (1989), Synthesis of chloramphenicol via an enzymatic enantioselective hydrolysis, *Synthesis* 444-446.

Chênevert, R., D'Astous, L. (1988), Enzyme-catalyzed hydrolysis of chlorophenoxypropionates, *Can. J. Chem.* **66**, 1219-1222.

Cheng, T. C., Harvey, S. P., Chen, G. L. (1996), Cloning and expression of a gene encoding a bacterial enzyme for decontamination of organophosphorus nerve agents and nucleotide sequence of the enzyme, *Appl. Environ. Microbiol.* **62**, 1636-1641.

Chen-Goodspeed, M., Sogorb, M. A., Wu, F., Hong, S. B., Raushel F. M. (2001a), Structural determinants of the substrate and stereochemical specificity of phosphotriesterase, *Biochemistry* **40**, 1335-1331.

Chen-Goodspeed, M., Sogorb, M. A., Wu, F., Raushel, F. M. (2001b), Enhancement, relaxation, and reversal of the stereoselectivity for phosphotriesterase by rational evolution of active site residues, *Biochemistry*, **40**, 1332-1339.

Cheon, Y.-H., Park, H.-S., Kim, J.-H., Kim, Y., Kim H.-S. (2004), Manipulation of the active site loops of D-hydantoinase, a (β/α)$_8$-barrel protein, for modulation of the substrate specificity, *Biochemistry* **43**, 7413-7420.

Cheong, C. S., Im, D. S., Kim, J., Kim, I. O. (1996), Lipase-catalyzed resolution of primary alcohol containing quaternary chiral carbon, *Biotechnol. Lett.* **18**, 1419-1422.

Chi, Y. M., Nakamura, K., Yano, T. (1988), Enzymic interesterification in supercritical carbon dioxide, *Agric. Biol. Chem.* **52**, 1541-1550.

Chiappe, C., Cordoni, A., Moro, G. L., Palese, C. D. (1998), Deracemization of (±)-*cis*-dialkyl substituted oxides via enantioconvergent hydrolysis catalyzed by microsomal epoxide hydrolase, *Tetrahedron: Asymmetry* **9**, 341-350.

Chibata, I., Tosa, T., Shibatani, T. (1992), The industrial production of optically active compounds by immobilized biocatalysts, in.: *Chirality in Industry* (Collins, A. N., Sheldrake, G. N., Crosby, J.; Eds.), pp. 352-370. Chichester, UK: Wiley.

Chin, J. T., Wheeler, S. L., Klibanov, A. M. (1994) On protein solubility in organic solvents. *Biotechnol. Bioeng.* **44**, 140-145.

Chinn, M. J., Iacazio, G., Spackman, D. G., Turner, N. J., Roberts, S. M. (1992), Regioselective enzymic acylation of methyl 4,6-*O*-benzylidene-α- and β-D-glucopyranoside, *J. Chem. Soc., Perkin Trans. I* 661-662.

Chiou, A. J., Wu, S. H., Wang, K. T. (1992), Enantioselective hydrolysis of hydrophobic amino acid derivatives by lipases, *Biotechnol. Lett.* **14**, 461-464.

Choi, J. H., Choi, Y. K., Kim, Y. H., Park, E. S., Kim, E. J., Kim, M.-J., Park J. (2004), Aminocyclopentadienyl ruthenium complexes as racemization catalysts for dynamic kinetic resolution of secondary alcohols at ambient temperature, *J. Org. Chem.* **69**, 1972-1977.

Choi, J. H., Kim, Y. H., Nam, S. H., Shin, S. T., Kim, M.-J., Park J. (2002), Aminocyclopentadienyl ruthenium chloride: catalytic racemization and dynamic kinetic resolution of alcohols at ambient temperature, *Angew. Chem. Int. Ed.* **41**, 2373-2376.

Choi, W. J., Huh, E. C., Park, H. J., Lee, E. Y., Choi, C. Y. (1998), Kinetic resolution for optically active epoxides by microbial enantioselective hydrolysis, *Biotechnol. Techn.* **12**, 225-228.

Choi, K. D., Jeohn, G. H., Rhee, J. S., Yoo, O. J. (1990), Cloning and nucleotide sequence of an esterase gene from *Pseudomonas fluorescens* and expression of the gene in *Escherichia coli*, *Agric Biol. Chem.* **54**, 2039-2045.

Choi, Y. K., Kim, M. J., Ahn, Y., Kim, M. J. (2001), Lipase/palladium-catalyzed asymmetric transformations of ketoximes to optically active amines, *Org. Lett.* **3**, 4099-4101.

Choi, Y. K., Suh, J. H., Lee, D., Lim, I. T., Jung, J. Y., Kim, M.-J. (1999), Dynamic kinetic resolution of acyclic allylic acetates using lipase and palladium, *J. Org. Chem.* **64**, 8423-8424.

Chong, J. M., Mar, E. K. (1991), Preparation of enantiomerically enriched α-hydroxystannanes via enzymatic resolution, *Tetrahedron Lett.* **32**, 5683-5686.

Chulalaksananukul, W., Condorét, J. S., Combes, D. (1993), Geranyl acetate synthesis by lipase-catalyzed transesterification in supercritical carbon dioxide, *Enzyme Microb. Technol.* **15**, 691-698.

Cipiciani, A., Fringuelli, F., Scappini, A. M. (1996), Enzymatic hydrolyses of acetoxy- and phenethylbenzoates by *Candida cylindracea* lipase, *Tetrahedron* **52**, 9869-9876.

Ciuffreda, P., Colombo, D., Ronchetti, F., Toma, L. (1990), Regioselective acylation of 6-deoxy-L- and -D-hexosides through lipase-catalyzed transesterification. Enhanced reactivity of the 4-hydroxy function in the L series, *J. Org. Chem.* **55**, 4187-4190.

Cleij, M., Archelas, A., Furstoss, R. (1998), Microbiological transformations. part 42: A two-liquid-phase preparative scale process for an epoxide hydrolase catalyzed resolution of *para*-bromo-α-methyl styrene oxide. Occurrence of a surprising enantioselectivity enhancement, *Tetrahedron: Asymmetry* **9**, 1839-1842.

Cohen, M. A., Paratt, J. S., Turner, N. J. (1992), Enantioselective hydrolysis of nitriles and amides using an immobilized whole cell system, *Tetrahedron: Asymmetry* **3**, 1543-1546.

Cohen, S. G., Lo, L. W. (1970), On the active site of α-chymotrypsin. Cyclized and noncyclized substrates with tetrasubstituted α-carbon atoms, *J. Biol. Chem.* **245**, 5718-5727.

Cohen, S. G., Milanovic, A., Schultz, R. M., Weinstein, S. Y. (1969), On the active site of α-chymotrypsin. Absolute configurations and kinetics of hydrolysis of cyclized and noncyclized substrates, *J. Biol. Chem.* **244**, 2664-2674.

Cohen, S. G., Schultz, R. M., Weinstein, S. Y. (1966), Stereospecificity in hydrolysis by α-chymotrypsin of esters of α,α-disubstituted acetic and β,β-disubstituted propionic acids, *J. Am. Chem. Soc.* **88**, 5315-5319.

Cohen, S. G., Crossley, J. (1964), Kinetics of hydrolysis of dicarboxylic esters and their a-acetamido dderivatives by α-chymotrypsin, *J. Am. Chem. Soc.* **86**, 4999-5003.

Cohen, S. G., Crossley, J., Khedouri, E. (1963), Action of α-chymotrypsin on diethyl N-acetylaspartate and on diethyl N-methyl-N-acetylaspartate, *Biochemistry* **2**, 820-823.

Cohen, S. G., Weinstein, S. Y. (1964), Hydrolysis of D(–)-ethyl β-phenyl-β-hydroxypropionate and D(–)-ethyl β-phenyl-β-acetamidopropionate by α-chymotrypsin, *J. Am. Chem. Soc.* **86**, 725-728.

Coleman, M. H., Macrae, A. R. (1977), Rearrangement of fatty acid esters in fat reaction reactants, *German Patent.* DE 2 705 608.

Collins, A. N., Sheldrake, G. N., Crosby, J. (Eds.) (1992), *Chirality in industry - the commercial manufacture and application of optically-active compounds*, New York: Wiley.

Colton, I. J., Ahmed, S. N., Kazlauskas, R. J. (1995), Isopropanol treatment increases enantioselectivity of *Candida rugosa* lipase toward carboxylic acid esters, *J. Org. Chem.* **60**, 212-217.

Comins, D. L., Salvador, J. M. (1993), Efficient synthesis and resolution of *trans*-2-(1-aryl-1-methylethyl)cyclohexanols: practical alternatives to 8-phenylmenthol, *J. Org. Chem.* **58**, 4656-4661.

Coope, J. F., Main, B. G. (1995), Biocatalytic resolution of a tertiary quinuclidinol ester using pig liver esterase, *Tetrahedron: Asymmetry* **6**, 1393-1398.

Copley, S. D. (2003), Enzymes with extra talents: moonlighting functions and catalytic promiscuity, *Curr. Opin. Chem. Biol.* **7**, 265-272.

Cotterill, I. C., Sutherland, A. G., Roberts, S. M., Grobbauer, R., Spreitz, J., Faber, K. (1991), Enzymatic resolution of sterically demanding bicyclo[3.2.0]heptanes: evidence for a novel hydrolase in crude porcine pancreatic lipase and the advantages of using organic media for some of the biotransformations, *J. Chem. Soc., Perkin Trans. 1* 1365-1368.

Cotterill, I. C., Dorman, G., Faber, K., Jaouhari, R., Roberts, S. M., Scheinmann, F., Spreitz, J., Sutherland, A. G., Winders, J. A., Wakefield, B. J. (1990), Chemoenzymic, enantiocomplementary, total asymmetric synthesis of leukotrienes-B3 and -B4, *J. Chem. Soc., Chem. Commun.* 1661-1663.

Cotterill, I. C., Finch, H., Reynolds, D. P., Roberts, S. M., Rzepa, H. S., Short, K. M., Slawin, A. M. Z., Wallis, C. J., Williams, D. J. (1988a), Enzymatic resolution of bicyclo[4.2.0]oct-2-en-7-ol and the preparation of some polysubstituted bicyclo[3.3.0.]octan-2-ones via highly strained tricyclo[4.2.0.01,5]1octan-8-ones, *J. Chem. Soc., Chem. Commun.* 470-472.

Cotterill, I. C., Macfarlane, E. L. A., Roberts, S. M. (1988b), Resolution of bicyclo[3.2.0]hept-2-en-6-ols and bicyclo[4.2.0]oct-2-en-endo-7-ol using lipases, *J. Chem. Soc., Perkin Trans. 1* 3387-3389.

Cousins, R. P. C., Mahmoudian, M., Youds, P. M. (1995), Enzymic resolution of oxathiolane intermediates - an alternative approach to the anti-viral agent lamivudine (3TC), *Tetrahedron: Asymmetry* **6**, 393-396.

Crameri, A., Raillard, S. A., Bermudez, E., Stemmer, W. P. (1998), DNA shuffling of a family of genes from diverse species accelerates directed evolution., *Nature* **391**, 288-291.

Crich, J. Z., Brieva, R., Marquart, P., Gu, R. L., Flemming, S., Sih, C. J. (1993), Enzymic asymmetric synthesis of α-amino acids. Enantioselective cleavage of 4-substituted oxazolin-5-ones and thiazolin-5-ones, *J. Org. Chem.* **58**, 3252-3258.

Crosby, J., Moilliet, J., Paratt, J. S., Turner, N. J. (1994), Regioselective hydrolysis of aromatic dinitriles using a whole cell catalyst, *J. Chem. Soc., Perkin Trans. 1* 1679-1687.

Crosby, J. A., Parratt, J. S., Turner, N. J. (1992), Enzymic hydrolysis of prochiral dinitriles, *Tetrahedron: Asymmetry* **3**, 1547-1550.

Crotti, P., Dibussolo, V., Favero, L., Minutolo, F., Pineschi, M. (1996), Efficient application of lipase-catalyzed transesterification to the resolution of γ-hydroxy ketones, *Tetrahedron: Asymmetry* **7**, 1347-1356.

Crout, D. H. G., Gaudet, V. S. B., Hallinan, K. O. (1993), Application of biotransformations in organic synthesis: preparation of enantiomerically pure esters of *cis*- and *trans*-epoxysuccinic acid, *J. Chem. Soc., Perkin Trans. 1*, 805-812.

Crout, D. H. G., MacManus, D. A., Critchley, P. (1991), Stereoselective galactosyl transfer to *cis*-cyclohexa-3,5-diene-1,2-diol, *J. Chem. Soc., Chem. Commun.* 376-378.

Crout, D. H. G., Gaudet, V. S. B., Laumen, K., Schneider, M. P. (1986), Enzymatic hydrolysis of (±)-trans-1,2-diacetoxycycloalkanes - a facile route to optically-active cycloalkane-1,2-diols, *J. Chem. Soc., Chem. Commun.* 808-810.

Csomós, P., Kanerva, L. T., Bernáth, G., Fülöp, F. (1996), Biocatalysis for the preparation of optically active β-lactam precursors of amino acids, *Tetrahedron: Asymmetry* **7**, 1789-1796.

Csuk, R., Dorr, P. (1994), Enantioselective enzymatic hydrolyses of building blocks for the synthesis of carbocyclic nucleosides, *Tetrahedron: Asymmetry* **5**, 269-276.

Cuperus, F. P., Bouwer, S. T., Kramer, G. F. H., Derksen, J. T. P. (1994), Lipases used for the production of peroxycarboxylic acids, *Biocatalysis* **9**, 89-96.

Cvetovich, R. J., Chartrain, M., Hartner, F. W., Roberge, C., Amato, J. S., Grabowski, E. J. (1996), An asymmetric synthesis of L-694,458, a human leukocyte elastase inhibitor, via novel enzyme resolution of β-lactam esters, *J. Org. Chem.* **61**, 6575-6580.

Cygler, M., Grochulski, P., Kazlauskas, R. J., Schrag, J. D., Bouthillier, F., Rubin, B., Serreqi, A. N., Gupta, A. K. (1994), Molecular basis for the chiral preferences of lipases, *J. Am. Chem. Soc.* **116**, 3180-3186.

Cygler, M., Schrag, J. D., Sussman, J. L., Harel, M., Silman, I., Gentry, M. K., Doctor, B. P. (1993), Relationship between sequence conservation and three-dimensional structure in a large family of esterases, lipases, and related proteins, *Protein Sci.* **2**, 366-382.

Cygler, M., Schrag, J. D., Ergan, F. (1992), Advances in structural understanding of lipases, *Biotechnol. Genet. Eng. Rev.* **10**, 143-184.

Dabulis, K., Klibanov, A. M. (1993), Dramatic enhancement of enzymatic activity in organic solvents by lyoprotectants, *Biotechnol. Bioeng.* **41**, 566-571.

Daffe, V., Fastrez, J. (1980), Enantiomeric enrichment in the hydrolysis of oxazolones catalyzed by cyclodextrins or proteolytic enzymes, *J. Am. Chem. Soc.* **102**, 3601-3605.

Dalrymple, B. P., Swadling, Y., Cybinski, D. H., Xue, G. P. (1996), Cloning of a gene encoding cinnamoyl ester hydrolase from the ruminal bacterium *Butyrivibrio fibrisolvens* E14 by a novel method, *FEMS Lett.* **143**, 115-120.

Danieli, B., Barra, C., Carrea, G., Riva, S. (1996a), Oxynitrilase-catalyzed transformation of substituted aldehydes. The case of (±)-2-phenylpropionaldehyde and of (±)-3-phenylbutyraldehyde, *Tetrahedron: Asymmetry* **7**, 1675-1682.

Danieli, B., Lesma, G., Passarella, D., Silvani, A. (1996b), Stereoselective enzymatic hydrolysis of dimethyl *meso*-piperidine-3,5-dicarboxylates, *Tetrahedron: Asymmetry* **7**, 345-348.

Danieli, B., Luisetti, M., Riva, S., Bertinotti, A., Ragg, E., Scaglioni, L., Bombardelli, E. (1995), Regioselective enzyme-mediated acylation of polyhydroxy natural compounds. A remarkable, highly efficient preparation of 6'-*O*-acetyl and 6'-*O*-carboxyacetyl Ginsenoside Rg1, *J. Org. Chem.* **60**, 3637-3642.

Danishefsky, S. J., Cabal, M. P., Chow, K. (1989), Novel stereospecific silyl group transfer reaction - practical routes to the prostaglandins, *J. Am. Chem. Soc.* **111**, 3456-3457.

D'Arrigo, P., de Ferra, L., Piergianni, V., Selva, A., Servi, S., Strini, A. (1996), Preparative transformation of natural phospholipids catalysed by phospholipase D from *Streptomyces*, *J. Chem. Soc., Perkin Trans. 1* 2651-2656.

De Amici, M., De Micheli, C., Carrea, G., Riva, S. (1996), Nitrile oxides in medicinal chemistry. 6. Enzymatic resolution of a set of bicyclic D2-isoxazolines, *Tetrahedron: Asymmetry* **7**, 787-796.

De Amici, M., De Micheli, C., Carrea, G., Spezia, S. (1989), Chemoenzymatic synthesis of chiral isoxazole derivatives, *J. Org. Chem.* **54**, 2646-2650.

de Bont, J. A. M. (1993), Bioformation of optically pure epoxides, *Tetrahedron: Asymmetry* **4**, 1331-1340.

de Jesus, P. C., Rezende, M. C., Nascimento, M. G. (1995), Enzymic resolution of alcohols via lipases immobilized in microemulsion-based gels, *Tetrahedron: Asymmetry* **6**, 63-66.

de Raadt, A., Klempier, N., Faber, K., Griengl, H. (1992), Chemoselective enzymatic hydrolysis of aliphatic and alicyclic nitriles, *J. Chem. Soc., Perkin Trans. I* 137-140.

de Vries, E. J., Janssen, D. B. (2003), Biocatalytic conversion of epoxides, *Curr. Opin. Biotechnol.* **14**, 414-420.

Deardorff, D. A., Matthews, A. J., McMeekin, D. S., Craney, C. L. (1986), A highly enantioselective hydrolysis of *cis*-3,5-diacetoxycyclopent-1-ene - an enzymatic preparation of 3(*R*)-acetoxy-5(*S*)-hydroxycyclopent-1-ene, *Tetrahedron Lett.* **27**, 1255-1256.

DeFrank, J. J., Cheng, T. C. (1991) Purification and properties of an organophosphorus acid anhydrase from a halophilic bacterial isolate, *J. Bacteriol.* **173**, 1938-1943.

Degueil-Castaing, M., de Jeso, B., Drouillard, S., Maillard, B. (1987), Enzymatic reactions in organic synthesis. 2. Ester interchange of vinyl esters, *Tetrahedron Lett.* **28**, 953-954.

Demirjan, D. C., Shah, P. C., Moris-Varas, F. (1999), Screening for novel enzymes, *Topics Curr. Chem.* **200**, 1-29.

De Jong, R. M., Tiesinga, J. J., Rozeboom, H. J., Kalk, K. H., Tang, L., Janssen, D. B., Dijkstra, B. W. (2003), Structure and mechanism of a bacterial haloalcohol dehalogenase: a new variation of the short-chain dehydrogenase/reductase fold without an NAD(P)H binding site, *Embo. J.* **22**, 4933-4944.

De Jong, R. M., Dijkstra, B. W. (2003), Structure and mechanism of bacterial dehalogenases: different ways to cleave a carbon-halogen bond, *Curr. Opin. Struct. Biol.* **13**, 722-730.

Derewenda, U., Swenson, L., Green, R., Wei, Y., Dodson, G. G., Yamaguchi, S., Haas, M. J., Derewenda, Z. S. (1994a), An unsusal buried polar cluster in a family of fungal lipases, *Nature Struct. Biol.* **1**, 36-47.

Derewenda, U., Swenson, L., Green, R., Wei, Y., Yamaguchi, S., Joerger, R., Haas, M. J., Derewenda, Z. S. (1994b), Current progress in crystallographic studies of new lipases from filamentous fungi, *Protein Engineer.* **7**, 551-557.

Derewenda, U., Swenson, L., Wei, Y., Green, R., Kobos, P. M., Joerger, R., Haas, M. J., Derewenda, Z. S. (1994c), Conformational lability of lipases observed in the absence of an oil-water interface: crystallografic studies of enzymes from the fungi *Humicola lanuginosa* and *Rhizopus delemar*, *J. Lipid Res.* **35**, 524-535.

Derewenda, U., Brzozowski, A. M., Lawson, D. M., Derewenda, Z. S. (1992), Catalysis at the interface: the anatomy of a conformational change in a triglyceride lipase, *Biochemistry* **31**, 1532-1541.

Derewenda, Z. S. (1994), Structure and function of lipases, *Adv. Prot. Chem.* **45**, 1-52.

Desai, S. B., Argade, N. P., Ganesh, K. N. (1996), Remarkable chemo-, regio-, and enantioselectivity in lipase-catalyzed hydrolysis - efficient resolution of (±)-*threo*-ethyl 3-(4-methoxyphenyl)-2,3-diacetoxypropionate leading to chiral intermediates of (+)-diltiazem, *J. Org. Chem.* **61**, 6730-6732.

DeSantis, G., Wong, K., Farwell, B., Chatman, K., Zhu, Z., Tomlinson, G., Huang, H., Tan, X., Bibbs, L., Chen, P., Kretz, K., Burk, M. J. (2003), Creation of a productive, highly enantioselective nitrilase through gene site saturation mutagenesis (GSSM), *J. Am. Chem. Soc.* **125**, 11476-11477.

DeSantis, G., Zhu, Z., Greenberg, W. A., Wong, K., Chaplin, J., Hanson, S. R., Farwell, B., Nicholson, L. W., Rand, C. L., Weiner, D. P., Robertson, D. E., Burk, M. J. (2002), An enzyme library approach to biocatalysis: development of nitrilases for enantioselective production of carboxylic acid derivatives, *J. Am. Chem. Soc.* **124**, 9024-9025.

DeTar, D. F. (1981), Computation of peptide-protein interactions. Catalysis by chymotrypsin: prediction of relative substrate reactivities, *J. Am. Chem. Soc.* **103**, 107-110.

Dinh, P. M., Howarth, J. A., Hudnott, A. R., Williams, J. M. J., Harris, W. (1996), Catalytic racemization of alcohols: applications to enzymic resolution reactions, *Tetrahedron Lett.* **37**, 7623-7626.

Dirlam, N. L., Moore, B. S., Urban, F. J. (1987), Novel synthesis of the aldose reductase inhibitor sorbinil via amidoalkylation, intramolecular oxazolidin-5-one alkylation, and chymotrypsin resolution, *J. Org. Chem.* **52**, 3587-3591.

Djerourou, A.-H., Blanco, L. (1991), Synthesis of optically active 2-sila-1,3-propanediols derivatives by enzymatic transesterification, *Tetrahedron Lett.* **32**, 6325-6326.

Doddema, H. J., Janssens, R. J. J., de Jong, J. P. J., van der Lugt, J. P., Oostrom, H. H. M. (1990), Enzymatic reactions in supercritical carbon dioxide and integrated product-recovery, in.: *Proc.*

5th Eur. Congr. Biotechnol. (Christiansen, L., Munck,. L., Villadsen, J., Eds.), pp. 239-242. Copenhagen: Munksgaard.

Doderer, K., Lutz-Wahl, S., Hauer, B., Schmid, R. D. (2003), Spectrophotometric assay for epoxide hydrolase activity toward any epoxide, *Anal. Biochem.* **321**, 131-134.

Dodson, G., Wlodawer, A. (1998), Catalytic triads and their relatives, *Trends Biochem. Sci.* **23**, 347-352.

Dodson, G. G., Lawson, D. M., Winkler, F. K. (1992), Structural and evolutionary relationships in lipase mechanism and activation, *Faraday Discuss.* 95-105.

Dörmõ, N., Gubicza, L. (2002) Application of pervaporation for removal of water produced during enzymatic esterification in ionic liquids, *Desalination* **149**, 267-268.

Doswald, S., Estermann, H., Kupfer, E., Stadler, H., Walther, W., Weisbrod, T., Wirz, B., Wostl, W. (1994), Large scale preparation of chiral building blocks for the P3 site of renin inhibitors, *Bioorg. Med. Chem.* **2**, 403-410.

Drauz, K., Waldmann, H. (1995), *Enzyme Catalysis in Organic Synthesis*, 1st edition, **Vol. 1 & 2**, Weinheim: VCH.

Drauz, K., Waldmann, H. (2002), *Enzyme Catalysis in Organic Synthesis*, 2nd edition, **Vol. 1-3**, Weinheim: VCH.

Drenth, J., Hol, W. G., Jansonius, J. N., Koekoek, R. (1972), Subtilisin Novo. The three-dimensional structure and its comparison with subtilisin BPN', *Eur. J. Biochem.* **26**, 177-181.

Drioli, S., Felluga, F., Forzato, C., Nitti, P., Pitacco, G., Valentin, E. (1998), Synthesis of (+)- and (−)-Phaseolinic acid by combination of enzymatic hydrolysis and chemical transformations with revision of the absolute configuration of the natural product, *J. Org. Chem.* **63**, 2385-2388.

Drioli, S., Felluga, F., Forzato, C., Nitti, P., Pitacco, G. (1996), A facile route to (+)- and (−)-*trans*-tetrahydro-5-oxo-2-pentylfuran-3-carboxylic acid, precursors of (+)- and (−)-methylenolactocin, *Chem. Commun.* 1289-1290.

Ducret, A., Giroux, A., Trani, M., Lortie, R. (1995), Enzymatic preparation of biosurfactants from sugars or sugar alcohols and fatty acids in organic media under reduced pressure, *Biotechnol. Bioeng.* **48**, 214-221.

Dudal, Y., Lortie, R. (1995), Influence of water activity on the synthesis of triolein catalyzed by immobilized *Mucor miehei* lipase, *Biotechnol. Bioeng.* **45**, 129-134.

Dufour, E., Storer, A. C., Ménard, R. (1995), Engineering nitrile hydratase activity into a cysteine protease be a single mutation, *Biochemistry* **34**, 16382-16388.

Duggleby, H. J., Tolley, S. P., Hill, C. P., Dodson, E. J., Dodson, G., Moody, P. C. (1995), Penicillin acylase has a single-amino-acid catalytic centre, *Nature* **373**, 264-268.

Duhamel, P., Renouf, P., Cahard, D., Yebga, A., Poirier, J.-M. (1993), Asymmetric synthesis by enzymatic hydrolysis of prochiral dienol diacetate, *Tetrahedron: Asymmetry* **4**, 2447-2450.

Dumont, T., Barth, D., Corbier, C., Branlant, G., Perrut, M. (1992), Enzymatic reaction kinetic: Comparison in an organic solvent and in supercritical carbon dioxide, *Biotechnol. Bioeng.* **39**, 329-333.

Dumortier, L., Liu, P., Dobbelaere, S., van der Eycken, J., Vandewalle, M. (1992), Lipase-catalyzed enantiotoposelective hydrolysis of *meso*-compounds derived from 2-cyclohexene-1,4-diol: synthesis of (−)-conduritol C, (−)-conduritol E, (−)-conduritol F and a homochiral derivative of conduritol A, *Synlett* 243-245.

Dunn, G., Montgomery, M. G., F. Mohammed, F., Coker, A., Cooper, J. B., Robertson, T., Garcia, J.-L., Bugg, T.D.H., Wood, S. P. (2005), The structure of the C–C bond hydrolase MhpC provides insights into its catalytic mechanism, *J. Mol. Biol.* **346**, 253-265.

Duran, R., Nishiyama, M., Horinouchi, S., Beppu, T. (1993), Characterization of nitrile hydratase genes cloned by DNA screening from *Rhodococcus erythropolis*, *Biosci. Biotechnol. Biochem.* **57**, 1323-1328.

Ebata, H., Toshima, K., Matsumura, S. (2001), A strategy for increasing molecular weight of polyester by lipase-catalyzed polymerization, *Chem. Lett.*, 798-799.

Ebbers, E. J., Ariaans, G. J. A., Houbiers, J. P. M., Bruggink, A., Zwanenburg, B. (1997), Controlled racemization of optically active organic compounds: Prospects for asymmetric transformation, *Tetrahedron* **53**, 9417-9476.

Eberling, J., Braun, P., Kowalczyk, D., Schultz, M., Kunz, H. (1996), Chemoselective removal of protecting groups from *O*-glycosyl amino acid and peptide (methoxyethoxy)ethyl esters using lipases and papain, *J. Org. Chem.* **61**, 2638-2646.

Eckstein, M., Sesing, M., Kragl, U., Adlercreutz, P. (2002) At low water activity α-chymotrypsin is more active in an ionic liquid than in non-ionic organic solvents, *Biotechnol. Lett.* **24**, 867-872.

Ebiike, H., Terao, Y., Achiwa, K. (1991), Acyloxymethyl as an activating group in lipase-catalyzed enantioselective hydrolysis. A versatile approach to chiral 4-aryl-1,4-dihydro-2,6-dimethyl-3,5-pyridinedicarboxylates, *Tetrahedron Lett.* **32**, 5805-5808.

Edwards, M. L., Matt, J. E., Wenstrup, D. L., Kemper, C. A., Persichetti, P. A., Margolin, A. L. (1996), Synthesis and enzymatic resolution of a carbocyclic analogue of ribofuranosylamine, *Org. Prep. Proc. Intl.* **28**, 193-201.

Effenberger, F., Böhme, J. (1994), Enzyme-catalyzed enantioselective hydrolysis of racemic naproxen nitrile, *Bioorg. Med. Chem.* **2**, 715-721.

Effenberger, F., Gutterer, B., Ziegler, T., Eckhardt, E., Aichholz, R. (1991), Enantioselective esterification of racemic cyanohydrins and enantioselective hydrolysis or transesterification of cyanohydrin esters by lipases, *Liebigs Ann. Chem.* 47-54.

Egloff, M. P., Marguet, F., Bueno, G., Verger, R., Cambillau, C., van Tilbeurgh, H. (1995a), The 2.46 Å resolution structure of the pancreatic lipase-colipase complex inhibited by a C11 alkyl phosphonate, *Biochemistry* **34**, 2751-2762.

Egloff, M. P., Sarda, L., Verger, R., Cambillau, C., van Tilbeurgh, H. (1995b), Crystallographic study of the structure of colipase and of the interaction with pancreatic lipase, *Protein Sci.* **4**, 44-57.

Egri, G., Baitzgacs, E., Poppe, L. (1996), Kinetic resolution of 2-acylated-1,2-diols by lipase-catalyzed enantiomer selective acylation, *Tetrahedron: Asymmetry* **7**, 1437-1448.

Ehrler, J., Seebach, D. (1990), Enantioselective saponification of substituted, achiral 3-acyloxy-propionates with lipases: synthesis of chiral derivatives of "Tris(hydroxymethyl)methane", *Liebigs Ann. Chem.* 379-388.

Eichhorn, E., Roduit, J.-P., Shaw, N., Heinzmann, K., Kiener, A. (1997), Preparation of (*S*)-piperazine-2-carboxylic acid, (*R*)-piperazine-2-carboxylic acid, and (*S*)-piperidine-2-carboxylic acid by kinetic resolution of the corresponding racemic carboxamides with stereoselective amidases in whole bacterial cells, *Tetrahedron: Asymmetry* **8**, 2533-2536.

Ema, T., Maeno, S., Takaya, Y., Sakai, T., Utaka, M. (1996), Kinetic resolution of racemic 2-substituted 3-cyclopenten-1-ols by lipase-catalyzed transesterifications - a rational strategy to improve enantioselectivity, *J. Org. Chem.* **61**, 8610-8616.

Ember, L. (1997), Detoxifying nerve agents, *Chem Eng News*, Sept 15: 26-29.

Endo, I., Odaka, M., Yohda, M. (1999), An enzyme controlled by light: the molecular mechanism of photoreactivity in nitrile hydratase, *Trends. Biotechnol.* **17**, 244-248.

Endo, I., Watanabe, I. (1989), Nitrile hydratase of *Rhodococcus* sp. N-774. Purification and amino acid sequence, *FEBS Lett.* **243**, 61-64.

Enzelberger, M. M., Bornscheuer, U. T., Gatfield, I., Schmid, R. D. (1997), Lipase-catalysed resolution of γ- and δ-lactones, *J. Biotechnol.* **56**, 129-133.

Erbeldinger, M., Mesiano, A. J., Russell, A. J. (2000) Enzymatic catalysis of formation of Z-aspartame in ionic liquid – An alternative to enzymatic catalysis in organic solvents. *Biotechnol. Prog.* 2000 **16**, 1129-1131.

Erickson, J. C., Schyns, P., Cooney, C. L. (1990), Effect of pressure on an enzymic reaction in a supercritical fluid, *AIChE J.* **36**, 299-301.

Esteban, A. I., Juanes, O., Conde, S., Goya, P., De Clercq, E., Martinez, A. (1995), New 1,2,6-thiadiazine dioxide acylonucleosides: synthesis and antiviral evalutation, *Bioorg. Med. Chem.* **3**, 1527-1535.

Esteban, A. I., Juanes, O., Martinez, A., Conde, S. (1994), Regioselective lipase-mediated acylation-deacylation in thiadiazine diacyclonucleosides, *Tetrahedron* **50**, 13865-13870.

Estell, D. A., Graycar, T. P., Miller, J. V., Powers, D. B., Burnier, J. P., Ng, P. G., Wells, J. A. (1986), Probing steric and hydrophobic effects on enzyme-substrate interactions by protein engineering, *Science* **233**, 659-663.

Estermann, H., Prasad, K., Shapiro, M. J., Bolsterli, J. J., Walkinshaw, M. D. (1990), Enzyme-catalyzed asymmetrization of bis(hydroxymethyl)-tetrahydrofurans, *Tetrahedron Lett.* **31**, 445-448.

Evans, C. T., Roberts, S. M., Shoberu, K. A., Sutherland, A. G. (1992), Potential use of carbocyclic nucleosides for the treatment of AIDS: chemo-enzymic syntheses of the enantiomers of carbovir, *J. Chem. Soc., Perkin Trans. I* 589-592.

Exl, C., Hoenig, H., Renner, G., Rogi-Kohlenprath, R., Seebauer, V., Seufer-Wasserthal, P. (1992), How large are the active sites of the lipases from *Candida rugosa* and from *Pseudomonas cepacia*? *Tetrahedron: Asymmetry* **3**, 1391-1394.

Faber, K. (2001), Non-sequential processes for the transformation of a racemate into a single stereoisomeric product: proposal for stereochemical classification, *Chem. Eur. J.* **7**, 5004-5010.

Faber, K. (2004), *Biotransformations in Organic Chemistry*, 5th ed., Berlin: Springer.

Faber, K., Mischitz, M., Kroutil, W. (1996), Microbial epoxide hydrolases, *Acta Chem. Scand.* **50**, 249-258.

Faber, K., Ottolina, G., Riva, S. (1993), Selectivity-enhancement of hydrolase reactions, *Biocatalysis* **8**, 91-132.

Faber, K., Riva, S. (1992), Enzyme-catalyzed irreversible acyl transfer, *Synthesis* 895-910.

Fabre, J., Paul, F., Monsan, P., Blonski, C., Périé, J. (1994), Enzymic synthesis of amino acid ester of butyl α-D-glucopyranoside, *Tetrahedron Lett.* **35**, 3535-3536.

Fabre, J., Betbeder, D., Paul, F., Monsan, P., Périé, J. (1993a), Regiospecific enzymic acylation of butyl-α-D-glucopyranoside, *Carbohydr. Res.* **243**, 407-411.

Fabre, J., Betbeder, D., Paul, F., Monsan, P., Périé, J. (1993b), Versatile enzymatic diacid ester synthesis of butyl α-D-glucopyranoside, *Tetrahedron* **49**, 10877-10882.

Fadel, A., Arzel, P. (1997a), Asymmetric construction of benzylic quaternary carbons by lipase-mediated enantioselective transesterification of prochiral α,α-disubstituted 1,3-propanediols, *Tetrahedron: Asymmetry* **8**, 283-291.

Fadel, A., Arzel, P. (1997b), Enzymatic asymmetrisation of prochiral α,α-disubstituted-malonates and -1,3-propanediols: formal asymmetric syntheses of (−)-aphanorphine and (+)-eptazocine, *Tetrahedron:Asymmetry* **8**, 371-374.

Fadnavis, N. W., Koteshwar, K. (1997), Remote control of stereoselectivity: lipase catalyzed enantioselective esterification of racemic α-lipoic acid, *Tetrahedron: Asymmetry* **8**, 337-339.

Falk-Heppner, M., Keller, M., Prinzbach, H. (1989), Unsaturated 1,4-cis and 1,4-trans-diaminotetradeoxycycloheptites - Enantiomerically pure, polyfunctionalized tropa derivatives by enzymatic hydrolysis, *Angew. Chem. Int. Ed.* **28**, 1253-1255.

Fallon, R. D., B.Stieglitz, Jr., I. T. (1997), A *Pseudomonas putida* capable of stereoselective hydrolysis of nitriles, *Appl. Microbiol. Biotechnol.* **47**, 156-161.

Farb, D., Jencks, W. D. (1980), Different forms of pig liver esterase, *Arch. Biochem. Biophys.* **203**, 214-226.

Farias, R. N., Torres, M., Canela, R. (1997), Spectrophotometric determination of the positional specificity of nonspecific and 1,3-specific lipases, *Anal. Biochem.* **252**, 186-189.

Felfer, U., Goriup, M., Koegl, M. F., Wagner, U., Larissegger-Schnell, B., Faber, K., Kroutil, W. (2005), The substrate spectrum of mandelate racemase: Minimum structural requirements for substrates and substrate model, *Adv. Synth. Catal.* **347**, 951-961.

Fellous, R., Lizzani-Cuvelier, L., Loiseau, M. A., Sassy, E. (1994), Resolution of racemic ε-lactones, *Tetrahedron: Asymmetry* **5**, 343-346.

Fernández, S., Ferrero, M., Gotor, V., Okamura, W. H. (1995), Selective acylation of A-ring precursors of vitamin D using enzymes in organic solvents, *J. Org. Chem.* **60**, 6057-6061.

Fernández, S., Brieva, R., Rebolledo, F., Gotor, V. (1992), Lipase -catalyzed enantioselective acylation of *N*-protected or unprotected 2-aminoalkan-1-ols, *J. Chem. Soc., Perkin Trans. 1* 2885-2889.

Fernandez-Lorente, G., Terreni, M., Mateo, C., Bastida, A., Fernandez-Lafuente, R., Dalmases, P., Huguet, J., Guisan, J. M. (2001), Modulation of lipase properties in macro-aqueous systems by controlled enzyme immobilization: enantioselective hydrolysis of a chiral ester by immobilized Pseudomonas lipase, *Enzyme Microb. Technol.* **28**, 389-396.

Fernandez, L., Beerthuyzen, M. M., Brown, J., Siezen, R. J., Coolbear, T., Holland, R., Kuipers, O. P. (2000), Cloning, characterization, controlled overexpression, and inactivation of the major tributyrin esterase gene of *Lactococcus lactis*, *Appl. Environ. Microbiol.* **66**, 1360-1368.

Ferraboschi, P., Molatore, A., Verza, E., Santaniello, E. (1996), The first example of lipase-catalyzed resolution of a stereogenic center in steroid side chains by transesterification in organic solvent, *Tetrahedron: Asymmetry* **7**, 1551-1554.

Ferraboschi, P., Casati, S., Manzocchi, A., Santaniello, E. (1995a), Studies on the regio- and enantioselectivity of the lipase-catalyzed transesterification of 1'- and 2'-naphthyl alcohols in organic solvent, *Tetrahedron: Asymmetry* **6**, 1521-1524.

Ferraboschi, P., Casati, S., Verza, E., Santaniello, E. (1995b), Regio- and enantioselective properties of the lipase-catalyzed irreversible transesterification of some 2-substituted-1,4- butanediols in organic solvents, *Tetrahedron: Asymmetry* **6**, 1027-1030.

Ferraboschi, P., Casati, S., Grisenti, P., Santaniello, E. (1994a), Selective enzymatic transformations of itaconic acid derivatives - an access to potentially useful building blocks, *Tetrahedron* **50**, 3251-3258.

Ferraboschi, P., Grisenti, P., Casati, S., Santaniello, E. (1994b), A new synthesis of (*R*)- and (*S*)-mevalonolactone from the enzymatic resolution of (*R,S*)-2-(3-methyl-2-butenyl)-oxiranemethanol, *Synlett* 754-756.

Ferraboschi, P., Grisenti, P., Manzocchi, A., Santaniello, E. (1994c), Regio- and enantioselectivity of *Pseudomonas cepacia* lipase in the transesterification of 2-substituted-1,4-butanediols, *Tetrahedron: Asymmetry* **5**, 691-698.

Ferraboschi, P., Casati, S., Grisenti, P., Santaniello, E. (1993), A chemoenzymatic approach to enantiomerically-pure (*R*)- and (*S*)-2,3-epoxy-2-(4-pentenyl)-propanol, a chiral building block for the synthesis of (*R*)- and (*S*)-frontalin, *Tetrahedron: Asymmetry* **4**, 9-12.

Ferraboschi, P., Brembilla, D., Grisenti, P., Santaniello, E. (1991), Enzymatic resolution of 2-substituted oxiranemethanols, a class of synthetically useful building blocks bearing a quaternary center, *J. Org. Chem.* **56**, 5478-5480.

Ferrer, M., Golyshina, O. V., Chernikova, T. N., Khachane, A. N., Matrins dos Santos, V. A. P., Yakimov, M. M., Timmis, K. N., Golyshin, P. N. (2005), Novel microbial enzymes minded from the Urania deep-sea hypersaline anoxic basin, *Chem. Biol.* **12**, 895-904.

Ferrero, M., Fernández, S., Gotor, V. (1997), Selective alkoxycarbonylation of A-ring precursors of vitamin D using enzymes in organic solvents. Chemoenzymatic synthesis of 1α,25-dihydroxyvitamin D3 C-5 A-ring carbamate derivatives, *J. Org. Chem.* **62**, 4358-4363.

Fiandanese, V., Hassan, O., Naso, F., Scilimati, A. (1993), A highly efficient kinetic resolution of ω-trimethylsilyl polyunsaturated secondary alcohols by lipase catalyzed transesterification, *Synlett* 491-493.

Fiaud, J. C., Gil, R., Legros, J. Y., Aribi-Zouioueche, L., Koenig, W. A. (1992), Kinetic resolution of 3-*t*-butyl and 3-phenyl cyclobutylidinethanols through lipase-catalyzed acylation with succinic anhydride, *Tetrahedron Lett.* **33**, 6967-6970.

Fink, A. L., Hay, G. W. (1969), Enzymic deacylation of esterified mono- and disaccharides. IV. Products of esterase-catalyzed deacetylations, *Can. J. Biochem.* **47**, 353-359.

Fischer, A., Bommarius, A. S., Drauz, K., Wandrey, C. (1994), A novel approach to enzymic peptide synthesis using highly solubilizing N^α-protecting groups of amino acids, *Biocatalysis* **8**, 289-307.

Fitzpatrick, P. A., Steinmetz, A. C. U., Ringe, D., Klibanov, A. M. (1993), Enzyme crystal structure in a neat organic solvent, *Proc. Natl. Acad. Sci. USA* **90**, 8653-8657.

Fitzpatrick, P. A., Klibanov, A. M. (1991), How can the solvent affect enantioselectivity? *J. Am. Chem. Soc.* **113**, 3166-3171.

Fletcher, P. D. I., Robinson, B. H., Freedman, R. B., Olfield, C. (1985), Activity of lipase in water in oil microemulsions, *J. Chem. Soc. Faraday Trans. I* **81**, 2667-2679.

Forró, E.; Ferenc Fülöp, F. (2003), Lipase-catalyzed enantioselective ring opening of unactivated alicyclic-fused β-lactams in an organic solvent, *Org. Lett.* **5**, 1209-1212.

Fouque, E., Rousseau, G. (1989), Enzymatic resolution of medium-ring lactones - synthesis of (*S*)-(+)-phoracantholide-I, *Synthesis* 661-666.

Fowler, P. W., Macfarlane, E. L. A., Roberts, S. M. (1991), Highly diastereoselective interesterification reactions involving a racemic acetate and a racemic carboxylic acid catalysed by lipase enzymes, *J. Chem. Soc., Chem. Commun.* 453-455.

Franssen, M. C. R., Jongejan, H., Kooijman, H., Spek, A. L., Bell R. P. L., Wijnberg, J. B. P. A., de Groot, A. (1999), Lipase-mediated resolution of octahydro-3,3,8a-trimethyl-naphthalenol, a key intermediate in the total synthesis of lactaranes and marasmanes, *Tetrahedron: Asymmetry*, **10**, 2729-2738.

Franssen, M. C. R., Jongejan, H., Kooijman, H., Spek, A. L., Mondril, N. L. F. L. C., de Santos, P. M. A. C. B., de Groot, A. (1996), Resolution of a tetrahydrofuran ester by *Candida rugosa* lipase (CRL) and an examination of CRL's stereochemical preference in organic media, *Tetrahedron: Asymmetry* **7**, 497-510.

Fregapane, G., Sarney, D. B., Vulfson, E. N. (1994), Facile chemo-enzymatic synthesis of monosaccharide fatty acid esters, *Biocatalysis* **11**, 9-18.

Fregapane, G., Sarney, D. B., Vulfson, E. N. (1991), Enzymic solvent-free synthesis of sugar acetal fatty acid esters, *Enzyme Microb. Technol.* **13**, 796-800.

Frykman, H., Öhrner, N., Norin, T., Hult, K. (1993), *S*-Ethyl thiooctanoate as acyl donor in lipase-catalyzed resolution of secondary alcohols, *Tetrahedron Lett.* **34**, 1367-1370.

Fuganti, C., Grasselli, P., Mendozza, M., Servi, S., Zucchi, G. (1997), Microbially-aided preparation of (*S*)-2-methoxycyclohexanone key intermediate in the synthesis of sanfetrinen, *Tetrahedron* **53**, 2617-2624.

Fuganti, C., Rosell, C. M., Servi, S., Tagliani, A., Terreni, M. (1992), Enantioselective recognition of the phenylacetyl moiety in the penicillin G acylase catalyzed hydrolysis of phenylacetate esters, *Tetrahedron: Asymmetry* **3**, 383-386.

Fuganti, C., Grasselli, P., Servi, S., Lazzarini, A., Casati, P. (1988), Substrate specificity and enantioselectivity of penicillinacylase catalyzed hydrolysis of phenacetyl esters of synthetically useful carbinols, *Tetrahedron* **44**, 2575-2582.

Fukuda, K., Kiyokawa, Y., Yanagiuchi, T., Wakai, Y., Kitamoto, K., Inoue, Y., Kimura, A. (2000), Purification and characterization of isoamyl acetate-hydrolyzing esterase encoded by the IAH1 gene of *Saccharomyces cerevisiae* from a recombinant *Escherichia coli*, *Appl. Microbiol. Biotechnol.* **53**, 596-600.

Fukusaki, E.-I., Satoda, S., Senda, S., Omata, T. (1998), Lipase-catalyzed kinetic resolution of 4-hydroxy-5-dodecynonitrile and its application to facile synthesis of a cupreous chafer beetle sex pheromone, *J. Ferment. Bioeng.* **86**, 508-509.

Fukusaki, E., Senda, S., Nakazono, Y., Omata, T. (1992a), Synthesis of the enantiomers of (Z-5-(1-octenyl)oxycyclopentan-2-one, a sey pheromone of the cupreous chafer beetle, *Anomala cuprea* Hope, *Biosci. Biotechnol. Biochem.* **56**, 1160-1161.

Fukusaki, E., Senda, S., Nakazono, Y., Yuasa, H., Omata, T. (1992b), Lipase-catalyzed kinetic resolution of 2,3-epoxy-8-methyl-1-nonanol, the key intermediate in the synthesis of the gypsy moth pheromone, *J. Ferment. Bioeng.* **73**, 280-283.

Fukusaki, E., Senda, S., Nakazono, Y., Yuasa, H., Omata, T. (1992c), Preparation of carboxyalkyl acrylate by lipase-catalyzed regioselective hydrolysis of corresponding methyl ester, *Bioorg. Med. Chem. Lett.* **2**, 411-414.

Fukusaki, E., Senda, S., Nakazono, Y., Omata, T. (1991), Lipase-catalyzed kinetic resolution of methyl 4-hydroxy-5- tetradecynoate and its application to a facile synthesis of Japanese beetle pheromone, *Tetrahedron* **47**, 6223-6230.

Fülling, G., Sih, C. J. (1987), Enzymatic second-order asymmetric hydrolysis of ketorolac esters: in situ racemization, *J. Am. Chem. Soc.* **109**, 2845-2846.

Furui, M., Furutani, T., Shibatani, T., Nakamoto, Y., Mori, T. H. (1996), A membrane bioreactor combined with crystallizer for production of optically active (2R,3S)-3-(4-methoxyphenyl)glycidic acid methyl ester, *J. Ferment. Bioeng.* **81**, 21-25.

Furukawa, M., Kodera, Y., Uemura, T., Hiroto, M., Matsushima, A., Kuno, H., Matsushita, H., Inada, Y. (1994), Alcoholysis of ε-decalactone with polyethylene glycol-modified lipase in 1,1,1-trichloroethane, *Biochem. Biophys. Res. Commun.* **199**, 41-45.

Gais, H. J., von der Weiden, I. (1996), Preparation of enantiomerically pure α-hydroxymethyl S-*tert*-butyl sulfones by *Candida antarctica* lipase catalyzed resolution, *Tetrahedron: Asymmetry* **7**, 1253-1256.

Gais, H. J., Elferink, V. H. M. (1995), Hydrolysis and formation of C–O-bonds, in.: *Enzyme Catalysis in Organic Synthesis* (Drauz, K., Waldmann, H.; Eds.), **Vol. 1**, pp. 165-271 (ref. on pp 343-348). Weinheim: VCH.

Gais, H. J., Hemmerle, H., Kossek, S. (1992), Enzyme-catalyzed asymmetric synthesis. 10. Pseudomonas cepacia lipase mediated synthesis of enantiomerically pure (2R,3S)- and (2S,3R)-2,3-O-cyclohexylideneerythritol monoacetate from 2,3-O-cyclohexylideneerythritol, *Synthesis (Stuttgart)* 169-173.

Gais, H.-J., Bülow, G., Zatorski, A., Jentsch, M., Maidonis, P., Hemmerle, H. (1989), Enzyme-catalyzed asymmetric synthesis. Enantioselectivity of pig liver esterase catalyzed hydrolyses of 4-substituted *meso*-cyclopentane 1,2-diesters, *J. Org. Chem.* **54**, 5115-5122.

Gais, H.-J., Zeissler, A., Maidonis, P. (1988), Diastereoselective D-galactopyranosyl transfer to *meso* diols catalyzed by β–galactosidases, *Tetrahedron Lett.* **29**, 5743-5744.

Gais, H.-J., Lukas, K. L., Ball, W. A., Braun, S., Lindner, H. J. (1986), Enzyme-catalyzed asymmetric synthesis. 4. Synthesis of homochiral building-blocks for the enantioselective total synthesis of cyclopentanoids with an esterase-catalyzed asymmetric reaction as key step, *Liebigs Ann. Chem.* 687-716.

Gais, H.-J., Lukas, K. L. (1984), Enantioselective and enantioconvergent syntheses of building-blocks for the total synthesis of cyclopentanoid natural products, *Angew. Chem. Int. Ed.* **23**, 142-143.

Gala, D., de Benedetto, D. J., Clark, J. E., Murphy, B. L., Schumacher, D. P., Steinman, M. (1996), Preparations of antifungal Sch 42427/SM 9164: preparative chromatographic resolution, and total asymmetric synthesis by enzymatic preparation of chiral α-hydroxy arylketones, *Tetrahedron Lett.* **37**, 611-614.

Gamalevich, G. D., Serebryakov, E. P., Vlasyuk, A. L. (1998), Lipase-mediated stereodivergent synthesis of both enantiomers of 4-(2,6-dimethylheptyl)benzoic acid, *Mendeleev Commun.* 8-9.

Gandhi, N. N. (1997), Applications of lipase, *J. Am. Oil Chem. Soc.* **74**, 621-634.

Ganske, F., Bornscheuer, U. T. (2005), Lipase-catalyzed glucose fatty acid ester synthesis in ionic liquids, *Org. Lett.* **7**, 3097-3098.

Garbay-Jaureguiberry, C., McCort-Tranchepain, I., Barbe, B., Ficheux, D., Roques, B. P. (1992), Improved synthesis of [*p*-phosphono and *p*-sulfo]methylphenylalanine. Resolution of [*p*-phosphono-, *p*-sulfo-, *p*-carboxy- and *p*-*N*-hydroxycarboxamido-]methylphenylalanine, *Tetrahedron: Asymmetry* **3**, 637-650.

Garcia, M. J., Azerad, R. (1997), Production of ring-substituted D-phenylglycines by microbial or enzymatic hydrolysis/deracemization of the corresponding D,L-hydantoins, *Tetrahedron: Asymmetry* **8**, 85-92.

Garcia, M. J., Rebolledo, F., Gotor, V. (1993), Practical enzymic route to optically active 3-hydroxyamides. Synthesis of 1,3-amino alcohols, *Tetrahedron: Asymmetry* **4**, 2199-2210.

Garcia, M. J., Rebolledo, F., Gotor, V. (1992), Enzymic synthesis of 3-hydroxybutyramides and their conversion to optically active 1,3-amino alcohols, *Tetrahedron: Asymmetry* **3**, 1519-1522.

Garcia-Alles, L. F., Moris, F., Gotor, V. (1993), Chemo-enzymatic synthesis of 2'-deoxynucleosides urethanes, *Tetrahedron Lett.* **34**, 6337-6338.

Garcia-Granados, A., Parra, A., Simeó, Y., Extremera, A. L. (1998), Chemical, enzymatic and microbiological synthesis of 8,12-eudesmanolides: synthesis of sivasinolide and yomogin analogues, *Tetrahedron* **54**, 14421-14436.

Gardossi, L., Bianchi, D., Klibanov, A. M. (1991), Selective acylation of peptides catalyzed by lipases in organic solvents, *J. Am. Chem. Soc.* **113**, 6328-6329.

Gaspar, J., Guerrero, A. (1995), Lipase-catalyzed enantioselective synthesis of naphthyl trifluoromethyl carbinols and their corresponding non-fluorinated counterparts, *Tetrahedron: Asymmetry* **6**, 231-238.

Gatfield, I. L. (1984), The enzymic synthesis of esters in nonaqueous systems, *Ann. N. Y. Acad. Sci.* **434**, 569-572.

Gaucher, A., Ollivier, J., Marguerite, J., Paugam, R., Salaun, J. (1994), Total asymmetric syntheses of (1*S*,2*S*)-norcoronamic acid, and of (1*R*,2*R*)- and (1*S*,2*S*)-coronamic acids from the diastereoselective cyclization of 2-(N-benzylideneamino)-4-chlorobutyronitriles, *Can. J. Chem.* **72**, 1312-1327.

Gavagnan, J. E., Fager, S. K., Fallon, R. D., Folsom, P. W., Herkes, F. E., Eisenberg, A., Hann, E. C., DiCosimo, R. (1998), Chemoenzymatic production of lactams from aliphatic α,ω-dinitriles, *J. Org. Chem.* **63**, 4792-4801.

Gelo-Pujic, M., Guibe-Jampel, e., Loupy, A., Galema, S. A., Mathe, D. (1996), Lipase-catalyzed esterification of some α-D-glucopyranosides in dry media under focused microwave irradiation, *J. Chem. Soc., Perkin Trans. 1* 2777-2780.

Genzel, Y.; Archelas, A.; Broxterman, Q. B.; Schulze, B.; Furstoss, R. (2001), A step toward a green chemistry preparation of enantiopure (*S*)-2-, -3-, and -4-pyridyloxirane via an epoxide hydrolase catalyzed kinetic resolution, *J. Org. Chem.* **66**, 538-543

Geresh, S., Gilboa, Y. (1991), Enzymic syntheses of alkyds. II. lipase-catalyzed polytransesterification of dichloroethyl fumarate with aliphatic and aromatic diols, *Biotechnol. Bioeng.* **37**, 883-888.

Geresh, S., Gilboa, Y. (1990), Enzymatic synthesis of alkyds, *Biotechnol. Bioeng.* **36**, 270-274.

Ghogare, A., Kumar, G. S. (1990), Novel route to chiral polymers involving biocatalytic transesterification of *O*-acryloyl oximes, *J. Chem. Soc., Chem. Commun.* 134-135.

Ghogare, A., Kumar, G. S. (1989), Oxime esters as novel irreversible acyl transfer agents for lipase catalysis in organic media, *J. Chem. Soc., Chem. Commun.* 1533-1535.

Ghosh, A. K., Kincaid, J. F., Haske, M. G. (1997), A convenient enzymatic route to optically active 1-aminoindan-2-ol: versatile ligands for HIV-1 protease inhibitors and asymmetric syntheses, *Synthesis* 541-544.

Ghosh, A. K., Chen, Y. (1995), Synthesis and optical resolution of high affinity P2-ligands for HIV-1- protease inhibitors, *Tetrahedron Lett.* **36**, 505-508.

Ghosh, D., Wawrzak, Z., Pletnev, V. Z., Li, N. Y., Kaiser, R., Pangborn, W., Jornvall, H., Erman, M., Duax, W. L. (1995), Structure of uncomplexed and linoleate-bound *Candida cylindracea* cholesterol esterase, *Structure* **3**, 279-288.

Gil, G., Ferre, E., Meou, A., Petit, J. L., Triantaphylides, C. (1987), Lipase-catalyzed ester formation in organic solvents. Partial resolution of primary allenic alcohols, *Tetrahedron Lett.* **28**, 1647-1648.

Gilbert, E. J. (1993), *Pseudomonas* lipases: Biochemical properties and molecular cloning, *Enzyme Microb. Technol.* **15**, 634-645.

Giver, L., Gershenson, A., Freskgard, P.-O., Arnold, F. H. (1998), Directed evolution of a thermostable esterase, *Proc. Natl. Acad. Sci. USA* **95**, 12809-12813.

Glueck, S. M., Pirker, M., Nestl, B. M., Ueberbacher, B. T., Larissegger-Schnell, B., Csar, K., Hauer, B., Stürmer, R., Kroutil, W., Faber, K. (2005), Biocatalytic racemization of aliphatic, arylaliphatic, and aromatic α-hydroxycarboxylic acids, *J. Org. Chem.*, **70**, 4028-4032.

Goddard, J. P., Reymond, J.-L. (2004), Recent advances in enzyme assays, *Trends Biotechnol.* **22**, 363-370.

Goderis, H. L., Ampe, G., Feyten, M. P., Fouwe, B. L., Guffens, W. M., Cauwenbergh, S. M. V., Tobback, P. P. (1987), Lipase-catalyzed ester exchange reactions in organic media with controlled humidity, *Biotechnol. Bioeng.* **30**, 258-266.

Goergens, U., Schneider, M. P. (1991a), Enzymatic preparation of enantiomerically pure and selectively protected 1,2- and 1,3-diols, *J. Chem. Soc., Chem. Commun.* 1066-1068.

Goergens, U., Schneider, M. P. (1991b), A facile chemoenzymatic route to enantiomerically-pure oxiranes: building blocks for biologically active compounds, *J. Chem. Soc., Chem. Commun.* 1064-1066.

Goetz, G., Iwan, P., Hauer, B., Breuer, M., Pohl, M. (2001), Continuous production of (*R*)-phenylacetylcarbinol in an enzyme-membrane reactor using a potent mutant of pyruvate decarboxylase from *Zymomonas mobilis*, *Biotechnol. Bioeng.* **74**, 317-325.

Goj, O., Burchardt, A., Haufe, G. (1997), A versatile approach to optically active primary 2-fluoro-2-phenylalkanols through lipase-catalyzed transformations, *Tetrahedron: Asymmetry* **8**, 399-408.

Goldberg, M., Thomas, D., Legoy, M. D. (1990), Water activity as a key parameter of synthesis reactions: the example of lipase in biphasic (liquid/solid) media, *Enzyme Microb. Technol.* **12**, 976-981.

Goldberg, M., Parvaresh, F., Thomas, D., Legoy, M. D. (1988), Enzymatic ester synthesis with continuous measurement of water activity, *Biochim. Biophys. Acta* **957**, 359-362.

Goldhust, A., Bohak, Z. (1989), Induction, purification and characterisation of the nitrilase of *Fusurium oxysporum*, *Biotechnol. Appl. Biochem.* **11**, 581-601.

Gopalan, V., Van der Jagt, D. J., Libell, D. P., Glew, R. H. (1992), Transglucosylation as a probe of the mechanism of action of mammalian cytosolic β-glucosidase, *J. Biol. Chem.* **267**, 9629-9638.

Gotor, V., Menendez, E., Mouloungui, Z., Gaset, A. (1993), Synthesis of optically active amides from β-furyl and β-phenyl esters by way of enzymic aminolysis, *J. Chem. Soc., Perkin Trans. 1* 2453-2456.

Gotor, V., Morís, F. (1992), Regioselective acylation of 2'-deoxynucleosides through an enzymatic reaction with oxime esters, *Synthesis* 626-628.

Gotor, V., Pulido, R. (1991), An improved procedure for regioselective acylation of carbohydrates: novel enzymatic acylation of α-D-glucopyranose and methyl α-D-glucopyranoside, *J. Chem. Soc., Perkin Trans. 1* 491-492.

Gotor, V., Astorga, C., Rebollodo, F. (1990), An improved method for the preparation of acylhydrazines: the first example of an enzymatic hydrazinolysis reaction, *Synlett* 387-388.

Gou, D.-M., Liu, Y.-C., Chen, C.-S. (1993), A practical chemoenzymatic synthesis of the taxol C-13 side chain N-benzoyl-(2R,2S)-3-phenylisoserine, *J. Org. Chem.* **58**, 1287-1289.

Gou, D.-M., Liu, Y.-C., Chen, C.-S. (1992), An efficient chemoenzymic access to optically active *myo*-inositol polyphosphates, *Carbohydr. Res.* **234**, 51-64.

Gradley, M. L., Deverson, C. J. F., Knowles, C. J. (1994), Asymmetric hydrolysis of R-(–), S(+)-2-methylbutyronitrile by *Rhodococcus rhodochrous* NCIMB 11216, *Arch. Microbiol.* **161**, 246-251.

Gradley, M. L., Knowles, C. J. (1994), Asymmetric hydrolysis of chiral nitriles by *Rhodococcus rhodochrous* NCIMB 11216 nitrilase, *Biotechnol. Lett.* **16**, 41-46.

Graham, L. D., Haggett, K. D., Jennings, P. A., Le Brocque, D. S., Whittaker, R. G., Schober, P. A. (1993), Random mutagenesis of the substrate-binding site of a serine protease can generate enzymes with increased activities and altered primary specificities, *Biochemistry* **32**, 6250-6258.

Granberg, K. L., Bäckvall, J.-E. (1992), Isomerization of (π-allyl)palladium complexes via nucleophilic displacement of palladium(0). A common mechanism in palladium(0)-catalyzed allylic substitution, *J. Am. Chem. Soc.* **114**, 6958-6963.

Greener, A., Callahan, M., Jerpseth, B. (1996), An efficient random mutagenesis technique using *E. coli* mutator strain, *Methods Mol. Biol.* **57**, 375-385.

Greenstein, J. P., Winitz, M. (1961), Enzymes involved in the determination, characterization, and preparation of amino acids, in.: *Chemistry of the Amino Acids*, **Vol. 2**, pp. 1753-1816. New York: Wiley.

Griffin, M., Trudgill, P. W. (1976), Purification and properties of cyclopentanone oxygenase of *Pseudomonas* NCIB 9872, *Eur. J. Biochem.* **63**, 199-209.

Griffith, D. A., Danishefsky, S. J. (1991), Total synthesis of allosamidin - an application of the sulfonamidoglycosylation of glycals, *J. Am. Chem. Soc.* **113**, 5863-5864.

Griffith, O. H., Volwerk, J. J., Kuppe, A. (1991), Phosphatidylinositol-specific phospholipases C from *Bacillus cereus* and *Bacillus thuringiensis*, *Methods Enzymol.* **197**, 493-502.

Griffiths, A. D., Tawfik, D. S. (2003), Directed evolution of an extremely fast phosphotriesterase by *in vitro* compartmentalization, *Embo J.* **22**, 24-35.

Griffiths, A. D., Tawfik, D. S. (1998), Man-made enzymes - from design to in vitro compartmentalisation, *Curr. Opin. Biotechnol.* **11**, 338-353.

Grochulski, P., Bouthillier, F., Kazlauskas, R. J., Serreqi, A. N., Schrag, J. D., Ziomek, E., Cygler, M. (1994), *Candida rugosa* lipase-inhibitor complexes simulating the acylation and deacylation transition states, *Biochemistry* **33**, 3494-3500.

Grochulski, P., Li, Y., Schrag, J. D., Bouthillier, F., Smith, P., Harrison, D., Rubin, B., Cygler, M. (1993), Insight into interfacial activation from an 'open' structure of *Candida rugosa* lipase, *J. Biol. Chem* **268**, 12843-12847.

Groeger, U., Drauz, K., Klenk, H. (1992), Amino acid transformation. 8. Enzymic resolution of N-alkyl amino acids, *Angew. Chem., Int. Ed.* **31**, 195-197.

Groeger, U., Drauz, K., Klenk, H. (1990), Transformation of amino acids. 4. Isolation of an L-stereospecific N-acyl-L-proline acylase and its use as a catalyst in organic synthesis, *Angew. Chem., Int. Ed..* **29**, 417-418.

Grogan, G., Rippe, C., Willetts, A. (1997), Biohydrolysis of substituted styrene oxides by *Beauveria densa* CMC 3240, *J. Mol. Catal. B: Enzym.* **3**, 253-257.

Grogan, G., Roberts, S. M., Willetts, A. J. (1996), Novel aliphatic epoxide hydrolase activities from dematiaceous fungi, *FEMS Microbiol. Lett.* **141**, 239-243.

Grognux, J., Reymond, J.-L. (2004), Classifying enzymes from selectivity fingerprints, *ChemBioChem* **5**, 826-831.

Gros, P., Betzel, C., Dauter, Z., Wilson, K. S., Hol, W. G. J. (1989), Molecular dynamics refinement of a thermitase-eglin-c complex at 1.98 A resolution and comparison of two crystal forms that differ in calcium content, *J. Mol. Biol.* **210**, 347-367.

Gross, R. A., Kumar, A., Kalra, B. (2001), Polymer synthesis by in vitro enzyme catalysis, *Chem. Rev.* **101**, 2097-2124.

Gruber-Khadjawi, M., Hönig, H., Illaszewicz, C. (1996), Chemoenzymatic methods for the preparation of optically active cyclic polyazido alcohols from easily available chiral starting materials, *Tetrahedron: Asymmetry* **7**, 807-814.

Gu, J., Li, Z. (1992), Enzyme reaction in aqueous-organic system. Kinetic resolution of racemic oxiranecarboxylic esters with lipase, *Chin. Chem. Lett.* **3**, 173-174.

Gu, Q. M., Sih, C. J. (1992), Improving the enantioselectivity of the *Candida cylindracea* lipase via chemical modification, *Biocatalysis* **6**, 115-126.

Gu, Q.-M., Chen, C.-S., Sih, C. J. (1986), A facile enzymatic resolution process for the preparation of (+)-S-2-(6-methoxy-2-naphthyl)propionic acid (naproxen), *Tetrahedron Lett.* **27**, 1763-1766.

Gu, R. L., Lee, I. S., Sih, C. J. (1992), Chemo-enzymic asymmetric synthesis of amino acids. Enantioselective hydrolyses of 2-phenyl-oxazolin-5-ones, *Tetrahedron Lett.* **33**, 1953-1956.

Guanti, G., Riva, R. (1995), Synthesis of optically active N-benzyl-2,4-bis(hydroxymethyl) substituted azetidines by lipase catalyzed acetylations, *Tetrahedron: Asymmetry* **6**, 2921-2924.

Guanti, G., Banfi, L., Brusco, S., Narisano, E. (1994a), Enzymatic preparation of homochiral 2-(N-carbobenzyloxypiperid-4-yl)-1,3-propanediol monoacetate - a facile entry to both enantiomers of 3-hydroxymethylquinuclidine, *Tetrahedron: Asymmetry* **5**, 537-540.

Guanti, G., Banfi, L., Riva, R. (1994b), Enzymatic asymmetrization of some prochiral and *meso* diols through monoacetylation with pig pancreatic lipase (PPL), *Tetrahedron: Asymmetry* **5**, 9-12.

Guanti, G., Banfi, L., Narisano, E. (1992), Chemoenzymic preparation of asymmetrized tris(hydroxymethyl)methane (THYM) and of asymmetrized bis(hydroxymethyl)acetaldehyde (BHYMA) as new highly versatile chiral building blocks, *J. Org. Chem.* **57**, 1540-1554.

Guanti, G., Banfi, L., Narisano, E. (1990a), Asymmetrized tris(hydroxymethyl)methane as a highly stereodivergent chiral building block: preparation of all four stereoisomers of protected 2-hydroxymethyl-1,3-butanediol, *Tetrahedron Lett.* **31**, 6421-6424.

Guanti, G., Banfi, L., Narisano, E. (1990b), Enzymes in organic synthesis: remarkable influence of a system on the enantioselectivity in PPL catalyzed monohydrolysis of 2-substituted 1,3-diacetoxypropanes, *Tetrahedron: Asymmetry* **1**, 721-724.

Guanti, G., Narisano, E., Podgorski, T., Thea, S., Williams, A. (1990c), Enzyme catalyzed monohydrolysis of 2-aryl-1,3-propanediol diacetates. A study of structural effects of the aryl moiety on the enantioselectivity, *Tetrahedron* **46**, 7081-7092.

Guanti, G., Banfi, L., Narisano, E. (1989), Enzymes as selective reagents in organic synthesis: enantioselective preparation of asymmetrized tris(hydroxymethyl)methane, *Tetrahedron Lett.* **30**, 2697-2698.

Guanti, G., Banfi, L., Narisano, E., Riva, R., Thea, S. (1986), Enzymes in asymmetric synthesis: Effect of reaction media on the PLE catalysed hydrolysis of diesters, *Tetrahedron Lett.* **27**, 4639-4642.

Gübitz, G., Cavaco-Paulo, A. (2003) New substrates for reliable enzymes: enzymatic modification of polymers, *Curr. Opin. Biotechnol.* **14**, 577-582.

Guevel, R., Paquette, L. A. (1993), An enzyme-based synthesis of (S)-(-)-3-methyl-2-[(phenylsulfonyl)methyl]butyl phenyl sulfide and the stereochemical course of its alkylation, *Tetrahedron: Asymmetry* **4**, 947-956.

Guibé-Jampel, E., Rousseau, G., Salaün, J. (1987), Enantioselective hydrolysis of racemic diesters by porcine pancreatic lipase, *J. Chem. Soc., Chem. Commun.* 1080-1081.

Guo, Z.-W. (1993), Novel plots of data from combined multistep enzymatic resolutions of enantiomers, *J. Org. Chem.* **58**, 5748-5752.

Guo, Z. W., Wu, S. H., Chen, C. S., Girdaukas, G., Sih, C. J. (1990), Sequential biocatalytic kinetic resolutions, *J. Am. Chem. Soc.* **112**, 4942-4945.

Guo, Z. W., Sih, C. J. (1989), Enantioselective inhibition: strategy for improving the enantioselectivity of biocatalytic systems, *J. Am. Chem. Soc.* **111**, 6836-6841.

Guo, Z. W., Ngooi, T. K., Scilimati, A., Fülling, G., Sih, C. J. (1988), Macrocyclic lactones via biocatalysis in nonaqueous media, *Tetrahedron Lett.* **29**, 5583-5586.

Guo, Z. W., Sih, C. J. (1988), Enzymic synthesis of macrocyclic lactones, *J. Am. Chem. Soc.* **110**, 1999-2001.

Gupta, A. K., Kazlauskas, R. J. (1993), Substrate modification to increase the enantioselectivity of hydrolases. A route to optically-active cyclic allylic alcohols, *Tetrahedron: Asymmetry* **4**, 879-888.

Gustafsson, J., Sandstroem, J., Sterner, O. (1995), Enantioselective esterification of 5-hydroxy-2-methyl-6- (diethoxymethyl)-1-cyclohexenecarbaldehyde. Synthesis of both enantiomers of 6-methylbicyclo[4.1.0]hept-2-ene-1,2-dicarbaldehyde, *Tetrahedron: Asymmetry* **6**, 595-602.

Gutman, A. L., Meyer, E., Kalerin, E., Polyak, F., Sterling, J. (1992), Enzymatic resolution of racemic amines in a continuous reactor in organic solvents, *Biotechnol. Bioeng.* **40**, 760-767.

Gutman, A. L. (1990), Enzymatic synthesis of chiral lactones and polyesters, *Mat. Res. Soc. Symp. Proc.* **174**, 217-222.

Gutman, A. L., Bravdo, T. (1989), Enzyme-catalyzed enantioconvergent lactonization of γ-hydroxy diesters in organic solvents, *J. Org. Chem.* **54**, 4263-4265.

Gutman, A. L., Zuobi, K., Boltansky, A. (1987), Enzymic lactonization of γ-hydroxy esters in organic solvents. Synthesis of optically pure γ-methylbutyrolactones and γ-phenylbutyrolactone, *Tetrahedron Lett.* **28**, 3861-3864.

Ha, H.-J., Yoon, K.-N., Lee, S.-Y., Park, Y.-S., Lim, M.-S., Yim, Y.-G. (1998), Lipase PS (*Pseudomonas cepacia*) mediated resolution of γ-substituted γ-((acetoxy)methyl)-γ-butyrolactones: complete stereochemical reversion by substituents, *J. Org. Chem.* **63**, 8062-8066.

Ha, H. J., Yi, D. G., Yoon, K. N., Song, C. E. (1996), Lipase-catalyzed preparation of optically active γ-substituted methyl-γ-butyrolactones, *Bull. Korean Chem. Soc.* **17**, 544-547.

Haas, M. J., Joerger, R. D. (1995), Lipases of the genera *Rhizopus* and *Rhizomucor*: versatile catalysts in nature and the laboratory, in.: *Food Biotechnology: Microorganisms.* (Khachatourians, G. G., Hui, H. Y.; Eds.), pp. 549-588. Weinheim: VCH.

Haase, B., Schneider, M. P. (1993), Enzyme assisted synthesis of enantiomerically pure δ-lactones, *Tetrahedron: Asymmetry* **4**, 1017-1026.

Halling, P. J. (1996), Predicting the behaviour of lipases in low-water media, *Biochem. Soc. Trans.* **25**, 170-174.

Halling, P. J. (1994), Thermodynamic predictions for biocatalysis in non-conventional media: theory, tests and recommendations for experimental design and analysis, *Enzyme Microb. Technol.* **16**, 178-206.

Halling, P. J. (1992), Salt hydrates for water activity control with biocatalysts in organic media, *Biotechnol. Lett.* **6**, 271-276.

Halling, P. J. (1990), High affinity binding of water by proteins is similar in air and organic solvents, *Biochim. Biophys. Acta* **1040**, 225-228.

Hamada, H., Shiromoto, M., Funahashi, M., Itoh, T., Nakamura, K. (1996), Efficient synthesis of optically pure 1,1,1-trifluoro-2-alkanols through lipase-catalyzed acylation in organic media, *J. Org. Chem.* **61**, 2332-2336, correction, p. 9635.

Hamaguchi, S., Asada, M., Hasegawa, J., Watanabe, K. (1985), Biological resolution of racemic 2-oxazolidinones. Part V. Stereospecific hydrolysis of 2-oxazolidinone esters and separation of products with an immobilized lipase column, *Agric Biol. Chem.* **49**, 1661-1667.

Hammond, D. A., Karel, A., Klibanov, A. M., Krukonis, V. J. (1985), Enzymatic reactions in supercritical gases, *Appl. Biochem. Biotechnol.* **11**, 393-400.

Han, D., Kwon, D. Y., Rhee, J. S. (1987), Determination of lipase activity in AOT-isooctane reversed micelles, *Agric. Biol. Chem.* **51**, 615-618.

Han, D., Rhee, J. S. (1986), Characteristics of lipase-catalyzed hydrolysis of olive oil in AOT-isooctane reversed micelles, *Biotechnol. Bioeng.* **28**, 1250-1255.

Hancock, S. M., Corbett, K., Fordham-Skelton, A. P., Gatehouse, J. A., Davis, B. G. (2005), Developing promiscuous glycosidases for glycoside synthesis: residues W433 and E432 in *Sulfolobus solfataricus* β-glycosidase are important glucoside- and galactoside-specificity determinants, *ChemBioChem* **6**, 866-875.

Handa, S., Earlam, G. J., Geary, P. J., Hawes, J. E., Phillips, G. T., Pryce, R. J., Ryback, G., Shears, J. H. (1994), The enantioselective synthesis of an important intermediate to the antiviral (-)-Carbovir, *J. Chem. Soc. Perkin Trans. 1* 1885-1886.

Handelsman, J. (2005), Sorting out metagenomes, *Nat. Biotechnol.* **23**, 38-39.

Handelsman, J. (2004), Metagenomics: Application of genomics to uncultured microorganisms, *Microbiol. Mol. Biol. Rev.* **68**, 669-685.

Hanefeld, U., Li, Y., Sheldon, R. A., Maschmeyer, T. (2000), CAL-B catalyzed enantioselective synthesis of cyanohydrins - a facile route to versatile building blocks, *Synlett*, 1775-1776.

Hansen, T. V., Waagen, V., Partali, V., Anthonsen, H. W., Anthonsen, T. (1995) Co-solvent enhancement of enantioselectivity in lipase-catalyzed hydrolysis of racemic esters. A process for production of homochiral C-3 building blocks using lipase B from *Candida antarctica*, *Tetrahedron: Asymmetry* **6**, 499-504.

Haraldsson, G. G. (1992), The application of lipases in organic synthesis, Supplement B: The chemistry of acid derivatives, (Patai, S.; Eds.), **Vol. 2**, pp. 1395-1473. Chichester: John Wiley & Sons.

Harel, M., Aharoni, A., Gaidukov, L., Brumshtein, B., Khersonsky, O., Meged, R., Dvir, H., Ravelli, R. B. G., McCarthy, A., Toker, L., Silman, I., Sussman, J. L., Tawfik, D. S. (2004), Structure and evolution of the serum paraoxonase family of detoxifying and anti-atherosclerotic enzymes, *Nature Struct. Mol. Biol.* **11**, 412-419.

Häring, D., Schüler, E., Adam, W., Saha-Möller, C. R., Schreier, P. (1999), Semisynthetic enzymes in asymmetric synthesis: enantioselective reduction of racemic hydroperoxides catalyzed by seleno-subtilisin, *J. Org. Chem.*, **64**, 832-835.

Häring, D., Schreier, P. (1998a), From detergent additive to semisynthetic peroxidase - simplified and up-scaled synthesis of seleno-subtilisin, *Biotechnol. Bioeng.*, **59**, 786-791.

Häring, D., Schreier, P. (1998b), Novel biocatalysts by chemical modification of known enzymes: cross-linked microcrystals of the semisynthetic peroxidase, *Angew. Chem. Int. Ed.*, **37**, 2471-2473.

Harper, D. (1985), Characterization of a nitrilase from *Nocardia* sp. (*Rhodococcus* group) NCIB 11215, using *p*-hydroxbenzonitrile as sole carbon source, *Int. J. Biochem.* **17**, 677-683.

Harper, D. (1977), Microbial metabolism of aromatic nitriles: enzymology of C-N cleavage by *Norcardia* sp. NCIB 11216, *J. Biochem.* **165**, 309-319.

Harris, K. J., Gu, Q. M., Shih, Y. E., Girdaukas, G., Sih, C. J. (1991), Enzymatic preparation of (3*S*,6*R*) and (3*R*,6*S*)-3-hydroxy-6- acetoxycyclohex-1-ene, *Tetrahedron Lett.* **32**, 3941-3944.

Hartleib, J., Ruterjans, H. (2001), Insights into the reaction mechanism of the diisopropyl fluorophosphatase from *Loligo vulgaris* by means of kinetic studies, chemical modification and site-directed mutagenesis, *Biochim. Biophys. Acta*, **1546**, 312-324.

Hatakeyama, S., Kawamura, M., Takano, S., Irie, H. (1994), Enantiospecific synthesis of (+)-paeonilactone C and (+)-paeoniflorigenone from *R*-(−)-carvone, *Tetrahedron Lett.* **35**, 7993-7996.

Haughton, L., Williams, J. M. J. (2001), Enzymatic hydrolysis and selective racemisation reactions of α-chloro esters, *Synthesis*, 943-946.

Hayes, D. G., Gulari, E. (1994), Improvement of enzyme activity and stability for reverse micellar-encapsulated lipases in the presence of short-chain and polar alcohols, *Biocatalysis* **11**, 223-231.

Hayes, D. G., Gulari, E. (1992), Formation of polyol-fatty acid esters by lipases in reverse micellar media, *Biotechnol. Bioeng.* **40**, 110-118.

Hayes, D. G., Gulari, E. (1991), 1-Monoglyceride production from lipase-catalyzed esterification of glycerol and fatty acid in reverse micelles, *Biotechnol. Bioeng.* **38**, 507-517.

Hedström, G., Backlund, M., Slotte, J. P. (1993), Enantioselective synthesis of ibuprofen esters in AOT/isooctane microemulsions by *Candida cylindracea* lipase, *Biotechnol. Bioeng.* **42**, 618-624.

Heim, J., Schmidt-Dannert, C., Atomi, H., Schmid, R. D. (1998), Functional expression of a mammalian acetylcholinesterase in *Pichia pastoris*: comparison to acetylcholinesterase, expressed and reconstituted from *Escherichia coli*, *Biochim. Biophys. Acta* **1396**, 306-319.

Hein, G. E., Niemann, C. (1962a), Steric course and specificity of α-chymotrypsin-catalyzed reactions. I., *J. Am. Chem. Soc.* **84**, 4487-4494.

Hein, G. E., Niemann, C. (1962b), Steric course and specificity of α-chymotrypsin-catalyzed reactions. II., *J. Am. Chem. Soc.* **84**, 4495-4503.

Heiss, L., Gais, H.-J. (1995), Polyethylene glycol monomethyl ether-modified pig liver esterase: preparation, characterization and catalysis of enantioselective hydrolysis in water and acylation in organic solvents, *Tetrahedron Lett.* **36**, 3833-3836.

Hemmerle, H., Gais, H.-J. (1987), Asymmetric hydrolysis and esterfication catalyzed by esterases from porcine pancreas in the synthesis of both enantiomers of cyclopentanoid building blocks, *Tetrahedron Lett.* **28**, 3471-3474.

Hemrika, W., Renirie, R., Dekker, H. L., Barnett, P., Wever, R. (1997), From phosphatases to vanadium peroxidases: A similar architecture of the active site, *Proc. Natl. Acad. Sci. USA*, **94**, 2145-2149.

Henegar, K. E., Ashford, S. W., Baughman, T. A., Sih, J. C., Gu, R. L. (1997), Practical asymmetric synthesis of (S)-4-ethyl-7,8-dihydro-4-hydroxy-1H-pyrano[3,4-f]indolizine-3,6,10(4H)-trione, a key intermediate for the synthesis of irinotecan and other camptothecin analogs, *J. Org. Chem.* **62**, 6588-6597.

Henke, E., Bornscheuer, U. T., Schmid, R. D., Pleiss, J. (2003), A molecular mechanism of enantiorecognition of tertiary alcohols by carboxylesterases, *ChemBioChem* **4**, 485-493.

Henke, E., Pleiss, J., Bornscheuer, U. T. (2002), Activity of lipases and esterases towards tertiary alcohols: Insights into structure-function relationchips, *Angew. Chem. Int. Ed.* **41**, 3211-3213.

Henke, E., Bornscheuer, U. T. (1999), Directed evolution of an esterase from *Pseudomonas fluorescens*. Random mutagenesis by error-prone PCR or a mutator strain and identification of mutants showing enhanced enantioselectivity by a resorufin-based fluorescence assay, *Biol. Chem.* **380**, 1029-1033.

Henkel, B., Kunath, A., Schick, H. (1995), Chemo-enzymatic synthesis of (5R,6S)-6-acetoxyhexadecan-5-olide - the major component of the mosquito oviposition attractant pheromone, *Liebigs Ann.* 921-923.

Henkel, B., Kunath, A., Schick, H. (1994), Synthesis of (3R,5S)-3-hydroxy-7-phenyl-6-heptyn-5-olide by an enantioselective enzyme-catalyzed lactonization of a racemic 3,5-dihydroxy ester, *Tetrahedron: Asymmetry* **5**, 17-18.

Henkel, B., Kunath, A., Schick, H. (1993), Enzyme-catalyzed lactonization of methyl (±)-(E)-3,5-dihydroxy-7-phenyl-6-heptenoates. A comparison of the behaviour of *syn*- and *anti*-compounds, *Tetrahedron: Asymmetry* **4**, 153-156.

Henkel, B., Kunath, A., Schick, H. (1992), Enantioselective lactonization of methyl 3,5-dihydroxyalkanoates. An access to (3R,5S,6E)-3-hydroxy-7-phenyl-6-hepten-5-olide by enzyme-catalyzed kinetic resolution in organic solvents, *Liebigs Ann. Chem.* 809-811.

Hennen, W. J., Sweers, H. M., Wang, Y.-F., Wong, C.-H. (1988), Enzymes in carbohydrate synthesis: Lipase-catalyzed selective acylation and deacylation of furanose and pyranose derivatives, *J. Org. Chem.* **53**, 4939-4945.

Hermoso, J., Pignol, D., Kerfelec, B., Crenon, I., Chapus, C., Fontecillacamps, J. C. (1996), Lipase activation by nonionic detergents - the crystal structure of the porcine lipase-colipase-tetraethylene glycol monooctyl ether complex, *J. Biol. Chem.* **271**, 18007-18016.

Hernáiz, M. J., Sánchez-Montero, J. M., Sinisterra, J. V. (1996), Improved stability of the lipase from *Candida rugosa* in different purification degrees by chemical modification, *Biotechnol. Techn.* **10**, 917-922.

Hernáiz, M. J., Sánchez-Montero, J. M., Sinisterra, J. V. (1994), Comparison of the enzymatic activity of commercial and semipurified lipases of *Candida cylindracea* in the hydrolysis of the esters of (*R*,*S*)-2-aryl propionic acids, *Tetrahedron* **50**, 10749-10760.

Herradón, B. (1994), Influence of the nature of the solvent on the enantioselectivity of lipase-catalyzed transesterification: a comparison between PPL and PCL, *J. Org. Chem.* **59**, 2891-2893.

Heymann, E., Peter, K. (1993), A note on the identity of porcine liver carboxylesterase and prolyl-beta-naphthylamidase, *Biol. Chem. Hoppe Seyler* **374**, 1033-1036.

Heymann, E., Junge, W. (1979), Characterization of the isoenzymes of pig-liver esterase. 1. Chemical studies, *Eur. J. Biochem.* **95**, 509-518.

Hill, C. M., Li, W.-S., Thoden, J. B., Holden, H. M., Raushel, F. M (2003), Enhanced degradation of chemical warfare agents through molecular engineering of the phosphotriesterase active site, *J. Am. Chem. Soc.* **125**, 8990-8991.

Hiratake, J., Yamamoto, K., Yamamoto, Y., Oda, J. (1989), Highly regioselective ring-opening of α-substituted cyclic acid anhydrides catalyzed by lipase, *Tetrahedron Lett.* **30**, 1555-1556.

Hirche, F., Schierhorn, A., Scherer, G., Ulbrich-Hofmann, R. (1997), Enzymatic introduction of N-heterocyclic and As-containing head groups into glycerophospholipids, *Tetrahedron Lett.* **38**, 1369-1370.

Hirose, Y., Kariya, K., Nakanishi, Y., Kurono, Y., Achiwa, K. (1995), Inversion of enantioselectivity in hydrolysis of 1,4-dihydropyridines by point mutation of lipase PS, *Tetrahedron Lett.* **36**, 1063-1066.

Hirose, Y., Kariya, K., Sasaki, I., Kurono, Y., Ebiike, H., Achiwa, K. (1992), Drastic solvent effect on lipase-catalyzed enantioselective hydrolysis of prochiral 1,4-dihydropyridines, *Tetrahedron Lett.* **33**, 7157-7160.

Hobson, A. H., Buckley, C. M., Aamand, J. L., Joergensen, S. T., Diderichsen, B., McConnell, D. J. (1993), Activation of a bacterial lipase by its chaperone, *Proc. Natl. Acad. Sci. USA* **90**, 5682-5686.

Hoegh, I., Patkar, S., Halkier, T., Hansen, M. T. (1995), Two lipases from *Candida antarctica* - cloning and expression in *Aspergillus oryzae*, *Can. J. Botan.* **73**, S869-S875.

Hoenke, C., Kluewer, P., Hugger, U., Krieger, R., Prinzbach, H. (1993), Biocatalysis en route to diamino-di(tri)deoxycyclohexitols and diamino-tetradeoxylcycloheptitols, *Tetrahedron Lett.* **34**, 4761-4764.

Hof, R. P., Kellog, R. M. (1996a), Lipase AKG mediated resolutions of α,α-disubstituted 1,2-diols in organic solvents - remarkably high regio- and enantioselectivity, *J. Chem. Soc., Perkin Trans. 1* 2051-2060.

Hof, R. P., Kellog, R. M. (1996b), Synthesis and lipase-catalyzed resolution of 5-(hydroxymethyl)-1,3-dioxolan-4-ones: masked glycerol analog as potential building blocks for pharmaceuticals, *J. Org. Chem.* **61**, 3423-3427.

Hof, R. P., Kellog, R. M. (1994), Lipase catalyzed resolutions of some α,α- disubstituted 1,2-diols in organic solvents; near absolute regio and chiral recognition, *Tetrahedron: Asymmetry* **5**, 565-568.

Hoff, B. H., Anthonsen, H. W., Anthonsen, T. (1996), The enantiomer ratio strongly depends on the alkyl part of the acyl donor in transesterification with lipase B from *Candida antarctica*, *Tetrahedron: Asymmetry* **7**, 3187-3192.

Höft, E., Hamann, H.-J., Kunath, A., Adam, W., Hoch, U., Saha-Möller, C. R., Schreier, P. (1995), Enzyme-catalyzed kinetic resolution of racemic secondary hydroperoxides, *Tetrahedron: Asymmetry* **6**, 603-608.

Högberg, H.-E., Edlund, H., Berglund, P., Hedenström, E. (1993), Water activity influences enantioselectivity in a lipase-catalyzed resolution by esterification in an organic solvent, *Tetrahedron: Asymmetry* **4**, 2123-2126.

Holdgrün, X. K., Sih, C. J. (1991), A chemoenzymic synthesis of optically active dihydropyridines, *Tetrahedron Lett.* **32**, 3465-3468.

Holla, E. W., Sinnwell, V., Klaffke, W. (1992), Two syntheses of 3-azido-3-deoxy-D-mannose, *Synlett* 413-415.

Holla, E. W. (1989), Enzymic synthesis of selectively protected glycals, *Angew. Chem. Intl. Ed.* **28**, 220-221.

Holmberg, E., Holmquist, M., Hedenström, E., Berglund, P., Norin, T., Hult, K. (1992), Reaction conditions for the resolution of 2-methylalkanoic acids in esterification and hydrolysis with lipase from *Candida cylindracea*, *Appl. Microbiol. Biotechnol.* **35**, 572-578.

Holmberg, E., Szmulik, P., Norin, T., Hult, K. (1989), Hydrolysis and esterification with lipase from *Candida cylindracea*. Influence of the reaction conditions and acid moiety on the enantiomeric excess, *Biocatalysis* **2**, 217-223.

Holmberg, K. (1994), Organic and bioorganic reactions in microemulsions, *Adv. Coll. Interface Sci.* **51**, 137-174.

Holmberg, K., Lassen, B., Stark, M. B. (1989), Enzymatic glycerolysis of a triglyceride in aqueous and nonaqueous microemulsions, *J. Am. Oil Chem. Soc.* **66**, 1796-1800.

Holmquist, M., Haeffner, F., Norin, T., Hult, K. (1996), A structural basis for enantioselective inhibition of *Candida rugosa* lipase by long-chain aliphatic alcohols, *Protein Sci.* **5**, 83-88.

Holzwarth, H. C., Pleiss, J., Schmid, R. D. (1997), Computer-aided modelling of stereoselective triglyceride hydrolysis catalyzed by *Rhizopus oryzae* lipase, *J. Mol. Catal. B: Enzym.* **3**, 73-82.

Hom, S. S. M., Scott, E. M., Atchison, R. E., Picataggio, S., Mielenz, J. R. (1991), Characterization and over-expression of a cloned *Pseudomonas* lipase gene, in.: *GBF Monographs* (Schmid, R. D., Alberghina, L., Verger, R.; Eds.), **Vol. 16**, pp. 267-270. Weinheim: VCH.

Honda, T., Ogino, T. (1998), Enantiodivergent synthesis of the key intermediate for a marine natural furanoterpene by chemoenzymatic process, *Tetrahedron: Asymmetry* **9**, 2663-2669.

Honda, J., Kandori, H., Okada, T., Nagamune, T., Shichida, Y., Sasabe, H., Endo, I. (1994), Spectroscopic observation of the intramolecular electron transfer in the photoactivation processes of nitrile hydratase, *Biochemistry* **33**, 3577-3583.

Hong, S.-B., Raushel, F. M. (1999), Stereochemical constraints on the substrate specificity of phosphotriesterase, *Biochemistry* **38**, 1159-1165.

Hongo, H., Iwasa, K., Kabuto, C., Matsuzaki, H., Nakano, H. (1997), Preparation of optically active photopyridone by lipase-catalyzed asymmetric resolution, *J. Chem. Soc., Perkin Trans. 1* 1747-1754.

Hönig, H., Shi, N., Polanz, G. (1994), Enzymatic resolutions of heterocyclic alcohols, *Biocatalysis* **9**, 61-69.

Hönig, H., Seufer-Wasserthal, P. (1990), A general method for the separation of enantiomeric *trans*-2-substituted cyclohexanols, *Synthesis* 1137-1140.

Horsman, G. P., Liu, A. M. F., Henke, E., Bornscheuer, U. T., Kazlauskas, R. J. (2003), Mutations in distant residues moderately increase the enantioselectivity of *Pseudomonas fluorescens* esterase toward methyl 3-bromo-2-methylpropanoate and ethyl 3-phenylbutyrate, *Chem. Eur. J.* **9**, 1933-1939.

Hoshino, O., Fuchino, H., Kikuchi, M. (1994), Lipase-catalyzed resolution of acetates of racemic phenolic aporphines and homoaporphines in organic solvent, *Heterocycles* **39**, 553-560.

Hosokawa, M., Takahashi, K., Kikuchi, Y., Hatano, M. (1995), Preparation of therapeutic phospholipids through porcine pancreatic phospholipase A_2-mediated esterification and Lipozyme-mediated hydrolysis, *J. Am. Oil Chem. Soc.* **72**, 1287-1291.

Hosoya, N., Hatayama, A., Irie, R., Sasaki, H., Katsuki, T. (1994), Rational design of Mn-Salen epoxidation catalysts: Preliminary results, *Tetrahedron* **50**, 4311-4322.

Hou, C. T. (1993), Screening of microbial esterases for asymmetric hydrolysis of 2-ethylhexyl butyrate, *J. Ind. Microbiol.* **11**, 73-81.

Houng, J.-Y., Hsieng, C. L., Chen, S. T. (1996a), Lipase-catalyzed kinetic resolution of ethyl D,L-2-amino-4-phenylbutyrate by hydrolysis, *Biotechnol. Techn.* **10**, 353-358.

Houng, J.-Y., Wu, M.-L., Chen, S.-T. (1996b), Kinetic resolution of amino acid esters catalyzed by lipases, *Chirality* **8**, 418-422.

Hoye, T. R., Ye, Z. X., Yao, L. J., North, J. T. (1996), Synthesis of the C-2-symmetric, macrocyclic alkaloid, (+)-xestospongin A its C(9)-epimer, (-)-xestospongin C - impact of substrate rigidity and reaction conditions on the efficiency of the macrocyclic dimerzation reaction, *J. Am. Chem. Soc.* **118**, 12074-12081.

Hoyle, A. J., Bunch, A. W., Knowles, C. J. (1998), The nitrilases of *Rhodococcus rhodochrous* NCIMB 11216, *Enzyme Microb. Technol.* **23**, 475-482.

Hsu, S.-H., Wu, S.-S., Wang, Y.-F., Wong, C.-H. (1990), Lipase-catalyzed irreversible transesterification using enol esters: XAD-8 immobilized lipoprotein lipase-catalyzed resolution of secondary alcohols, *Tetrahedron Lett.* **31**, 6403-6406.

Hu, S., Tat, D., Martinez, C. A., Yazbeck, D. R., Tao, J. (2005), An efficient and practical chemoenzymatic preparation of optically active secondary amines, *Org. Lett.* **7**, in press.

Huang, W., Jia, J., Cummings, J., Nelson, M., Schneider, G., Lindqvist, Y. (1997), Crystal structure of nitrile hydratase reveals a novel iron centre in a novel fold, *Structure* **15**, 691-699.

Huber, R. E., Gaunt, M. T., Hurlburt, K. L. (1984), Binding and reactivity at the "glucose" site of galactosyl-β-galactosidase (*Escherichia coli*), *Arch. Biochem. Biophys.* **234**, 151-160.

Hubner, B., Haensler, M., Hahn, U., (1999), Modification of ribonuclease T1 specificity by random mutagenesis of the substrate binding segment, *Biochemistry* **38**, 1371-1376.

Huerta, F. F., Minidis, A. B. E., Baeckvall, J.-E. (2001), Racemization in asymmetric synthesis. Dynamic kinetic resolution and related processes in enzyme and metal catalysis, *Chem. Soc. Rev.*, 321-331.

Huerta, F. F., Laxmi, Y. R. S., Bäckvall, J.-E. (2000), Dynamic kinetic resolution of α-hydroxy acid esters, *Org. Lett.* **2**, 1037-1040.

Huge-Jensen, B., Andreasen, F., Christensen, T., Christensen, M., Thim, L., Boel, E. (1989), *Rhizomucor miehei* triglyceride lipase is processed and secreted from transformed *Aspergillus oryzae*, *Lipids* **24**, 781-785.

Huge-Jensen, B., Galluzzo, D. R., Jensen, R. G. (1987), Partial purification and characterization of free and immobilized lipases from *Mucor miehei*, *Lipids* **22**, 559-565.

Hughes, D. L., Song, Z., Smith, G. B., Bergan, J. J., Dezeny, G. C., Reider, P. J., Grabowski, E. J. J. (1993), A practical chemoenzymic synthesis of an LTD4 antagonist, *Tetrahedron: Asymmetry* **4**, 865-874.

Hughes, D. L., Bergan, J. J., Amato, J. S., Bhupathy, M., Leazer, J. L., McNamara, J. M., Sidler, D. R., Reider, P. J., Grabowski, E. J. J. (1990), Lipase-catalyzed asymmetric hydrolysis of esters having remote chiral/prochiral centers, *J. Org. Chem.* **55**, 6252-6259.

Hughes, D. L., Bergan, J. J., Amato, J. S., Reider, P. J., Grabowski, E. J. J. (1989), Synthesis of chiral dithioacetals: a chemoenzymic synthesis of a novel LTD4 antagonist, *J. Org. Chem.* **54**, 1787-1788.

Hui, D. Y., Kissel, J. A. (1990), Sequence identity between human pancreatic cholesterol esterase and bile salt-stimulated milk lipase, *FEBS Lett.* **276**, 131-134.

Hulshof, L. A., Roskam, J. H. (1989), Phenylglycidate stereoisomers, conversion products thereof with e.g. 2-nitrophenol and preparation of diltiazem, *Eur. Pat. Appl.* EP 0343714.

Hultin, P. G., Jones, J. B. (1992), Dilemma regarding the active site model for porcine pancreatic lipase, *Tetrahedron Lett.* **33**, 1399-1402.

Hyatt, J. A., Skelton, C. (1997), A kinetic resolution route to the (*S*)-chromanmethanol intermediate for synthesis of the natural tocols, *Tetrahedron: Asymmetry* **8**, 523-526.

Iacazio, G., Roberts, S. M. (1993), Investigation of the regioselectivity of some esterifications involving methyl 4,6-*O*-benzylidene D-glycopyranosides and *Pseudomonas fluorescens* lipase, *J. Chem. Soc., Perkin Trans. 1* 1099-1101.

Igarashi, Y., Otsutomo, S., Harada, M., Nakano, S., Watanabe, S. (1997), Lipase mediated resolution of indene bromohydrin, *Synthesis* 549-552.

Iglesias, L. E., Sánchez, V. M., Rebolledo, F., Gotor, V. (1997), *Candida antarctica* B lipase catalysed resolution of (±)-1-(heteroaryl) ethylamines, *Tetrahedron: Asymmetry* **8**, 2675-2677.

Iglesias, L. E., Baldessari, A., Gros, E. G. (1996), Lipase-catalyzed chemospecific *O*-acylation of 3-mercapto-1-propanol and 4-mercapto-1-butanol, *Bioorg. Med. Chem. Lett.* **6**, 853-856.

Ihara, M., Suzuki, M., Fukumoto, K., Kabuto, C. (1990), Asymmetric total synthesis of atisine via intramolecular double Michael reaction, *J. Am. Chem. Soc.* **112**, 1164-1171.

Iimori, T., Azumaya, I., Hayashi, Y., Ikegami, S. (1997), A practical preparation of optically active *endo*-bicyclo[3.3.0]octen-2-ols, *Chem. Pharm. Bull.* **45**, 207-208.

Iizumi, T., Nakamura, K., Shimada, Y., Sugihara, A., Tominaga, Y., Fukase, T. (1991), Cloning, nucleotide sequence and expression in *Escherichia coli* of a lipase and its activator genes from *Pseudomonas* sp. KWI-56, *Agric. Biol. Chem.* **55**, 2349-2357.

Ikeda, M., Clark, D. S. (1998), Molecular cloning of extremely thermostable esterase gene from hyperthermophilic archaeon *Pyrococcus furiosus* in *Escherichia coli*, *Biotechnol. Bioeng.* **57**, 624-629.

Ikeda, I., Klibanov, A. M. (1993), Lipase-catalyzed acylation of sugars solubilized in hydrophobic solvents by complexation, *Biotechnol. Bioeng.* **42**, 788-791.

Ikushima, Y., Saito, N., Arai, M., Blanch, H. W. (1995), Activation of a lipase triggered by interactions with supercritical carbon dioxide in the near critical region, *J. Phys. Chem.* **99**, 8941-8944.

Ikushima, Y., Saito, N., Yokohama, T., Hatakeda, K., Ito, S., Arai, M., Blanch, H. W. (1993), Solvent effects on an enzymatic ester synthesis in supercritical carbon dioxide, *Chem. Lett.* 109-112.

Imperiali, B., Prins, T. J., Fisher, S. L. (1993), Chemoenzymatic synthesis of 2-amino-3-(2,2'-bipyridinyl)propanoic acids, *J. Org. Chem.* **58**, 1613-1616.

Inagaki, M., Hiratake, J., Nishioka, T., Oda, J. (1992), One-pot synthesis of optically active cyanohydrin acetates from aldehydes via lipase -catalyzed kinetic resolution coupled with in situ formation and racemization of cyanohydrins, *J. Org. Chem.* **57**, 5643-5649.

Inagaki, M., Hiratake, J., Nishioka, T., Oda, J. (1991), Lipase-catalyzed kinetic resolution with *in situ* racemization: one-pot synthesis of optically active cyanohydrin acetates from aldehydes, *J. Am. Chem. Soc.* **113**, 9360-9361.

Inagaki, M., Hiratake, J., Nishioka, T., Oda, J. (1989), Lipase-catalyzed stereoselective acylation of [1,1'-binaphthyl]-2,2'-diol and deacylation of its esters in organic solvent, *Agric. Biol. Chem.* **53**, 1879-1884.

Ingvorsen, K., Yde, B., Godtfredsen, S. E., Tsuchiya, R. T. (1988), Microbial hydrolysis of organic nitriles and amides, in.: *Ciba Geigy Symp.: Cyanide Compounds in Biology*, **Vol. 140**, pp. 16-31. New York: John Wiley & Sons.

Isowa, Y., Ohmori, M., Ichikawa, T., Mori, K., Nonaka, Y., Kihara, K., Oyama, K., Satoh, H., Nishimura, S. (1979), The thermolysin-catalyzed condensation reactions of *N*-substituted aspartic acid and glutamic acids with phenylalanine alkyl esters., *Tetrahedron Lett.* 2611-2612.

Ito, T., Shimizu, M., Fujisawa, T. (1998), Preparation and use of novel (S)-β-chlorodifluoromethyl-β-propiolactone as a chiral fluorinated building block, *Tetrahedron*, **54**, 5523-5530.

Itoh, T., Akasaki, E., Kudo, K., Shirakami, S. (2001) Lipase-catalyzed enantioselective acylation in the ionic liquid solvent system: reaction of enzyme anchored to the solvent, *Chem. Lett.* 262-263.

Itoh, T., Akasaki, E., Nishimura, Y. (2002) Efficient lipase-catalyzed enantioselective acylation under reduced pressure conditions in an ionic liquid solvent system, *Chem. Lett.*, 154-155.

Itoh, T., Nishimura, Y., Ouchi, N., Hayase, S. (2003), 1-Butyl-2,3-dimethylimidazolium tetrafluoroborate: the most desirable ionic liquid solvent for recycling use of enzyme in lipase-catalyzed transesterification using vinyl acetate as acyl donor, *J. Mol. Catal. B: Enzym.* **26**, 41-45.

Itoh, T., Shiromoto, M., Inoue, H., Hamada, H., Nakamura, K. (1996a), Simple preparation of optically pure bis(trifluoromethyl)-alkanediols through lipase-catalyzed reaction, *Tetrahedron Lett.* **37**, 5001-5002.

Itoh, T., Takagi, Y., Murakami, T., Hiyama, Y., Tsukube, H. (1996b), Crown ethers as regulators of enzymatic reactions: enhanced reaction rate and enantioselectivity in lipase-catalyzed hydrolysis of 2-cyano-1-methylethyl acetate, *J. Org. Chem.* **61**, 2158-2163.

Itoh, T., Uzu, A., Kanda, N., Takagi, Y. (1996c), Preparation of 3-alkyl-4-hydroxy-2-butenyl acetate through highly regioselective lipase catalyzed hydrolysis of the corresponding diacetates, *Tetrahedron Lett.* **37**, 91-92.

Itoh, T., Chika, J., Takagi, Y., Nishiyama, S. (1993a), An efficient enantioselective total synthesis of antitumor lignans: synthesis of enantiomerically pure 4-hydroxyalkanenitriles via an enzymic reaction, *J. Org. Chem.* **58**, 5717-5723.

Itoh, T., Hiyama, Y., Betchaku, A., Tsukube, H. (1993b), Enhanced reaction rate and enantioselectivity in lipase-catalyzed hydrolysis by addition of a crown ether, *Tetrahedron Lett.* **34**, 2617-2620.

Itoh, T., Ohara, H., Takagi, Y., Kanada, N., Uneyama, K. (1993c), Preparation of a new chiral building block for synthesizing broadly varied types of tertiary alcohols, *Tetrahedron Lett.* **34**, 4215-4218.

Itoh, T., Takagi, Y., Nishiyama, S. (1991), Enhanced enantioselectivity of an enzymatic reaction by the sulfur functional group. A simple preparation of optically active ß-hydroxy nitriles using a lipase, *J. Org. Chem.* **56**, 1521-1524.

Itoh, T., Ohta, T. (1990), A simple method for the preparation of optically active α-hydroxystannanes by the enantioselective hydrolysis using a lipase, *Tetrahedron Lett.* **31**, 6407-6408.

Itoh, T., Ohta, T., Sano, M. (1990), An efficient preparation of the optically active γ–hydroxy stannanes using lipase-catalyzed hydrolysis, *Tetrahedron Lett.* **31**, 6387-6390.

Iwasaki, Y., Mishima, N., Mizumoto, K., Nakano, H., Yamane, T. (1995), Extracellular production of phospholipase D of *Streptomyces antibioticus* using recombinant *Escherichia coli*, *J. Ferment. Bioeng.* **79**, 417-421.

Iwasaki, Y., Niwa, S., Nakano, H., Nagasawa, T., Yamane, T. (1994), Purification and properties of phosphatidylinositol-specific phospholipase C from *Streptomyces antibioticus*, *Biochim. Biophys. Acta* **1214**, 221-228.

Izumi, T., Aratani, S. (1993), Lipase-catalyzed resolution of 1-hydroxymethyl-2-methylferrocene, *J. Chem. Technol. Biotechnol.* **57**, 33-36.

Izumi, T., Hino, T., Ishihara, A. (1993), Enzymic kinetic resolution of [3](1,1')ferrocenophane derivatives, *J. Chem. Technol. Biotechnol.* **56**, 45-49.

Izumi, T., Tamura, F., Sasaki, K. (1992), Enzymic kinetic resolution of [4](1,2)ferrocenophane derivatives, *Bull. Chem. Soc. Jpn.* **65**, 2784-2788.

Jackson, M. A., King, J. W. (1997), Lipase-catalyzed glycerolysis of soy bean oil in supercritical carbon dioxide, *J. Am. Oil Chem. Soc.* **74**, 103-106.

Jackson, M. A., King, J. W., List, G. R., Neff, W. E. (1997), Lipase-catalyzed randomization of fats and oils in flowing supercritical carbon dioxide, *J. Am. Oil Chem. Soc.* **74**, 635-639.

Jackson, R. C.; Handschumacher, R. E. (1970), *Escherichia coli* L-asparaginase. Catalytic activity and subunit nature, *Biochemistry* **9**, 3585-3590.

Jaeger, K.-E., Reetz, M. T. (1998), Microbial lipases form versatile tools for biotechnology, *Trends Biotechnol.* **16**, 396-403.

Jaeger, K. E., Liebeton, K., Zonta, A., Schimossek, K., Reetz, M. T. (1996), Biotechnological application of *Pseudomonas aeruginosa* lipase: efficient kinetic resolution of amines and alcohols, *Appl. Microbiol. Biotechnol.* **46**, 99-105.

Jaeger, K.-E., Ransac, S., Dijkstra, B. W., Colson, C., van Heuvel, M., Misset, O. (1994), Bacterial lipases, *FEMS Microbiol. Rev.* **15**, 29-63.

Jahn, M., Chen, H., Muellegger, J., Marles, J., Warren, R. A. J., Withers, S. G. (2004), Thioglycosynthases: double mutant glycosidases that serve as scaffolds for thioglycoside synthesis, *Chem. Commun.*, 274-275.

Jahn, M., Marles, J., Warren, R. A. J., Withers, S. G. (2003), Thioglycoligases: mutant glycosidases for thioglycoside synthesis, *Angew. Chem. Intl. Ed.* **42**, 352-354.

Jahn, M., Withers, S. G. (2003), New approaches to enzymatic oligosaccharide synthesis: glycosynthases and thioglycoligases, *Biocatal. Biotransform.* **21**, 59-166.

Janes, L. E.; Cimpoia, A.; Kazlauskas, R. J. (1999) Protease-mediated separation of cis and trans diastereomers of 2-(R,S)-benzyloxymethyl-4-(S)-carboxylic acid-1,3-dioxolane methyl ester: intermediates for the synthesis of dioxolane nucleosides, *J. Org. Chem.* **64**, 9019-9029.

Janes, L. E., Löwendahl, A. C., Kazlauskas, R. J. (1998), Quantitative screening of hydrolase librarires using pH indicators: Identifying active and enantioselective hydrolases, *Chem. Eur. J.* **4**, 2324-2331.

Janes, L. E., Kazlauskas, R. J. (1997a), Empirical rules for the enantiopreference of lipase from *Aspergillus niger* toward secondary alcohols and carboxylic acids, especially α-amino acids, *Tetrahedron: Asymmetry* **8**, 3719-3733.

Janes, L. E., Kazlauskas, R. J. (1997b), Quick E. A fast spectrophotometric method to measure the enantioselectivity of hydrolases, *J. Org. Chem.* **62**, 4560-4561.

Janssen, D. B. (2004), Evolving haloalkane dehalogenases, *Curr. Opin. Chem. Biol.* **8**, 150-159.

Janssen, A. E. M., Klabbers, C., Franssen, M. C. R., van't Riet, K. (1991a), Enzymic synthesis of carbohydrate esters in 2-pyrrolidone, *Enzyme Microb. Technol.* **13**, 565-572.

Janssen, A. J. M., Klunder, A. J. H., Zwanenburg, B. (1991b), Resolution of secondary alcohols by enzyme-catalyzed transesterification in alkyl carboxylates as the solvent, *Tetrahedron* **47**, 7645-7662.

Janssen, A. E. M., Lefferts, A. G., van't Riet, K. (1990), Enzymatic synthesis of carbohydrate esters in aqueous media, *Biotechnol. Lett.* **12**, 711-716.

Janssens, R. J. J., Lugt, J. P. V. D., Oostrom, W. H. M. (1992), The integration of biocatalysis and downstream processing in supercritical carbon dioxide, in.: *High Press. Biotechnol.* (Balny, C., Hayashi, R., Heremans, K., Masson, P.; Eds.), **Vol. 224**, pp. 447-449.

Jao, S.-C., Huang, L.-F., Tao, Y. S., Li, W.-S. (2004), Hydrolysis of organophosphate triesters by *Escherichia coli* aminopeptidase P, *J. Mol. Catal. B: Enzym.* **27**, 7-12.

Jerina, D. M., Daly, J. W. (1974), Arene oxides: a new aspect of drug metabolism, *Science* **185**, 573-582.

Jeromin, G. E., Welsch, V. (1995), Diketene a new esterification reagent in the enzyme-aided synthesis of chiral alcohols and chiral acetoacetic acid esters, *Tetrahedron Lett.* **36**, 6663-6664.

Jiménez, O., Bosch, M. P., Guerrero, A. (1997), Lipase-catalyzed enantioselective synthesis of methyl (*R*)- and (*S*)-2-tetradecyloxiranecarboxylate through sequential kinetic resolution, *J. Org. Chem.* **62**, 3496-3499.

Johnson, C. J., Bis, S. J. (1992), Enzymic asymmetrization of *meso*-2-cycloalken-1,4-diols and their diacetates in organic and aqueous media, *Tetrahedron Lett.* **33**, 7287-7290.

Johnson, C. R., Bis, S. J. (1995), Enzymatic asymmetrization of 6-amino-2-cycloheptene-1,4-diol derivatives: synthesis of tropane alkaloids (+)- and (–)-calystegine A(3), *J. Org. Chem.* **60**, 615-623.

Johnson, C. R., Xu, Y., Nicolaou, K. C., Yang, Z., Guy, R. K., Dong, J. G., Berova, N. (1995), Enzymic resolution of a key stereochemical intermediate for the synthesis of (–)-taxol, *Tetrahedron Lett.* **36**, 3291-3294Johnson, C. R., Golebiowski, A., Steensma, D. H., Scialdone, M. A. (1993), Enantio- and diastereoselective transformations of cycloheptatriene to sugars and related products, *J. Org. Chem.* **58**, 7185-7194.

Johnson, C. R., Harikrishnan, L. S., Golebiowski, A. (1994), Enantioselective synthesis of 7-cycloocten-1,3,5,6-tetraol derivatives by enzymic asymmetrization, *Tetrahedron Lett.* **35**, 7735-7738.

Johnson, C. R., Sakaguchi, H. (1992), Enantioselective transesterifications using immobilized, recombinant *Candida antarctica* lipase B: resolution of 2-iodo-2-cycloalken-1- ols, *Synlett* 813-816.

Johnson, C. R., Plé, P. A., Adams, J. P. (1991), Enantioselective synthesis of (+)- and (–)-conduritol C from benzene via microbial oxidation and enzymic asymmetrization, *J. Chem. Soc., Chem. Commun.* 1006-1007.

Johnson, C. R., Senanayake, C. H. (1989), Polyhydroxylated seven-membered chiral building blocks. Asymmetric synthesis of compactin analogues, *J. Org. Chem.* **54**, 735-736.

Johnson, C. R., Penning, T. D. (1986), Triply convergent synthesis of (-)-prostaglandin-E_2, *J. Am. Chem. Soc.* **108**, 5655-5656.

Johnston, B. R., Morgan, B., Oehlschlager, A. C., Ramaswamy, S. (1991), A convenient synthesis of both enantiomers of seudenol and their conversion to 1-methyl-2-cyclohexen-1-ol, *Tetrahedron: Asymmetry* **2**, 377-380.

Jones, R. M., Collier, L. S., Neidle, E. L., Williams, P. A. (1999), *areABC* genes determine the catabolism of aryl esters in *Acinetobacter* sp. strain ADP1, *J. Bacteriol.* **181**, 4568-4575.

Jones, B. C. N. M., Simons, C., Nishimura, H., Zemlicka, J. (1995), Resolution of (±)-cytallene - a highly active anti-HIV agent with axial dissymmetry, *Nucleos. Nucleot.* **14**, 431-434.

Jones, J. B. (1993), Probing the specificity of synthetically useful enzymes, *Can. J. Chem.* **71**, 1273-1282.

Jones, J. B., Hinks, R. S., Hultin, P. G. (1985), Enzymes in organic synthesis. 33. Stereoselective pig liver esterase-catalyzed hydrolyses of meso cyclopentyl-, tetrahydrofuranyl-, and tetrahydrothiophenyl-1,3-diesters, *Can. J. Chem.* **63**, 452-456.

Jones, J. B., Beck, J. F. (1976), Asymmetric syntheses and resolutions using enzymes, in.: *Applications of Biochemical Systems in Organic Chemistry* (Jones, J. B., Sih, C. J., Perlman, D.; Eds.), , pp. 107-401. New York: Wiley.

Jones, J. B., Marr, P. W. (1973), Potential of α-chymotrysin for the resolution of alicyclic acids and esters, *Tetrahedron Lett.* 3165-3168.

Jones, M. M., Williams J. M. J. (1998), Dynamic kinetic resolution in the hydrolysis of an α-bromo ester, *Chem. Commun.*, 2519-2520.

Jones, M., Page, M. I. (1991), An esterase with β-lactamase activity, *J. Chem. Soc., Chem. Commun.*, 316-317.

Jorgensen, S., Skov, K. W., Diderichsen, B. (1991), Cloning, sequence, and expression of a lipase gene from *Pseudomonas cepacia*: lipase production in heterologous hosts requires two *Pseudomonas* genes, *J. Bacteriol.* **173**, 559-567.

Jouglet, B., Rousseau, G. (1993), Enzymatic resolution of *N*-hydroxylmethyl γ-butyrolactams. An access to optically active γ-butyrolactams, *Tetrahedron Lett.* **34**, 2307-2310.

Junge, W., Heymann, E. (1979), Characterization of the isoenzymes of pig liver esterase. 2. Kinetic studies, *Eur. J. Biochem.* **95**, 519-525.

Justiz, O. H., Fernandez-Lafuente, R., Guisan, J. M., Negri, P., Pagani, G., Pregnolato, M., Terreni, M. (1997), One-pot chemoenzymic synthesis of 3'-functionalized cephalosporines (Cefazolin) by three consecutive biotransformations in fully aqueous medium., *J. Org. Chem.* **62**, 9099-9106.

Kaftzik, N., Wasserscheid, P., Kragl, U. (2002) Use of ionic liquids to increase the yield and enzyme stability in the β-galactosidase catalysed synthesis of *N*-acetyllactosamine, *Org. Proc. Res. Dev.* **6**, 553-557.

Kaga, H., Siegmund, B., Neufellner, E., Faber, K., Paltauf, F. (1994), Stabilization of *Candida* lipase against acetaldehyde by adsorption onto Celite, *Biotechnol. Tech.* **8**, 369-374.

Kagan, H. B., Fiaud, J. C. (1988), Kinetic resolution, *Top. Stereochem.* **18**, 249-330.

Kakeya, H., Sakai, N., Sano, A., Yokoyama, M., Sugai, T., Ohta, H. (1991a), Microbial hydrolysis of 3-substituted glutaronitriles, *Chem. Lett.* 1823-1824.

Kakeya, H., Sakai, N., Sugai, T., Ohta, H. (1991b), Microbial hydrolysis as a potent method for the preparation of optically active nitriles, amides and carboxylic acids, *Tetrahedron Lett.* **32**, 1343-1346.

Kakeya, H., Sakai, N., Sugai, T., Ohta, H. (1991c), Preparation of optically active α-hydroxy acid derivatives by microbial hydrolysis of cyanohydrins and its application to the synthesis of (*R*)-4-dodecanolide, *Agric. Biol. Chem.* **55**, 1877-1881.

Kalaritis, P., Regenye, R. W., Partridge, J. J., Coffen, D. L. (1990), Kinetic resolution of 2-substituted esters catalyzed by a lipase ex. *Pseudomonas fluorescens*, *J. Org. Chem.* **55**, 812-815.

Kamat, S., Critchley, G., Beckman, E. J., Russell, A. J. (1995), Biocatalytic synthesis of acrylates in organic solvents and supercritical fluids: III. Does carbon dioxide covalently modify enzymes? *Biotechnol. Bioeng.* **46**, 610-620.

Kamat, S., Iwaskezyz, B., Beckman, E. J., Russell, A. J. (1993), Biocatalysis synthesis of acrylates in organic solvents and supercritical fluids: II. Tuning enzyme activity by changing pressure, *Proc. Natl. Acad. Sci. USA* **90**, 2940-2944.

Kamat, S., Burrera, J., Beckman, E. J., Russell, A. J. (1992), Biocatalysis synthesis of acrylates in organic solvents and supercritical fluids: I. Optimization of enzyme environment, *Biotechnol. Bioeng.* **40**, 158-166.

Kaminska, J., Gornicka, I., Sikora, M., Gora, J. (1996), Preparation of homochiral (*S*)- And (*R*)-1-(2-furyl)ethanols by lipase-catalyzed transesterification, *Tetrahedron: Asymmetry* **7**, 907-910.

Kamphuis, J., Boesten, W. H. J., Kapten, B., Hermes, H. F. M., Sonke, T., Broxterman, Q. B., van den Tweel, W. J. J., Schoemaker, H. E. (1992), The production and uses of optically pure natural and unnatural amino acids, in.: *Chirality in Industry* (Collins, A. N., Sheldrake, G. N., Crosby, J.; Eds.), pp. 187-208. Chichester, UK: Wiley.

Kanerva, L. T., Vänttinen, E. (1997), Optimized double kinetic resolution for the preparation of (*S*)-solketal, *Tetrahedron: Asymmetry* **8**, 923-933.

Kanerva, L. T., Csomos, P., Sundholm, O., Bernath, G., Fulop, F. (1996), Approach to highly enantiopure β-amino acid esters by using lipase catalysis in organic media, *Tetrahedron: Asymmetry* **7**, 1705-1716.

Kanerva, L. T., Sundholm, O. (1993a), Enzymic acylation in the resolution of methyl *threo*-2-hydroxy-3-(4-methoxyphenyl)-3-(2-X-phenylthio)propionates in organic solvents, *J. Chem. Soc., Perkin Trans. I* 2407-2410.

Kanerva, L. T., Sundholm, O. (1993b), Lipase catalysis in the resolution of racemic intermediates of diltiazem synthesis in organic solvents, *J. Chem. Soc., Perkin Trans. I* 1385-1389.

Kang, S.-K., Jeon, J.-H., Yamaguchi, T., Kim, J.-S., Ko, B.-S. (1995), Enzymic synthesis of (S)-(–)-1-(2-thienyl)propyl acetate, *Tetrahedron: Asymmetry* **6**, 2139-2142.

Kaptein, B., Boesten, W. H. J., Broxterman, Q. B., Peters, P. J. H., Schoemaker, H. E., Kamphuis, J. (1993), Enzymatic resolution of α,α-disubstituted α-amino-acid estes and amides, *Tetrahedron: Asymmetry* **4**, 1113-1116.

Kataoka, M., Honda, K., Shimizu, S. (2000), 3,4-Dihydrocoumarin hydrolase with haloperoxidase activity from *Acinetobacter calcoaceticus* F46, *Eur. J. Biochem.* **267**, 3-10.

Kataoka, M., Shimizu, K., Sakamoto, K., Yamada, H., Shimizu, S. (1995), Lactonohydrolase-catalyzed optical resolution of pantoyl lactone: selection of a potent enzyme producer and optimization of culture and reaction conditions for practical resolution, *Appl. Microbiol. Biotechnol.* **44**, 333-338.

Kato, K., Katayama, M., Fujii, S., Kimoto, H. (1996), Effective preparation of optically active 4,4,4-trifluoro-3-(indole-3-) butyric acid, a novel plant growth regulator, using lipase from *Pseudomonas fluorescens*, *J. Ferment. Bioeng.* **82**, 355-360.

Kato, K., Katayama, M., Gautam, R. K., Fujii, S., Fukaya, H., Kimoto, H. (1995a), A facile preparation of enantiomers of ethyl 4,4,4-trifluoro-3- (indole-3-)butyrate, a novel plant growth regulator, *J. Ferment. Bioeng.* **79**, 171-173.

Kato, K., Katayama, M., Gautam, R. K., Fujii, S., Kimoto, H. (1995b), Preparation and lipase-catalyzed optical resolution of 2,2,2-trifluoro-1-(naphthyl)ethanols, *Biosci. Biotechnol. Biochem.* **59**, 271-276.

Kato, K., Katayama, M., Gautam, R. K., Fujii, S., Kimoto, H. (1994), Lipase-catalyzed optical resolution of 2,2,2-trifluoro-1-(1-naphthyl)ethanol, *Biosci. Biotechnol. Biochem.* **58**, 1353-1354.

Kato, Y., Tsuda, T., Asano, Y. (1999), Nitrile hydratase involved in aldoxime metabolism from *Rhodococcus* sp. strain YH3-3 purification and characterization, *Eur. J. Biochem.* **263**, 662-670.

Katsuki, T., Martin, V. S. (1996), Asymmetric epoxidation of allylic alcohols: The Katsuki-Sharpless epoxidation reaction, *Org. React. (N.Y.)* **48**, 1-299.

Kawakami, K., Yoshida, S. (1995), Sol-gel entrapment of lipase using a mixture of tetramethoxysilane and methyltrimethoxysilane as the alkoxide precursor: esterification activity in organic media, *Biotechnol. Tech.* **9**, 701-704.

Kawanami, Y., Moriya, H., Goto, Y., Tsukao, K., Hashimoto, M. (1996), Lipase-catalyzed kinetic resolution of *trans*-2,5-disubstituted pyrrolidine derivatives, *Tetrahedron* **52**, 565-570.

Kawanami, Y., Moriya, H., Goto, Y. (1994), Lipase-catalyzed kinetic resolution of *trans*-2,5-disubstituted pyrrolidine derivatives, *Chem. Lett.* 1161-1162.

Kawasaki, M., Nakamura, K., Kawabata, S. (1999), Lipase-catalyzed enantioselective deacetylation of ortho-substituted phenyl acetates with 1-butanol in organic solvents, *J. Mol. Catal. B: Enzym.*, **6**, 447-451.

Kawashiro, K., Sugahara, H., Tsukioka, T., Sugiyama, S., Hayashi, H. (1996), Effect of ester moiety of substrates on enantioselectivity of protease catalysis in organic media, *Biotechnol. Lett.* **18**, 1381-1386.

Kazlauskas, R. J. (2006), Chromogenic assay of hydrolases for activity and enantioselectivity, in: *Enzyme Assays and Fingerprinting* (Reymond, J.-L., Ed.), Weinheim: Wiley-VCH.

Kazlauskas, R. J., Bornscheuer, U. T. (1998), Biotransformations with lipases, in.: *Biotechnology-Series* (Rehm, H. J., Reed, G., Pühler, A., Stadler, P. J. W., Kelly, D. R.; Eds.), **Vol. 8a**, pp. 37-191. Weinheim: VCH-Wiley.

Kazlauskas, R. J., Weissfloch, A. N. E. (1997), A structure-based rationalization of the enantiopreference of subtilisin toward secondary alcohols and isosteric primary amines, *J. Mol. Catal. B: Enzym.* **3**, 65-72.

Kazlauskas, R. J. (1991), (S)-(-)- and (R)-(+)-1,1'-bi-2-naphthol, *Org. Synth.* **70**, 60-67.

Kazlauskas, R. J., Weissfloch, A. N. E., Rappaport, A. T., Cuccia, L. A. (1991), A rule to predict which enantiomer of a secondary alcohol reacts faster in reactions catalyzed by cholesterol esterase, lipase from *Pseudomonas cepacia*, and lipase from *Candida rugosa*, *J. Org. Chem.* **56**, 2656-2665.

Kazlauskas, R. J. (1989), Resolution of binaphthols and spirobiindanols using cholesterol esterase, *J. Am. Chem. Soc.* **111**, 4953-4959.

Ke, T., Klibanov, A. M. (1998), Insights into the solvent dependence of chymotryptic prochiral selectivity, *J. Am. Chem. Soc.* **120**, 4259-4263.

Ke, T., Wescott, C. R., Klibanov, A. M. (1996), Prediction of the solvent dependence of enzymatic prochiral selectivity by means of structure-based thermodynamic calculations, *J. Am. Chem. Soc.* **118**, 3366-3374.

Keil, O., Schneider, M. P., Rasor, J. P. (1995), New hydantoinases from thermophilic microorganisms. Synthesis of enantiomerically pure D-amino acids, *Tetrahedron: Asymmetry* **6**, 1257-1260.

Kelly, D. R., Wan, P. W. H., Tang, J. (1998), Flavin monooxygenases - uses as catalysts for Baeyer-Villiger ring expansion and heteroatom oxidation, in.: *Biotechnology* (Rehm, H. J., Reed, G., Pühler, A., Stadler, P. J. W., Kelly, D. R.; Eds.), **Vol. 8a**, pp. 535-587. Weinheim: Wiley-VCH.

Kelly, N. M., Reid, R. G., Willis, C. L., Winton, P. L. (1996), Chemo-enzymatic synthesis of isotopically labelled L-valine, L-isoleucine and *allo*-isoleucine, *Tetrahedron Lett.* **37**, 1517-1520.

Kerscher, V., Kreiser, W. (1987), Enantiomerenreine glycerin-derivate durch enzymatische hydrolyse prochiraler ester, *Tetrahedron Lett.* **28**, 531-534.

Khalaf, N., Govardhan, C. P., Lalonde, J. J., Persichetti, R. A., Wang, Y. F., Margolin, A. L. (1996), Cross-linked enzyme crystals as highly active catalysts in organic solvents, *J. Am. Chem. Soc.* **118**, 5494-5495.

Khalameyzer, V., Bornscheuer, U. T. (1999), Overexpression and characterization of an esterase from Streptomyces diastatochromogenes, *Biotechnol. Lett.* **21**, 101-104.

Khalameyzer, V., Fischer, I., Bornscheuer, U. T., Altenbuchner, J. (1999), Screening, nucleotide sequence and biochemical characterization of an esterase from *Pseudomonas fluorescens* with high activity toward lactones, *Appl. Environm. Microbiol.* **65**, 477-482.

Khan, S. A., Halling, P. J., Bell, G. (1990), Measurement and control of water activity with an aluminum oxide sensor in organic two-phase reaction mixtures for enzymic catalysis, *Enzyme Microb. Technol.* **12**, 453-458.

Khlebnikov, V., Mori, K., Terashima, K., Tanaka, Y., Sato, M. (1995), Lipase-catalyzed resolution of sterically crowded 1,2-diols, *Chem. Pharm. Bull.* **43**, 1659-1662.

Khmelnitsky, Y. L., Welch, S. H., Clark, D. S., Dordick, J. S. (1994), Salts dramatically enhance activity of enzymes suspended in organic solvents, *J. Am. Chem. Soc* **116**, 2647-2648.

Khushi, T., O'Toole, K. J., Sime, J. T. (1993), Biotransformation of phosphonate esters, *Tetrahedron Lett.* **34**, 2375-2378.

Kiefer, M., Vogel, R., Helmchen, G., Nuber, B. (1994), Resolution of (1,1'-binaphthalene)-2,2'-dithiol by enzyme-catalyzed hydrolysis of a racemic diacyl derivative, *Tetrahedron* **50**, 7109-7114.

Kielbasinski, P., Zurawinski, R., Pietrusiewicz, K. M., Zablocka, K., Mikolajczyk, M. (1994), Enzymatic heteroatom chemistry. 1. Enzymatic resolution of racemic phosphinoylacetates having a stereogenic phosphorus atom, *Tetrahedron Lett.* **35**, 7081-7084.

Kikkawa, S., Takahashi, K., Katada, T., Inada, Y. (1989), Esterification of chiral secondary alcohols with fatty acid in organic solvents by polyethylene glycol-modified lipase, *Biochem. Int.* **19**, 1125-1131.

Kikuchi, M., Koshiyama, I., Fukushima, D. (1983), A new enzyme, proline acylase (*N*-acyl-L-proline amidohydrolase) from *Pseudomonas* species, *Biochim. Biophys. Acta* **744**, 180-188.

Kim, Y.-W., Lee, S. S., Warren, R. A. J., Withers, S. G. (2004), Directed evolution of a glycosynthase from *Agrobacterium* sp. increases its catalytic activity dramatically and expands its substrate repertoire, *J. Biol. Chem.* **279**, 42787-42793.

Kim, M.-J., Chung, Y. I., Choi, Y. K., Lee, H. K., Kim, D., Park, J. (2003), (*S*)-Selective dynamic kinetic resolution of secondary alcohols by the combination of subtilisin and an aminocyclopentadienylruthenium complex as the catalysts, *J. Am. Chem. Soc.* **125**, 11494 -11495.

Kim, K. W., Song, B., Choi, M. Y., Kim, M. J. (2001) Biocatalysis in ionic liquids: markedly enhanced enantioselectivity of lipase, *Org. Lett.* **3**, 1507-1509.

Kim, K. K., Song, H. K., Shin, D. H., Hwang, K. Y., Choe, S., Yoo, O. J., Suh, S. W. (1997a), Crystal structure of carboxylesterase from *Pseudomonas fluorescens*, an α/β-hydrolase with broad substrate specificity, *Structure* **5**, 1571-1584.

Kim, K. K., Song, H. K., Shin, D. H., Hwang, K. Y., Suh, S. W. (1997b), The crystal structure of a triacylglycerol lipase from *Pseudomonas cepacia* reveals a highly open conformation in the absence of a bound inhibitor, *Structure* **5**, 173-185.

Kim, M.-J., Choi, G.-B., Kim, J.-Y., Kim, H.-J. (1995), Lipase-catalyzed transesterification as a practical route to homochiral acyclic anti-1,2-diols. A new synthesis of (+)- and (−)-endo-brevicomin, *Tetrahedron Lett.* **36**, 6253-6256.

Kim, H., Ziani-Cherif, C., Oh, J., Cha, J. K. (1995), New [4+3] cycloaddition approach to cis-2,8-disubstiuted oxocanes, *J. Org. Chem.* **60**, 792-793.

Kim, M. J., Cho, H. (1992), *Pseudomonas* lipases as catalysts in organic synthesis: specificity of lipoprotein lipase, *J. Chem. Soc., Chem. Commun.* 1411-1413.

Kim, M. J., Choi, Y. K. (1992), Lipase-catalyzed enantioselective transesterification of *O*-trityl 1,2-diols. Practical synthesis of (*R*)-tritylglycidol, *J. Org. Chem.* **57**, 1605-1607.

Kim, S. B., Choi, C. Y. (1995), Effect of solid salt hydrates on the asymmetric esterification of 2-chloropropionic acid: control of water activity in organic solvent, *Biotechnol. Lett.* **17**, 1075-1076.

Kim, T., Chung, K. (1989), Some characteristics of palm kernel olein hydrolysis by *Rhizopus arrhizus* lipase in reversed micelle of AOT in isooctane, and additive effects, *Enzyme Microb. Technol.* **11**, 528-532.

Kingerywood, J., Johnson, J. S. (1996), Resolution of anti-3-oxotricyclo[2.2.1.0]heptane-7-carboxylic acid by *Candida antartica* lipase A, *Tetrahedron Lett.* **37**, 3975-3976.

Kirk, O., Conrad, L. S. (1999), Metal-free haloperoxidases: fact or artifact? *Angew. Chem., Int. Ed.* **38**, 977-979.

Kirk, O., Christensen, M. W., Damhus, T., Godtfredsen, S. E. (1994), Enzyme catalyzed degradation and formation of peroxycarboxylic acids, *Biocatalysis* **11**, 65-77.

Kita, Y., Takebe, Y., Murata, K., Naka, T., Akai, S. (1996), 1-Ethoxyvinyl acetate as a novel, highly reactive, and reliable acyl donor for enzymatic resolution of alcohols, *Tetrahedron Lett.* **37**, 7369-7372.

Kitaguchi, H., Fitzpatrick, P. A., Huber, J. E., Klibanov, A. M. (1989), Enzymic resolution of racemic amines: crucial role of the solvent, *J. Am. Chem. Soc.* **111**, 3094-3095.

Kitayama, T. (1996), Asymmetric syntheses of pheromones for *Bactrocera nigrotibialis*, *Andrena wilkella*, and *Andrena haemorrhoa* F from a chiral nitro alcohol, *Tetrahedron* **52**, 6139-6148.

Kitazume, T., Murata, K., Kokusho, Y., Iwasaki, S. (1988), Enzymes active in organic media. Synthesis of optically active trifluoromethylated compounds via asymmetric addition reactions, *J. Fluorine Chem.* **39**, 75-86.

Kitazume, T., Ikeya, T., Murata, K. (1986), Synthesis of optically active trifluorinated compounds: asymmetric Michael addition with hydrolytic enzymes, *J. Chem. Soc., Chem. Commun.* 1331-1333.

Klempier, N., Harter, G., de Raadt, A., Griengl, H., Braunegg, G. (1996), Chemoselective hydrolysis of nitriles by *Rhodococcus rhodochrous* NCIMB 11216, *Food Technol. Biotechnol.* **34**, 67-70.

Klempier, N., de Raadt, A., Faber, K., Griengl, H. (1991), Selective transformation of nitriles into amides and carboxylic acids by an immobilized nitrilase, *Tetrahedron Lett.* **32**, 341-344.

Klempier, N., Geymayer, P., Stadler, P., Faber, K., Griengl, H. (1990), Biocatalytic preparation of bicyclo[3.2.0]heptane derivatives, *Tetrahedron: Asymmetry* **1**, 111-118.

Klempier, N., Faber, K., Griengl, H. (1989), Chemoenzymic large-scale preparation of homochiral bicyclo[3.2.0]hept-2-en-6-one, *Biotechnol. Lett.* **11**, 685-688.

Klibanov, A. M. (1990), Asymmetric transformations catalyzed by enzymes in organic solvents, *Acc. Chem. Res.* **23**, 114-120.

Klibanov, A. M. (1989), Enzymatic catalysis in anhydrous organic solvents, *Trends Biochem. Sci.* **14**, 141-144.

Klibanov, A. M. (1997), Why are enzymes less active in organic solvents than in water? *Trends Biotechnol.* **15**, 97-101.

Kloosterman, M., de Nijs, M. P., Weijnen, J. G. J., Schoemaker, H. E., Meijer, E. M. (1989), Regioselective hydrolysis of carbohydrate secondary acyl esters by lipases, *J. Carbohydr. Chem.* **8**, 333-341.

Kloosterman, M., Elferink, V. H. M., van Iersel, J., Roskam, J. H., Meijer, E. M., Hulshof, L. A., Sheldon, R. A. (1988), Lipases in the preparation of β-blockers, *Trends Biotechnol.* **6**, 251-256.

Kloosterman, M., Mosmuller, E. W. J., Schoemaker, H. E., Meijer, E. M. (1987), Application of lipases in the removal of protective groups on glycerides and glycosides, *Tetrahedron Lett.* **28**, 2989-2992.

Klotz-Berendes, B., Kleemiß, W., Jegelka, U., Schäfer, H. J., Kotila, S. (1997), Enzymatic synthesis of optically active mono-alkylated malonic monoesters, *Tetrahedron: Asymmetry* **8**, 1821-1823.

Knani, D., Gutman, A. L., Kohn, D. H. (1993), Enzymatic polyesterification in organic enzyme-catalyzed synthesis of linear polyesters. I. condensation polymerization of hydroxyesters. II. ring-opening polymerization of ε-caprolactone, *J. Polym. Sci. A: Polymer Chem.* **31**, 1221-1232.

Knani, D., Kohn, D. H. (1993), Enzymatic polymerization in organic media. II. Enzyme-catalyzed synthesis of lateral-substituted alipahtic polyesters and copolyesters, *J. Polym. Sci. A: Polymer Chem.* **31**, 2887-2897.

Knez, Z., Habulin, M. (1992), Lipase catalyzed transesterification in supercritical carbon dioxide, in.: *Biocatalysis in Non-Conventional Media* (Tramper, J.; Eds.), **Vol. 8**, pp. 401-406. Amsterdam: Elsevier.

Knezovic, S., Sunjic, V., Levai, A. (1993), Enantioselective hydrolysis of some 3-(2-nitrophenoxy)butanoates catalyzed by *Pseudomonas fluorescens* and *Pseudomonas* sp. lipase, *Tetrahedron: Asymmetry* **4**, 313-320.

Kobayashi, M., Shimizu, S. (2000), Nitrile hydrolases, *Curr. Opin. Chem. Biol.* **4**, 95-102.

Kobayashi, M., Suzuki, T., Fujita, T., Masuda, M., Shimizu, S. (1995), Occurence of enzymes involved in biosynthesis of indole-3-acetic acid from indole-3-acetonitrile in plant-associated bacteria, *Agrobacterium* and *Rhizobium*, *Proc. Natl. Acad. Sci. USA* **92**, 714-718.

Kobayashi, M., Shimizu, S. (1994), Versatile nitrilases: Nitrile-hydrolysing enzymes, *FEMS Microbiol. Lett.* **120**, 217-224.

Kobayashi, M., Izui, H., Nagasawa, T., Yamada, H. (1993), Nitrilase in biosynthesis of the plant hormone indole-3-acetic acid from indole-3-acetonitrile: Cloning of the *Alcaligenes* gene and site-directed mutagenesis of cysteine residues, *Proc. Natl. Acad. Sci. USA* **90**, 247-251.

Kobayashi, M., Nagasawa, T., Yamada, H. (1992), Enzymatic syntheses of acrylamide: a success story not yet over, *Trends Biotechnol.* **10**, 402-408.

Kobayashi, M., Nishiyama, M., Nagasawa, T., Horinouchi, S., Beppu, T., Yamada, H. (1991), Cloning, nucleotide sequence and expression in *Escherichia coli* of two cobalt-containing nitrile hydratase genes from *Rhodococcus rhodochrous* J1, *Biochim. Biophys. Acta* **1129**, 23-33.

Kobayashi, M., Yanaka, N., Nagasawa, T., Yamada, H. (1990a), Monohydrolysis of an aliphatic dinitrile compound by nitrilase from *Rhodococcus rhodochrous* K22, *Tetrahedron* **46**, 5587-5590.

Kobayashi, M., Yanaka, N., Nagasawa, T., Yamada, H. (1990b), Nitrilase-catalyzed production of pyrazinoic acid, an antimycobacterial agent, from cyanopyrazine by resting cells of *Rhodococcus rhodochrous* J1, *J. Antibiot. (Tokyo)* **43**, 1316-1320.

Kobayashi, M., Nagasawa, T., Yamada, H. (1989), Nitrilase of *Rhodococcus rhodochrous* J1: purification and characterization, *Eur. J. Biochem.* **182**, 349-356.

Kobayashi, M., Nagasawa, T., Yamada, H. (1988), Regiospecific hydrolysis of dinitrile compounds by nitrilase from *Rhodococcus rhodochrous* J1, *Appl. Microbiol. Biotechnol.* **29**, 231-233.

Kobayashi, S., Shoda, S., Uyama, H. (1994), Enzymatic polymerization, *J. Syn. Org. Chem. Jpn.* **52**, 754-764.

Kobayashi, S., Uyama, H. (1993), Enzymatic polymerization of cyclic acid anhydrides and glycols by a lipase catalyst, *Makromol. Chem., Rapid Commun.* **14**, 841-844.

Kobayashi, S., Kamiyama, K., Ohno, M. (1990), Chiral synthon obtained with pig liver esterase: Introduction of chiral centers into cyclohexene skeleton, *Chem. Pharm. Bull.* **38**, 350-354.

Kodera, Y., Nishimura, H., Matsushima, A., Hiroto, M., Inada, Y. (1994), Lipase made active in hydrophobic media by coupling with polyethylene glycol, *J. Am. Oil Chem. Soc.* **71**, 335-338.

Kodera, Y., Furukawa, M., Yokoi, M., Kuno, H., Matsushita, H., Inada, Y. (1993), Lactone synthesis from 16-hydroxyhexadecanoic acid ethyl ester in organic solvents catalyzed with polyethylene glycol-modified lipase, *J. Biotechnol.* **31**, 219-224.

Koga, T., Nagao, A., Terao, J., Sawada, K., Mukai, K. (1994), Synthesis of a phosphatidyl derivative of Vitamin E and its antioxidant activity in phospholipid bilayers, *Lipids* **29**, 83-89.

Kohno, M., Funatsu, J., Mikami, B., Kugimiya, W., Matsuo, T., Morita, Y. (1996), The crystal structure of lipase II from *Rhizopus niveus* at 2.2 angstrom resolution, *J. Biochem.* **120**, 505-510.

Koichi, Y., Suginaka, K., Yamamoto, Y. (1995), Lipase-promoted asymmetric transesterification of 4-alkyloxetane-2-ones with ring opening, *J. Chem. Soc. Perkin Trans. 1*, 1645-1646.

Komeda, H., Hori, Y., Kobayashi, M., Shimizu, S. (1996), Transcriptional regulation of the *Rhodococcus rhodochrous* J1 nitA gene encoding a nitrilase, *Proc. Natl. Acad. Sci. USA* **93**, 10572-10577.

Kometani, T., Isobe, T., Goto, M., Takeuchi, Y., Haufe, G. (1998), Enzymatic resolution of 2-fluoro-2-arylacetic acid derivatives, *J. Mol. Catal. B: Enzym.*, **5**, 171-174.

Konarzycka-Bessler, M., Bornscheuer, U. T. (2003), A high-throughput-screening method for the determination of the synthetic activity of hydrolases, *Angew. Chem. Int. Ed.* **42**, 1418-1420.

Konigsberger, K., Luna, H., Prasad, K., Repic, O., Blacklock, T. J. (1996), Separation of *cis/trans*-cyclohexanecarboxylates by enzymic hydrolysis: preference for diequatorial isomers, *Tetrahedron Lett.* **37**, 9029-9032.

Koseki, T., Furuse, S., Iwano, K., Sakai, H., Matsuzawa, H. (1997), An *Aspergillus awamori* acetylesterase: purification of the enzyme, and cloning and sequencing of the gene, *Biochem. J.* **326**, 485-490.

Koshiro, S., Sonomoto, K., Tanaka, A., Fukui, S. (1985), Stereoselective esterification of *dl*-menthol by polyurethane-entrapped lipase in organic solvent, *J. Biotechnol.* **2**, 47-57.

Koskinen, A. M. P., Klibanov, A. M. (Eds.) (1996), *Enzymatic Reactions in Organic Media*, Glasgow: Chapman & Hall.

Kötting, J., Eibl, H. (1994), Lipases and phospholipases in organic synthesis, in.: *Lipases: Their Structure, Biochemistry, and Application* (Woolley, P., Petersen, S. B.; Eds.), pp. 289-313. Cambridge: Cambridge University Press.

Kovacs, J. A. (2004), Synthetic analogous of cysteinate-ligated non-heme iron and non-corrinoid cobalt enzymes, *Chem. Rev.* **104**, 825-848.

Koyama, N., Doi, Y. (1996), Miscibility, thermal properties, and enzymatic degradability of binary blends of poly[(R)-3-hydroxybutyric acid] with poly(ε-caprolactone-co-lactide), *Macromolecules* **29**, 5843-5851.

Kragl, U., Eckstein, M., Kaftzik, N. (2002) Enzyme catalysis in ionic liquids, *Curr. Opin. Biotechnol.* **13**, 565-571.

Krebsfänger, N., Zocher, F., Altenbuchner, J., Bornscheuer, U. T. (1998a), Characterization and enantioselectivity of a recombinant esterase from *Pseudomonas fluorescens*, *Enzyme Microb. Technol.* **22**, 641-646.

Krebsfänger, N., Schierholz, K., Bornscheuer, U. T. (1998b), Enantioselectivity of a recombinant esterase from *Pseudomonas fluorescens* towards alcohols and carboxylic acids, *J. Biotechnol.* **60**, 105-111.

Kreiner, M., Moore, B. D., Parker, M. C. (2001), Enzyme-coated micro-crystals: a 1-step method for high activity biocatalyst preparation, *Chem. Commun.* 1096-1097.

Krief, A., Surleraux, D., Ropson, N. (1993), Novel enantioselective synthesis of optically active (1R)-cis- and (1R)-trans-chrysanthemic acids, *Tetrahedron: Asymmetry* **4**, 289-292.

Kroutil, W., Genzel, Y., Pietzsch, M., Syldatk, C., Faber, K. (1998a), Purification and characterization of a highly selective epoxide hydrolase from *Nocardia* sp. EH1, *J. Biotechnol.* **61**, 143-150.

Kroutil, W., Orru, R. V. A., Faber, K. (1998b), Stabilization of *Nocardia* EH1 epoxide hydrolase by immobilization, *Biotechnol. Lett.* **20**, 373-377.

Kroutil, W., Kleewein, A., Faber, K. (1997a), A computer program for analysis, simulation and optimization of asymmetric catalytic processes proceeding through two consecutive steps. Type 1: asymmetrization-kinetic resolutions, *Tetrahedron: Asymmetry* **8**, 3251-3261.

Kroutil, W., Kleewein, A., Faber, K. (1997b), A computer program for analysis, simulation and optimization of asymmetric catalytic processes proceeding through two consecutive steps. Type 2: sequential kinetic resolutions, *Tetrahedron: Asymmetry* **8**, 3263-3274.

Kroutil, W., Mischitz, M., Faber, K. (1997c), Deracemization of (±)-2,3-disubstituted oxiranes *via* biocatalytic hydrolysis using bacterial epoxide hydrolases: kinetics of an enantioconvergent process, *J. Chem. Soc., Perkin Trans. 1* 3629-3636.

Kroutil, W., Osprian, I., Mischitz, M., Faber, K. (1997d), Chemoenzymatic synthesis of (S)-(−)-frontalin using bacterial epoxide hydrolases, *Synthesis (Stuttgart)* 156-158.

Kroutil, W., Mischitz, M., Plachota, P., Faber, K. (1996), Deracemization of (±)-cis-2,3-epoxyheptane *via* enantioconvergent biocatalytic hydrolysis using *Nocardia* EH1-epoxide hydrolase, *Tetrahedron Lett.* **37**, 8379-8382.

Kruizinga, W. H., Bolster, J., Kellogg, R. M., Kamphuis, J., Boesten, W. H. J., Meijer, E. M., Schoemaker, H. E. (1988), Synthesis of optically pure α-alkylated α-amino acids and a single-step method for enantiomeric excess determination, *J. Org. Chem.* **53**, 1826-1827.

Kuboki, A., Okazaki, H., Sugai, T., Ohta, H. (1997), An expeditions route to *N*-glycolylneuraminic acid based on enzyme-catalyzed reaction, *Tetrahedron* **53**, 2387-2400.

Kudo, I., Murakami, M., Hara, S., Inoue, K. (1993), Mammalian non-pancreatic phospholipases A2, *Biochim Biophys Acta* **1170**, 217-231.

Kuge, Y., Shioga, K., Sugaya, T., Tomioka, S. (1993), Enzymatic optical resolution of dibenzoxepins and its application to an optically active antiallergic agent with thromboxane A2 receptor antagonist activity, *Biosci. Biotechnol. Biochem.* **57**, 1157-1160.

Kullmann, W. (1987), *Enzymatic peptide synthesis*, Boca Raton: CRC-Press.

Kundu, N., Roy, S., Maenza, F. (1972), Esterase activity of chymotrypsin on oxygen-substituted tyrosine substrates, *Eur. J. Biochem* **28**, 311-315.

Kunz, H., Kowalczyk, D., Braun, P., Braum, G. (1994), Enzymatic hydrolysis of hydrophilic diethyleneglycol and polyethyleneglycol esters of peptides and glycopeptides by lipases, *Angew. Chem. Int. Ed.* **33**, 336-339.

Kurtzman, A. L., Govindarajan, S., Vahle, K., Jones, J. T., Heinrichs, V., Patten, P. A. (2001), Advances in directed protein evolution by recursive genetic recombination: applications to therapeutic proteins, *Curr. Opin. Biotechnol.* **12**, 361- 370.

Kvittingen, L., Sjursnes, B. J., Anthonsen, T., Halling, P. (1992), Use of salt hydrates to buffer optimal water level during lipase catalyzed synthesis in organic media: a practical procedure for organic chemists, *Tetrahedron* **48**, 2793-2802.

Laane, C. (1987), Medium-engineering for bioorganic synthesis, *Biocatalysis* **1**, 17-22.

Laane, C., Boeren, S., Vos, K., Verger, C. (1987), Rules for optimization of biocatalysis in organic solvents, *Biotechnol. Bioeng.* **30**, 81-87.

Lacourciere, G. M., Armstrong, R. N. (1993), The catalytic mechanism of microsomal epoxide hydrolase involves an ester intermediate, *J. Am. Chem. Soc.* **115**, 10466-10467.

Ladner, W. E., Whitesides, G. M. (1984), Lipase-catalyzed hydrolysis as a route to esters of chiral epoxy alcohols, *J. Am. Chem. Soc.* **106**, 7250-7251.

Laib, T., Ouazzani, J., Zhu, J. (1998), Horse liver esterase catalyzed enantioselective hydrolysis of *N,O*-diacetyl-2-amino-1-arylethanol, *Tetrahedron: Asymmetry* **9**, 169-178.

Lallemand, J. Y., Leclaire, M., Levet, R., Aranda, G. (1993), Easy access to an optically pure precursor of forskolin, *Tetrahedron: Asymmetry* **4**, 1775-1778.

Lalonde, J., Margolin, A. (2002), Immobilization of Enzymes, in.: *Enzyme Catalysis in Organic Synthesis* (Drauz, K., Waldmann, H.; Eds.), 2nd ed., **Vol. 2**, pp. 163-184. Weinheim: Wiley-VCH.

Lalonde, J. J. (1995), The preparation of homochiral drugs and peptides using cross-linked enzyme crystals, *Chimica oggi* **9**, 31-35.

Lalonde, J. J., Govardhan, C., Khalaf, N., Martinez, A. G., Visuri, K., Margolin, A. L. (1995), Cross-linked crystals of *Candida rugosa* lipase: highly efficient catalysts for the resolution of chiral esters, *J. Am. Chem. Soc.* **117**, 6845-6852.

Lalonde, J. J., Bergbreiter, D. E., Wong, C.-H. (1988), Enzymatic kinetic resolution of α-nitro α-methyl carboxylic acids, *J. Org. Chem.* **53**, 2323-2327.

Lam, L. K. P., Brown, C. M., de Jeso, B., Lym, L., Toone, E. J., Jones, J. B. (1988), Enzymes in organic synthesis. 42. Investigation of the effects of the isozymal composition of pig liver esterase on its stereoselectivity in preparative-scale ester hydrolyses of asymmetric synthetic value, *J. Am. Chem. Soc.* **110**, 4409-4411.

Lam, L. K. P., Hui, R. A. H. F., Jones, J. B. (1986), Enzymes in organic synthesis. 35. Stereoselective pig liver esterase catalyzed hydrolyses of 3-substituted glutarate diesters. Optimization of enantiomeric excess via reaction conditions control, *J. Org. Chem.* **51**, 2047-2050.

Lamare, S., Legoy, M. D. (1995), Working at controlled water activity in a continuous process: the gas/solid system as a solution, *Biotechnol. Bioeng.* **45**, 387-397.

Lambusta, D., Nicolosi, G., Patti, A., Piattelli, M. (1996), Lipase-mediated resolution of racemic 2-hydroxymethyl-1-methylthioferrocene, *Tetrahedron Lett.* **37**, 127-130.

Lambusta, D., Nicolosi, G., Patti, A., Piatelli, M. (1993), Enzyme-mediated regioprotection-deprotection of hydroxyl groups in (+)-catechin, *Synthesis* 1155-1158.

Lampe, T. F. J., Hoffmann, H. M. R., Bornscheuer, U. T. (1996), Lipase mediated desymmetrization of *meso* 2,6- di(acetoxymethyl)tetrahydropyran-4-one derivatives. An innovative route to enantiopure 2,4,6-trifunctionalized C-glycosides, *Tetrahedron: Asymmetry* **7**, 2889-2900.

Lang, D., Hofmann, B., Haalck, L., Hecht, H. J., Spener, F., Schmid, R. D., Schomburg, D. (1996), Crystal structure of a bacterial lipase from *Chromobacterium viscosum* ATCC 6918 refined at 1.6 Å resolution, *J. Mol. Biol.* **259**, 704-717.

Lange, S., Musidlowska, A., Schmidt-Dannert, C., Schmitt, J., Bornscheuer, U. T. (2001), Cloning, functional expression, and characterization of recombinant pig liver esterase, *ChemBioChem* **2**, 576-582.

Langrand, G., Baratti, J., Buono, G., Triantaphylides, C. (1986), Lipase catalyzed reactions and strategy for alcohol resolution, *Tetrahedron Lett.* **27**, 29-32.

Langrand, G., Secchi, M., Buono, G., Baratti, J., Triantaphylides, C. (1985), Lipase-catalyzed ester formation in organic solvents. An easy preparative resolution of α-substituted cyclohexanols, *Tetrahedron Lett.* **26**, 1857-1860.

Lankiewicz, L., Kasprzykowski, F., Grzonka, Z., Kettmann, U., Hermann, P. (1989), Resolution of racemic amino acids with thermitase, *Bioorg. Chem.* **17**, 275-280.

Larissegger-Schnell, B., Kroutil, W., Faber, K. (2005), Chemo-enzymatic synthesis of (*R*)- and (*S*)-2-hydroxy-4-phenylbutanoic acid via enantio-complementary deracemization of (±)-2-hydroxy-4-phenyl-3-butenoic acid using a racemase-lipase two-enzyme system, *Synlett*, 1936-1938.

Larsson, A. L. E., Persson, B. A., Backvall, J.-E. (1997), Enzymic resolution of alcohols coupled with ruthenium-catalyzed racemization of the substrate alcohol, *Angew. Chem. Int. Ed.* **36**, 1211-1212.

Larsson, K. M., Adlercreutz, P., Mattiasson, B. (1990), Enzymatic catalysis in microemulsions: enzyme reuse and product recovery, *Biotechnol. Bioeng.* **36**, 135-141.

Laszlo, J. A., Compton, D. L. (2001) α-Chymotrypsin catalysis in imidazolium-based ionic liquids, *Biotechnol. Bioeng.* **75**, 181-186.

Lau, R. M., van Rantwijk, F., Sheddon, K. R., Sheldon, R. A. (2000) Lipase-catalyzed reactions in ionic liquids, *Org. Lett.* **2**, 4189-4191.

Laughlin, L. T., Tzeng, H.-F., Lin, S., Armstrong, R. N. (1998), Mechanism of microsomal epoxide hydrolase. Semifunctional site-specific mutants affecting the alkylation half-reaction, *Biochemistry* **37**, 2897-2904.

Laumen, K., Ghisalba, O. (1994), Preparative-scale chemo-enzymic synthesis of optically pure D-*myo*-inositol 1-phosphate, *Biosci. Biotechnol. Biochem.* **58**, 2046-2049.

Laumen, K., Breitgoff, D., Seemayer, R., Schneider, M. P. (1989), Enantiomerically pure cyclohexanols and cyclohexane-1,2-diol derivatives, chiral auxiliaries and substitutes for (-)-8-phenylmenthol. A facile enzymic route, *J. Chem. Soc., Chem. Commun.* 148-150.

Laumen, K., Breitgoff, D., Schneider, M. P. (1988a), Enzymic preparation of enantiomerically pure secondary alcohols. Ester synthesis by irreversible acyl transfer using a highly selective ester hydrolase from *Pseudomonas* sp.; an attractive alternative to ester hydrolysis, *J. Chem. Soc., Chem. Commun.* 1459-1461.

Laumen, K., Schneider, M. P. (1988b), A highly selective ester hydrolase from *Pseudomonas* sp. for the enzymic preparation of enantiomerically pure secondary alcohols chiral auxiliaries in organic synthesis, *J. Chem. Soc., Chem. Commun.* 598-600.

Laumen, K. E. (1987), Esterhydrolasen – Anwendung in der organischen Synthese: chirale Bausteine aus Estern prochiraler und racemischer Alkohole. *Ph.D. Thesis*, Bergische Universität, Wuppertal, Germany.

Laumen, K., Schneider, M. P. (1986), A facile chemoenzymatic route to optically pure building blocks for cyclopentanoid natural products, *J. Chem. Soc. Chem. Commun.* 1298-1299.

Laumen, K., Schneider, M. (1984), Enzymatic hydrolysis of prochiral cis-1,4-diacyl-2-cyclopentene diols - preparation of (1*S*,4*R*) and (1*R*,4*S*)-4-hydroxy-2-cylopentenyl derivatives, versatile building blocks for cyclopentanoic natural products, *Tetrahedron Lett.* **25**, 5875-5878.

Lavandera, I., Fernández, S., Magdalena, J., Ferrero, M., Kazlauskas, R. J., Gotor, V. (2005), An inverse substrate orientation for the regioselective acylation of 3',5'-diaminonucleosides catalyzed by *Candida antarctica* lipase B?, *ChemBioChem* **6**, 1181-1190.

Lawson, W. B. (1967), The conformation of substrates during hydrolysis at the active site of chymotrypsin, *J. Biol. Chem.* **242**, 3397-3401.

Layh, N., Parratt, J., Willetts, A. (1998), Characterization and partial purification of an enatioselective arylacetonitrilase from *Pseudomonas fluorescens* DSM 7155, *J. Mol. Catal. B: Enzymat.* **5**, 467-474.

Layh, N., Hirrlinger, B., Stolz, A., Knackmuss, H.-J. (1997), Enrichment strategies for nitrile-hydrolysing bacteria, *Appl. Microbiol. Biotechnol.* **47**, 668-674.

Layh, N., Knackmuss, H.-J., Stolz, A. (1995), Enantioselective hydrolysis of ketoprofen amide by *Rhodococcus* sp. C3II and *Rhodococcus erythropolis* MP 50, *Biotechnol. Lett.* **17**, 187-192.

Layh, N., Stolz, A., Böhme, J., Effenberger, F., Knackmuss, H.-J. (1994), *J. Biotechnol.* **33**, 175-182.

Layh, N., Stolz, A., Förster, S., Effenberger, F., Knackmuss, H.-J. (1992), Enantioselective hydrolysis of O-acetylmandelonitrile to O-acetylmandelic acid by bacterial nitrilases, *Arch. Microbiol.* **158**, 405-411.

Leanna, M. R., Morton, H. E. (1993), *N*-(Boc)-L-(2-Bromoallyl)-glycine: a versatile intermediate for the synthesis of optically active unnatural amino acids, *Tetrahedron Lett.* **34**, 4485-4488.

Lee, E. Y., Yoo, S. S., Kim, S. K., Lee, S. J., Oh, Y. K., Park, S. (2004), Production of (*S*)-styrene oxide by recombinant *Pichia pastoris* containing epoxide hydrolase from *Rhodotorula glutinis*, *Enzyme Microb. Technol.* **35**, 624-631.

Lefker, B. A., Hada, W. A., McGarry, P. J. (1994), An efficient synthesis of enantiomerically enriched aryllactic esters, *Tetrahedron Lett.* **35**, 5205-5208.

LeGrand, D. M., Roberts, S. M. (1992), Enzyme-catalyzed hydrolysis of 3,5-*cis*-diacetoxy-4-*trans*-benzyloxymethylcyclopentene and the synthesis of aristeromycin precursors, *J. Chem. Soc., Perkin Trans. 1* 1751-1752.

Lejeune, K. E., Mesiano, A. J., Bower, S. B., Grimsley, J. K., Wild, J. R., Russell, A. J. (1997), Dramatically stabilized phosphotriesterase-polymers for nerve agent degradation, *Biotechnol. Bioeng.* **54**, 105-114.

Lemke, K., Lemke, M., Theil, F. (1997), A three-dimensional predictive active site model for lipase from *Pseudomonas cepacia*, *J. Org. Chem.* **62**, 6268-6273.

Lemke, K., Theil, F., Kunath, A., Schick, H. (1996), Lipase-catalyzed kinetic resolution of phenylethan-1,2-diol by sequential transesterification - the influence of the solvent, *Tetrahedron: Asymmetry* **7**, 971-974.

Lemoult, S. C., Richardson, P. F., Roberts, S. M. (1995), Lipase-catalyzed Baeyer-Villiger reactions, *J. Chem. Soc., Perkin Trans. 1* 89-91.

Leroy, E., Bensel, N., Reymond, J. L. (2003), A low background high-throughput screening (HTS) fluorescence assay for lipases and esterases using acyloxymethylethers of umbelliferone, *Bioorg. Med. Chem. Lett.* **13**, 2105-2108.

Levayer, F., Rabiller, C., Tellier, C. (1995), Enzyme catalysed resolution of 1,3-diarylpropan-1,3-diols, *Tetrahedron: Asymmetry* **6**, 1675-1682.

Levy-Schil, S., Soubrier, F., Crutz-Le Coq, A. M., Faucher, D., Crouzet, J., Petre, D. (1995), Aliphatic nitrilase from a soil-isolated *Comamonas testosteroni* sp.: gene cloning and overexpression, purification and primary structure, *Gene* **161**, 15-20.

Ley, S. V., Mio, S., Meseguer, B. (1996), Dispiroketals in synthesis (part 20): preparation of chiral 2,2'-bis(halomethyl) and 2,2'bis(phenylthiomethyl)dihydropyrans, *Synlett* 787.

Li, C., Montgomery, M. G., Mohammed, F., Li, J.-J., Wood, S. P., Bugg, T. D. H. (2005), Catalytic mechanism of C–C hydrolase MhpC from *Escherichia coli*: kinetic analysis of His263 and Ser110 site-directed mutants, *J. Mol. Biol.* **346**, 241-251.

Li, W.-S., Li, Y., Hill, C. M., Lum, K. T., Raushel, F. M. (2002), Enzymatic synthesis of chiral organophosphothioates from prochiral precursors, *J. Am. Chem. Soc.* **124**, 3498-3499.

Li, W.-S., Lum, K. T., Chen-Goodspeed, M., Sogorb, M. A., Raushel, F. M. (2001) Stereoselective detoxification of chiral sarin and soman analogues by phosphotriesterase. *Bioorg. Med. Chem.* **9**, 2083-2091.

Li, Y., Aubert, S. D., Maes, E. G., Raushel, F. M. (2004), Enzymatic resolution of chiral phosphinate esters, *J. Am. Chem. Soc.* **126**, 8888-8889.

Li, Y., Aubert, S. D., Raushel, F. M. (2003), Operational control of stereoselectivity during the enzymatic hydrolysis of racemic organophosphorus compounds, *J. Am. Chem. Soc.* **125**, 7526-7527.

Li, Y.-F., Hammerschmidt, F. (1993), Enzymes in organic chemistry, part 1: enantioselective hydrolysis of α-(acyloxy)phosphonates by esterolytic enzymes, *Tetrahedron: Asymmetry* **4**, 109-120.

Li, Y.-X., Straathof, A. J. J., Hanefeld., U. (2002), Enantioselective formation of mandelonitrile acetate - investigation of a dynamic kinetic resolution, *Tetrahedron: Asymmetry*, **13**, 739-743.

Liang, S., Paquette, L. A. (1990), Biocatalytic-based synthesis of optically-pure (C-6)-functionalized 1-(tert-butyldimethylsilyloxy) 2-methyl-(*E*)-2-heptenes, *Tetrahedron: Asymmetry* **1**, 445-452.

Liang, X., Bols, M. (1999), Chemoenzymatic synthesis of enantiopure 1-azafagonine, *J. Org. Chem.*, **64**, 8485-8488.

Lin, H., Tao, H., Cornish, V. W. (2004), Directed evolution of a glycosynthase via chemical complementation, *J. Am. Chem. Soc.* **126**, 15051-15059.

Lin, Y. Y., Palmer, D. N., Jones, J. B. (1974), The specificity of the nucleophilic site of α-chymotrypsin and its potential for the resolution of alcohols. Enzyme-catalyzed hydrolyses of some (+)-, (−)-, and (±)-2-butyl, -2-octyl, and -α-phenethyl esters, *Can. J. Chem.* **52**, 469-476.

Linderman, R. J., Walker, E. A., Haney, C., Roe, R. M. (1995), Determination of the regiochemistry of insect epoxide hydrolase catalyzed epoxide hydration of juvenile hormone by ^{18}O-labelling studies, *Tetrahedron* **51**, 10845-10856.

Ling, L., Ozaki, S. (1993), Enzyme aided synthesis of D-*myo*-inositol 1,4,5-trisphosphate, *Tetrahedron Lett.* **34**, 2501-2504.

Ling, L., Watanabe, Y., Akiyama, T., Ozaki, S. (1992), A new efficient method of resolution of *myo*-inositol derivatives by enzyme catalyzed regio- and enantio-selective esterification in organic solvent, *Tetrahedron Lett.* **33**, 1911-1914.

Linker, T. (1997), The Jacobsen-Katsuki epoxidation and its controversial mechanism, *Angew. Chem. Int. Ed.* **36**, 2060-2062.

Linko, Y. Y., Seppälä, J. (1996), Producing high molecular weight biodegradable polyesters, *Chemtech* **26**, 25-31.

Linko, Y.-Y., Wang, Z.-L., Seppälä, J. (1995a), Lipase-catalyzed linear aliphatic polyester synthesis in organic solvent, *Enzyme Microb. Technol.* **17**, 506-511.

Linko, Y.-Y., Wang, Z.-L., Seppälä, J. (1995b), Lipase-catalyzed synthesis of poly(1,4-butyl sebacate) from sebacic acid or its derivatives with 1,4-butanediol, *J. Biotechnol.* **40**, 133-138.

Littlechild, J., Garcia-Rodriguez, E., Dalby, A., Isupov, M. (2002), Structural and functional comparisons between vanadium haloperoxidase and acid phosphatase enzymes, *J. Mol. Recognit.* **15**, 291-296.

Liu, A. M. F., Somers, N. A., Kazlauskas, R. J., Brush, T. S., Zocher, F., Enzelberger, M. M., Bornscheuer, U. T., Horsman, G. P., Mezzetti, A., Schmidt-Dannert, C., Schmid, R. D. (2001), Mapping the substrate selectivity of new hydrolases using colorimetric screening:lipases from *Bacillus thermocatenulatus* and *Ophiostoma piliferum*, esterases from *Pseudomonas fluorescens* and *Streptomyces diastatochromogenes*, *Tetrahedron: Asymmetry* **12**, 545-556.

Liu, K. K. C., Nozaki, K., Wong, C. H. (1990), Problems of acyl migration in lipase-catalyzed enantioselective transformation of *meso*-1,3-diol systems, *Biocatalysis* **3**, 169-177.

Liu, Y.-C., Chen, C.-S. (1989), An efficient synthesis of optically active D-*myo*-inositol 1,4,5-triphosphate, *Tetrahedron Lett.* **30**, 1617-1620.

Ljunger, G., Adlercreutz, P., Mattiasson, B. (1994), Lipase catalyzed acylation of glucose, *Biotechnol. Lett.* **16**, 1167-1172.

Lobell, M., Schneider, M. P. (1993), Lipase-catalyzed formation of lactones via irreversible intramolecular acyl transfer, *Tetrahedron: Asymmetry* **4**, 1027-1030.

Lokotsch, W., Fritsche, K., Syldatk, C. (1989), Resolution of D,L-menthol by interesterification with triacetin using the free and immobilized lipase of *Candida cylindracea*, *Appl. Microbiol. Biotechnol.* **31**, 467-472.

López, R., Montero, E., Sanchez, F., Canada, J., Fernandez-Mayoralas, A. (1994), Regioselective acetylations of alkyl β-D-xylopyranosides by use of lipase PS in organic solvents and application to the chemoenzymatic synthesis of oligosaccharides, *J. Org. Chem.* **59**, 7027-7032.

López-Serrano, P., Cao, L., van Rantwijk, F., Sheldon, R. A. (2002), Cross-linked enzyme aggregates with enhanced activity: application to lipases, *Biotechnol. Lett.* **24**, 1379-1383.

Lord, M. D., Negri, J. T., Paquette, L. A. (1995), Oxonium ion-initiated pinacolic ring expansion reactions - application to the enantioselective synthesis of the spirocyclic sesquiterpene ethers dactyloxene-B and -C, *J. Org. Chem.* **60**, 191-195.

Lorenz, P., Eck, J. (2004), Screening for Noval Industrial Biocatalysts, *Eng. Life Sci.* **4**, 501-504.

Lorenz, P., Liebeton, K., Niehaus, F., Schleper, C., Eck, J. (2003), The impact of non-cultivated biodiversity on enzyme discovery and evolution, *Biocatal. Biotransform.* **21**, 87-91.

Lorenz, W. W., Wiegel, J. (1997), Isolation, analysis, and expression of two genes from *Thermoanaerobacterium* sp. strain JW/SL YS485: a β-xylosidase and a novel acetyl xylan esterase with cephalosporin C deacetylase activity, *J. Bacteriol.* **179**, 5436-5441.

Lotti, M., Grandori, R., Fusetti, F., Longhi, S., Brocca, S., Tramontano, A., Alberghina, L. (1993), Cloning and analysis of *Candida cylindracea* lipase sequences, *Gene* **124**, 45-55.

Lovey, R. G., Saksena, A. K., Girijavallabhan, V. M. (1994), PPL-catalyzed enzymatic asymmetrization of a 5-substituted prochiral 1,3-diol with remote chiral functionality - improvements toward synthesis of the eutomers of SCH 45012, *Tetrahedron Lett.* **35**, 6047-6050.

Lozano, P., de Diego, T., Carrie, D., Valutier, M., Iborra, J. L. (2003) Enzymatic ester synthesis in ionic liquids, *J. Mol. Catal. B: Enzym.* **21**, 9-13.

Lozano, P., de Diego, T., Carrie, D., Vaultier, M., Iborra, J. L. (2001a) Over-stabilization of *Candida antarctica* lipase B by ionic liquids in ester synthesis, *Biotechnol. Lett.* **23**, 1529-1533.

Lozano, P., de Diego, T., Guegan, J. P., Vaultier, M., Iborra, J. L. (2001b) Stabilization of α-chymotrypsin by ionic liquids in transesterification reactions, *Biotechnol. Bioeng.* **75**, 563-569.

Luić, M., Tomić, S., Leščić, I., Ljubović, E., Šepac, D., Šunjić, V., Vitale, L., Saenger, W., Kojić-Prodić, B. (2001) Complex of *Burkholderia cepacia* lipase with transition state analogue of 1-phenoxy-2-acetoxybutane, biocatalytic, structural and modelling study, *Eur. J. Biochem.* **268**, 3964-3973.

Lundell, K., Raijola, T., Kanerva, L. T. (1998), Enantioselectivity of *Pseudomonas cepacia* and *Candida rugosa* lipases for the resolution of secondary alcohols: The effect of *Candida rugosa* isoenzymes, *Enzyme Microb. Technol.* **22**, 86-93.

Lundh, H., Nordin, O., Hedenström, E., Högberg, H. E. (1995), Enzyme catalysed irreversible transesterifications with vinyl acetate - are they really irreversible? *Tetrahedron: Asymmetry* **6**, 2237-2244.

Lutz, D., Huffer, M., Gerlach, D., Schreier, P. (1992), Carboxylester-lipase-mediated reactions, in.: *Flavor precursors: thermal and enzymatic conversions* (Teranishi, R., Takeoka, G. R., Guentert, M.; Eds.), pp. 32-45. Washington DC: American Chemical Society.

Lutz, S., Ostermeier, M., Benkovic, S. J. (2001), Rapid generation of incremental trunction libraries for protein engineering using α-phosphothioate nucleotides, *Nucl. Acids Res.* **29**, 1-7.

MacDonald, R. T., Pulapura, S. K., Svirkin, Y. Y., Gross, R. A., Kaplan, D. L., Akkara, J., Swift, G., Wolk, S. (1995), Enzyme-catalyzed ε-caprolactone ring-opening polymerization, *Macromolecules* **28**, 73-78.

Macfarlane, E. L. A., Roberts, S. M., Turner, N. J. (1990), Enzyme-catalyzed inter-esterification procedure for the preparation of esters of a chiral secondary alcohol in high enantiomeric purity, *J. Chem. Soc., Chem. Commun.* 569-571.

MacKeith, R. A., McCague, R., Olivo, H. F., Roberts, S. M., Taylor, S. J. C., Xiong, H. (1994), Enzyme-catalyzed kinetic resolution of 4-endo-hydroxy-2- oxabicyclo[3.3.0]oct-7-en-3-one and employment of the pure enantiomers for the synthesis of antiviral and hypocholestemic agents, *Bioorg. Med. Chem.* **2**, 387-394.

MacKeith, R. A., McCague, R., Olivo, H. F., Palmer, C. F., Roberts, S. M. (1993), Conversion of (–)-4-hydroxy-2-oxabicyclo[3.3.0]oct-7-en-3-one into the anti-HIV agent carbovir, *J. Chem. Soc., Perkin Trans. 1* 313-314.

Maelicke, A. (1991), Acetylcholine esterase: the structure, *Trends Biochem. Sci.* **16**, 355-356.

Mackenzie, L. F., Wang, Q., Warren, R. A. J., Withers, S. G. (1998), Glycosynthases: mutant glycosidases for oligosaccharide synthesis, *J. Am. Chem. Soc.* **120**, 5583-5584.

Magnusson, A. O., Takwa, M., Hamberg, A., Hult. K. (2005), An S-selective lipase was created by rational redesign and the enantioselectivity increased with temperature, *Angew. Chem. Intl. Ed.* **44**, 4582-45.

Magnusson, A., Hult, K., Holmquist, M. (2001), Creation of an enantioselective hydrolase by engineered substrate-assisted catalysis, *J. Am. Chem. Soc.* **123**, 4354-4355.

Majeric, M., Sunjic, V. (1996), Preparation of (S)-2-ethylhexyl-p-methoxycinnamate by lipase catalyzed sequential kinetic resolution, *Tetrahedron: Asymmetry* **7**, 815-824.

Makita, A., Nihira, T., Yamada, Y. (1987), Lipase-catalyzed synthesis of macrocyclic lactones in organic solvents, *Tetrahedron Lett.* **28**, 805-808.

Maleczka, R. E. J., Paquette, L. A. (1991), Adaptation of oxyanionic sigmatropy to the convergent enantioselective synthesis of ambergris-type odorants, *J. Org. Chem.* **56**, 6538-6546.

Malézieux, B., Jaouen, G., Salaün, J., Howell, J. A. S., Palin, M. G., et al. (1992), Enzymatic generation of planar chirality in the (arene)tricarbonyl-chromium series, *Tetrahedron: Asymmetry* **3**, 375-376.

Manco, G., Adinolfi, E., Pisani, F. M., Ottolina, G., Carrea, G., Rossi, M. (1998), Overexpression and properties of a new thermophilic and thermostable esterase from *Bacillus acidocaldarius* with sequence similarity to hormone-sensitive lipase subfamily, *Biochem. J.* **332**, 203-212.

Margolin, A. L. (1996), Novel crystalline catalysts, *Trends Biotechnol.* **14**, 223-230.

Margolin, A. L. (1993a), Enzymes in the synthesis of chiral drugs, *Enzyme Microb. Technol.* **15**, 266-280.

Margolin, A. L. (1993b), Synthesis of optically pure mechanism-based inhibitors of γ-aminobutyic acid aminotransferase (GABA-T) via enzyme-catalyzed resolution, *Tetrahedron Lett.* **34**, 1239-1242.

Margolin, A. L., Delinck, D. L., Whalon, M. R. (1990), Enzyme-catalyzed regioselective acylation of castanospermine, *J. Am. Chem. Soc.* **112**, 2849-2854.

Margolin, A. L., Crenne, J.-Y., Klibanov, A. M. (1987), Stereoselective oligomerizations catalyzed by lipases in organic solvents, *Tetrahedron Lett.* **28**, 1607-1610.

Marier-Greiner, U. H., Obermaier-Skrobranek, B. M., Estermaier, L. M., Kammerloher, W., Freund, C., Wulfing, C., Burkert, U. I., Matern, D. H., Breuer, M., Eulitz, M., et al. (1991), Isolation and properties of a nitrile hydratase from the soil fungus *Myrothecium verrucaria* that is highly specific for the fertilizer cyanamide and cloning of its gene, *Proc. Natl. Acad. Sci. USA* **88**, 4260-4264.

Marr, R., Gamse, T., Schilling, T., Klingsbichel, E., Schwab, H., Michor, H. (1996), Enzymatic catalysis in supercritical carbon dioxide: comparison of different lipases and a novel esterase, *Biotechnol. Lett.* **18**, 79-84.

Marshalko, S. J., Schweitzer, B. I., Beardsley, G. P. (1995), Chiral chemical synthesis of DNA containing (S)-9-(1,3-dihydroxy-2-propoxymethyl)-guanine (dhpg) and effects on thermal stability, duplex structure, and thermodynamics of duplex formation, *Biochemistry* **34**, 9235-9248.

Martin, S. F., Hergenrother, P. J. (1998), Enzymatic synthesis of a modified phospholipid and its evalutation as a substrate for *B. cereus* phospholipase C, *Bioorg. Med. Chem. Lett.* **8**, 593-596.

Martinek, K., Levashov, A. V., Khmelnitski, Y. L., Klyachko, N., Berezin, I. V. (1982), Colloidal solution of water in organic solvents: A microheterogeneous medium for enzymatic reactions, *Science* **218**, 889-891.

Martinez, C., Nicolas, A., van Tilbeurgh, H., Egloff, M.-P., Cudrey, C., Verger, R., Cambillau, C. (1994), Cutinase, a lipolytic enzyme with a preformed oxyanion hole, *Biochemistry* **33**, 83-89.

Martinez, C., Geus, P. D., Lauwereys, M., Matthyssens, G., Cambillau, C. (1992), Fusarium solani cutinase is a lipolytic enzyme with a catalytic serine accessible to the solvent, *Nature* **356**, 615-618.

Martinkova, L., Klempier, N., Bardakji, J., Kandelbauer, A., Ovesna, M., Podarilova, T., Kuzma, M., Prepechalova, I., Griengl, H., Kren, V. (2001), Biotransformation of 3-substituted methyl (*R*,*S*)-4-cyanobutanoates with nitrile- and amide-converting biocatalysts, *J. Mol. Catal. B: Enzym.* **14**, 95-99.

Martinkova, L., Klempier, N., Prepechalova, I., Prikrylova, V., Ovesna, M., Griengl, H., Kren, V. (1998), Chemoselective biotransformation of nitriles by *Rhodococcus equi* A4, *Biotechnol. Lett.* **20**, 63-66.

Martínková, L., Stolz, A., Knackmuss, H.-J. (1996), Enantioselectivity of the nitrile hydratase from *Rhodococcus equi* A4 towards substituted (*R*,*S*)-2-arylpropionitriles, *Biotechnol. Lett.* **18**, 1073-1076.

Martín-Matute, B., Edin, M., Bogár, K., Kaynak, F. B., Bäckvall, J.-E. (2005), Combined ruthenium(II) and lipase catalysis for efficient dynamic kinetic resolution of secondary alcohols. Insight into the racemization mechanism, *J. Am. Chem. Soc.* **127**, 8817-8825.

Martín-Matute, B., Edin, M., Bogár, K., Bäckvall, J.-E. (2004), Highly compatible metal and enzyme catalysts for efficient dynamic kinetic resolution of alcohols at ambient temperature, *Angew. Chem. Intl. Ed.* **43**, 6535-6539.

Martins, J. F., Sampaio, T. C., Carvalho, I. B., da Ponte, M. N., Barreiros, S. (1992), Lipase catalyzed esterification of glycidol in chloroform and in supercritical carbon dioxide, in.: *High Press. Biotechnol.* (Balny, C., Hayashi, R., Heremans, K., Masson, P.; Eds.), **Vol. 224**, pp. 411-415.

Martres, M., Gil, G., Meon, A. (1994), Preparation of optically active aziridine carboxylates by lipase-catalyzed alcoholysis, *Tetrahedron Lett.* **35**, 8787-8790.

Marty, A., Chulalaksananukul, W., Willemot, R. M., Condorét, J. S. (1992), Kinetics of lipase-catalyzed esterification in supercritical carbon dioxide, *Biotechnol. Bioeng.* **39**, 273-280.

Marty, A., Chulalaksananukul, W., Condoret, J. S., Willemot, R. M., Durand, G. (1990), Comparison of lipase-catalysed esterification in supercritical carbon dioxide and in n-hexane, *Biotechnol. Lett.* **12**, 11-16.

Mateo, C., Palomo, J. M., van Langen, L. M., van Rantwijk, F., Sheldon, R. A. (2004), A new, mild cross-linking methodology to prepare cross-linked enzyme aggregates, *Biotech. Bioeng.* **86**, 273-276.

Mathew, C. D., Nagasawa, T., Kobayashi, M., Yamada, H. (1988), Nitrilase-catalyzed production of nicotinic acid from 3-cyanopyridine in *Rhodococcus rhodochrous* J1, *Appl. Environm. Microbiol.* **54**, 1030-1032.

Matsumae, H., Furui, M., Shibatani, T., Tosa, T. (1994), Production of optically active 3-phenylglycidic acid ester by the lipase from *Serratia marcescens* on a hollow-fiber membrane reactor, *J. Ferment. Bioeng.* **78**, 59-63.

Matsumae, H., Furui, M., Shibatani, T. (1993), Lipase-catalyzed asymmetric hydrolysis of 3-phenylglycidic acid ester, the key intermediate in the synthesis of diltiazem hydrochloride, *J. Ferment. Bioeng.* **75**, 93-98.

Matsumoto, K., Fuwa, S., Shimojo, S., Kitajima, H. (1996), Preparation of optically active diol derivatives by the enzymatic hydrolysis of cyclic carbonates, *Bull. Chem. Soc. Jpn.* **69**, 2977-2987.

Matsumoto, K., Fuwa, S., Kitajima, H. (1995), Enzyme-mediated enantioselective hydrolysis of cyclic carbonates, *Tetrahedron Lett.* **36**, 6499-6502.

Matsumoto, K. (1992), Production of 6-APA, 7-ACA, and 7-ADCA by immobilized penicillin and cephalosporin amidases, in.: *Industrial Applications of Immobilized Biocatalysts* (Tanaka, A., Tosa, T., Kobayashi, T.; Eds.), pp. New York: Marcel Dekker.

Matsuo, N., Ohno, N. (1985), Preparation of optically active 1-acetoxy-2-aryloxypropionitriles and its application to a facile synthesis of (*S*)-(-)-Propranolol, *Tetrahedron Lett.* **26**, 5533-5534.

Matsushima, M., Inoue, H., Ichinose, M., Tsukada, S., Miki, K., Kurokawa, K., Takahashi, T., Takahashi, K. (1991), The nucleotide and deduced amino acid sequence of porcine liver proline-β-naphthylamidase. Evidence for the idendity with carboxylesterase, *FEBS Lett.* **293**, 37-41.

Matta, M. S., Rohde, M. F. (1972), α-Chymotrypsin and rigid substrates. Reactivity of some *p*-nitrophenyl 1,2,3,4-tetrahydro-2-naphthoates and indan-2-carboxylates, *J. Am. Chem. Soc.* **94**, 8573-8578.

Matthews, B. W. (1988), Structural basis of the action of thermolysin and related zinc peptidases, *Acc. Chem. Res.* **21**, 333-340.

Matthews, B. W., Sigler, P. B., Henderson, R., Blow, D. M. (1967), Three-dimensional structure of tosyl-α-chymotrypsin., *Nature* **214**, 652-656.

Mattson, A., Orrenius, C., Oehrner, N., Unelius, C. R., Hult, K., Norin, T. (1996), Kinetic resolution of chiral auxiliaries with C_2-symmetry by lipase-catalyzed alcoholysis and aminolysis, *Acta Chem. Scand.* **50**, 918-921.

Mattson, A., Öhrner, N., Norin, T., Hult, K. (1993), Resolution of diols with C_2 symmetry by lipase-catalyzed transesterification, *Tetrahedron: Asymmetry* **4**, 925-930.

May, O., Verseck, S., Bommarius, A., Drauz, K. (2002), Development of dynamic kinetic resolution processes for biocatalytic production of natural and nonnatural L-amino acids, *Org. Proc. Res. Dev.* **6**, 452-457.

May, O., Nguyen, P. T., Arnold, F. H. (2000), Inverting enantioselectivity by directed evolution of hydantoinase for improved production of L-methionine, *Nat. Biotechnol.* **18**, 317-320.

Mazdiyasni, H., Konopacki, D. B., Dickman, D. A., Zydowsky, T. M. (1993), Enzyme-catalyzed synthesis of optically-pure β-sulfonamidopropionic acids. Useful starting materials for P-3 site modified renin inhibitors, *Tetrahedron Lett.* **34**, 435-438.

McCague, R., Taylor, S. J. C. (1997), Four case studies in the development of biotransformation-based processes, in.: *Chirality in Industry II* (Collins, A. N., Sheldrake, G. N., Crosby, J.; Eds.), pp. 183-206. Chichester: John Wiley & Sons.

McNeill, G. P., Shimizu, S., Yamane, T. (1991), High-yield enzymic glycerolysis of fats and oils, *J. Am. Oil Chem. Soc.* **68**, 1-5.

McPhalen, C. A., Schnebli, H. P., James, M. N. G. (1985), Crystal and molecular structure of the inhibitor eglin from leeches in complex with subtilisin Carlsberg, *FEBS Lett.* **188**, 55-58.

Meng, D., Sorensen, E. J., Bertinato, P., Danishefsky, S. J. (1996), Studies toward a synthesis of epothilone A: Use of hydropyran templates for the management of acyclic stereochemical relationships, *J. Org. Chem.* **61**, 7998-7999.

Merlo, V., Reece, F. J., Roberts, S. M., Gregson, M., Storer, R. (1993), Synthesis of optically active 5'-noraristeromycin: enzyme-catalyzed kinetic resolution of 9-(4-hydroxycyclopent-2-enyl)purines, *J. Chem. Soc., Perkin Trans. I* 1717-1718.

Mertoli, P., Nicolosi, G., Patti, A., Piattelli, M. (1996), Convenient lipase-assisted preparation of both enantiomers of suprofen, a non-steroidal anti-inflammatory drug, *Chirality* **8**, 377-380.

Meth-Cohn, O., Wang, M. X. (1997a), An in-depth study of the biotransformation of nitriles into amides and/or acids using *Rhodococcus rhodochrous* AJ270, *J. Chem. Soc., Perkin Trans. I* 1099-1104.

Meth-Cohn, O., Wang, M. X. (1997b), Rationalisation of the regioselective hydrolysis of aliphatic dinitriles with *Rhodococcus rhodochrous* AJ270, *Chem. Commun.* 1041-1042.

Mezzetti, A., Schrag, J., Cheong, C. S., Kazlauskas, R. J. (2005) Mirror-image packing in enantiomer discrimination: Molecular basis for the enantioselectivity of *Burkholderia cepacia* lipase toward 2-methyl-3-phenyl-1-propanol , *Chem. Biol.* **12**, 427-437.

Michor, H., Marr, R., Gamse, T., Schilling, T., Klingsbichel, E., Schwab, H. (1996), Enzymic catalysis in supercritical carbon dioxide: comparison of different lipases and a novel esterase, *Biotechnol. Lett.* **18**, 79-84.

Miller, C. A. (2000), Advances in enzyme discovery technology: capturing diversity, *Inform* **11**, 489-495.

Miller, D. A., Blanch, H. W., Prausnitz, J. M. (1990), Enzymic interesterification of triglycerides in supercritical carbon dioxide, *Ann. N. Y. Acad. Sci.* **613**, 534-537.

Milton, J., Brand, S., Jones, M. F., Rayner, C. M. (1995), Enzymatic resolution of α-acetoxysulfides - a new approach to the synthesis of homochiral *S,O*-acetals, *Tetrahedron: Asymmetry* **6**, 1903-1906.

Minning, S., Schmidt-Dannert, C., Schmid, R. D. (1998), Functional expression of *Rhizopus oryzase* lipase in *Pichia pastoris*: High-level production and some properties, *J. Biotechnol.* **66**, 147-156.

Misawa, E., ChanKwoChion, C. K. C., Archer, I. V., Woodland, M. P., Zhou, N.-Y., Carter, S. F., Widdowson, D. A., Leak, D. J. (1998), Characterization of a catabolic epoxide hydrolase from *Corynebacterium* sp., *Eur. J. Biochem.* **253**, 173-183.

Mischitz, M., Faber, K. (1996), Chemo-enzymatic synthesis of (2*R*,5*S*)- and (2*R*,5*R*)-5-(1-hydroxy-1-methylethyl)-2-methyl-2-vinyl-tetrahydrofuran ('linalool oxide'): Preparative application of a highly selective epoxide hydrolase, *Synlett* 978-979.

Mischitz, M., Mirtl, C., Saf, R., Faber, K. (1996), Regioselectivity of *Rhodococcus* sp. NCIMB11216 epoxide hydrolase: Applicability of *E*-values for description of enantioselectivity depends on substrate structure, *Tetrahedron: Asymmetry* **7**, 2041-2046.

Mischitz, M., Faber, K., Willetts, A. (1995a), Isolation of a highly enantioselective epoxide hydrolase from *Rhodococcus* sp. NCIMB11216, *Biotechnol. Lett.* **17**, 893-898.

Mischitz, M., Kroutil, W., Wandel, U., Faber, K. (1995b), Asymmetric microbial hydrolysis of epoxides, *Tetrahedron: Asymmetry* **6**, 1261-1272.

Mischitz, M., Faber, K. (1994), Asymmetric opening of an epoxide by azide catalyzed by an immobilized enzyme preparation from *Rhodococcus* sp., *Tetrahedron Lett.* **35**, 81-84.

Mitrochkine, A., Gil, G., Reglier, M. (1995a), Synthesis of enantiomerically pure *cis*- and *trans*-2-amino-1-indanol, *Tetrahedron: Asymmetry* **6**, 1535-1538.

Mitrochkine, A., Mycke, F., Martres, M., Gil, G., Heumann, A., Reglier, M. (1995b), Synthesis of enantiomerically pure (1*S*,2*R*)-epoxyindan and *cis*-(1*R*,2*S*)-2-amino-1-indanol, *Tetrahedron: Asymmetry* **6**, 59-62.

Miyanaga, A., Fushinobu, S., Ito, K., Wakagi, T. (2001), Crystal structure of cobalt-containing nitrile hydratase, *Biochem. Biophys. Res. Commun.* **288**, 1169-1174.

Miyaoka, H., Sagawa, S., Nagaoka, H., Yamada, Y. (1995), Efficient synthesis of enantiomerically pure 5,5-dimethyl-4-hydroxy-2-cyclopentenone, *Tetrahedron: Asymmetry* **6**, 587-594.

Miyazaki, K., Wintrode, P. L., Grayling, R. A., Rubingh, D. N., Arnold, F. H. (2000), Directed evolution study of temperature adaptation in a psychophilic enzyme, *J. Mol. Biol.* **297**, 1015-1026.

Miyazawa, T., Kurita, S., Ueji, S., Yamada, T., Kuwata, S. (1992), Resolution of racemic carboxylic acids via the lipase-catalyzed irreversible transesterification using vinyl esters: effects of alcohols as nucleophiles and organic solvents on enantioselectivity, *Biotechnol. Lett.* **14**, 941-946.

Mizuguchi, E., Nagai, H., Uchida, H., Achiwa, K. (1994), Lipase-catalyzed enantioselective reaction-based on remote recognition of the stereogenic carbon-atom away from the reaction site, *J. Synth. Org. Chem. Jpn.* **52**, 638-648.

Mohar, B., Stimac, A., Kobe, J. (1994), Chiral building blocks for carbocyclic *N*- and *C*-ribonucleosides through biocatalytic asymmetrization of *meso*-cyclopentane-1,3-dimethanols, *Tetrahedron: Asymmetry* **5**, 863-878.

Mohr, P., Waespe-Sarcevic, N., Tamm, C., Gawronska, K., Gawronski, J. K. (1983), A study of stereoselective hydrolysis of symmetrical diesters with pig liver esterase, *Helv. Chim. Acta* **66**, 2501-2511.

Molinari, F., Brenna, O., Valenti, M., Aragozzini, F. (1996), Isolation of a novel carboxylesterase from *Bacillus coagulans* with high enantioselectivity toward racemic esters of 1,2-O-isopropylideneglycerol, *Enzyme Microb. Technol.* **19**, 551-556.

Mölm, D., Risch, N. (1995), Resolution of racemic 1-azaadamantane derivatives by pig liver esterase catalysis, *Liebigs Ann.* 1901-1902.

Monfort, N., Archelas, A., Furstoss, R. (2002), Enzymatic transformations. Part 53: Epoxide hydrolase-catalysed resolution of key synthons for azole antifungal agents, *Tetrahedron: Asymmetry* **13**, 2399-2401.

Moore, B. D., Stevenson, L., Watt, A., Flitsch, S., Turner, N. J., Cassidy, C., Graham, D. (2004), Rapid and ultra sensitive determination of enzyme activities using surface-enhanced resonance raman scattering, *Nat. Biotechnol.* **22**, 1133-1138.

Moore, J. C., Arnold, F. H. (1996), Directed evolution of a para-nitrobenzyl esterase for aqueous-organic solvents, *Nat. Biotechnol.* **14**, 458-467.

Moore, S. A., Kingston, R. L., Loomes, K. M., Hernell, O., Blackberg, L., Baker, H. M., Baker, E. N. (2001) The structure of truncated recombinant human bile salt-stimulated lipase reveals bile salt-independent conformational flexibility at the active-site loop and provides insights into heparin binding, *J. Mol. Biol.* **312**, 511-523.

Moree, W. J., Sears, P., Kawashiro, K., Witte, K., Wong, C. H. (1997), Exploitation of subtilisin BPN' as catalyst for the synthesis of peptides containing noncoded amino acids, peptide mimetics and peptide conjugates, *J. Am. Chem. Soc.* **119**, 3942-3947.

Morgan, B., Zaks, A., Dodds, D. R., Liu, J., Jain, R., Megati, S., Njoroge, F. G., Girijavallabhan, V. M. (2000), Enzymatic kinetic resolution of piperidine atropisomers: synthesis of a key intermediate of the farnesyl protein transferase inhibitor, SCH66336, *J. Org. Chem.* **65**, 5451-5459.

Moorlag, H., Kellogg, R. M., Kloosterman, M., Kaptein, B., Kamphuis, J., Schoemaker, H. E. (1990), Pig-liver-esterase-catalyzed hydrolyses of racemic α-substituted α-hydroxy esters, *J. Org. Chem.* **55**, 5878-5881.

Moravcová, J., Vanclová, Z., Capková, J., Kefurt, K., Stanek, J. (1997), Enzymic hydrolysis of methyl 2,3-di-O-acetyl-5-deoxy-α and β-D-xylofuranosides - an active-site model of pig liver esterase, *J. Carbohydr. Chem.* **16**, 1011-1028.

Morgan, B., Dodds, D. R., Zaks, A., Andrews, D. R., Klesse, R. (1997a), Enzymatic desymmetrization of prochiral 2-substituted-1,3-propanediols: a practical chemoenzymatic synthesis of a key precursor of SCH5108, a broad spectrun orally active antifungal agent, *J. Org. Chem.* **62**, 7736-7743.

Morgan, B., Stockwell, B. R., Dodds, D. R., Andrews, D. R., Sudhakar, A. R., Nielsen, C. M., Mergelsberg, I., Zumbach, A. (1997b), Chemoenzymatic approaches to SCH 56592, a new azole antifungal, *J. Am. Oil Chem. Soc.* **74**, 1361-1370.

Morgan, B., Oehlschlager, A. C., Stokes, T. M. (1992), Enzyme reactions in apolar solvents. 5. The effect of adjacent unsaturation on the PPL-catalyzed kinetic resolution of secondary alcohols, *J. Org. Chem.* **57**, 3231-3236.

Morgan, B., Oehlschlager, A. C., Stokes, T. M. (1991), Enzyme reactions in apolar solvents. The resolution of branched and unbranched 2-alkanols by porcine pancreatic lipase, *Tetrahedron* **47**, 1611-1620.

Morgan, J., Pinhey, J. T., Sherry, C. J. (1997), Reaction of organolead triacetates with 4-ethoxycarbonyl-2-methyl-4,5-dihydro-1,3-oxazol-5-one. The synthesis of α-aryl-and α-vinyl-*N*-

acetylglycines and their ethyl esters and their enzymic resolution, *J. Chem. Soc., Perkin Trans. 1* 613-619.

Mori, K. (1995), Biochemical methods in enantioselective synthesis of bioactive natural products, *Synlett* 1097-1109.

Mori, K., Puapoomchareon, P. (1991), Preparation of optically pure 2,4,4-trimethyl-2-cyclohexen-1-ol, a new and versatile chiral building block in terpene synthesis, *Liebigs Ann. Chem.* 1053-1056.

Mori, K., Hazra, B. G., Pfeiffer, R. J., Gupta, A. K., Lindgren, B. S. (1987), Synthesis and bioactivity of optically-active forms of 1-methyl-2-cyclohexen-1-ol, an aggregation pheromone of *Dendroctonus pseudotsugae*, *Tetrahedron* **43**, 2249-2254.

Morihara, K., Oka, T. (1983), Enzymic semisynthesis of human insulin by transpeptidation method with *Achromobacter* protease: comparison with the coupling method, *Pept. Chem.* **20**, 231-236.

Morimoto, T., Murakami, N., Nagatsu, A., Sakakibara, J. (1994), Enzymatic regioselective acylation of 3-*O*-β-D-galactopyranosyl-*sn*-glycerol by *Achromobacter* sp lipase, *Chem. Pharm. Bull.* **42**, 751-753.

Morís, F., Gotor, V. (1994), Selective aminoacylation of nucleosides through an enzymatic reaction with oxime aminoacyl esters, *Tetrahedron* **50**, 6927-6934.

Morís, F., Gotor, V. (1993a), Enzymic acylation and alkoxycarbonylation of α-, *xylo*-, *anhydro*-, and *arabino*-nucleosides, *Tetrahedron* **49**, 10089-10098.

Morís, F., Gotor, V. (1993b), A useful and versatile procedure for the acylation of nucleosides through an enzymic reaction, *J. Org. Chem.* **58**, 653-660.

Morís, F., Gotor, V. (1992a), Lipase-mediated alkoxycarbonylation of nucleosides with oxime carbonates, *Tetrahedron* **48**, 9869-9876.

Morís, F., Gotor, V. (1992b), A novel and convenient route to 3'-carbonates from unprotected 2'-deoxynucleosides through an enzymatic reaction, *J. Org. Chem.* **57**, 2490-2492.

Morisseau, C., Nellaiah, H., Archelas, A., Furstoss, R., Baratti, J. C. (1997), Asymmetric hydrolysis of racemic *para*-nitrostyrene oxide using an epoxide hydrolase preparation from *Aspergillus niger*, *Enzyme Microb. Technol.* **20**, 446-452.

Morley, K. L., Kazlauskas, R. J. (2005), Improving enzyme properties: when are closer mutations better? *Trends Biotechnol.* **23**, 231-237.

Morrone, R., Nicolosi, G., Patti, A., Piattelli, M. (1995), Resolution of racemic flurbiprofen by lipase-mediated esterification in organic solvent, *Tetrahedron: Asymmetry* **6**, 1773-1778.

Moussou, P., Archelas, A., Baratti, J., Furstoss, R. (1998a), Determination of the regioselectivity during epoxide hydrolase oxirane ring opening: A new method from racemic epoxides, *J. Mol. Catal. B: Enzym.* **5**, 213-217.

Moussou, P., Archelas, A., Baratti, J., Furstoss, R. (1998b), Microbiological transformations. Part 39: Determination of the regioselectivity occurring during oxirane ring opening by epoxide hydrolases: a theoretical analysis and a new method for its determination, *Tetrahedron: Asymmetry* **9**, 1539-1547.

Moussou, P., Archelas, A., Furstoss, R. (1998c), Microbiological transformations. 40. Use of fungal epoxide hydrolases for the synthesis of enantiopure alkyl epoxides, *Tetrahedron* **54**, 1563-1572.

Muchmore, D. C. (1993), Enantiomeric enrichment of (*R*,*S*)-3-quinuclidinol, *US Patent* US 5215918.

Mugford, P. F., Magloire, V. P., Kazlauskas, R. J. (2005), Unexpected subtilisin-catalyzed hydrolysis of a sulfinamide bond in preference to a carboxamide bond in *N*-acyl sulfinamides, *J. Am. Chem. Soc.* **127**, 6536-6537.

Mulchandani, A., Kaneva, I., Chen, W. (1999), Detoxification of organophosphate pesticides by immobilized *Escherichia coli* expressing organophosphorus hydrolase on cell surface, *Biotechnol. Bioeng.*, **63**, 216-223.

Mulvihill, M. J., Gage, J. L., Miller, M. J. (1998), Enzymatic resolution of aminocyclopentenols as precursors to D- and L-carbocyclic nucleosides, *J. Org. Chem.* **63**, 3357-3363.

Mulzer, J., Greifenberg, S., Beckstett, A., Gottwald, M. (1992), Selective acetate hydrolysis of diastereomers with porcine pancreatic lipase (PPL) as an access to useful chiral building blocks, *Liebigs Ann. Chem.* 1131-1135.

Murakami, T., Nojiri, M., Nakayama, H., Odaka, M., Yohda, M., Dohmae, N., Takio, K., Nagamune, T., Endo, I. (2000), Post-translational modification is essential for catalytic activity of nitrile hydratase, *Protein. Sci.* **9**, 1024-1030.

Musidlowska-Persson, A., Bornscheuer, U. T. (2003a), Recombinant porcine intestinal carboxylesterase: Cloning from the pig liver esterase gene by site-directed mutagenesis, functional expression and characterization, *Prot. Eng.* **16**, 1139-1145.

Musidlowska-Persson, A., Bornscheuer, U. T. (2003b), Site directed mutagenesis of recombinant pig liver esterase yields in mutant with altered enantioselectivity, *Tetrahedron: Asymmetry* **14**, 1341-1344.

Musidlowska, A., Lange, S., Bornscheuer, U. T. (2001), By overexpression in the yeast *Pichia pastoris* to enhanced enantioselectivity: New aspects in the application of pig liver esterase, *Angew. Chem. Int. Ed.* **40**, 2851-2853.

Mustranta, A. (1992), Use of lipases in the resolution of racemic ibuprofen, *Appl. Microbiol. Biotechnol.* **38**, 61-66.

Na, A., Eriksson, C., Eriksson, S.-G., Österberg, E., Holmberg, K. (1990), Synthesis of phosphatidylcholine with (n-3) fatty acids by phospholipase A2 in microemulsion, *J. Am. Oil Chem. Soc.* **67**, 766-770.

Naemura, K., Murata, M., Tanaka, R., Yano, M., Hirose, K., Tobe, Y. (1996), Enantioselective acylation of primary and secondary alcohols catalyzed by lipase QL from *Alcaligenes* sp.: a predictive active site model for lipase QL to identify which enantiomer of an alcohol reacts faster in this acylation, *Tetrahedron: Asymmetry* **7**, 3285-3294.

Naemura, K., Fukuda, R., Murata, M., Konishi, M., Hirose, K., Tobe, Y. (1995), Lipase YS-catalyzed enantioselective acylation of alcohols: a predictive active site model for lipase YS to identify which enantiomer of an alcohol reacts faster in this acylation, *Tetrahedron: Asymmetry* **6**, 2385-2394.

Naemura, K., Fukuda, R., Konishi, M., Hirose, K., Tobe, Y. (1994), Lipase YS catalysed acylation of alcohols: a predictive active site model for lipase YS to identify which enantiomer of a primary or secondary alcohol reacts faster in this acylation, *J. Chem. Soc. Perkin Trans. I* 1253-1256.

Naemura, K., Fukuda, R., Takahashi, N., Konishi, M., Hirose, Y., Tobe, Y. (1993a), Enzyme-catalyzed asymmetric acylation and hydrolysis of cis-2,5-disubstituted tetrahydrofuran derivatives - contribution to the development of models for reactions catalyzed by porcine liver esterase and porcine pancreatic lipase, *Tetrahedron: Asymmetry* **4**, 911-918.

Naemura, K., Ida, H., Fukuda, R. (1993b), Lipase YS-catalyzed enantioselective transesterification of alcohols of bicarbocyclic compounds, *Bull. Chem. Soc. Jpn.* **66**, 573-577.

Naemura, K., Takahashi, N., Tanaka, S. (1992), Resolution of the diols of bicyclo[2.2.1]heptane, bicyclo[2.2.2]octane and bicyclo[3.2.1]octane by enzymatic hydrolysis, and their absolute-configurations, *J. Chem. Soc. Perkin Trans. I* 2337-2343.

Naemura, K., Miyabe, H., Shingai, Y. (1991), Synthesis and enantiomer recognition of crown ethers containing cyclohexane-1,2-diol derivatives as the chiral center and enzymatic resolution of the chiral subunits, *J. Chem. Soc. Perkin Trans. I* 957-959.

Nagai, H., Shiozawa, T., Achiwa, K., Terao, Y. (1993), Convenient syntheses of optically active β-lactams by enzymatic resolution, *Chem. Pharm. Bull.* **41**, 1933-1938.

Nagamune, T., Kurata, H., Hirata, M., Honda, J., Koike, H., Ikeuchi, M., Inoue, Y., Hirata, A., Endo, I. (1990), Purification of inactivated photoresponsive nitrile hydratase, *Biochem. Biophys. Res. Commun.* **168**, 437-442.

Nagao, Y., Kume, M., Wakabayashi, R. C., Nakamura, T., Ochiai, M. (1989), Efficient preparation of new chiral synthons useful for (+)-carbacyclin synthesis by utilizing enzymatic hydrolysis of prochiral σ-symmetric diesters, *Chem. Lett.* 239-242.

Nagasawa, T., Yamada, H. (1995), Microbial production of commodity chemicals, *Pure Appl. Chem.* **67**, 1241-1256.

Nagasawa, T., Takeuchi, K., Yamada, H. (1991), Characterization of a new cobalt-containing nitrile hydratase purified from urea-induced cells of *Rhodococcus rhodochrous* J1, *Eur. J. Biochem.* **196**, 581-589.

Nagasawa, T., Mauger, J., Yamada, H. (1990a), Novel nitrilase, aryloacetonitrilase, of *Alcaligenes faecalis* JM3, purification and characterization, *Biochemistry* **194**, 765-772.

Nagasawa, T., Nakamura, T., Yamada, H. (1990b), Production of acrylic acid and methacrylic acid using *Rhodococcus rhodochrous* J1 nitrilase, *Appl. Microbiol. Biotechnol.* **34**, 322-324.

Nagasawa, T., Mathew, C. D., Mauger, J., Yamada, H. (1988), Nitrile hydratase-catalyzed production of nicotinamide from 3-cyanopyridine in *Rhodococcus rhodochrous* J1, *Appl. Environm. Microbiol.* **54**, 1766-1769.

Nagasawa, T., Nanba, H., Ryuno, K., Takeuchi, K., Yamada, H. (1987), Nitrile hydratase of *Pseudomonas chlororaphis* B23. Purification and characterization., *Eur. J. Biochem.* **162**, 691-698.

Nagasawa, T., Ryuno, K., Yamada, H. (1986), Nitrile hydratase of *Brevibacterium* R312-purification and characterization, *Biochem. Biophys. Res. Commun.* **139**, 1305-1312.

Nagashima, S., Nakasako, M., Dohmae, N., Tsujimura, M., Takio, K., Odaka, M., Yohda, M., Kamiya, N., Endo, I. (1998), Novel non-heme iron center of nitrile hydratase with a claw setting of oxygen atoms, *Nat. Struct. Biol.* **5**, 347-351.

Nagata, M. (1996), Enzymatic degradation of aliphatic polyesters copolymerized with various diamines, *Macromol. Rapid Commun.* **17**, 583-587.

Nagumo, S., Arai, T., Akita, H. (1996), Enzymatic hydrolysis of *meso-(syn-syn)*-1,3,5-tricetoxy-2,4-dimethylpentane and acetylation of *meso-(syn-syn)*-3-benzyloxy-2,4-dimethylpentane-1,5-diol by lipase, *Chem. Pharm. Bull.* **44**, 1391-1394.

Nair, M. S., Anilkumar, A. T. (1996), Versatile chiral intermediates for terpenoid synthesis using lipase-catalyzed acylation, *Tetrahedron: Asymmetry* **7**, 511-514.

Nakada, M. (1995), The first total synthesis of antitumor macrolide, Rhizoxin, *Mem. Sch. Sci. Eng. Waseda Univ.* **59**, 167-195.

Nakamura, K., Takebe, Y., Kitayama, T., Ohno, A. (1991), Effect of solvent structure on enantioselectivity of lipase-catalyzed transesterification, *Tetrahedron Lett.* **32**, 4941-4944.

Nakamura, K., Inoue, Y., Kitayama, T., Ohno, A. (1990a), Stereoselective preparation of (R)-4-nitro-2-butanol and (R)-5-nitro-2-pentanol mediated by a lipase, *Agric. Biol. Chem.* **54**, 1569-1570.

Nakamura, K., Ishihara, K., Ohno, A., Uemura, M., Nishimura, H., Hayashi, Y. (1990b), Kinetic resolution of (η^6-arene)chromium complexes by a lipase, *Tetrahedron Lett.* **31**, 3603-3604.

Nakamura, K. (1990), Kinetics of lipase-catalyzed esterification in supercritical CO_2, *Trends Biotechnol.* **8**, 288-291.

Nakamura, K., Chi, Y. M., Yamada, Y., Yano, T. (1986), Lipase activity and stability in supercritical carbon dioxide, *Chem. Eng. Commun.* **45**, 207-212.

Nakanishi, Y., Watanabe, H., Washizu, K., Narahashi, Y., Kurono, Y. (1991), Cloning, sequencing and regulation of the lipase gene from *Pseudomonas* sp. M-12-33, in.: *Lipases* (Schmid, R. D., Alberghina, L., Verger, R.; Eds.), **Vol. 16**, pp. 263-266. Weinheim: VCH.

Nakano, H., Ide, Y., Tsuda, T., Yang, J., Ishii, S., Yamane, T. (1999), Improvement in the organic solvent stability of *Pseudomonas* lipase by random mutation, in.: *Ann. N. Y. Acad. Sci.* (Laskin, A. I., Li, G.-X., Yu, Y.-T.; Eds.), **Vol. 864**, pp. 431-434. New York: N. Y. Acad. Sci.

Nakano, H., Okuyama, Y., Iwasa, K., Hongo, H. (1996), Lipase-catalyzed resolution of 2-azabicyclo[2.2.1]hept-5-en-3-ones, *Tetrahedron: Asymmetry* **7**, 2381-2386.

Nara, S. J., Harjani, J. R., Salunkhe, M. M., Mane, A. T., Wadgaonkar, P. P. (2003) Lipase-catalyzed polyester synthesis in 1-butyl-3-methylimidazolium hexafluorophosphate ionic liquid, *Tetrahedron Lett.* **44**, 1371-1373.

Nara, S. J., Harjani, J. R,., Salunkhe, M. M. (2002) Lipase-catalyzed transesterification in ionic liquids and organic solvents: a comprehensive study, *Tetrahedron Lett.* **43**, 2979-2982.

Nardini, M., Lang, D. A., Liebeton, K., Jaeger, K.-E., Dijkstra, B. W. (2000) Crystal structure of *Pseudomonas aeruginosa* lipase in the open conformation. The prototype for family I.1 of bacterial lipases, *J. Biol. Chem.* **275**, 31219-31225.

Nardini, M., S., R. I., Rozeboom, H. J., Kalk, K. H., Rink, R., Janssen, D. B., Dijkstra, B. W. (1999), The X-ray structure of epoxide hydrolase from *Agrobacterium radiobacter* AD1, *J. Biol. Chem.* **274**, 14579-14596.

Natoli, M., Nicolosi, G., Piattelli, M. (1992), Regioselective alcoholysis of flavonoid acetates with lipase in an organic solvent, *J. Org. Chem.* **57**, 5776-5778.

Natoli, M., Nicolosi, G., Piattelli, M. (1990), Enzyme-catalyzed alcoholysis of flavone acetates in organic solvent, *Tetrahedron Lett.* **31**, 7371-7374.

Neidhart, D. J., Petsko, G. A. (1988), The refined crystal structure of subtilisin Carlsberg at 2.5 Å resolution, *Protein Eng.* **2**, 271-276.

Nellaiah, H., Morisseau, C., Archelas, A., Furstoss, R., Baratti, J. C. (1996), Enantioselective hydrolysis of *p*-nitrostyrene oxide by an epoxide hydrolase preparation from *Aspergillus niger*, *Biotechnol. Bioeng.* **49**, 70-77.

Ness, J. E., Welch, M., Giver, L., Bueno, M., Cherry, J. R., Borchert, T., Stemmer, W. P. C., Minshull, J. (1999), DNA shuffling of subgenomic sequences of subtilisin, *Nat. Biotechnol.* **17**, 893-896.

Nettekoven, M., Psiorz, M., Waldmann, H. (1995), Synthesis of enantiomerically pure 4-alkylsubstituted tryptophan derivatives by a combination of organometallic reactions with enantioselective enzymatic transformations, *Tetrahedron Lett.* **36**, 1425-1428.

Neuwald, A. F. (1997), An unexpected structural relationship between integral membrane phosphatases and soluble haloperoxidases, *Prot. Sci.* **6**, 1764-1767.

New, R. R. C. (1993), Biological and biotechnological applications of phospholipids, in.: *Phospholipid Handbook* (Cevc, G.; Ed.), pp. 855-878. New York: Marcel Dekker.

Neylon, C. (2004), Chemical and biochemical strategies for the randomization of protein encoding DNA sequences: library construction methods for directed evolution, *Nucl. Acids Res.* **32**, 1448-1459.

Ng-Youn-Chen, M. C., Serreqi, A. N., Huang, Q., Kazlauskas, R. J. (1994), Kinetic resolution of pipecolic acid using partially-purified lipase from *Aspergillus niger*, *J. Org. Chem.* **59**, 2075-2081.

Nicolosi, G., Patti, A., Piattelli, M., Sanfilippo, C. (1995a), Desymmetrization of cis-1,2-dihydroxycycloalkanes by stereoselective lipase mediated esterification, *Tetrahedron: Asymmetry* **6**, 519-524.

Nicolosi, G., Patti, A., Piattelli, M., Sanfilippo, C. (1995b), Enzymatic access to homochiral 1-acetoxy-2-hydroxycyclohexane-3,5-diene through lipase-assisted acetylation in organic solvent, *Tetrahedron Lett.* **36**, 6545-6546.

Nicolosi, G., Patti, A., Piattelli, M. (1994a), Lipase -mediated separation of the stereoisomers of 1-(1-hydroxyethyl)-2-(hydroxymethyl)ferrocene, *J. Org. Chem.* **59**, 251-254.

Nicolosi, G., Patti, A., Piattelli, M., Sanfilippo, C. (1994b), Desymmetrization of *meso*-hydrobenzoin via stereoselective enzymatic esterification, *Tetrahedron: Asymmetry* **5**, 283-288.

Nicolosi, G., Piattelli, M., Sanfilippo, C. (1993), Lipase-catalyzed regioselective protection of hydroxyl groups in aromatic dihydroxyaldehydes and ketones, *Tetrahedron* **49**, 3143-3148.

Nicolosi, G., Morrone, R., Patti, A., Piattelli, M. (1992), Lipase mediated desymmetrization of 1,2-bis(hydroxymethyl)ferrocene in organic medium: production of both enantiomers of 2-acetoxymethyl-1-hydroxymethyl ferrocene, *Tetrahedron: Asymmetry* **3**, 753-758.

Nicotra, F., Riva, S., Secundo, F., Zucchelli, L. (1989), An interesting example of complementary regioselective acylation of secondary hydroxyl groups by different lipases, *Tetrahedron Lett.* **30**, 1703-1704.

Nilsson, J., Blackberg, L., Carlsson, P., Enerback, S., Hernell, O., Bjursell, G. (1990), cDNA cloning of human-milk bile-salt-stimulated lipase and evidence for its identity to pancreatic carboxylic ester hydrolase, *Eur. J. Biochem.* **192**, 543-550.

Nilsson, K. G. I. (1987), A simple strategy for changing the regioselectivity of glycosidase-catalyzed formation of disaccharides, *Carbohydr. Res.* **167**, 95-103.

Nishino, H., Mori, T., Okahata, Y. (2002), Enzymatic silicone oligomerization catalyzed by a lipid-coated lipase, *Chem. Commun.*, 2684-2685.

Nishio, T., Kamimura, M., Murata, M., Terao, Y., Achiwa, K. (1989), Production of optically active esters and alcohols from racemic alcohols by lipase-catalyzed stereoselective transesterification in nonaqueous reaction system, *J. Biochem.* **105**, 510-512.

Nishizawa, K., Ohgami, Y., Matsuo, N., Kisida, H., Hirohara, H. (1997), Studies on hydrolysis of chiral, achiral and racemic alcohol esters with *Pseudomonas cepacia* lipase - mechanism of stereospecificity of the enzyme, *J. Chem. Soc., Perkin Trans. II* 1293-1298.

Nishizawa, M., Shimizu, M., Ohkawa, H., Kanaoka, M. (1995), Stereoselective production of (+)-*trans*-chrysanthemic acid by a microbial esterase: cloning, nucleotide sequence, and overexpression of the esterase gene of *Arthrobacter globiformis* in *Escherichia coli*, *Appl. Environ. Microbiol.* **61**, 3208-3215.

Nishizawa, M., Gomi, H., Kishimoto, F. (1993), Purification and some properties of carboxylesterase from Arthobacter globiformis; Stereoselective hydrolysis of ethyl chrysanthemate, *Biosci. Biotech. Biochem.* **57**, 594-598.

Nobes, G. A. R., Kazlauskas, R. J., Marchessault, R. H. (1996), Lipase-catalyzed ring-opening polyerization of lactones: A novel route to poly(hydroxyalkanoate)s, *Macromolecules* **29**, 4829-4833.

Noble, M. E. M., Cleasby, A., Johnson, L. N., Egmond, M. R., Frenken, L. G. J. (1994), Analysis of the structure of *Pseudomonas glumae* lipase, *Protein Eng.* **7**, 559-562.

Noble, M. E. M., Cleasby, A., Johnson, L. N., Egmond, M. R., Frenken, L. G. J. (1993), The crystal structure of triacylglycerol lipase from *Pseudomonas glumae* reveals a partially reductant catalytic aspartate, *FEBS Lett.* **331**, 123-128.

Node, M., Inoue, T., Araki, M., Nakamura, D., Nishide, K. (1995), An asymmetric synthesis of C_2-symmetric dimethyl 3,7-dioxo-cis-bicyclo[3.3.0]octane-2,6-dicarboxylate by lipase-catalyzed demethoxycarbonylation, *Tetrahedron Lett.* **36**, 2255-2256.

Noguchi, T., Honda, J., Nagamune, T., Sasabe, H., Inoue, Y., Endo, I. (1995), Photosensitive nitrile hydratase intrinsically possesses nitric oxide bound to the non-heme iron center: evidence by Fourier transform infrared spectroscopy, *FEBS Lett.* **358**, 9-12.

Nojiri, M., Nakayama, H., Odaka, M., Yohda, M., Takio, K., Endo, I. (2000), Cobalt-substituted Fe-type nitrile hydratase of *Rhodococcus* sp. N-771, *FEBS Lett.* **465**, 173-177.

Nojiri, M., Yohda, M., Odaka, M., Matsushita, Y., Tsujimura, M., Yoshida, T., Dohmae, N., Takio, K., Endo, I. (1999), Functional expression of nitrile hydratase in *Escherichia coli*: Requirement of a nitrile hydratase activator and post-translational modification of a ligand cysteine, *J. Biochem. (Tokyo)* **125**, 696-704.

Norin, M., Hult, K., Mattson, A., Norin, T. (1993), Molecular modelling of chymotrypsin-substrate interactions: calculation of enantioselectivity, *Biocatalysis* **7**, 131-147.

Noritomi, H., Almarsson, Ö., Barletta, G. L., Klibanov, A. M. (1996), The influence of the mode of enzyme preparation on enzymatic enantioselectivity in organic solvents and its temperature dependence, *Biotechnol. Bioeng.* **51**, 95-99.

Nozaki, K., Uemura, A., Yamashita, J., Yasumoto, M. (1990), Enzymatic regioselective acylation of the 3' hydroxyl groups of the 2'-deoxy-5-fluorouridine (FUdR) and 2'-deoxy-5-trifluoromethyluridine (CF_3UdR), *Tetrahedron Lett.* **31**, 7327-7328.

Oakley, A. J., Klvana, M., Otyepka, M., Nagata, Y., Wilce, M. C., Damborsky, J. (2004), Crystal structure of haloalkane dehalogenase LinB from *Sphingomonas paucimobilis* UT26 at 0.95 A resolution: dynamics of catalytic residues, *Biochemistry* **43**, 870-878.

O'Brian, P. J., Herschlag, D. (1999), Catalytic promiscuity and the evolution of new enzymatic activities, *Chem. Biol.* **6**, R91-R105.

O'Brian, P. J., Herschlag, D. (1998), Sulfatase activity of *E. coli* alkaline phosphatase demonstrates a functional link to arylsulfatases, an evolutionarily related enzyme family, *J. Am. Chem. Soc.* **120**, 12369-12370.

Oberhauser, T., Bodenteich, M., Faber, K., Penn, G., Griengl, H. (1987), Enzymic resolution of norbornane-type esters, *Tetrahedron* **43**, 3931-3944.

Odaka, M., Fujii, K., Hoshino, M., Noguchi, T., Tsujimura, M., Nagashima, S., Yohda, M., Nagamune, T., Inoue, Y., Endo, I. (1997), Activity regulation of photoreactive nitrile hydratase by nitric oxide, *J. Am. Chem. Soc.* **119**, 3785-3791.

Oddon, G., Uguen, D. (1997), Toward a total syntheis of an aglycone of spiramycine, two complementary accesses to a C-5/C-9 fragment, *Tetrahedron Lett.* **38**, 4411-4414.

Oesch, F. (1974), Purification and specificity of a human microsomal epoxide hydratase, *Biochem. J.* **139**, 77-88.

Oesch, F. (1973), Mammalian epoxide hydrases: inducible enzymes catalysing the inactivation of carcinogenic and cytotoxic metabolites derived from aromatic and olefinic compounds, *Xenobiotica* **3**, 305-340.

Oetting, J., Holzkamp, J., Meyer, H. H., Pahl, A. (1997), Total synthesis of the piperidinol alkaloid (−)-(2R,3R,6S)-cassine, *Tetrahedron: Asymmetry* **8**, 477-484.

Ogawa, J., Shimizu, S. (2002), Industrial microbial enzymes: their discovery by screening and use in large-scale production of useful chemicals in Japan, *Curr. Opin. Biotechnol.* **13**, 367-375.

Ogawa, J., Shimizu, S. (1997), Diversity and versatility of microbial hydantoin-transforming enzymes, *J. Mol. Catal. B: Enzym.* **2**, 163-176.

Oguntimein, G. B., Erdmann, H., Schmid, R. D. (1993), Lipase catalysed synthesis of sugar ester in organic solvents, *Biotech. Lett.* **15**, 175-180.

O'Hagan, D., Zaidi, N. A. (1994a), Enzyme-catalyzed condensation polymerization of hydroxyundecanoic acid with lipase from *Candida cylindracea*, *Polymer* **35**, 3576-3578.

O'Hagan, D., Zaidi, N. A. (1994b), The resolution of tertiary α-acetylene acetate esters by the lipase from *Candida cylindracea*, *Tetrahedron: Asymmetry* **5**, 1111-1118.

O'Hagan, D., Rzepa, H. S. (1994), Stereoelectronic influence of fluorine in enzyme resolutions of α-fluoroesters, *J. Chem. Soc., Perkin Trans. II* 3-4.

O'Hagan, D., Zaidi, N. A. (1993), Polymerization of 10-hydroxydecanoic acid with lipase from *Candida cylindracea*, *J. Chem. Soc., Perkin Trans. I* 2389-2390.

O'Hagan, D., Zaidi, N. A. (1992), Hydrolytic resolution of tertiary acetylenic acetate esters with the lipase from *Candida cylindracea*, *J. Chem. Soc. Perkin Trans. I* 947-948.

Ohno, M., Otsuka, M. (1990), Chiral synthons by ester hydrolysis catalysed by pig liver esterase, *Org. React.* **37**, 1-55.

Ohno, M., Kobayashi, S., Iimori, T., Wang, Y. F., Izawa, T. (1981), Synthesis of (S)- and (R)-4-[(methoxycarbonyl)methyl]-2-azetidinone by chemicoenzymatic approach, *J. Am. Chem. Soc.* **103**, 2405-2406.

Öhrner, N., Orrenius, C., Mattson, A., Norin, T., Hult, K. (1996), Kinetic resolutions of amine and thiol analogues of secondary alcohols catalyzed by the *Candida antarctica* lipase B, *Enzyme Microb. Technol.* **19**, 328-331.

Öhrner, N., Martinelle, M., Mattson, A., Norin, T., Hult, K. (1992), Displacement of the equilibrium in lipase catalyzed transesterification in ethyl octanoate by continuous evaporation of ethanol, *Biotechnol. Lett.* **14**, 263-268.

Öhrner, K., Mattson, A., Norin, T., Hult, K. (1990), Enantiotopic selectivity of pig liver esterase isoenzymes, *Biocatalysis* **4**, 81-88.

Ohsawa, K., Shiozawa, T., Achiwa, K., Terao, Y. (1993), Synthesis of optically active nucleoside analogs containing 2,3-dideoxyapiose in the presence of a catalytic amount of trimethylsilyl iodide, *Chem. Pharm. Bull.* **41**, 1906-1909.

Ohta, H. (1996), Stereochemistry of enzymatic hydrolysis of nitriles, *Chimia* **50**, 434-436.

Ohta, H., Kimura, Y., Sugano, Y., Sugai, T. (1989), Enzymatic kinetic resolution of cyanohydrin acetates and its application to the synthesis of (S)-(−)-frontalin, *Tetrahedron* **45**, 5469-5476.

Ohtani, T., Nakatsukasa, H., Kamezawa, M., Tachibana, H., Naoshima, Y. (1998), Enantioselectivity of *Candida antarctica* lipase for some synthetic substrates including aliphatic secondary alcohols, *J. Mol. Catal. B: Enzym.* **4**, 53-60.

Ohtani, T., Kikuchi, K., Kamezawa, M., Hamatani, H., Tachibana, H., Totani, T., Naoshima, Y. (1996), Chemoenzymatic synthesis of (+)-(4E,15E)-docosa-4,15-dien-1-yn-3-ol, a component of the marine sponge *Cribrochalina vasculum*, *J. Chem. Soc., Perkin Trans. I* 961-962.

Okumura, S., Iwai, M., Tominaga, Y. (1984), Synthesis of ester oligomer by *Aspergillus niger* lipase, *Agric. Biol. Chem.* **48**, 2805-2808.

Oladepo, D. K., Halling, P. J., Larsen, V. F. (1995), Effect of different supports on the reaction rate of *Rhizomucor miehei* lipase in organic media, *Biocatal. Biotransform.* **12**, 47-54.

Oladepo, D. K., Halling, P. J., Larsen, V. F. (1994), Reaction rates in organic media show similar dependence on water activity with lipase catalyst immobilized on different supports, *Biocatalysis* **8**, 283-287.

Olivieri, R., Fascetti, E., Angelini, L., Degen, L. (1981), Microbial transformation of racemic hydantoins to D-amino acids., *Biotechnol. Bioeng.* **23**, 2173-2183.

Olivieri, R., Fascetti, E., Angelini, L., Degen, L. (1979), Enzymatic conversion of *N*-carbamoyl-D-amino acids to D-amino acids, *Enzyme Microb. Technol.* **1**, 201-204.

Ollis, D. L., Cheah, E., Cygler, M., Dijkstra, B., Frolow, F., Franken, S., Harel, M., Remington, S. J., Silman, I. (1992), The α/β hydrolase fold, *Protein Eng.* **5**, 197-211.

Onakunle, O. A., Knowles, C. J., Bunch, A. W. (1997), The formation and substrate specificity of bacterial lactonases capable of enantioselective resolution of racemic lactones, *Enzyme Microb. Technol.* **21**, 245-251.

Onumonu, A. N., Colocoussi, A., Mathews, C., Woodland, M. P., Leak, D. J. (1994), Microbial alkene epoxidation - merits and limitations, *Biocatalysis* **10**, 211-218.

Ooi, Y., Mitsuo, N., Satoh, T. (1985), Enzymatic synthesis of glycosides of racemic alcohols using β-galactosidase and separation of the diasteromers by high-performance liquid chromatograhy using a conventional column, *Chem. Pharm. Bull.* **33**, 5547-5550.

Orrenius, C., Norin, T., Hult, K., Carrea, G. (1995a), The *Candida antarctica* lipase B catalyzed kinetic resolution of seudenol in non-aqueous media of controlled water activity, *Tetrahedron: Asymmetry* **6**, 3023-3030.

Orrenius, C., Oehrner, N., Rotticci, D., Mattson, A., Hult, K., Norin, T. (1995b), *Candida antarctica* lipase B catalyzed kinetic resolutions: substrate structure requirements for the preparation of enantiomerically enriched secondary alcohols, *Tetrahedron: Asymmetry* **6**, 1217-1220.

Orrenius, C., Mattson, A., Norin, T. (1994), Preparation of 1-pyridinylethanols of high enantiomeric purity by lipase catalyzed transesterifications, *Tetrahedron: Asymmetry* **5**, 1363-1366.

Orru, R. V. A., Archelas, A., Furstoss, R., Faber, K. (1998a), Epoxide hydrolases and their synthetic applications, *Adv. Biochem. Eng./Biotechnol.* **63**, 145-167.

Orru, R. V. A., Mayer, S. F., Kroutil, W., Faber, K. (1998b), Chemoenzymatic deracemization of (±)-2,2-disubstituted oxiranes, *Tetrahedron* **54**, 859-874.

Orru, R. V. A., Osprian, I., Kroutil, W., Faber, K. (1998c), An efficient large-scale synthesis of (*R*)-(−)-mevalonolactone using simple biological and chemical catalysts, *Synthesis* 1259-1263.

Orru, R. V. A., Kroutil, W., Faber, K. (1997), Deracemization of (±)-2,2-disubstituted epoxides *via* enantioconvergent chemoenzymic hydrolysis using *Nocardia* EH1 epoxide hydrolase and sulfuric acid, *Tetrahedron Lett.* **38**, 1753-1754.

Orsat, B., Alper, P. B., Moree, W., Mak, C.-P., Wong, C.-H. (1996), Homocarbonates as substrates for the enantioselective enzymatic protection of amines, *J. Am. Chem. Soc.* **118**, 712-713.

Osprian, I., Kroutil, W., Mischitz, M., Faber, K. (1997), Biocatalytic resolution of 2-methyl-2-(aryl)alkyloxiranes using novel bacterial epoxide hydrolases, *Tetrahedron: Asymmetry* **8**, 65-71.

Osswald, S., Wajant, H., Effenberger, F. (2002), Characterization and synthetic applications of recombinant AtNIT1 from *Arabidopsis thaliana*, *Eur. J. Biochem.* **269**, 680-687.

Ostermeier, M., Nixon, A. E., Benkovic, S. J. (1999), Incremental truncation as a strategy in the engineering of novel biocatalysts, *Bioorg. Med. Chem.* **7**, 2139-2144.

Otto, R. T., Bornscheuer, U. T., Syldatk, C., Schmid, R. D. (1998a), Synthesis of aromatic *n*-alkyl-glucoside esters in a coupled ß-glucosidase and lipase reaction, *Biotechnol. Lett.* **20**, 437-440.

Otto, R., Bornscheuer, U. T., Syldatk, C., Schmid, R. D. (1998b), Lipase-catalyzed synthesis of arylaliphatic esters of β-D(+)-glucose, alkyl- and arylglucosides and characterization of their surfactant properties, *J. Biotechnol.* **64**, 231-237.

Ottolina, G., Bovara, R., Riva, S., Carrea, G. (1994), Activity and selectivity of some hydrolases in enantiomeric solvents, *Biotechnol. Lett.* **16**, 923-928.

Oyama, K. (1992), Industrial production of aspartame, in.: *Chirality in Industry* (Collins, A. N., Sheldrake, G. N., Crosby, J., Eds.), pp. 237-247. Chichester: Wiley.

Ozaki, E., Sakimae, A. (1997), Purification and characterization of recombinant esterase from *Pseudomonas putida* MR-2068 and its application to the optical resolution of dimethyl methylsuccinate, *J. Ferment. Bioeng.* **83**, 535-539.

Ozaki, E., Sakimae, A., Numazawa, R. (1995), Nucleotide sequence of the gene for a thermostable esterase from *Pseudomonas putida* MR-2068, *Biosci. Biotech. Biochem.* **59**, 1204-1207.

Ozegowski, R., Kunath, A., Schick, H. (1996), Synthesis of enantiomerically enriched 2,3- and 3,4-dimethylpentan-5-olides by lipase-catalyzed regio- and enantioselective alcoholysis of *cis*- and *trans*-2,3-dimethylpentanedioic anhydrides, *Liebigs Ann. Chem.* 1443-1448.

Ozegowski, R., Kunath, A., Schick, H. (1995a), The different behaviour of *syn*- and *anti*-2,3-dimethylbutanedioic anhydride in the lipase-catalyzed enantioselective alcoholysis, *Tetrahedron: Asymmetry* **6**, 1191-1194.

Ozegowski, R., Kunath, A., Schick, H. (1995b), Lipase-catalyzed sequential esterification of (±)-2-methylbutanedioic anhydride - a biocatalytical access to an enantiomerically pure 1-monoester of (*S*)-2-methylbutanedioic acid, *Liebigs Ann.* 1699-1702.

Ozegowski, R., Kunath, A., Schick, H. (1994a), Enzymes in organic synthesis. 19. The enzyme-catalyzed sequential esterification of (±)-*anti*-2,4-dimethylglutaric anhydride - an efficient route to enantiomerically enriched mono- and diesters of *anti*-2,4-dimethylglutaric acid, *Liebigs Ann.* 215-217.

Ozegowski, R., Kunath, A., Schick, H. (1994b), Lipase-catalyzed conversion of (±)-2-methylglutaric anhydride into (*S*)-2-methyl- and (*R*)-4-methyl-δ-valerolactone via a regio- and enantioselective sequential esterification, *Liebigs Ann. Chem.* 1019-1023.

Pai, Y. C., Fang, J. M., Wu, S. H. (1994), Resolution of homoallylic alcohols containing dithioketene acetal functionalities - synthesis of optically active γ-lactones by a combination of chemical and enzymatic methods, *J. Org. Chem.* **59**, 6018-6025.

Pallavicini, M., Valoti, E., Villa, L., Piccolo, O. (1994), Lipase-catalyzed resolution of glycerol-2,3-carbonate, *J. Org. Chem.* **59**, 1751-1754.

Palmer, D. R. J., Garrett, J. B., Sharma, V., Meganathan, R., Babbitt, P. C., Gerlt, J. A. (1999), Unexpected divergence of enzyme function and sequence: "*N*-acylamino acid racemase" is *o*-succinylbenzoate synthase, *Biochemistry* **38**, 4252-4258.

Pàmies, O., Bäckvall, J.-E. (2004), Chemoenzymatic dynamic kinetic resolution, *Trends Biotechnol.* **22**, 130-135.

Pàmies, O., Bäckvall, J.-E. (2003), Combination of enzymes and metal catalysts. A powerful approach in asymmetric catalysis, *Chem. Rev.* **103**, 3247-3261.

Pàmies, O., Bäckvall, J.-E. (2002), Chemoenzymatic dynamic kinetic resolution of β-halo alcohols. An efficient route to chiral epoxides, *J. Org. Chem.* **67**, 9006-9010.

Pàmies, O., Bäckvall, J.-E. (2001), Dynamic kinetic resolution of β-azido alcohols. An efficient route to chiral aziridines and β-amino alcohols, *J. Org. Chem.* **66**, 4022-4025; correction: (2002), *J. Org. Chem.* **67**, 1418.

Pamperin, D., Schulz, C., Hopf, H., Syldatk, C., Pietzsch, M. (1998), Chemoenzymatic synthesis of optically pure planar chiral (*S*)-(–)-5-formyl-4-hydroxyl[2.2]paracyclophane, *Eur. J. Org. Chem.* 1441-1445.

Panunzio, M., Camerini, R., Mazzoni, A., Donati, D., Marchioro, C., Pachera, R. (1997), Lipase-catalysed resolution of *cis*-1-ethoxycarbonyl-2-hydroxy-cyclohexane: enantioselective total synthesis of 10-ethyl-trinem, *Tetrahedron: Asymmetry* **8**, 15-17.

Panza, L., Brasca, S., Riva, S., Russo, G. (1993a), Selective lipase-catalyzed acylation of 4,6-*O*-benzylidene-D-glucopyranosides to synthetically useful esters, *Tetrahedron: Asymmetry* **4**, 931-932.

Panza, L., Luisetti, M., Crociati, E., Riva, S. (1993b), Selective acylation of 4,6-*O*-benzylidene glycopyranosides by enzymic catalysis, *J. Carbohydr. Chem.* **12**, 125-130.

Parida, S., Dordick, J. S. (1991), Substrate structure and solvent hydrophobicity control lipase catalysis and enantioselectivity in organic media, *J. Am. Chem. Soc.* **113**, 2253-2259.

Park, S., Morley, K. L., Horsman, G. P., Holmquist, M., Hult, K., Kazlauskas, R. J. (2005a), Focusing mutations into the *P. fluorescens* esterase binding site increases enantioselectivity more effectively than distant mutations, *Chem. Biol.* **12**, 45-54.

Park, O.-J., Lee, S.-H., Park, T.-Y., Lee, S.-W., Cho, K.-H. (2005b), Enzyme-catalyzed preparation of methyl (*R*)-*N*-(2,6-dimethylphenyl)alaninate: a key intermediate for (*R*)-metalaxyl, *Tetrahedron: Asymmetry* **16**, 1221-1225.

Park, S., Kazlauskas, R. J. (2003) Biocatalysis in ionic liquids – advantages beyond green technology, *Curr. Opin. Biotechnol.* **14**, 432-437.

Park, S., Viklund, F., Kazlauskas, R. J., Hult, K. (2003a) Vacuum-driven lipase-catalyzed direct condensation of L-ascorbic acid and fatty acids in ionic liquids: synthesis of natural surface-active antioxidants, *Green Chem.* **5**, 715-719.

Park, S., Forró, E., Grewal, H., Fülöp, F., Kazlauskas, R. J. (2003b), Molecular basis for the enantioselective ring opening of β-lactams catalyzed by *Candida antarctica* lipase B, *Adv. Synth. Catal.* **345**, 986-995.

Park, H. G., Chang, H. N., Dordick, J. S. (1994), Enzymatic synthesis of various aromatic polyesters in anhydrous organic solvents, *Biocatalysis* **11**, 263-271.

Parker, M.-C., Besson, T., Lamare, S., Legoy, M.-D. (1996), Microwave radiation can increase the rate of enzyme-catalyzed reactions in organic media, *Tetrahedron Lett.* **37**, 8383-8386.

Parmar, V. S., Kumar, A., Bisht, K., Mukherjee, S., Prasad, A. K., Sharma, S. K., Wengel, J., Olsen, C. E. (1997), Novel chemoselective de-esterification of esters of polyacetoxy aromatic acids by lipases, *Tetrahedron* **53**, 2163-2176.

Parmar, V. S., Pati, H. N., Sharma, S. K., Singh, A., Malhotra, S., al., e. (1996), Hydrolytic reactions on polyphenolic perpropanoates by porcine pancreatic lipase immobilized in microemulsion-based gels, *Bioorg. Med. Chem. Lett.* **6**, 2269-2274.

Parmar, V. S., Prasad, A. K., Sharma, N. K., Varan, A., Pati, H. N., Sharma, S. K., Bisht, K. S. (1993a), Lipase-catalyzed selective deacylation of peracetylated benzopyranones, *J. Chem. Soc., Chem. Commun.* 27-29.

Parmar, V. S., Sinha, R., Bisht, K. S., Gupta, S., Prasad, A. K., Taneja, P. (1993b), Regioselective esterification of diols and triols with lipases in organic solvents, *Tetrahedron* **49**, 4107-4116.

Parmar, V. S., Prasad, A. K., Sharma, N. K., Singh, S. K., Pati, H. N., Gupta, S. (1992), Regioselective deacylation of polyacetoxy aryl-methyl ketones by lipases in organic solvents, *Tetrahedron* **48**, 6495-6498.

Partali, V., Waagen, V., Alvik, T., Anthonsen, T. (1993), Enzymic resolution of butanoic esters of 1-phenylmethyl and 1-(2-phenylethyl) ethers of 3-chloro-1,2-propanediol, *Tetrahedron: Asymmetry* **4**, 961-968.

Patel, R. N., Banerjee, A., Szarka, L. J. (1997), Stereoselective acetylation of racemic 7-[N,N'-bis(benzyloxycarbonyl)-N-(guanidinoheptanoyl)]-α-hydroxyglycine, *Tetrahedron: Asymmetry* **8**, 1767-1771.

Patel, R. N., Banerjee, A., Ko, R. Y., Howell, J. M., Li, W.-S., Comezoglu, F. T., Partyka, R. A., Szarka, L. (1994), Enzymic preparation of (3R-cis)-3-(acetyloxy)-4-phenyl-2-azetidinone - a taxol side-chain synthon, *Biotechnol. Appl. Biochem.* **20**, 23-33.

Patel, R. N., Howell, J. M., McNamee, C. G., Fortney, K. F., Szarka, L. J. (1992a), Stereoselective enzymatic hydrolysis of α-[(acetylthio)methyl]benzenepropanoic acid and 3-acetylthio-2-methylpropanoic acid, *Biotechnol. Appl. Biochem.* **16**, 34-47.

Patel, R. N., Liu, M., Banerjee, A., Szarka, L. J. (1992b), Stereoselective enzymatic hydrolysis of (exo,exo)-7-oxabicyclo[2.2.1]heptane-2,3-dimethanol diacetate ester in a biphasic system, *Appl. Microbiol. Biotechnol.* **37**, 180-183.

Patel, R. N., Howell, J. M., Banerjee, A., Fortney, K. F., Szarka, L. J. (1991), Stereoselective enzymatic esterification of 3-benzoylthio-2-methylpropanoic acid, *Appl. Microbiol. Biotechnol.* **36**, 29-34.

Patti, A., Sanfilippo, C., Piattelli, M., Nicolosi, G. (1996), Enzymatic desymmetrization of conduritol D. Preparation of homochiral intermediates for the synthesis of cyclitols and aminocyclitols, *Tetrahedron: Asymmetry* **7**, 2665-2670.

Payne, M. S., Wu, S., Fallon, R. D., Tudor, G., Stieglitz, B., Turner, I. M., Jr., Nelson, M. J. (1997), A stereoselective cobalt-containing nitrile hydratase, *Biochemistry* **36**, 5447-5454.

Pearson, A. J., Srinivasan, K. (1992), Approaches to the synthesis of heptitol derivatives via iron-mediated stereocontrolled functionalization of cycloheptatrienone, *J. Org. Chem.* **57**, 3965-3973.

Pearson, A. J., Lai, Y. S. (1988), A synthesis of the (+)-Prelog-Djerassi lactone, *J. Chem. Soc., Chem. Commun.* 442-443.

Pearson, A. J., Bansal, H. S., Lai, Y. S. (1987), Enzymatic hydrolysis of 3,7-diacetoxycycloheptene derivatives, *J. Chem. Soc., Chem. Commun.* 519-520.

Pedragosa-Moreau, S., Archelas, A., Furstoss, R. (1996a), Microbiological transformations. 31: Synthesis of enantiopure epoxides and vicinal diols using fungal epoxide hydrolase mediated hydrolysis, *Tetrahedron Lett.* **37**, 3319-3322.

Pedragosa-Moreau, S., Archelas, A., Furstoss, R. (1996b), Microbiological transformations. 32. Use of epoxide hydrolase mediated biohydrolysis as a way to enantiopure epoxides and vicinal diols: Application to substituted styrene oxides, *Tetrahedron* **52**, 4593-4606.

Pedragosa-Moreau, S., Morisseau, C., Zylber, J., Archelas, A., Baratti, J., Furstoss, R. (1996c), Microbiological transformations. 33. Fungal epoxide hydrolases applied to the synthesis of enantiopure *para*-substituted styrene oxides. A mechanistic approach, *J. Org. Chem.* **61**, 7402-7407.

Pedragosa-Moreau, S., Archelas, A., Furstoss, R. (1995), Epoxydes énantiopurs: obtention par voie chimique ou par voie enzymatique, *Bull. Soc. Chim. Fr.* **132**, 769-800.

Pedragosa-Moreau, S., Archelas, A., Furstoss, R. (1993), Microbiological transformations. 28. Enantiocomplementary epoxide hydrolyses as a preparative access to both enantiomers of styrene oxide, *J. Org. Chem.* **58**, 5533-5536.

Pelenc, V. P., Paul, F. M. B., Monsan, P. F. (1993), Method for the enyzmatic production of α-glucosides and esters of α-glucosides, and utilization of products thus obtained, *Int. Patent Appl.* WO 93/04185.

Pelletier, I., Altenbuchner, J. (1995), A bacterial esterase is homologous with non-heme haloperoxidases and displays brominating activity, *Microbiology* **141**, 459-468.

Pellissier, H. (2003), Dynamic kinetic resolution, *Tetrahedron* **59**, 8291-8327.

Pereira, R. A., Graham, D., Rainey, F. A., Cowan, D. A. (1998), A novel thermostable nitrile hydratase, *Extremophiles* **2**, 347-357.

Perona, J. J., Craik, C. S. (1995), Structural basis of substrate specificity in the serine proteases, *Protein Sci.* **4**, 337-360.

Persichetti, R. A., Lalonde, J. J., Govardhan, C. P., Khalaf, N. K., Margolin, A. L. (1996), *Candida rugosa* lipase - enantioselectivity enhancements in organic solvents, *Tetrahedron Lett.* **37**, 6507-6510.

Persson, M., Bornscheuer, U. T. (2003) Increased stability of an esterase from *Bacillus stearothermophilus* in ionic liquids as compared to organic solvents, *J. Mol. Catal. B: Enzym.* **22**, 21-27.

Persson, B. A., Larsson, A. L. E., Le Ray, M., Bäckvall, J.-E. (1999a), Ruthenium- and enzyme-catalyzed dynamic kinetic resolution of secondary alcohols, *J. Am. Chem. Soc.* **121**, 1645-1650.

Persson, B. A.; Huerta, F. F.; Bäckvall, J.-E. (1999b), Dynamic kinetic resolution of secondary diols via coupled ruthenium and enzyme catalysis, *J. Org. Chem.* **64**, 5237-5240.

Perugino, G., Cobucci-Ponzano, B., Rossi, M., Moracci, M. (2005), Recent advances in the oligosaccharide synthesis promoted by catalytically engineered glycosidases, *Adv. Synth. Catal.* **347**, 941-950.

Perugino, G., Trincone, A., Rossi, M., Moracci, M. (2004), Oligosaccharide synthesis by glycosynthases, *Trends Biotechnol.* **22**, 31-37.

Perugino, G., Trincone, A., Giordano, A., van der Oost, J., Kaper, T., Rossi, M., Moracci, M. (2003), Activity of hyperthermophilic glycosynthases is significantly enhanced at acidic pH, *Biochemistry* **42**, 8484-8493.

Petersen, E. I., Valinger, G., Sölkner, B., Stubenrauch, G., Schwab, H. (2001), A novel esterase from *Burkholderia gladioli* shows high deacetylation activity on cephalosporins is related to β-lactamases and DD-peptidases, *J. Biotechnol.* **89**, 11-25.

Petschen, I., Malo, E. A., Bosch, M. P., Guerrero, A. (1996), Highly enantioselective synthesis of long chain alkyl trifluoromethyl carbinols and β-thiotrifluoromethyl carbinols through lipases, *Tetrahedron: Asymmetry* **7**, 2135-2143.

Phytian, S. J. (1998), Esterases, in.: *Biotechnology-Series* (Rehm, H. J., Reed, G., Pühler, A., Stadler, P. J. W., Kelly, D. R.; Eds.), **Vol. 8a**, pp. 193-241. Weinheim: VCH-Wiley.

Picard, M., Gross, J., Lübbert, E., Tölzer, S., Krauss, S., van Pée, K.-H., Berkessel, A. (1997), Metal-free bacterial haloperoxidases as unusual hydrolases: activation of H_2O_2 by the formation of peracetic acid, *Angew. Chem. Intl. Ed.* **36**, 1196-1199.

Piersma, S. R., Nojiri, M., Tsujimura, M., Noguchi, T., Odaka, M., Yohda, M., Inoue, Y., Endo, I. (2000), Arginine 56 mutation in the beta subunit of nitrile hydratase: importance of hydrogen bonding to the non-heme iron center, *J. Inorg. Biochem.* **80**, 283-288.

Pietzsch, M., Vielhauer, O., Pamperin, D., Ohse, B., Hopf, H. (1999), On the kinetics of the enzyme-catalyzed hydrolysis of axial chiral alkyl allenecarboxylates: preparation of optically active *R*-(−)-2-ethyl-4-phenyl-2,3-hexadiene-carboxylic acid and its optically pure *S*-(+)-methylester, *J. Mol. Catal. B: Enzym.* **6**, 51-57.

Pinot, F., Grant, D. F., Beetham, J. K., Parker, A. G., Borhan, B., Landt, S., Jones, A. D., Hammock, B. D. (1995), Molecular and biochemical evidence for the involvement of the Asp-333-His-523 pair in the catalytic mechanism of soluble epoxide hydrolase, *J. Biol. Chem.* **270**, 7968-7974.

Pisch, S., Bornscheuer, U., Meyer, H. H., Schmid, R. D. (1997), Properties of unusual phospholipids IV: Chemoenzymatic synthesis of phospholipids bearing acetylenic fatty acids, *Tetrahedron* **53**, 14627-14634.

Pleiss, J., Fischer, M., Peiker, M., Thiele, C., Schmid, R. D. (2000), Lipase engineering database: Understanding and exploiting sequence-structure-function relationships, *J. Mol. Catal. B: Enzym.* **10**, 491-508.

Pleiss, J., Fischer, M., Schmid, R. D. (1998), Anatomy of lipase binding sites: the scissile fatty acid binding site, *Chem. Phys. Lipids* **93**, 67-80.

Pletnev, V., Addlagatta, A., Wawrzak, Z., Duax, W. (2003) Three-dimensional structure of homodimeric cholesterol esterase-ligand complex at 1.4 Å resolution, *Acta Crystallogr., Sect.D* **59**, 50-56.

Pogorevc, M., Strauss, U. T., Hayn, M., Faber, K. (2000), Novel carboxyl esterase preparations for the resolution of linalyl acetate, *Chem. Monthly* **131**, 639-644.

Pohl, T., Waldmann, H. (1996), Enzymatic synthesis of a characteristic phosphorylated and glycosylated peptide fragment of the large subunit of mammalian RNA polymerase II, *Angew. Chem. Intl. Ed.* **35**, 1720-1723.

Pohl, T., Waldmann, H. (1995), Enhancement of the enantioselectivity of penicillin G acylase from *E. coli* by "substrate tuning", *Tetrahedron Lett.* **36**, 2963-2966.

Pohlenz, H. D., Boidol, W., Schuttke, I., Streber, W. R. (1992), Purification and properties of an *Arthrobacter oxydans* P52 carbamate hydrolase specific for the herbicide phenmedipham and nucleotide sequence of the corresponding gene, *J. Bacteriol.* **174**, 6600-6607.

Polla, M., Frejd, T. (1991), Synthesis of optically active cyclohexenol derivatives via enzyme-catalyzed ester hydrolysis of 4-acetoxy-3-methyl-2-cyclohexenone, *Tetrahedron* **41**, 5883-5894.

Poppe, L., Novák, L., Kajtár-Peredy, M., Szántay, C. (1993), Lipase catalyzed enantiomer selective hydrolysis of 1,2-diol acetates, *Tetrahedron: Asymmetry* **4**, 2211-2217.

Poppe, L., Novak, L. (1992), *Selective biocatalysis*, Weinheim: VCH.

Pottie, M., van der Eycken, J., Vandewalle, M. (1991), Synthesis of optically active derivatives of erythritol, *Tetrahedron: Asymmetry* **2**, 329-330.

Pozo, M., Gotor, V. (1993a), Chiral carbamates through an enzymic alkoxycarbonylation reaction, *Tetrahedron* **49**, 4321-4326.

Pozo, M., Gotor, V. (1993b), Kinetic resolution of vinyl carbonates through a lipase-mediated synthesis of their carbonate and carbamate derivatives, *Tetrahedron* **49**, 10725-10732.

Pozo, M., Pulido, R., Gotor, V. (1992), Vinyl carbonates as novel alkoxycarbonylation reagents in enzymatic synthesis of carbonates, *Tetrahedron* **48**, 6477-6484.

Prasad, A. K., Sorensen, M. D., Parmar, V. S., Wengel, J. (1995), Tri-*O*-acyl 2-deoxy-D-ribofuranose: an effective enzyme-assisted one-pot synthesis from 2-deoxy-D-ribose and transformation into 2'-deoxynucleosides, *Tetrahedron Lett.* **36**, 6163-6166.

Prepechalova, I., Martinkova, L., Stolz, A., Ovesna, M., Bezouska, K., Kopecky, J., Kren, V. (2001), Purification and characterization of the enantioselective nitrile hydratase from *Rhodococcus equi* A4, *Appl. Microbiol. Biotechnol.* **55**, 150-156.

Provencher, L., Wynn, H., Jones, J. B., Krawczyk, A. R. (1993), Enzymes in organic synthesis 51. Probing the dimensions of the large hydrophobic pocket of the active site of pig liver esterase, *Tetrahedron: Asymmetry* **4**, 2025-2040.

Puertas, S., Brieva, R., Rebolledo, F., Gotor, V. (1993), Lipase catalyzed aminolysis of ethyl propiolate and acrylic esters. Synthesis of chiral acrylamides, *Tetrahedron* **49**, 4007-4014.

Pulido, R., Gotor, V. (1993), Enzymatic regioselective alkoxycarbonylation of hexoses and pentoses with carbonate oxime esters, *J. Chem Soc. Perkin Trans. 1* 589-592.

Pulido, R., López, F. O., Gotor, V. (1992), Enzymatic regioselective acylation of hexoses and pentoses using oxime esters, *J. Chem Soc. Perkin Trans. 1* 2981-2988.

Qi, D., Tann, C.-M. Haring, D., Distefano M. D. (2001), Generation of new enzymes via covalent modification of existing proteins, *Chem. Rev.*, **101**, 3081-3111.

Quartey, E. G. K., Hustad, J. A., Faber, K., Anthonsen, T. (1996), Selectivity enhancement of PPL-catalyzed resolution by enzyme fractionation and medium engineering - syntheses of both enantiomers of tetrahydropyran-2-methanol, *Enzyme Microb. Technol.* **19**, 361-366.

Quax, W. J., Broekhuizen, C. P. (1994), Development of a new *Bacillus* carboxyl esterase for use in the resolution of chiral drugs, *Appl. Microbiol. Biotechnol.* **41**, 425-431.

Quirós, M., Sanchez, V. M., Brieva, R., Rebbolledo, F., Gotor, V. (1993), Lipase-catalyzed synthesis of optically active amides in organic media, *Tetrahedron: Asymmetry* **4**, 1105-1112.

Quyen, D. T., Schmidt-Dannert, C., Schmid, R. D. (1999), High-level formation of active *Pseudomonas cepacia* lipase after heterologous expression of the encoding gene and its modified chaperone in *Escherichia coli* and rapid in vitro refolding, *Appl. Environm. Microbiol.* **65**, 787-794.

Rabiller, C. G., Koenigsberger, K., Faber, K., Griengl, H. (1990), Enzymic recognition of diastereomeric esters, *Tetrahedron* **46**, 4231-4240.

Rakels, J. L. L., Caillat, P., Straathof, A. J. J., Heijnen, J. J. (1994a), Modification of the enzyme enantioselectivity by product inhibition, *Biotechnol. Prog.* **10**, 403-409.

Rakels, J. L. L., Wolff, A., Straathof, A. J. J., Heijnen, J. J. (1994b), Sequential kinetic resolution by two enantioselective enzymes, *Biocatalysis* **9**, 31-47.

Ramaswamy, S., Oehlschlager, A. C. (1991), Chemico-enzymatic syntheses of racemic and chiral isomers of 7-methyl-1,6-dioxaspiro[4.5]decane, *Tetrahedron* **47**, 1157-1162.

Ramaswamy, S., Hui, R. A. H. F., Jones, J. B. (1986), Enantiomerically selective pig liver esterase-catalysed hydrolyses of racemic allenic esters, *J. Chem. Soc., Chem. Commun.* 1545-1546.

Ramos-Tombo, G. M., Schär, H. P., Busquets, H. P., Fernandez, X., Ghisalba, O. (1986), Synthesis of both enantiomeric forms of 2-substituted 1,3-propanediol monoacetates starting from a common prochiral precursor, using enzymatic transformations in aqueous and in organic media, *Tetrahedron Lett.* **27**, 5707-5710.

Ransac, S., Carriére, F., Rogalska, E., Verger, R., Marguet, F., Buono, G., Melo., E. P., Cabral, J. M. S., Egloff, M. P. E., van Tilbeurgh, H., Cambillau, C. (1996), The kinetics specificities and structural features of lipases, in.: *Molecular Dynamics of Biomembranes* (de Kamp, J. A. F.; Ed.), **Vol. H 96**, pp. 254-304. Berlin: Springer.

Rantakylae, M., Aaltonen, O. (1994), Enantioselective esterification of ibuprofen in supercritical carbon dioxide by immobilized lipase, *Biotechnol. Lett.* **16**, 825-830.

Rathbone, D. A., Holt, P. J., Lowe, C. R., Bruce, N. C. (1997), Molecular analysis of the *Rhodococcus* sp. strain H1 her gene and characterization of its product, a heroin esterase, expressed in *Escherichia coli*, *Appl. Environm. Microbiol.* **63**, 2062-2066.

Raushel, F. M. (2002), Bacterial detoxification of organophosphate nerve agents, *Curr. Opin. Microbiol.* **5**, 288-295.

Raushel, F. M., Holden, H. M. (2000) Phosphotriesterase: an enzyme in search of its natural substrate, *Adv. Enzymol. Relat. Areas Mol. Biol.* **74**, 51-93.

Rees, G. D., Robinson, B. H., Stephenson, G. R. (1995), Preparative-scale kinetic resolutions catalyzed by microbial lipases immobilized in AOT-stabilized microemulsion-based organogels: cryoenzymology as a tool for improving enantioselectivity, *Biochim. Biophys. Acta* **1259**, 73-81.

Rees, G. D., Jenta, T. R. J., Nascimento, M. G., Catauro, M., Robinson, B. H., Stephenson, G. R., Olphert, R. D. G. (1993), Use of water-in-oil microemulsions and gelatin-containing microemulsion based gels for lipase -catalysed ester synthesis in organic solvents, *Indian J. Chem., Sect. B* **32B**, 30-34.

Rees, G. D., de Nascimento, M. G., Jenta, T. R. J., Robinson, B. H. (1991), Reverse enzyme synthesis in microemulsion-based organo-gels, *Biochim. Biophys. Acta* **1073**, 493-501.

Reetz, M. T., Bocola, M., Carballeira, J. D.; Zha, D., Vogel, A. (2005), Expanding the range of substrate acceptance of enzymes: combinatorial active-site saturation test, *Angew. Chem. Int. Ed.* **44**, 4192-4196.

Reetz, M. T. (2004), Controlling the enantioselectivity of enzymes by directed evolution: practical and theoretical ramifications, *Proc. Natl. Acad. Sci. USA* **101**, 5716-5722.

Reetz, M. T., Torre, C., Eipper, A., Lohmer, R., Hermes M., Brunner, B., Maichele, A., Bocola, M., Arand, M., Cronin, A., Genzel, Y., Archelas, A., Furstoss, R. (2004), Enhancing the enantioselectivity of an epoxide hydrolase by directed evolution, *Org. Lett.* **6**, 177-180.

Reetz, M. T., Wiesenhofer, W., Francio, G., Leitner, W. (2002) Biocatalysis in ionic liquids: batchwise and continuous flow processes using supercritical carbon dioxide as the mobile phase, *Chem. Commun*, 992-993.

Reetz, M. T. (2002), New methods for the high-throughput screening of enantioselective catalysts and biocatalsts, *Angew. Chem. Int. Ed.* **41**, 1335-1338.

Reetz, M. T., Wilensek, S., Zha, D., Jaeger, K.-E. (2001), Directed evolution of an enantioselective enzyme through combinatorial multiple-cassette mutagenesis, *Angew. Chem. Int. Ed.* **40**, 3589-3591.

Reetz, M. T., Becker, M. H., Klein, H. W., Stöckigt, D. (1999), A method for high-throughput screening of enantioselective catalysts, *Angew. Chem. Int. Ed.* **38**, 1758-1761.

Reetz, M. T. (1997), Entrapment of biocatalysts in hydrophobic sol-gel materials for use in organic chemistry, *Adv. Mater.* **9**, 943-954.

Reetz, M. T., Zonta, A., Schimossek, K., Liebeton, K., Jaeger, K.-E. (1997), Creation of enantioselective biocatalysts for organic chemistry by in vitro evolution, *Angew. Chem. Int. Ed.* **36**, 2830-2832.

Reetz, M. T., Zonta, A., Simpelkamp, J. (1996a), Efficient immobilization of lipases by entrapment in hydrophobic sol-gel materials, *Biotechnol. Bioeng.* **49**, 527-534.

Reetz, M. T., Zonta, A., Simpelkamp, J., Konen, W. (1996b), *In situ* fixation of lipase-containing hydrophobic sol-gel materials on sintered glass - highly efficient heterogeneous biocatalysts, *Chem. Commun.* 1397-1398.

Reetz, M. T., Schimossek, K. (1996), Lipase-catalyzed dynamic kinetic resolution of chiral amines: use of palladium as the racemization catalyst, *Chimia* **90**, 668-669.

Reetz, M., Zonta, A., Simpelkamp, J. (1995), Efficient heterogeneous biocatalysts by entrapment of lipases in hydrophobic sol-gel materials, *Angew. Chem. Int. Ed.* **34**, 301-303.

Reetz, M. T., Dreisbach, C. (1994), Highly efficient lipase-catalyzed kinetic resolution of chiral amines, *Chimia* **48**, 570.

Reid, T. W., Fahrney,D. (1967), The pepsin-catalyzed hydrolysis of sulfite esters, *J. Am. Chem. Soc.* **89**, 3941-3943.

Reiter, B., Glieder, A., Talker, D., Schwab, H. (2000), Cloning and characterization of EstC from *Burkholderia gladioli*, *Appl. Microbiol. Biotechnol.* **54**, 778-785.

Renirie, R., Hemrika, W., Wever, R. (2000), Peroxidase and phosphatase activity of active-site mutants of vanadium chloroperoxidase from the fungus *Curvularia inaequalis*. Implications for the catalytic mechanisms, *J. Biol. Chem.* **275**, 11650-11657.

Renold, P., Tamm, C. (1995), Comparison of the hydrolysis of cyclic *meso*-diesters with pig liver esterease (PLE) and rabbit liver esterase (RLE), *Biocatal. Biotransform.* **12**, 37-46.

Renouf, P., Poirier, J.-M., Duhamel, P. (1997), Asymmetric hydrolysis of pro-chiral 3,3-disubstituted 2,4-diacetoxycyclohexa-1,4-dienes, *J. Chem. Soc., Perkin Trans. 1* 1739-1745.

Reymond, J.-L., Wahler, D. (2002), Substrate arrays as enzyme fingerprinting tools, *ChemBioChem* **8**, 701-708.

Ricca, J.-M., Crout, D. H. G. (1993), Selectivity and specificity in substrate binding to proteases: novel hydrolytic reactions catalysed by α-chymotrypsin suspended in organic solvents with low water content and mediated by ammonium hydrogen carbonate, *J. Chem. Soc., Perkin Trans. 1* 1225-1233.

Ricks, E. E., Estrada-Valdes, M. C., McLean, T. L., Iacobucci, G. A. (1992), Highly enantioselective hydrolysis of (*R,S*)-phenylalanine isopropyl ester by subtilisin Carlsberg. Continuous synthesis of (*S*)-phenylalanine in a hollow fiber/liquid membrane reactor, *Biotechnol. Prog.* **8**, 197-203.

Rigby, J. H., Sugathapala, P. (1996), Lipase-mediated resolution and higher-order cycloaddition of substituted tricarbonyl (n6-cycloheptatriene)chromium(0) complexes, *Tetrahedron Lett.* **37**, 5293-5296.

Ringia, E. A. T., Garrett, J. B., Thoden, J. B., Holden, H. M., Rayment, I., Gerlt, J. A. (2004), Evolution of enzymatic activity in the enolase superfamily: functional studies of the promiscuous o-succinylbenzoate synthase from *Amycolatopsis*, *Biochemistry* **43**, 224-229.

Rink, R., Spelberg, J. H. L., Pieters, R. J., J. Kingma, Nardini, M., Kellogg, R. M., Dijkstra, B. W., Janssen, D. B. (1999), Mutation of tyrosine residues involved in the alkylation half reaction of epoxide hydrolase from *Agrobacterium radiobacter* AD1 results in improved enantioselectivity, *J. Am. Chem. Soc.* **121**, 7417-7418.

Rink, R., Fennema, M., Smids, M., Dehmel, U., Janssen, D. B. (1997), Primary structure and catalytic mechanism of the epoxide hydrolase from *Agrobacterium radiobacter* AD1, *J. Biol. Chem.* **272**, 14650-14657.

Riva, S. (1996), Regioselectivity of hydrolases in organic media, in.: *Enzymatic Reactions in Organic Media* (Koskinen, A. M. P., Klibanov, A. M.; Eds.), pp. 140-169. Glasgow: Chapman & Hall.

Riva, S., Danieli, B., Luisetti, M. (1996), A two-step efficient chemoenzymatic synthesis of flavonoid glycoside malonates, *J. Nat. Prod.* **59**, 618-621.

Riva, S., Chopineau, J., Kieboom, A. P. G., Klibanov, A. M. (1988), Protease-catalyzed regioselective esterification of sugars and related compounds in anhydrous dimethylformamide, *J. Am. Chem. Soc.* **110**, 584-589.

Roberts, S. M., Wan, P. W. H. (1998), Enzyme-catalysed Baeyer-Villiger oxidations, *J. Mol. Catal. B: Enzym.* **4**, 111-136.

Roberts, S. M. (Ed.) (1999), *Biocatalysts for Fine Chemical Cynthesis*, Weinheim: Wiley-VCH.

Roberts, S. M., Shoberu, K. A. (1991), Enzymic resolution of *cis*- and *trans*-4-hydroxycyclopent-2-enylmethanol derivatives and a novel preparation of carbocyclic 2',3'-dideoxydidehydronucleosides and aristeromycin, *J. Chem. Soc., Perkin Trans. 1* 2605-2607.

Roberts, S. M. (1989), Use of enzymes as catalysts to promote key transformations in organic synthesis, *Phil. Trans. R. Soc. London B* **324**, 577-587.

Robertson, D. E., Chaplin, J. A., DeSantis, G., Podar, M., Madden, M., et al. (2004), Exploring nitrilase sequence space for enantioselective catalysis, *Appl. Environ. Microbiol.* **70**, 2429-2436.

Robertson, D. E., Steer, B. A. (2004), Recent progress in biocatalyst discovery and optimization, *Curr. Opin. Chem. Biol.* **8**, 141-149.

Robinson, G. K., Alston, M. J., Knowles, C. J., Cheetham, P. S. J., Motion, K. R. (1994), An investigation into the factors influencing lipase-catalyzed intramolecular lactonization in microaqueous systems, *Enzyme Microb. Technol.* **16**, 855-863.

Rocco, V. P., Danishefsky, S. J., Schulte, G. K. (1991), Substrate specificity in enzymatically mediated transacetylation reactions of calicheamicinone intermediates, *Tetrahedron Lett.* **32**, 6671-6674.

Rosell, C. M., Vaidya, A. M., Halling, P. J. (1996), Continuous *in situ* water activity control for organic phase biocatalysis in a packed bed hollow fiber reactor, *Biotechnol. Bioeng.* **49**, 284-289.

Rosenquist, Å., Kvarnström, I., Classon, B., Samuelsson, B. (1996), Synthesis of enantiomerically pure bis(hydroxymethyl)-branched cyclohexenyl and cyclohexyl purines as potential inhibitors of HIV, *J. Org. Chem.* **61**, 6282-6288.

Rotticci, D., Orrenius, C., Hult, K., Norin, T. (1997), Enantiomerically enriched bifunctional sec-alcohols prepared by *Candida antarctica* lipase B catalysis. Evidence of non-steric interactions, *Tetrahedron: Asymmetry* **8**, 359-362.

Rúa, M. L., Diaz-Maurino, T., Fernandez, V. M., Otero, C., Ballesteros, A. (1993), Purification and characterization of two distinct lipases from *Candida cylindracea*, *Biochim. Biophys. Acta* **1156**, 181-189.

Rubio, E., Fernandez-Mayorales, A., Klibanov, A. M. (1991), Effect of the solvent on enzyme regioselectivity, *J. Am. Chem. Soc.* **113**, 695-696.

Rudolf, M. T., Schultz, C. (1996), Lipase-catalyzed regio- and enantioselective esterification of *rac*-1,2-*O*-cyclohexylidene-*myo*-inositol, *Liebigs Ann. Chem.* 533-537.

Rupley, J. A., Gratton, E., Careri, G. (1983), Water and globular proteins, *Trends Biochem. Sci.* **8**, 18-22.

Ruppert, S., Gais, H. J. (1997), Activity enhancement of pig liver esterase in organic solvents by colyophilization with methoxypolyethylene glycol: kinetic resolution of alcohols, *Tetrahedron: Asymmetry* **8**, 3657-3664.

Rüsch gen. Klaas, M., Warwel, S. (1998), A three-step-one-pot chemo-enzymatic synthesis of epoxyalkanolacylates, *Synth. Comm.* **28**, 251-260.

Rüsch gen. Klaas, M., Warwel, S. (1996), Chemoenzymatic epoxidation of unsaturated fatty acid esters and plant oils, *J. Am. Oil Chem. Soc.* **73**, 1453-1457.

Sahai, P., Vishwakarma, R. A. (1997), Phospholipase-A_2-mediated stereoselective synthesis of (*R*)-1-*O*-alkylglycero-3-phosphate and alkyl-acyl analogues: application for synthesis of radio-labelled biosynthetic precursors of cell surface glycoconjugates of *Leishmania donovani*, *J. Chem. Soc., Perkin Trans. 1* 1845-1849.

Sakagami, H., Kamikubo, T., Ogasawara, K. (1997), A facile synthesis of *R*-4-amino-3-hydroxybutanoic acid (gabob) from 3-hydroxypyridine, *Synlett* 221-222.

Sakagami, H., Samizu, K., Kamikubo, T., Ogasawara, K. (1996a), Enantiocontrolled synthesis and absolute configuration of (+)-crooksidine, an indole alkaloid from *Haplophyton crooksii* using chiral piperideinol block, *Synlett* 163-164.

Sakagami, H. K., Kamikubo, T., Ogasawara, K. (1996b), Novel reduction of 3-hydroxypyridine and its use in the enantioselective synthesis of (+)-pseudoconhydrine and (+)-*N*-methylpseudoconhydrine, *Chem. Commun.* 1433-1434.

Sakai, T., Kawabata, I., Kishimoto, T., Ema, T., Utaka, M. (1997), Enhancement of the enantioselectivity in lipase-catalyzed kinetic resolutions of 3-phenyl-2H-azirine-2-methanol by lowering the temperature to -40 degree, *J. Org. Chem.* **62**, 4906-4907.

Saksena, A. K., Girijavallabhan, V. M., Lovey, R. G., Pike, R. E., Wang, H., Ganguly, A. K., Morgan, B., Zaks, A., Puar, M. S. (1995), Highly stereoselective access to novel 2,2,4-trisubstituted tetrahydrofurans by halocyclization: practical chemoenzymic synthesis of SCH 51048, a broad-spectrum orally active antifungal agent, *Tetrahedron Lett.* **36**, 1787-1790.

Sakurai, T., Margolin, A. L., Russell, A. J., Klibanov, A. M. (1988), Control of enzyme enantioselectivity by the reaction medium, *J. Am. Chem. Soc.* **110**, 7236-7237.

Salazar, L., Sih, C. J. (1995), Optically active dihydropyridines via lipase-catalyzed enantioselective hydrolysis, *Tetrahedron: Asymmetry* **6**, 2917-2920.

Sánchez, V. M., Rebolledo, F., Gotor, V. (1997), *Candida antarctica* lipase catalyzed resolution of ethyl (±)-3-aminobutyrate, *Tetrahedron: Asymmetry* **8**, 37-40.

Sanchez-Montero, J. M., Hamon, V., Thomas, D., Legoy, M. D. (1991), Modulation of lipase hydrolysis and synthesis reactions using carbohydrates, *Biochim. Biophys. Acta* **1078**, 345-350.

Sanfillipo, C., Patti, A., Piattelli, M., Nicolosi, G. (1997), An efficient enzymatic preparation of (+)- and (−)-conduritol E, a cyclitol with C2 symmetry, *Tetrahedron: Asymmetry* **8**, 1569-1573.

Santaniello, E., Ferraboschi, P., Grisenti, P., Manzocchi, A. (1992), The biocatalytic approach to the preparation of enantiomerically-pure chiral building blocks, *Chem. Rev.* **92**, 1071-1140.

Sarney, D. B., Vulfson, E. N. (1995), Application of enzymes to the synthesis of surfactants, *Trends Biotechnol.* **13**, 164-172.

Sarney, D. B., Kapeller, H., Fregapane, G., Vulfson, E. N. (1994), Chemo-enzymatic synthesis of disaccharide fatty acid esters, *J. Am. Oil Chem. Soc.* **71**, 711-714.

Sato, M., Ohuchi, H., Abe, Y., Kaneko, C. (1992), Regio-, stereo-, and enantioselective synthesis of cyclobutanols by means of the photoaddition of 1,3-dioxin-4-ones and lipase catalyzed acylation, *Tetrahedron: Asymmetry* **3**, 313-328.

Sato, S., Murakata, T., Ochifuji, M., Fukushima, M., Suzuki, T. (1994), Development of immobilized enzyme entrapped within inorganic matrix and its catalytic activity in organic medium, *J. Chem. Eng. Jpn.* **27**, 732-736.

Sattler, A., Haufe, G. (1995), Synthesis of (9R)- and (9S)-10-fluorodecan-9-olide (fluorophoracantholide I) first lipase-catalyzed enantioselective esterification of β-fluoroalcohols, *Tetrahedron: Asymmetry* **6**, 2841-2848.

Savile, C. K., Kazlauskas R. J. (2005) How substrate solvation contributes to the enantioselectivity of subtilisin toward secondary alcohols, *J. Am. Chem. Soc.*, **127**, in press.

Savile, C. K., Magloire, V. P., Kazlauskas, R. J. (2005) Subtilisin-catalyzed resolution of N-acyl arylsulfinamides, *J. Am. Chem. Soc.* **2005**, *127*, 2104-2113.

Savjalov, W. W. (1901), Zur Theorie der Eiweissverdauung, *Pflügers Arch. Ges. Physiol.* **85**, 171.

Sayle, R. A., Milner-White, E. J. (1995), RASMOL: biomolecular graphics for all, *Trends Biochem. Sci.* **20**, 374-376.

Scharff, E. I., Koepke, J., Fritzsch, G., Lucke, C., Ruterjans, H. (2001), Crystal structure of diisopropylfluorophosphatase from *Loligo vulgaris*, *Structure* **9**, 493-502.

Schechter, I., Berger, A. (1967), On the active site of proteases. I. Papain., *Biochem. Biophys. Res. Commun.* **27**, 157-162.

Scheckermann, C., Schlotterbeck, A., Schmidt, M., Wray, V., Lang, S. (1995), Enzymatic monoacylation of fructose by two procedures, *Enzyme Microb. Technol.* **17**, 157-162.

Scheib, H., Pleiss, J., Kovac, A., Paltauf, F., Schmid, R. D. (1999), Stereoselectivity of *Mucorales* lipases toward triradylglycerols - a simple solution to a complex problem, *Protein Sci.* **8**, 215-221.

Scheib, H., Pleiss, J., Stadler, P., Kovac, A., Potthoff, A. P., Haalck, L., Spener, F., Paltauf, F., Schmid, R. D. (1998), Rational design of *Rhizopus oryzae* lipase with modified stereoselectivity toward triradylglycerols, *Protein Eng.* **11**, 675-682.

Schellenberger, V., Jakubke, H. D. (1991), Protease-catalyzed kinetically controlled peptide synthesis, *Angew. Chem. Int. Ed.* **30**, 1437-1449.

Schlotterbeck, A., Lang, S., Wray, V., Wagner, F. (1993), Lipase-catalyzed monoacylation of fructose, *Biotechnol. Lett.* **15**, 61-64.

Schmid, A., Dordick, J. S., Hauer, B., Kiener, A., Wubbolts, M., Witholt, B. (2001), Industrial biocatalysis today and tomorrow, *Nature* **409**, 258-268.

Schmid, R. D., Verger, R. (1998), Lipases - interfacial enzymes with attractive applications, *Angew. Chem. Int. Ed.* **37**, 1608-1633.

Schmidt, M., Barbayianni, E., Fotakopoulou, I., Höhne, M., Constantinou-Kokotou, V., Bornscheuer, U. T., Kokotos, G. (2005), Enzymatic removal of carboxyl protecting groups. 1. Cleavage of the tert-butyl moiety, *J. Org. Chem.* **70**, 3737-3740.

Schmitke, J. L., Wescott, C. R., Klibanov, A. M. (1996), The mechanistic dissection of the plunge in enzymatic activity upon transition from water to anhydrous solvents, *J. Am. Chem. Soc.* **118**, 3360-3365.

Schneider, M. P., Goergens, U. (1992), An efficient route to enantiomerically pure antidepressants: Tomoxetine, Nisoxetine and Fluoxetine, *Tetrahedron: Asymmetry* **3**, 525-528.

Schneider, M., Engel, N., Hönicke, P., Heinemann, G., Görisch, H. (1984), Hydrolytic enzymes in organic synthesis. 3. Enzymatic syntheses of chiral building-blocks from prochiral *meso*-substrates - preparation of methyl(hydrogen)-1,3-cycloalkanedicarboxylates, *Angew. Chem. Int. Ed.* **23**, 67-68.

Schnell, B., Faber, K., Kroutil, W. (2003), Enzymatic racemization and its application to synthetic biotransformations, *Adv. Synth. Catal.* **345**, 653-666.

Schoemaker, H. E., Mink, D., Wubbolts, M. G. (2003), Dispelling the myths - Biocatalysis in industrial synthesis, *Science* **299**, 1694-1697.

Schoevaart, R., Wolbers, M. W., Golubovic, M., Ottens, M., Kieboom, A. P., van Rantwijk, F., van der Wielen, L. A., Sheldon, R. A. (2004), Preparation, optimization, and structures of cross-linked enzyme aggregates (CLEAs), *Biotechnol. Bioeng.* **87**, 754-762.

Schöfer, S. H., Kaftzik, N., Wasserscheid, P., Kragl, U. (2001) Enzyme catalysis in ionic liquids: lipase catalysed kinetic resolution of 1-phenylethanol with improved enantioselectivity, *Chem. Commun.*, 425-426.

Schoffers, E., Golebiowski, A., Johnson, C. R. (1996), Enantioselective synthesis through enzymatic asymmetrization, *Tetrahedron* **52**, 3769-3826.

Schrag, J. D., Li, Y., Cygler, M., Lang, D., Burgdorf, T., Hecht, H. J., Schmid, R., Schomburg, D., Rydel, T. J., Oliver, J. D., Strickland, L. C., Dunaway, C. M., Larson, S. B., Day, J., McPherson, A. (1997), The open conformation of a *Pseudomonas* lipase, *Structure* **5**, 187-202.

Schrag, J. D., Li, Y., Wu, S., Cygler, M. (1991), Ser-His-Glu triad forms the catalytic site of the lipase from *Geotrichum candidum*, *Nature* **351**, 761-764.

Schricker, B., Thirring, K., Berner, H. (1992), α-Chymotrypsin catalyzed enantioselective hydrolysis of alkenyl α-amino acid esters, *Bioorg. Med. Chem. Lett.* **2**, 387-390.

Schudok, M., Kretzschmar, G. (1997), Enzyme catalyzed resolution of alcohols using ethoxyvinyl acetate, *Tetrahedron Lett.* **38**, 387-388.

Schueller, C. M., Manning, D. D., Kiessling, L. L. (1996), Preparation of (*R*)-(+)-7-oxabicyclo[2.2.1]hept-5-ene-exo-2-carboxylic acid, a precursor to substrates for the ring opening metathesis polymerization, *Tetrahedron Lett.* **37**, 8853-8856.

Schultz, M., Hermann, P., Kunz, H. (1992), Enzymatic cleavage of tert-butyl esters: themitase catalyzed deprotection of peptides and *O*-glycopeptides, *Synlett* 37-38.

Schutt, H., Schmidt-Kastner, G., Arens, A., Preiss, M. (1985), Preparation of optically active D-arylglycines for use as side chains for semisynthetic penicillins and cephalosporins using immobilized subtilisins in two-phase systems, *Biotechnol. Bioeng.* **27**, 420-433.

Scilimati, A., Ngooi, T. K., Sih, C. J. (1988), Biocatalytic resolution of (±)-hydroxyalkanoic esters. A strategy for enhancing the enantiomeric specificity of lipase-catalyzed ester hydrolysis, *Tetrahedron Lett.* **29**, 4927-4930.

Scott, D. L., White, S. P., Otwinowski, Z., Yuan, W., Gelb, M. H., Sigler, P. B. (1990), Interfacial catalysis: the mechanism of phospholipase A2, *Science* **250**, 1541-1546.

Sears, P., Wong, C.-H. (2001) Toward automated synthesis of oligosaccharides and glycoproteins, *Science* **291**, 2344-2350.

Secundo, F., Riva, S., Carrea, G. (1992), Effects of medium and of reaction conditions on the enantioselectivity of lipases in organic solvents and possible rationales, *Tetrahedron: Asymmetry* **3**, 267-280.

Seebach, D., Eberle, M. (1986), Enantioselective cleavage of *meso*-nitrodiol diacetates by an esterase concentrate from fresh pig liver: preparation of useful nitroaliphatic building blocks for EPC syntheses, *Chimia* **40**, 315-318.

Seemayer, R., Bar, N., Schneider, M. P. (1992), Enzymic preparation of isomerically pure 1,4:3,6-dianhydro-D- glucitol monoacetates - precursors for isoglucitol 2- and 5-mononitrates, *Tetrahedron: Asymmetry* **3**, 1123-1126.

Seibert, C. M., Raushel, F. M. (2005), Structural and catalytic diversity within the amidohydrolase superfamily, *Biochemistry* **44**, 6383-6391.

Seidegard, J., de Pierre, J. W. (1983), Microsomal epoxide hydrolase. Properties, regulation and function, *Biochim. Biophys. Acta* **695**, 251-270.

Seki, M., Furutani, T., Miyake, T., Yamanaka, T., Ohmizu, H. (1996), A novel synthesis of a key intermediate for penems and carbapenems utilizing lipase-catalyzed kinetic resolution, *Tetrahedron: Asymmetry* **7**, 1241-1244.

Serebryakov, E. P., Gamalevich, G. D., Strakhov, A. V., Vasil'ev, A. A. (1995), Enhancement and reversal of enantioselectivity of the enzymic hydrolysis of (*RS*)-3-(4-methoxycarbonyl)phenyl-2-methylprop-1-yl acetate upon its transformation into the η^6- arene(tricarbonyl)chromium complex, *Mendeleev Commun.* 175-176.

Serreqi, A. N., Kazlauskas, R. J. (1995), Kinetic resolution of sulfoxides with pendant acetoxy groups using cholesterol esterase: substrate mapping and an empirical rule for chiral phenols, *Can. J. Chem.* **73**, 1357-1367.

Serreqi, A. N., Kazlauskas, R. J. (1994), Kinetic resolution of phosphines and phosphine oxides with phosphorus stereocenters by hydrolases, *J. Org. Chem.* **59**, 7609-7615.

Seu, Y. B., Kho, Y. H. (1992), Enzymic preparation of optically active 2-acetoxymethylglycidol, a new chiral building block in natural product synthesis, *Tetrahedron Lett.* **33**, 7015-7016.

Sharma, A., Chattopadhyay, S., Mamdapur, V. R. (1995), PPL catalyzed monoesterification of α,ω-dicarboxylic acids, *Biotechnol. Lett.* **17**, 939-942.

Shaw, N. M., Naughton, A., Robins, K., Tinschert, A., Schmid, E., Hischier,M.-L., Venetz, V., Werlen, J., Zimmermann, T., Brieden, W., de Riedmatten, P., Roduit, J.-P., Zimmermann, B., Neumüller, R. (2002), Selection, purification, characterisation, and cloning of a novel heat-stable stereo-specific amidase from *Klebsiella oxytoca*, and its application in the synthesis of enantiomerically pure (*R*)- and (*S*)-3,3,3-trifluoro-2-hydroxy-2-methylpropionic acids and (*S*)-3,3,3-trifluoro-2-hydroxy-2-methylpropionamide, *Org. Proc. Res. Dev.* **6**, 497-504.

Shaw, N. M., Robins, K. T., Kiener, A. (2003), Lonza: 20 Years of Biotransformations, *Adv. Synth. Catal.* **345**, 425-435.

Sheldon, R. A., Lau, R. M., Sorgedrager, M. J., van Rantwijk, F., Seddon, K. R. (2002) Biocatalysis in ionic liquids, *Green Chem.* **4**, 147-151.

Sheldon, R. A. (1993), Chirotechnology-industrial synthesis of optically-active compounds, New York: Marcel Dekker.

Shewale, J. G., Sudhakaran, V. K. (1997), Penicillin V acylase: its potential in the production of 6-aminopenicillanic acid, *Enzyme Microb. Technol.*, **20**, 402-410.

Shieh, W. R., Gou, D. M., Chen, C. S. (1991), Computer-aided substrate design for biocatalysis: an enzymatic access to optically active propranolol, *J. Chem. Soc., Chem. Commun.* 651-653.

Shimizu, S., Ogawa, J., Kataoka, M., Kobayashi, M. (1997), Screening of novel microbial enzymes for the production of biologically and chemically useful enzymes, *Adv. Biochem. Eng./Biotechnol.* **58**, 45-87.

Stecher, H., Faber, K. (1997), Biocatalytic deracemization techniques: dynamic resolutions and stereoinversions, *Synthesis* 1-16.

Steinreiber, A., Faber, K. (2001), Microbial epoxide hydrolases for preparative biotransformations, *Curr. Opin. Biotechnol.* **12**, 552-558.

Stemmer, W. P. C. (1994a), DNA shuffling by random fragmentation and reassembly: *In vitro* recombination for molecular evolution, *Proc. Natl. Acad. Sci. USA* **91**, 10747-10751.

Stemmer, W. P. C. (1994b), Rapid evolution of a protein *in vitro* by DNA shuffling, *Nature* **370**, 389-391.

Stevenson, D., Feng, R., Dumas, F., Groleau, D., Mihoc, A., Storer, A. (1992), Mechanistic and structural studies on *Rhodococcus* ATCC39484 nitrilase, *Biotechnol. Appl. Biochem.* **15**, 283-302.

Stinson, C. (1997), U.K.'s Chirotech goes it alone, *Chem. Eng. News* **27**, 15-16.

Stokes, T. M., Oehlschlager, A. C. (1987), Enzyme reaction in apolar solvents: the resolution of (±)-sulcatol with porcine pancreatic lipase, *Tetrahedron Lett.* **28**, 2091-2094.

Stolz, A., Trott, S., Binder, M., Bauer, R., Hirrlinger, B., Layh, N., Knackmuss, H.-J. (1998), Enantioselective nitrile hydratases and amidases from different bacterial isolates, *J. Mol. Catal. B: Enzym.* **5**, 137-141.

Straathof, A. J. J., Rakels, J. L. L., van Tol, J. B. A., Heijnen, J. J. (1995), Improvement of lipase-catalyzed kinetic resolution by tandem transesterification, *Enzyme Microb. Technol.* **17**, 623-628.

Stranix, B. R., Darling, G. D. (1995), Functional polymers from (vinyl)polystyrene. Enzyme immobilization through a cysteinyl-S-ethyl spacer, *Biotechnol. Tech.* **9**, 75-80.

Strauss, U. T., Faber, K. (1999), Deracemization of (±)-mandelic acid using a lipase-mandelate racemase two-enzyme system, *Tetrahedron: Asymmetry* **10**, 4079-4081.

Strauss, U. T., Felfer, U., Faber, K. (1999), Biocatalytic transformation of racemates into chiral building blocks in 100% chemical yield and 100% enantiomeric excess, *Tetrahedron: Asymmetry* **10**, 107-117.

Stuckey, J. A., Dixon, J. E. (1999), Crystal structures of a phospholipase D family member, *Nature Struct. Biol.*, **6**, 278-284.

Suemune, H., Tanaka, M., Obaishi, H., Sakai, K. (1988), Enzymic procedure for the synthesis of prostaglandin A2, *Chem. Pharm. Bull.* **36**, 15-21.

Sugahara, T., Ogasawara, K. (1996), An expedient route to (-)-*cis*-2-oxabicyclo[3.3.0]oct-6-en-3-one via a *meso*-asymmetrization, *Synlett* 319-320.

Sugahara, T., Satoh, I., Yamada, O., Takano, S. (1991), Efficient enzymic preparation of (+)- and (−)-Corey lactone derivatives, *Chem. Pharm. Bull.* **39**, 2758-2760.

Sugai, T., Yamazaki, T., Yokoyama, M., Ohta, H. (1997), Biocatalysis in organic synthesis: The use of nitrile- and amide-hydrolyzing microorganisms, *Biosci. Biotech. Biochem.* **61**, 1419-1427.

Sugai, T., Ikeda, H., Ohta, H. (1996), Biocatalytic approaches to both enantiomers of (2R,3S)-2-allyloxy-3,4,5,6-tetrahydro-2H-pyran-3-ol, *Tetrahedron* **52**, 8123-8134.

Sugai, T., Katoh, O., Ohta, H. (1995), Chemo-enzymatic synthesis of (R,R)-(−)-pyrenophorin, *Tetrahedron* **51**, 11987-11998.

Sugai, T., Ohta, H. (1991), A simple preparation of (R)-2-hydroxy-4-phenyl-butanoic acid, *Agric. Biol. Chem.* **55**, 293-294.

Sugai, T., Kakeya, H., Ohta, H. (1990a), Enzymatic preparation of enantiomerically enriched tertiary α-benzyloxy acid esters. application to the synthesis of (S)-(+)-frontalin, *J. Org. Chem.* **55**, 4643-4647.

Sugai, T., Kakeya, H., Ohta, H. (1990b), A synthesis of (R)-(−)-mevalonolactone by the combination of enzymatic and chemical methods, *Tetrahedron* **46**, 3463-3468.

Sugai, T., Ohsawa, S., Yamada, H., Ohta, H. (1990c), Preparation of enantiomerically enriched compounds using enzymes, VII. A synthesis of japanese beetle pheromone utilizing lipase-catalyzed enantioselective lactonization, *Synthesis* 1112-1114.

Sugai, T., Mori, K. (1988), Preparative bioorganic chemistry; 8. Efficient enzymatic preparation of (1R,4S)-(+)-4-hydroxy-2-cyclopentenyl acetate, *Synthesis* 19-22.

Suginaka, K., Hayashi, Y., Yamamoto, Y. (1996), Highly selective resolution of secondary alcohols and acetoacetates with lipases and diketenes in organic media, *Tetrahedron: Asymmetry* **7**, 1153-1158.

Sugiura, Y., Kuwahara, J., Nagasawa, T., Yamada, H. (1987), Nitrile hydratase: the first non-heme iron enzyme with a typical low-spin Fe(III)-active center, *J. Am. Chem. Soc.* **109**, 5848-5850.

Sundram, H., Golebiowski, A., Johnson, C. R. (1994), Chemoenzymatic synthesis of (2R,3R)-3-hydroxyproline from cyclopentadiene, *Tetrahedron Lett.* **35**, 6975-6976.

Suresh, C. G., Pundle, A. V., SivaRaman, H., Rao, K. N., Brannigan, J. A., McVey, C. E., Verma, C. S., Dauter, Z., Dodson, E. J., Dodson, G. G. (1999), Penicillin V acylase crystal structure reveals new Ntn-hydrolase family members, *Nature Struct. Biol.*, **6**, 414-416.

Sussman, J. L.Harel, M., Frolow, F., Oefner, C., Goldman, A., Toker, L., Silman, I. (1991), Atomic structure of acetylcholinesterase from *Torpedo californica*: a prototypic acetylcholine-binding protein., *Science* **253**, 872-879.

Sutherland, A., Willis, C. L. (1997), Enantioselective syntheses of α-amino-β-hydroxy acids, [^{15}N]-L-allothreonine and [^{15}N]-L-threonine, *Tetrahedron Lett.* **38**, 1837-1840.

Suzuki, Y., Marumo, S. (1972), Fungal metabolism of (±)-epoxyfarnesol and its absolute stereochemistry, *Tetrahedron Lett.* **19**, 1887-1890.

Svendsen, A., Borch, K., Barfoed, M., Nielsen, T. B., Gormsen, E., Patkar, S. A. (1995), Biochemical properties of cloned lipases from the *Pseudomonas* family, *Biochim. Biophys. Acta* **1259**, 9-17.

Svendsen, A. (1994), Sequence comparisons within the lipase family, in.: *Lipases* (Wooley., P., Peterson, S. B.; Eds.), pp. 1-21. Cambridge: Cambridge University Press.

Svensson, I., Wehtje, E., Adlercreutz, P., Mattiasson, B. (1994), Effects of water activity on reaction rates and equilibrium positions in enzymatic esterifications, *Biotechnol. Bioeng.* **44**, 549-556.

Svirkin, Y. Y., Xu, J., Gross, R. A., Kaplan, D. L., Swift, G. (1996), Enzyme-catalyzed stereoselective ring-opening polymerization of α-methyl-β-propiolactone, *Macromolecules* **29**, 4591-4597.

Swanson, P. E. (1999), Dehalogenases applied to industrial-scale biocatalysis, *Curr. Opin. Biotechnol.* **10**, 365-369.

Swaving, J., de Bont, J. A. N. (1998), Microbial transformation of epoxides, *Enzyme Microb. Technol.* **22**, 19-26.

Sweers, H. M., Wong, C.-H. (1986), Enzyme-catalyzed regioselective deacylation of protected sugars in carbohydrate synthesis, *J. Am. Chem. Soc.* **108**, 6421-6422.

Syed, R., Wu, Z. P., Hogle, J. M., Hilvert, D. (1993), Crystal structure of selenosubtilisin at 2.0-Å resolution *Biochemistry,* **32**, 6157-6164.

Syldatk, C., May, O., Altenbuchner, J., Mattes, R., Siemann, M. (1999), Microbial hydantoinases - industrial enzymes from the origin of life? *Appl. Microbiol. Biotechnol.* **51**, 293-309.

Syldatk, C., Müller, R., Pietzsch, M., Wagner, F. (1992a), Microbial and enzymatic production of L-amino acids from D,L-5-substituted hydantoins, in.: *Biocatalytic production of amino acids and derivatives* (Rozell, J. D., Wagner, F.; Eds.), pp. 131-176. New York: Hanser.

Syldatk, C., Müller, R., Siemann, M., Wagner, F. (1992b), Microbial and enzymatic production of D-amino acids from D,L-5-substituted hydantoins, in.: *Biocatalytic production of amino acids and derivatives* (Rozell, J. D., Wagner, F.; Eds.), pp. 75-127. New York: Hanser.

Sym, E. A. (1936), Action of esterase in the presence of organic solvents, *Biochem. J.* **30**, 609-617.

Taipa, M. A., Liebeton, K., Costa, J. V., Cabral, J. M. S., Jaeger, K. E. (1995), Lipase from *Chromobacterium viscosum*: biochemical characterization indicating homology to the lipase from *Pseudomonas glumae, Biochim. Biophys. Acta* **1256**, 396-402.

Takagi, Y., Teramoto, J., Kihara, H., Itoh, T., Tsukube, H. (1996), Thiacrown ether as regulator of lipase-catalyzed trans-esterification in organic media - practical optical resolution of allyl alcohols, *Tetrahedron Lett.* **37**, 4991-4992.

Takahashi, M., Ogasawara, K. (1996), Lipase-mediated resolution of *trans*-1-azidoindan-2-ol: a new route to optically pure *cis*-1-aminoindan-2-ol, *Synthesis* 636-638.

Takahashi, S., Ueda, M., Atomi, H., Beer, H. D., Bornscheuer, U. T., Schmid, R. D., Tanaka, A. (1998), Extracellular production of active *Rhizopus oryzae* lipase by *Saccharomyces cerevisiae, J. Ferment. Bioeng.* **86**, 164-168.

Takahashi, S., Ohashi, T., Kii, Y., Kumagai, H., Yamada, H. (1979), Microbial transformation of hydantoins to amino acids. III. Microbial transformation of hydantoins to N-carbamyl-D-amino acids, *J. Ferment. Technol.* **57**, 328-332.

Takahashi, T., Nishigai, M., Ikai, A., Takahashi, K. (1991), Electron microscopy and biochemical evidence that proline-β-naphthylamidase is composed of three identical subunits, *FEBS* **280**, 297-300.

Takahashi, T., Ikai, A., Takahashi, K. (1989), Purification and characterization of proline-β-naphthylamidase, a novel enzyme from pig intestinal mucosa, *J. Biol. Chem.* **264**, 11565-11571.

Takami, M., Hidaka, N., Suzuki, Y. (1994), Phospholipase D-catalyzed synthesis of phosphatidyl aromatic compounds, *Biosci. Biotech. Biochem.* **58**, 2140-2144.

Takano, S., Higashi, Y., Kamikubo, T., Moriya, M., Ogasawara, K. (1993a), Enantiodivergent preparation of chiral 2,5-cyclohexadienone synthons, *Synthesis* 948-950.

Takano, S., Moriya, M., Higashi, Y., Ogasawara, K. (1993b), Enantiodivergent route to conduritol C via lipase -mediated asymmetrization, *J. Chem. Soc., Chem. Commun.* 177-178.

Takano, S., Setoh, M., Ogasawara, K. (1993c), Enantiocomplementary resolution of 4-hydroxy-5-(4-methoxyphenoxy)-1-pentyne using the same lipase, *Tetrahedron: Asymmetry* **4**, 157-160.

Takano, S., Setoh, M., Yamada, O., Ogasawara, K. (1993d), Synthesis of optically active 4-benzyloxymethyl- and 4-(4-methoxyphenoxy)methyl-buten-2-olides via lipase mediated resolution, *Synthesis* 1253-1256.

Takano, S., Suzuki, M., Ogasawara, K. (1993e), Enantiocomplementary preparation of optically pure 2-trimethylsilylethynyl-2-cyclopentenol by homochiralization of racemic precursors: a new route to the key intermediate of 1,25-dihydroxycholecalciferol and vincamine, *Tetrahedron: Asymmetry* **4**, 1043-1046.

Takano, S., Setoh, M., Ogasawara, K. (1992a), Resolution of racemic O-(4-methoxyphenyl)glycidol, *Heterocycles* **34**, 173-180.

Takano, S., Yamane, T., Takahashi, M., Ogasawara, K. (1992b), Efficient chiral route to a key building block of 1,25-dihydroxyvitamin D3 via lipase -mediated resolution, *Synlett* 410-412.

Takano, S., Yamane, T., Takahashi, M., Ogasawara, K. (1992c), Enantiocomplementary synthesis of functionalized cycloalkenol building blocks using lipase, *Tetrahedron: Asymmetry* **3**, 837-840.

Takano, S., Inomata, K., Takahashi, M., Ogasawara, K. (1991), Expedient preparation and enantiomerization of optically pure dicyclopentadienone (tricyclo[5.2.1.0^{2,6}]deca-4,8-dien-3-one), *Synlett* 636-638´.

Takaoka, Y., Kajimoto, T., Wong, C.-H. (1993), Inhibition of N-acetylglucosaminyltransfer enzymes: chemical-enzymatic synthesis of new five-membered acetamido azasugars, *J. Org. Chem.* **58**, 4809-4812.

Takayama, S., Moree, W. J., Wong, C. H. (1996), Enzymatic resolution of amines and amino alcohols using pent-4-enoyl derivatives, *Tetrahedron Lett.* **37**, 6287-6290.

Tamai, S., Miyauchi, S., Morizane, C., Miyagi, K., Shimizu, H., Kume, M., Sano, S., Shiro, M., Nagao, Y. (1994), Enzymic hydrolyses of the σ-symmetric dicarboxylic diesters bearing a sulfinyl group as the prochiral center, *Chem. Lett.* 2381-2384.

Tamm, C. (1992), Pig liver esterase catalyzed hydrolysis: substrate specificity and stereoselectivity, *Pure Appl. Chem.* **64**, 1187-1191.

Tan, D. S., Guenter, M. M., Drueckhammer, D. G. (1995), Enzymatic resolution coupled with substrate racemization using a thioester substrate, *J. Am. Chem. Soc.* **117**, 9093-9094.

Tanaka, K., Shogase, Y., Osuga, H., Suzuki, H., Nakamura, K. (1995), Enantioselective synthesis of helical molecules: lipase-catalyzed resolution of bis(hydroxymethyl)[7]thiaheterohelicene, *Tetrahedron Lett.* **36**, 1675-1678.

Tanaka, M., Norimine, Y., Fujita, T., Suemune, H., Sakai, K. (1996), Chemoenzymatic synthesis of antiviral carbocyclic nucleosides - asymmetric hydrolysis of *meso*-3,5-bis(acetoxymethyl)cyclopentenes using *Rhizopus delemar* lipase, *J. Org. Chem.* **61**, 6952-6957.

Tanaka, M., Yoshioka, M., Sakai, K. (1992), Practical enzymic procedure for the synthesis of (–)-aristeromycin, *J. Chem. Soc., Chem. Commun.* 1454-1455.

Tanaka, N., Dumay, V., Liao, Q., Lange, A. J., Wever, R. (2002), Bromoperoxidase activity of vanadate-substituted acid phosphatases from *Shigella flexneri* and *Salmonella enterica* ser. typhimurium, *Eur. J. Biochem.* **269**, 2162-2167.

Taniguchi, T., Kanada, R. M., Ogasawara, K. (1997), Lipase-mediated kinetic resolution of tricyclic acyloins, endo-3-hydroxytricyclo[4.2.1.02,5]non-7-en-4-one and endo-3-hydroxytricyclo[4.2.2.02,5]dec-7-en-4-one, *Tetrahedron: Asymmetry* **8**, 2773-2780.

Taniguchi, T., Ogasawara, K. (1997), Lipase–triethylamine-mediated dynamic transesterification of a tricyclic acyloin having a latent *meso*-structure: a new route to optically pure oxodicyclopentadiene, *Chem. Commun.* 1399-1400.

Tanimoto, H., Oritani, T. (1996), Practical synthesis of Ambrox from farnesyl acetate involving lipase-catalyzed resolution, *Tetrahedron: Asymmetry* **7**, 1695-1704.

Tanyeli, C., Demir, A. S., Dikici, E. (1996), New chiral synthon from the PLE catalyzed enantiomeric separation of 6-acetoxy-3-methylcyclohex-2-en-1-one, *Tetrahedron: Asymmetry* **7**, 2399-2402.

Taschner, M. J., Black, D. J. (1988), The enzymatic Baeyer-Villiger Oxidation: Enantioselective synthesis of lactones from mesomeric cyclohexanones, *J. Am. Chem. Soc.* **110**, 6892-6893.

Tauber, M. M., Cavaco-Paulo, A., Robra, K.-H., Gübitz, G. M. (2000), Nitrile hydratase and amidase from *Rhodococcus rhodochrous* hydrolyze acrylic fibers and granular polyacrylonitriles, *Appl. Environ. Microbiol.* **66**, 1634-1638.

Taylor, S. J. C., McCague, R., Wisdom, R., Lee, C., Dickson, K., Ruecroft, G., O'Brien, F., Littlechild, J., Bevan, J., Roberts, S. M., Evans, C. T. (1993), Development of the biocatalytic resolution of 2-azabicyclo[2.2.1]hept-5- en-3-one as an entry to single-enantiomer carbocyclic nucleosides, *Tetrahedron: Asymmetry* **4**, 1117-1128.

Taylor, S. K., Atkinson, R. F., Almli, E. P., Carr, M. D., van Huis, T. J., Whittaker, M. R. (1995), The synthesis of three important lactones via an enzymatic resolution strategy that improves ee's and yields, *Tetrahedron: Asymmetry* **6**, 157-164.

Terao, Y., Tsuji, K., Murata, M., Achiwa, K., Nishio, T., Watanabe, N., Seto, K. (1989), Facile process for enzymic resolution of racemic alcohols, *Chem. Pharm. Bull.* **37**, 1653-1655.

Terao, Y., Murata, M., Achiwa, K. (1988), Highly-efficient lipase-catalyzed asymmetric synthesis of chiral glycerol derivatives leading to practical synthesis of *(S)*-propranolol, *Tetrahedron Lett.* **29**, 5173-5176.

Terradas, F., Teston-Henry, M., Fitzpatrick, P. A., Klibanov, A. M. (1993), Marked dependence of enzyme prochiral selectivity on the solvent, *J. Am. Chem. Soc.* **115**, 390-396.

Terreni, M., Pagani, G., Ubiali, D., Fernandez-Lafuente, R., Mateo, C., Guisan, J. M. (2001), Modulation of penicillin acylase properties via immobilization techniques: one-pot chemoenzymatic synthesis of Cephamandole from Cephalosporin C, *Bioorg. Med. Chem. Lett.* **11**, 2429-2432.

Terzyan, S., Wang, C.-S., Downs, D., Hunter, B., Zhang, X. (2000) Crystal structure of the catalytic domain of human bile salt activated lipase, *Protein Sci.* **9**, 1783-1790.

Tesch, C., Nikoleit, K., Gnau, V., Götz, F., Bormann, C. (1996), Biochemical and molecular characterization of the extracellular esterase from *Streptomyces diastatochromogenes*, *J. Bacteriol.* **178**, 1858-1865.

Testet-Lamant, V., Archaimbault, B., Durand, J., Rigaud, M. (1992), Enzymatic synthesis of structural analogs of PAF-acether by phospholipase D-catalysed transphosphatidylation, *Biochim. Biophys. Acta* **1123**, 347-350.

Theil, F. (1997), *Enzyme in der Organischen Synthese*, Heidelberg: Spektrum Akademischer Verlag.

Theil, F. (1995), Lipase-supported synthesis of biologically active compounds, *Chem. Rev.* **95**, 2203-2227.

Theil, F., Lemke, K., Ballschuh, S., Kunath, A., Schick, H. (1995), Lipase-catalyzed resolution of 3-(aryloxy)-1,2-propanediol derivatives - towards an improved active site model of *Pseudomonas cepacia* lipase (Amano PS), *Tetrahedron: Asymmetry* **6**, 1323-1344

Theil, F., Weidner, J., Ballschuh, S., Kunath, A., Schick, H. (1994), Kinetic resolution of acyclic 1,2-diols using a sequential lipase -catalyzed transesterification in organic solvents, *J. Org. Chem.* **59**, 388-393.

Theil, F., Kunath, A., Schick, H. (1992), "Double enantioselection" by a lipase-catalyzed transesterification of a meso-diol with a racemic carboxylic ester, *Tetrahedron Lett.* **33**, 3457-3460.

Theil, F., Schick, H. (1991), An improved procedure for the regioselective acetylation of monosaccharide derivatives by pancreatin-catalyzed transesterification in organic solvents, *Synthesis* 533-535.

Theil, F., Schick, H., Winter, G., Reck, G. (1991), Lipase-catalyzed transesterification of *meso*-cyclopentane diols, *Tetrahedron* **47**, 7569-7582.

Theil, F., Schick, H., Nedkov, P., Boehme, M., Haefner, B., Schwarz, S. (1988), Synthesis of enantiomerically pure prostaglandin intermediates by enzymic hydrolysis of (1SR,5RS,6RS,7RS)-7-Acetoxy-6-acetoxymethyl-2-oxabicyclo [3.3.0]octan- 3-one, *J. Prakt. Chem.* **330**, 893-899.

Therisod, M., Klibanov, A. M. (1987), Regioselective acylation of secondary hydroxyl groups in sugars catalyzed by lipases in organic solvents, *J. Am. Chem. Soc.* **109**, 3977-3981.

Therisod, M., Klibanov, A. M. (1986), Facile enzymatic preparation of monoacylated sugars in pyridine, *J. Am. Chem. Soc.* **108**, 5638-5640.

Thiem, J. (1995), Applications of enzymes in synthetic carbohydrate chemistry, *FEMS Microbiol. Rev.* **16**, 193-211.

Tholander, F., Kull, F., Ohlson, E., Shafqat, J., Thunnissen, M. M., Haeggstrom, J. Z. (2005), Leukotriene A4 hydrolase: Insights to the molecular evolution by homology modeling and mutational analysis of enzyme from *Saccharomyces cerevisiae*, *J. Biol. Chem.* **280**, online: 10.1074/jbc.M506821200.

Thuring, J. W. J. F., Klunder, A. J. H., Nefkens, G. H. L., Wegman, M. A., Zwanenburg, B. (1996a), Lipase catalyzed dynamic kinetic resolution of some 5-hydroxy-2(5H)-furanones, *Tetrahedron Lett.* **37**, 4759-4760.

Thuring, J. W. J. F., Nefkens, G. H. L., Wegman, M. A., Klunder, A. J. H., Zwanenburg, B. (1996b), Enzymatic kinetic resolution of 5-hydroxy-4-oxa-*endo*-tricyclo[5.2.1.0^{2,6}]dec-8-en-3-ones - a useful approach to D-ring synthons for strigol analogues with remarkable stereoselectivity, *J. Org. Chem.* **61**, 6931-6935.

van Tilbeurgh, H., Egloff, M.-P., Martinez, C., Rugani, N., Verger, R., Cambillau, C. (1993), Interfacial activation of the lipase-procolipase complex by mixed micelles revealed by X-ray crystallography, *Nature* **362**,

Tokiwa, Y., Suzuki, T., Ando, T. (1979), Synthesis of copolyamide-esters and some aspects involved in their hydrolysis by lipase, *J. Appl. Polym. Sci.* **24**, 1701-1711.

Tokuyama, S. (2001), Discovery and application of a new enzyme N-acylamino acid racemase, *J. Mol. Catal. B: Enzym.* **12**, 3-14.

Tokuyama, S., Hatano, K. (1996), Overexpression of the gene for N-acylamino acid racemase from *Amycolatopsis* sp. TS-1-60 in *Escherichia coli* and continuous production of optically active methionine by a bioreactor, *Appl. Microbiol. Biotechnol.* **44**, 774-777.

Tong, J. H., Petitclerc, C., D'Iorio, A., Benoiton, N. L. (1971), Resolution of ring-substituted phenylalanines by the action of α-chymotrypsin on their ethyl esters, *Can. J. Biochem.* **49**, 877-881.

Toone, E. J., Jones, J. B. (1991), Enzymes in organic synthesis. 49. Resolutions of racemic monocyclic esters with pig liver esterase, *Tetrahedron: Asymmetry* **2**, 207-222.

Toone, E. J., Werth, M. J., Jones, J. B. (1990), Enzymes in organic synthesis. 47. Active-site model for interpreting and predicting the specificity of pig liver esterase, *J. Am. Chem. Soc.* **112**, 4946-4952.

Torre, O., Alfonso, I., Gotor. V. (2004), Lipase catalysed Michael addition of secondary amines to acrylonitrile, *Chem. Commun.* 1724-1725.

Toyooka, N., Nishino, A., Momose, T. (1993), Ring differentiation of the *trans*-decahydronaphthalene system via chemoenzymic dissymmetrization of its s-symmetric glycol: synthesis of a highly functionalized chiral building block for the terpene synthesis, *Tetrahedron Lett.* **34**, 4539-4540.

Trani, M., Ducret, A., Pepin, P., Lortie, R. (1995), Scale-up of the enantioselective reaction for the enzymic resolution of (*R,S*)-ibuprofen, *Biotechnol. Lett.* **17**, 1095-1098.

Treilhou, M., Fauve, A., Pougny, J. R., Prome, J. C., Veschambre, H. (1992), Use of biological catalysts for the preparation of chiral molecules. 8. Preparation of propargylic alcohols. Application in the total synthesis of leukotriene B4, *J. Org. Chem.* **57**, 3203-3208.

Trollsås, M., Orrenius, C., Sahlén, F., Gedde, U. W., Norin, T., Hult, A., Hermann, D., Rudquist, P., Komitov, L., Lagerwall, S. T., Lindström, J. (1996), Preparation of a novel cross-linked polymer for second-order nonlinear optics, *J. Am. Chem. Soc.* **118**, 8542-8548.

Tsai, S. W., Dordick, J. S. (1996), Extraordinary enantiospecificity of lipase catalysis in organic media induced by purification and catalyst engineering, *Biotechnol. Bioeng.* **52**, 296-300.

Tsai, S. W., Wei, H. J. (1994a), Effect of solvent on enantioselective esterification of naproxen by lipase with trimethylsilyl methanol, *Biotechnol. Bioeng.* **43**, 64-68.

Tsai, S. W., Wei, H. J. (1994b), Enantioselective esterification of racemic naproxen by lipases in organic solvents, *Enzyme Microb. Technol.* **16**, 328-333.

Tsai, S. W., Wei, H. J. (1994c), Kinetics of enantioselective esterification of naproxen by lipase in organic solvents, *Biocatalysis* **11**, 33-45.

Tsuboi, S., Yamafuji, N., Utaka, M. (1997), Lipase-catalyzed kinetic resolution of 3-chloro-2-hydroxyalkanoates. Its application for the synthesis of (–)-disparlure, *Tetrahedron: Asymmetry* **8**, 375-379.

Tsugawa, R., Okumura, S., Ito, T., Katsuga, N. (1966), Production of L-glutamic acid from D,L-5-hydantoin propionic acid by microorganisms, *Agric. Biol. Chem.* **30**, 27-34.

Tsuji, K., Terao, Y., Achiwa, K. (1989), Lipase-catalyzed asymmetric synthesis of chiral 1,3-propanediols and its application to the preparation of optically-pure building block for renin inhibitors, *Tetrahedron Lett.* **30**, 6189-6192.

Tsujimura, M., Dohmae, N., Odaka, M., Chijimatsu, M., Takio, K., Yohda, M., Hoshino, M., Nagashima, S., Endo, I. (1997), Structure of the photoreactive iron center of the nitrile hydratase from *Rhodococcus* sp. N-771. Evidence of a novel post-translational modification in the cysteine ligand, *J. Biol. Chem.* **272**, 29454-29459.

Tsujimura, M., Odaka, M., Nagashima, S., Yohda, M., Endo, I. (1996), Photoreactive nitrile hydratase: the photoreaction site is located on the alpha subunit, *J Biochem (Tokyo, Jpn.)* **119**, 407-413.

Tulinsky, A., Blevins, R. A. (1987), Structure of a tetrahedral transition state complex of α-chymotrypsin at 1.8-Å resolution, *J. Biol. Chem.* **262**, 7737-7743.

Tuomi, W. V., Kazlauskas, R. J. (1999), Molecular basis for enantioselectivity of lipase from *Pseudomonas cepacia* toward primary alcohols. Modeling, kinetics and chemical modification of Tyr29 to increase or decrease enantioselectivity, *J. Org. Chem.* **64**, 2638-2647.

Turner, N. J. (2003), Directed evolution of enzymes for applied biocatalysis, *Trends Biotechnol.* **21**, 474-478.

Turner, N. J., Winterman, J. R., McCague, R., Parratt, J. S., Taylor, S. J. C. (1995), Synthesis of homochiral L-(S)-*tert*-leucine via a lipase catalyzed dynamic resolution process, *Tetrahedron Lett.* **36**, 1113-1116.

Uchiyama, T., Takashi, A., Ikemura, T., Watanabe, K. (2005), Substrate-induced gene-expression screening of environmental metagenome libraries for isolation of catabolic genes, *Nat. Biotechnol.* **23**, 88-93.

Udding, J. H., Fraanje, J., Goubitz, K., Hiemstra, H., Speckamp, W. N., Kaptein, B., Schoemaker, H. E., Kamphuis, J. (1993), Resolution of methyl *cis*-3-chloromethyl-2-tetrahydrofurancarboxylate via enzymic hydrolysis, *Tetrahedron: Asymmetry* **4**, 425-432.

Ueji, S., Fujino, R., Okubo, N., Miyazawa, T., Kurita, S., Kitadani, M., Muromatsu, A. (1992), Solvent-induced inversion of enantioselectivity in lipase-catalyzed esterification of 2-phenoxypropionic acids, *Biotechnol. Lett.* **14**, 163-168.

Uejima, A., Fukui, T., Fukusaki, E., Omata, T., Kawamoto, T., Sonomoto, K., Tanaka, A. (1993), Efficient kinetic resolution of organosilicon compounds by stereoselective esterification with hydrolases in organic solvent, *Appl. Microbiol. Biotechnol.* **38**, 482-486.

Uemasu, I., Hinze, W. L. (1994), Enantioselective esterification of 2-methylbutyric acid catalyzed via lipase immobilized in microemulsion-based organogels, *Chirality* **6**, 649-653.

Uemura, T., Furukawa, M., Kodera, Y., Hiroto, M., Matsushima, A., Kuno, H., Matsushita, H., Sakurai, K., Inada, Y. (1995), Polyethylene glycol-modified lipase catalyses asymmetric alcoholysis of δ-decalactone in *n*-decanol, *Biotechnol. Lett.* **17**, 61-66.

Uemura, M., Nishimura, H., Yamada, S., Hayashi, Y. (1994), Kinetic resolution of hydroxymethyl-substituted (arene)Cr(CO)3 and (diene)Fe(CO)3 by lipase, *Tetrahedron: Asymmetry* **5**, 1673-1682.

Uemura, M., Nishimura, H., Yamada, S., Nakamura, K., Hayashi, Y. (1993), Kinetic resolution of hydroxymethyl substituted iron (diene)Fe(CO)3 complexes by lipase, *Tetrahedron Lett.* **34**, 6581-6582.

Uemura, A., Nozaki, K., Yamashita, J., Yasumoto, M. (1989a), Lipase-catalyzed regioselective acylation of sugar moieties of nucleosides, *Tetrahedron Lett.* **30**, 3817-3818.

Uemura, A., Nozaki, K., Yamashita, J., Yasumoto, M. (1989b), Regioselective deprotection of 3,5-*O*-acylated pyrimidine nucleosides by lipase and esterase, *Tetrahedron Lett.* **30**, 3819-3820.

Uenishi, J., Hiraoka, T., Hata, S., Nishiwaki, K., Yonemitsu, O., Nakamura, K., Tsukube, H. (1998), Chiral pyridines: optical resolution of 1-(2-pyridyl)- and 1-[6-(2,2'-bipyridyl)]ethanols by lipase-catalyzed enantioselective acetylation, *J. Org. Chem.* **63**, 2481-2487.

Uenishi, J., Nishiwaki, K., Hata, S., Nakamura, K. (1994), An optical resolution of pyridyl and bipyridylethanols and a facile preparation of optically pure oligopyridines, *Tetrahedron Lett.* **35**, 7973-7976.

Ulbrich-Hofmann, R., Haftendorn, R., Dittrich, N., Hirche, F., Aurich, I. (1998), Phospholipid analogs - chemoenzymatic syntheses and properties as enzyme effectors, *Fett/Lipid* **100**, 114-120.

Um, P. J., Drueckhammer, D. G. (1998), Dynamic enzymatic resolution of thioesters, *J. Am. Chem. Soc.* **120**, 5605-5610.

Uppenberg, J., Öhrner, N., Norin, M., Hult, K., Patkar, S., Waagen, V., Anthonsen, T., Jones, T. A. (1995), Crystallographic and molecular modelling studies of lipase B from *Candida antarctica* reveal a stereospecificity pocket for secondary alcohols, *Biochemistry* **34**, 16838-16851.

Uppenberg, J., Hansen, M. T., Patkar, S., Jones, T. A. (1994), The sequence, crystal structure determination and refinement of two crystal forms of lipase B from *Candida antarctica*, *Structure* **2**, 293-308.

Urban, F. J., Breitenbach, R., Vincent, L. A. (1990), Synthesis of optically active 3(*R*)-[(alkylsulfonyl)oxy] thiolanes from 2(*R*)-hydroxy-4-(methylthio)butanoic acid or D-methionine, *J. Org. Chem.* **55**, 3670-3672.

Ushio, K., Nakagawa, K., Nakagawa, K., Watanabe, K. (1992), An easy access to optically-pure (*R*)-malic acid via enantioselective hydrolysis of diethyl malate by *Rhizopus* lipase, *Biotechnol. Lett.* **14**, 795-800.

Uyama, H., Takeya, K., Kobayashi, S. (1995), Enzymic ring-opening polymerization lactones to polyesters by lipase catalyst: unusually high reactivity of macrolides, *Bull. Chem. Soc. Jpn.* **68**, 56-61.

Uyama, H., Kobayashi, S. (1994), Lipase-catalyzed polymerization of divinyl adipate with glycols to polyesters, *Chem. Lett.* 1687-1690.

Uyama, H., Kobayashi, S. (1993), Enzymatic ring-opening polymerization of lactones catalyzed by lipases, *Chem. Lett.* 1149-1150.

Uyama, H., Takeya, K., Kobayashi, S. (1993), Synthesis of polyesters by enzymic ring-opening copolymerization using lipase catalyst, *Proc. Jpn. Acad.* **69B**, 203-207.

Uyttenbröck, W., Hendriks, D., Vriend, G., de Baere, I., Moens, L., Scharpé, S. (1993), Molecular characterization of an extracellular acid-resistant lipase produced by *Rhizopus javanicus*, *Biol. Chem. Hoppe Seyler* **374**, 245-254.

Valis, T. P., Xenakis, A., Kolisis, F. N. (1992), Comparative studies of lipase from *R. delemar* in various microemulsion systems, *Biocatalysis* **6**, 267-279.

Valivety, R. H., Halling, P. J., Macrae, A. R. (1992a), Reaction rate with suspended lipase catalyst shows similar dependence on water activity in different organic solvents, *Biochim. Biophys. Acta* **1118**, 218-222.

Valivety, R. H., Halling, P. J., Peilow, A. D., Macrae, A. R. (1992b), Lipases from different sources vary widely in dependence of catalytic activity on water activity, *Biochim. Biophys. Acta* **1122**, 143-146.

Van Almsick, A., Buddrus, J., Hönicke-Schmidt, P., Laumen, K., Schneider, M. P. (1989), Enzymatic preparation of optically active cyanohydrin acetates, *J. Chem. Soc., Chem. Commun.* 1391-1393.

Van de Velde, F., Konemann, L., van Rantwijk, F., Sheldon, R. A. (2000), The rational design of semisynthetic peroxidases, *Biotechnol. Bioeng.* **67**, 87-96.

Van de Velde, F., Könemann, L., van Rantwijk, F., Sheldon, R. A. (1998), Enantioselective sulfoxidation mediated by vanadium-incorporated phytase: a hydrolase acting as a peroxidase, *Chem. Commun.*, 1891-1892.

Van den Heuvel, N., Cuiper, A. D., van der Deen, H., Kellogg, R. M., Feringa, B. L. (1997), Optically active 6-acetyloxy-2H-pyran-3(6H)-one obtained by lipase catalyzed transesterification and esterification, *Tetrahedron Lett.* **38**, 1655-1658.

Van der Deen, H., Cuiper, A. D., Hof, R. P., Vanoeveren, A., Feringa, B. L., Kellogg, R. M. (1996), Lipase-catalyzed second-order asymmetric transformations as resolution and synthesis strategies for chiral 5-(acyloxy)-2(5H)-furanone and pyrrolinone synthons, *J. Am. Chem. Soc.* **118**, 3801-3803.

Van der Eycken, J., Vandewalle, M., Heinemann, G., Laumen, K., Schneider, M. P., Kredel, J., Sauer, J. (1989), Enzymic preparation of optically active bicyclo[2.2.1]heptene derivatives, building blocks for terpenoid natural products. An attractive alternative to enantioselective Diels-Alder syntheses, *J. Chem. Soc., Chem. Commun.* 306-308.

Vanhooke, J. L., Benning, M. M., Raushel, F. M., Holden, H. M. (1996), Three-dimensional structure of the zinc-containing phosphotriesterase with the bound substrate analog diethyl 4-methylbenzylphosphonate, *Biochemistry*, **35**, 6020-6025.

Van Rantwijk, F., Lau, RM., Sheldon, RA. (2003) Biocatalytic transformations in ionic liquids, *Trends Biotechnol.* **21**, 131-138.

Van Tol, J. B. A., Jongejan, J. A., Duine, J. A. (1995a), Description of hydrolase-enantioselectivity must be based on the actual kinetic mechanism: analysis of the kinetic resolution of glycidyl (2,3-epoxy-1-propyl) butyrate by pig pancreas lipase, *Biocatal. Biotransform.* **12**, 99-117.

Van Tol, J. B. A., Kraayveld, D. E., Jongejan, J. A., Duine, J. A. (1995b), The catalytic performance of pig pancreas lipase in enantioselective transesterification in organic solvents, *Biocatal. Biotransform.* **12**, 119-136.

Vänttinen, E., Kanerva, L. T. (1997), Optimized double kinetic resolution for the preparation of (*S*)-solketal, *Tetrahedron: Asymmetry* **8**, 923-933.

Vänttinen, E., Kanerva, L. T. (1995), Combination of the lipase-catalysed resolution with the Mitsunobu esterification in one pot, *Tetrahedron: Asymmetry* **6**, 1779-1786.

Vänttinen, E., Kanerva, L. T. (1994), Lipase-catalysed transesterification in the preparation of optically active solketal, *J. Chem. Soc., Perkin Trans I* 3459-3463.

Verger, R. (1997), 'Interfacial activation' of lipases: facts and artefacts, *Trends Biotechnol.* **15**, 32-38.

Vermuë, M. H., Tramper, J., de Jong, J. P. J., Oostrom, W. H. M. (1992), Enzyme transesterification in near-critical carbon dioxide: Effect of pressure, Hildebrand solubility parameter and water content, *Enzyme Microb.Technol.* **14**, 649-655.

Verseck, S., Bommarius, A., Kula , M.-R. (2001), Screening, overexpression and characterization of an *N*-acylamino acid racemase from *Amycolatopsis orientalis* subsp. *lurida*, *Appl. Microbiol. Biotechnol.* **55**, 354-361.

Veum, L., Kanerva, L. T., Halling, P. J., Maschmeyer, T., Hanefeld, U. (2005), Optimisation of the enantioselective synthesis of cyanohydrin esters, *Adv. Synth. Catal.* **347**, 1015-1021.

Veum, L., Kuster, M., Telalovic, S., Hanefeld, U., Maschmeyer, T. (2002), Enantioselective synthesis of protected cyanohydrins, *Eur. J. Org. Chem.* **2002**, 1516-1522.

Vogel, K., Cook, J., Chmielewski, J. (1996), Subtilisin-catalyzed religation of proteolyzed hen egg-white lysozyme: investigation of the role of disulfides, *Chem. Biol.* **3**, 295-299.

Von der Osten, C. H., Sinskey, A. J., III, C. F. B., Pederson, R. L., Wang, Y.-F., Wong, C.-H. (1989), Use of a recombinant bacterial fructose-1,6-diphosphate aldolase in aldol reactions: preparative synthesis of 1-deoxynojirimycin, 1-deoxymannojirimycin, 1,4-dideoxy-1,4-imino-D-arabinitol, and fagomine, *J. Am. Chem. Soc.* **111**, 3924-3927.

Vörde, C., Högberg, H. E., Hedenström, E. (1996), Resolution of 2-methylalkanoic esters - enantioselective aminolysis by (*R*)-1-phenylethylamine of ethyl 2-methyloctanoate catalysed by lipase B from *Candida antarctica*, *Tetrahedron: Asymmetry* **7**, 1507-1513.

Vorderwülbecke, T., Kieslich, K., Erdmann, H. (1992), Comparison of lipases by different assays, *Enzyme Microb. Technol.* **14**, 631-639.

Vriesema, B. K., ten Hoeve, W., Wynberg, H., Kellogg, R. M., Boesten, W. H. J., Meijer, E. M., Schoemaker, H. E. (1986), Resolution of 2-amino-5-(thiomethyl)pentanoic acid (homomethionine) with aminopeptidase from *Pseudomonas putida* or chiral phosphoric acids, *Tetrahedron Lett.* **27**, 2045-2048.

Vulfson, E. N. (1994), Industrial applications of lipases, in.: *Lipases: Their Structure Biochemistry and Application* (Wooley., P., Peterson, S. B.; Eds.), pp. 271-288. Cambridge: Cambridge University Press.

Waagen, V., Hollingstaeter, I., Partali, V., Thorstad, O., Anthonsen, T. (1993), Enzymatic resolution of butanoic esters of 1-phenyl, 1-phenylmethyl, 1-[2-phenylethyl] and 1-[2-phenoxyethyl]ethers of 3-methoxy-1,2-propanediol, *Tetrahedron: Asymmetry* **4**, 2265-2274.

Wagegg, T., Enzelberger, M. M., Bornscheuer, U. T., Schmid, R. D. (1998), The use of methoxy acetoxy esters significantly enhances reaction rates in the lipase-catalyzed preparation of optical pure 1-(4-chloro-phenyl) ethyl amines, *J. Biotechnol.* **61**, 75-78.

Wahler, D., Reymond, J.-L. (2002), The adrenaline test for enzymes, *Angew. Chem Int. Ed.* **41**, 1229-1232.

Wahler, D., Badalassi, F., Crotti, P., Reymond, J.-L. (2001), Enzyme fingerprints by fluorogenic and chromogenic substrate arrays, *Angew. Chem. Int. Ed.* **40**, 4457-4460.

Wahler, D., Reymond, J.-L. (2001), High-throughput screening for biocatalysts, *Curr. Opin. Chem. Biol.* **12**, 535-544.

Walde, P., Han, D., Luisi, P. L. (1993), Spectroscopic and kinetic studies of lipases solubilized in reverse micelles, *Biochemistry* **32**, 4029-4034.

Waldinger, C., Schneider, M., Botta, M., Corelli, F., Summa, V. (1996), Aryl propargylic alcohols of high enantiomeric purity via lipase catalyzed resolutions, *Tetrahedron: Asymmetry* **7**, 1485-1488.

Waldmann, H., Naegele, E. (1995), Synthesis of the palmitoylated and farnesylated C-terminal lipohexapeptide of the human N-ras protein by employing an enzymically removable urethane protecting group, *Angew. Chem. Intl. Ed.* **34**, 2259-2262.

Waldmann, H., Sebastian, D. (1994), Enzymic protecting group techniques, *Chem. Rev.* **94**, 911-937.

Waldmann, H., Braun, P., Kunz, H. (1991), New enzymic protecting group techniques for the construction of peptides and glycopeptides, *Biomed. Biochim. Acta* **50**, S243-S248.

Waldmann, H. (1989), A new access to chiral 2-furylcarbinols by enantioselective hydrolysis with penicillin acylase, *Tetrahedon Lett.* **30**, 3057-3058.

Wallace, J. S., Baldwin, B. W., Morrow, C. J. (1992), Separation of remote diol and triol stereoisomers by enzyme-catalyzed esterification in organic media or hydrolysis in aqueous media, *J. Org. Chem.* **57**, 5231-5239.

Wallace, J. S., Reda, K. B., Williams, M. E., Morrow, C. J. (1990), Resolution of a chiral ester by lipase-catalyzed transesterification with polyethylene glycol in organic media, *J. Org. Chem.* **55**, 3544-3546.

Wallace, J. S., Morrow, C. J. (1989a), Biocatalytic synthesis of polymers. Synthesis of an optically active, epoxy-substituted polyester by lipase-catalyzed polymerization, *J. Polym. Sci., Part A: Polym. Chem.* **27**, 2553-2567.

Wallace, J. S., Morrow, C. J. (1989b), Biocatalytic synthesis of polymers. II. Preparation of [AA-BB]x polyesters by porcine pancreatic lipase catalyzed polymerization, *J. Polym. Sci. A: Polymer Chem.* **27**, 3271-3284.

Walts, A. E., Fox, E. M. (1990), A lipase fraction for resolution of glycidyl esters to high enantiomeric excess, *US Patent* US 4 923 810.

Wandel, U., Mischitz, M., Kroutil, W., Faber, K. (1995), Highly selective asymmetric hydrolysis of 2,2-disubstituted epoxides using lyophilized cells of *Rhodococcus* sp. NCIMB11216, *J. Chem. Soc., Perkin Trans. 1* 735-736.

Wang, M.-X., Li, J.-J., Ji, G.-J., Li, J.-S. (2001), Enantioselective biotransformations of racemic 2-aryl-3-methylbutyronitriles using *Rhodococcus* sp. AJ270, *J. Mol. Catal. B: Enzym.* **14**, 77-83.

Wirz, B., Schmid, R., Foricher, J. (1992), Asymmetric enzymatic hydrolysis of prochiral 2-*O*-allylglycerol ester derivatives, *Tetrahedron: Asymmetry* **3**, 137-142.

Wirz, B., Walther, W. (1992), Enzymic preparation of chiral 3-(hydroxymethyl)piperidine derivatives, *Tetrahedron: Asymmetry* **3**, 1049-1054.

Wolff, A., Straathof, A. J. J., Heijnen, J. J. (1994), Enzymatic resolution of racemates contaminated by racemic product, *Biocatalysis* **11**, 249-261.

Wong, C.-H. (1995), Enzymatic and chemo-enzymatic synthesis of carbohydrates, *Pure Appl. Chem.* **67**, 1609-1616.

Wong, C.-H., Whitesides, G. M. (1994), *Enzymes in Synthetic Organic Chemistry*, Oxford: Pergamon Press.

Wong, T. S., Tee, K. L., Hauer, B., Schwaneberg, U. (2004), Sequence saturation mutagenesis (SeSaM): a novel method for directed evolution, *Nucl. Acids Res.* **32**, e26.

Woolley, P., Petersen, S. B. (1994), Lipases: Their structure, biochemistry, and application, Cambridge: Cambridge University Press.

Wright, C. S., Alden, R. A., Kraut, J. (1969), Structure of subtilisin BPN' at 2.5 A resolution, *Nature* **221**, 235-242.

Wu, D. R., Cramer, S. M., Belfort, G. (1993), Kinetic resolution of racemic glycidyl butyrate using a multiphase membrane enzyme reactor: experiments and model verification, *Biotechnol. Bioeng.* **41**, 979-990.

Wu, F., Li, W.-S., Chen-Goodspeed, M., Sogorb, M. A., Raushel, F. M. (2000), Rationally engineered mutants of phosphotriesterase for preparative scale isolation of chiral organophosphates, *J. Am. Chem. Soc.* **122**, 10206-10207.

Wu, S.-H., Chu, F.-Y., Chang, C.-H., Wang, K.-T. (1991), The synthesis of D-isoglutamine by a chemoenzymatic method, *Tetrahedron Lett.* **32**, 3529.

Wu, S. H., Guo, Z. W., Sih, C. J. (1990), Enhancing the enantioselectivity of *Candida* lipase-catalyzed ester hydrolysis via noncovalent enzyme modification, *J. Am. Chem. Soc.* **112**, 1990-1995.

Wu, S.-H., Zhang, L.-Q., Chen, C. S., Girdaukas, G., Sih, C. J. (1985), Bifunctional chiral synthons via biochemical methods. VII. Optically active 2,2'-dihydroxy-1,1'-binaphthyl, *Tetrahedron Lett.* **26**, 4323-4326.

Wu, Z. P., Hilvert, D. (1990), Selenosubtilisin as a glutathione peroxidase mimic, *J. Am. Chem. Soc.*, **112**, 5647-5648.

Wu, Z. P., Hilvert, D. (1989) Conversion of a protease into an acyl transferease: selenosubtilisin, *J. Am. Chem. Soc.* **111**, 4513-4514.

Wünsche, K., Schwaneberg, U., Bornscheuer, U. T., Meyer, H. H. (1996), Chemoenzymatic route to β-blockers via 3-hydroxy esters, *Tetrahedron: Asymmetry* **7**, 2017-2022.

Wyatt, J. M., Linton, E. A. (1988), The industrial potential of microbial nitrile biochemistry, in.: *Ciba Geigy Symp.: Cyanide Compounds in Biology*, **Vol. 140**, pp. 32-48. New York: John Wiley & Sons.

Xie, Z. F., Suemune, H., Sakai, K. (1993), Synthesis of chiral building blocks using *Pseudomonas fluorescens* lipase catalyzed asymmetric hydrolysis of *meso* diacetates, *Tetrahedron: Asymmetry* **4**, 973-980.

Xie, Z. F. (1991), *Pseudomonas fluorescens* lipase in asymmetric synthesis, *Tetrahedron: Asymmetry* **2**, 733-750.

Xie, Z. F., Suemune, H., Sakai, K. (1990), Stereochemical observation on the enantioselective hydrolysis using *Pseudomonas fluorescens* lipase, *Tetrahedron: Asymmetry* **1**, 395-402.

Xu, J., Gross, R. A., Kaplan, D. L., Swift, G. (1996), Chemoenzymatic synthesis and study of poly(α-methyl-β-propiolactone) stereocopolymers, *Macromolecules* **29**, 4582-4590.

Xu, J. H., Kawamoto, T., Tanaka, A. (1995a), Efficient kinetic resolution of dl-menthol by lipase catalyzed enantioselective esterification with acid anhydride in fed-batch reactor, *Appl. Microbiol. Biotechnol.* **43**, 402-407.

Xu, J. H., Kawamoto, T., Tanaka, A. (1995b), High-performance continuous operation for enantioselective esterification of menthol by use of acid anhydride and free lipase in organic solvent, *Appl. Microbiol. Biotechnol.* **43**, 639-643.

Yamada, H., Kobayashi, M. (1996), Nitrile hydratase and its application to industrial production of acrylamide, *Biosci. Biotechnol. Biochem.* **60**, 1391-400.

Yamada, O., Ogasawara, K. (1995), Lipase-mediated preparation of optically pure four-carbon di- and triols from a *meso*-precursor, *Synthesis* 1291-1294.

Yamaguchi, Y., Komatsu, A., Moroe, T. (1976), Optical resolution of menthols and related compounds. Part III. Preliminary fractionation of microbial menthyl ester hydrolases and esterolysis by commercial lipases, *J. Agric. Chem. Soc. Jpn.* **50**, 619-620.

Yamaki, T., Olikawa, T., Ito, K., Nakamura, T. (1997), Cloning and sequencing of a nitrile hydratase gene from *Pseudonocardia thermophila* JCM3095, *J. Ferment. Bioeng.* **83**, 474-477.

Yamamoto, K., Fujimatsu, I., Komatsu, K.-I. (1992), Purification and characterization of the nitrilase from *Alcaligenes faecalis* ATCC 8750 responsible for enantioselective hydrolysis of mandelonitrile, *J. Ferment. Bioeng.* **73**, 425-430.

Yamamoto, K., Komatsu, K.-I. (1991), Purification and characterization of nitrilase responsible for the enantioselective hydrolysis from *Acinetobacter* sp. AK 226, *Agric. Biol. Chem.* **55**, 1459-1466.

Yamamoto, K., Oishi, K., Fujimatsu, I., Komatsu, K.-I. (1991), Production of R-(–)-mandelic acid from mandelonitrile by *Alcaligenes faecalis* ATCC 8750, *Appl. Environm. Microbiol.* **57**, 3028-3032.

Yamamoto, K., Ueno, Y., Otsubo, K., Kawakami, K., Komatsu, K.-I. (1990), Production of *S*-(+)-Ibuprofen from a nitrile compound by *Acinetobacter* sp. strain AK226, *Appl. Environm. Microbiol.* **56**, 3125-3129.

Yamamoto, K., Nishioka, T., Oda, J., Yamamoto, Y. (1988), Asymmetric ring opening of cyclic acid anhydrides with lipase in organic solvents, *Tetrahedron Lett.* **29**, 1717-1720.

Yamamoto, Y., Iwasa, M., Sawada, S., Oda, J. (1990), Asymmetric synthesis of optically active 3-substituted δ-valerolactones using lipase in organic solvnets, *Agric. Biol. Chem.* **54**, 3269-3274.

Yamamura, K., Kaiser, E. T. (1976), Studies on the oxidase activity of copper(II) carboxypeptidase A, *J. Chem. Soc., Chem. Commun.,* 830-831.

Yamano, T., Tokuyama, S., Aoki, I., Nishiguchi, Y., Nakahama, K., Takanohashi, K. (1993), Synthesis of *d*-biotin chiral intermediates via a biochemical method, *Bull. Chem. Soc. Jpn.* **66**, 1456-1460.

Yamazaki, T., Ohnogi, T., Kitazume, T. (1990), Asymmetric synthesis of both enantiomers of 2-trifluoromethyl-4-aminobutyric acid, *Tetrahedron: Asymmetry* **1**, 215-218.

Yamazaki, Y., Hosono, K. (1990), Facile resolution of planar chiral organometallic alcohols with lipase in organic solvents, *Tetrahedron Lett.* **31**, 3895-3896.

Yamazaki, Y., Morohashi, N., Hosono, K. (1991), Lipase-mediated homotopic and heterotopic double resolution of a planar chiral organometallic alcohol, *Biotechnol. Lett.* **13**, 81-86.

Yang, H., Henke, E., Bornscheuer, U. T. (1999), The use of vinyl esters significantly enhanced enantioselectivities and reaction rates in lipase-catalyzed resolutions of arylaliphatic carboxylic acids, *J. Org. Chem.* **64**, 1709-1712.

Yang, H., Cao, S. G., Han, S. P., Feng, Y., Ding, Z. T., Sun, L. F., Cheng, Y. H. (1995a), Optical resolution of *(R,S)* 2-octanol with lipases in organic solvent, *Ann. N. Y. Acad. Sci.* **750**, 250-254.

Yang, F., Hoenke, C., Prinzbach, H. (1995b), Biocatalytic resolutions in total syntheses of purpurosamine and sannamine/sporamine type building blocks of aminoglycoside antibiotics, *Tetrahedron Lett.* **36**, 5151-5154.

Yao, S.P., Lu, D.-S., Wu, Q., Cai, Y., Xu, S.-H., Lin X.-F. (2004), A single-enzyme, two-step, one-pot synthesis of *N*-substituted imidazole derivatives containing a glucose branch via combined acylation/Michael addition reaction, *Chem. Commun.* 2006-2007.

Yao, Y., Lalonde, J. J. (2003), Unexpected enantioselectivity and activity of penicillin acylase in the resolution of methyl 2,2-dimethyl-1,3-dioxane-4-carboxylate, *J. Mol. Catal. B: Enzym.* **22**, 55-59.

Yasufuku, Y., Ueji, S. (1997), High temperature-induced high enantioselectivity of lipase for esterifications of 2-phenoxypropionic acids in organic solvent, *Bioorg. Chem.* **25**, 88-99.

Yasufuku, Y., Ueji, S. (1996), Improvement (5-fold) of enantioselectivity for lipase-catalyzed esterification of a bulky substrate at 57°C in organic solvent, *Biotechnol. Tech.* **10**, 625-628.

Yasufuku, Y., Ueji, S. (1995), Effect of temperature on lipase-catalyzed esterification in organic solvent, *Biotechnol. Lett.* **17**, 1311-1316.

Yee, N. K., Nummy, L. J., Byrne, D. P., Smith, L. L., Roth, G. P. (1998), Practical synthesis of enantiomerically pure trans-4,5-disubstituted 2-pyrrolidinone via enzymatic resolution. Preparation of the LTB4 inhibitor BIRZ-227, *J. Org. Chem.* **63**, 326-330.

Yennawar, H. P., Yennawar, N. H., Farber, G. K. (1995), A structural explanation for enzyme memory in nonaqueous solvents, *J. Am. Chem. Soc.* **117**, 577-585.

Yeo, S. H., Nihira, T., Yamada, Y. (1998), Purification and characterization of tert-butyl ester-hydrolyzing lipase from *Burkholderia* sp. YY62, *Biosci. Biotechnol. Biochem.* **62**, 2312-2317.

Yokoyama, M., Sugai, T., Ohta, H. (1993), Asymmetric hydrolysis of a disubstituted malonontrile by the aid of a microorganism, *Tetrahedron: Asymmetry* **4**, 1081-1084.

Yonezawa, T., Sakamoto, Y., Nogawa, K., Yamazaki, T., Kitazume, T. (1996), Highly efficient synthetic metod of optically active 1,1,1-trifluoro-2-alkanols by enzymatic hydrolysis of the corresponding 2-chloroacetates, *Chem. Lett.* 855-856.

You, L., Arnold, F. H. (1994), Directed evolution of subtilisin E in *Bacillus subtilis* to enhance the total activity in aqueous dimethylformamide, *Protein Eng.* **9**, 77-83.

Zaidi, N. A., O'Hagan, D., Pitchford, N. A., Howard, J. A. K. (1995), The solid state structure of the 34-membered macrocyclic diolide of 16-hydroxyhexadecanoic acid, formed by porcine pancreatic lipase mediated cyclisation in hexane, *J. Chem. Res., Synop.* 427.

Zaks, A., Klibanov, A. M. (1985), Enzyme-catalyzed processes in organic solvents, *Proc. Natl. Acad. Sci. USA* **82**, 3192-3196.

Zaks, A., Klibanov, A. M. (1984), Enzymatic catalysis in organic media at 100°C, *Science* **224**, 1249-1251.

Zhang, J., Reddy, J., Roberge, C., Senanayake, C., Greasham, R., Chartrain, M. (1995), Chiral bioresolution of racemic indene oxide by fungal epoxide hydrolase, *J. Ferment. Bioeng.* **80**, 244-246.

Zhang, L. H., Chung, J. C., Costello, T. D., Valvis, I., Ma, P., Kauffman, S., Ward, R. (1997), The enantiospecific synthesis of an isoxazoline. A RGD mimic platelet GPIIb/IIIa antagonist, *J. Org. Chem.* **62**, 2466-2470.

Zhang, X. M., Archelas, A., Furstoss, R. (1991), Microbiological transformations. 19. Asymmetric dihydroxylation of the remote double bond of geraniol: A unique stereochemical control allowing easy access to both enantiomers of geraniol-6,7-diol, *J. Org. Chem.* **56**, 3814-3817.

Zhao, L., Han, B., Huang, Z., Miller, M., Huang, H., Malashock, D. S., Zhu, Z., Milan, A., Robertson, D. E., Weiner, D. P., Burk, M. J. (2004), Epoxide hydrolase-catalyzed enantioselective synthesis of chiral 1,2-diols via desymmetrization of meso-epoxides, *J. Am. Chem. Soc.* **126**, 11156-11157.

Zhao, H., Malhotra, S. V. (2002) Enzymatic resolution of amino acid esters using ionic liquid *N*-ethyl pyridinium trifluoroacetate. *Biotechnol. Lett.* **24**, 1257-1260.

Zhao, H., Giver, L., Affholter, J. A., Arnold, F. H. (1998), Molecular evolution by staggered extension process (StEP) in vitro recombination, *Nat. Biotechnol.* **16**, 258-261.

Zhu, L.-M., Tedford, M. C. (1990), Applications of pig liver esterases (PLE) in asymmetric synthesis, *Tetrahedron* **46**, 6587-6611.

Zmijewski, M. J., Jr., Briggs, B. S., Thompson, A. R., Wright, I. G. (1991), Enantioselective acylation of a β-lactam intermediate in the synthesis of loracarbef using penicillin G amidase, *Tetrahedron Lett.* **32**, 1621-1622.

Zocher, F., Enzelberger, M. M., Bornscheuer, U. T., Hauer, B., Schmid, R. D. (1999), A colorimetric assay suitable for screening epoxide hydrolase activity, *Anal. Chim. Acta* **391**, 345-351.

Zock, J., Cantwell, C., Swartling, J., Hodges, R., Pohl, T., Sutton, K., Jr., P. R., McGilvray, D., Queener, S. (1994), The *Bacillus subtilis pnb*A gene encoding *p*-nitrobenzyl esterase: cloning, sequence and high-level expression in *Escherichia coli*, *Gene* **151**, 37-43.

Zou, J., Hallberg, B. M., Bergfors, T., Oesch, F., Arand, M., Mowbray, S. L., Jones, T. A. (2000), Structure of *Aspergillus niger* epoxide hydrolase at 1.8 A resolution: implications for the structure and function of the mammalian microsomal class of epoxide hydrolases, *Structure Fold Des.* **8**, 111-122.

Index

A

7-ACA, 217
Acetic acid assay, 89
Acetylcholine esterase, *see AChE*
AChE, 201
   from Electric eel, 201
   from *Torpedo californica*, 201
   physiological role, 201
   resolution of meso-diesters, 202
Acrylamide
   enzymatic synthesis of, 269
Activity
   AChE, 202
   effect of solvent, 30
   lipase, 87
   protease, 220
   staining, *see zymogram*
   water activity, 33
Acyl donors, 84
Acyl migration
   in meso-1,3-diols, 122
Acylase
   commercial reactions, 240
Adrenaline test, 91
Adsorption, *see immobilization*
Agrobacterium radiobacter
   epoxide hydrolase, 255
Alcaligenes faecalis
   nitrile hydrolyzing activity, 275
Alcaligenes sp. lipase
   resolution of
     secondary alcohols, 118
Alcohols
   in non-sugars, 180
   in sugars
     primary, 165
     secondary, 175
   resolution of
     axially-disymmetric, 135
     phosphorus stereocenter, 141
     primary alcohols, 121
     quaternary centers, 130
     remote stereocenters, 137
     secondary alcohols, 94
     spiro compounds, 135
     sulfur stereocenter, 141
     tertiary alcohols, 130
     with amidase, 224

     with esterase, 198, 202, 210
     with lipase, 94, 121
     with protease, 224
Aldol addition, 64
Aliphatic hydroxyls
   acylation/deacylation of diols, 181
Amano AH, *see PCL*
Amano D *see ROL*
Amano N *see ROL*
Amano P *see PCL*
Amano PS *see PCL*
Amberlite, 44
Amidase
   active site model, 237
   availability, 213
   commercial reactions, 240
   enantioselective reactions, 224
   hydrolysis of nitriles, 263
   occurrence, 213
   resolution of
     amino acids, 237
     carboxylic acids, 230
   structure, 222
   substrate binding nomenclature, 214
   synthesis of amides, 215
Amide
   synthesis of
     with amidase, 215
     with protease, 215
Amines
   protection/deprotection
     with lipase, 184
   resolution of
     with amidase, 224
     with lipase, 141
     with protease, 224
Amino acid acylase
   availability, 213
   general features, 220
   resolution of
     amides, 229
Amino acids
   resolution of, 230
7-Aminocephalosporanic acid *see 7-ACA*
6-Aminopenicillanic acid *see 6-APA*
Anhydrides
   as acyl donors, 86
   resolution of, 154

ANL
  availability, 73
  classification, 75
  in dynamic kinetic resolution, 11
  in polymerization, 189
  in protection/deprotection, 184
  resolution of
    amino acids, 183
    carboxylic acids, 149
    primary alcohols, 128
  secondary hydroxyls in sugars, 175
6-APA, 220
Arthrobacter globiformis esterase, 210
Aspartame, 216
Assays
  AChE, 202
  acylase, 56
  amidase, 56
  epoxide hydrolase, 56
  esterase, 56, 87
  fluorimetric, 56, 88
  for directed evolution, 54
  lipase, 56, 87
  nitrile hydratase, 267
  protease, 220
  spectrophotometric, 88
Asymmetric synthesis *see also lipase, esterase, epoxide hydrolase, nitrilase, amidase, protease*
  principle, 24
Availability
  amidases, 213
  esterase, 77, 193
  lipase, 70
  proteases, 213

B

Bacillus coagulans esterase, 210
Bacillus subtilis esterase *see also Carboxylesterase NP*
  directed evolution of, 59
Bacillus thermocatenulatus
  lipase *see BTL2*
BChE, 201
BTL2
  availability, 73
  classification, 75
β-Blockers, 120
BMIM *see ionic liquids*
Burkholderia cepacia
  lipase *see PCL*
Butyl-3-methylimidazolium
  tetrafluoroborate, 35
Butyrylcholine esterase *see BChE*

C

CAL-A
  availability, 73
  classification, 75
CAL-B
  availability, 73
  classification, 74
  properties, 76
  protection/deprotection of
    carboxyl groups, 186
  resolution of
    amides, 143
    amines, 142
    anhydrides, 154
    carboxylic acids
      α-stereocenter, 145
      β-stereocenter, 150, 151
    ferrocenes, 140
    primary alcohols, 128
    secondary alcohols, 100, 102, 105
    thiols, 144
  substrate binding site, 83
  synthesis of
    alkyl glycosides, 170
    bioesters, 170
    nucleosides, 171
    peroxycarboxylic acids, 191
Candida antarctica lipase *see CAL-A, CAL-B*
Candida cylindrea lipase *see CRL*
Candida rugosa lipase *see CRL*
Carbohydrates
  alkyl glycosides, 167
  primary hydroxyls, 165
  secondary hydroxyls, 175
  synthesis of
    surfactants/detergents, 166, 169
Carboxylesterase NP, 208
  resolution of
    carboxylic acids, 209
    naproxen, 208
Carboxylic acids
  resolution of
    with amidase, 230, 238
    with esterase, 197, 200, 202, 210
    with lipase, 145
    with protease, 230, 238
Catalytic promiscuity, 62
  functional group analogs, 62
CE
  availability, 73
  classification, 74
  resolution of
    phosphorus stereocenters, 141

secondary alcohols, 95, 203
sulfur stereocenters, 141
thiols, 144
cEH *see soluble epoxide hydrolase*
Celite, 43
Cephalosporin, 217
Cephalosporinase
directed evolution, 54
Chemoselective reactions, 163
Chicken liver esterase *see CLE*
Cholesterol esterase *see CE*
Chromobacterium viscosum lipase *see CVL*
Chrysanthemic acid
resolution of, 210
Chymotrypsin
active site structure, 236
availability, 213
general features, 218
resolution of
alcohols, 227
amino acids, 234
carboxylic acids, 234, 238
structure, 222
Citronellol
resolution of, 40
CLE
resolution with, 203
CLEA's, 46
CLEC, 45
CLL
availability, 73
classification, 75
Cobalt NHases, 267
Corynebacterium
epoxide hydrolase, 255
CRL
availability, 71, 73
classification, 74
effect of vinyl acetate, 85
PEG-modified, 46
polymerization, 189
primary hydroxyls in sugars, 165
properties, 76
protection/deprotection of
carboxyl groups, 186
reaction with
phenolic hydroxyls, 180
resolution of
amino acids, 183
carboxylic acids
α-stereocenter, 146
β-stereocenter, 150, 151
ferrocenes, 140
phosphorus stereocenters, 141
primary alcohols, 128

secondary alcohols, 102
proposed binding site, 97
silanes, 141
sulfur stereocenters, 141
secondary hydroxyls in sugars, 175
substrate binding site, 83
supercritical carbon dioxide, 40
synthesis of
alkyl glycosides, 170
cross-linked enzyme aggregates *see CLEA's*
Cross-linked enzyme crystals *see CLEC*
CVL
availability, 71, 73
classification, 75
in reverse micelles, 38
resolution of
silanes, 141
secondary hydroxyls in sugars, 177
Cytosolic epoxide hydrolase *see soluble epoxide hydrolase*

D

D-4-hydroxyphenylglycine, 11
Diisopropyl fluorophosphatases, 286
Diketene, 86
Diltiazem, 160
Directed evolution, 50
assay systems, 54
esterase, 59
examples, 57
focused, 61
hydantoinase, 59
library creation, 51
lipase, 57
organophosphorous hydrolase, 284
principle, 51
protease, 60
screening, 56
selection, 54
DKR *see dynamic kinetic resolution*
DNA-shuffling, 52
Double enantioselection, 153
D-phenylglycine, 11
Dynamic kinetic resolution, 9
by addition/elimination, 14
by deprotonation/protonation, 12
by nucleophilic substitution, 17
by oxidation/reduction, 17
of acyloins, 12
of carboxylic acids, 12
of cyanohydrins, 15
of hydantoins, 10
with palladium catalysts, 16, 18

E

E *see enantiomeric ratio*
Enantioconvergence, 252, 257, 259
Enantiomeric ratio
   definition, 5
Enantioselective reactions
   asymmetric synthesis, 25
      carboxylic acids, 197
      diols, 198, 202, 282
      epoxides, 254
      principle, 24
   commercial, 160
   dynamic kinetic resolution, 9
   in situ recycling
      principle, 6
   of triglycerides, 130
   quantitative analysis, 5
   recycling
      principle, 6
   resolution of
      amines, 224
      carboxylic acids, 198, 200, 209, 210, 230, 241
      epoxides, 253, 256, 258, 260
      nitriles, 274
      primary alcohols, 121, 210
      quaternary stereocenters, 130, 133
      secondary alcohols, 94, 95, 199, 202, 210, 224
         electronic effects, 96
      tertiary alcohols, 130
   sequential kinetic resolution, 7
      principle, 6
Enantioselectivity
   definition, 5
   influence of solvent, 31
   influence of temperature, 31
Enol esters, 85
Entrapment *see immobilization*
Epoxide
   resolution of, 249, 253, 256, 258, 260
Epoxide hydrolase, 249
   active site residues, 250
   Aspergillus niger, 258
   Beauveria sulfurescens, 258
   degradation of aromatics, 249
   Diploida gossipina, 259
   enantiopreference, 251
   mammalian, 253
      microsomal, 250
      resolution with, 253, 254
      soluble, 250
   mechanism, 250
   microbial
      bacterial, 254
      fungal, 257
      yeast, 260
   Nocardia sp., 256
   regioselectivity, 252
   resolution of
      epoxides, 256
   Rhodococcus, 256
   Rhodotorula sp., 260
epPCR, 52
Error-prone polymerase chain reaction *see epPCR*
Esterase, 191
   assay, 87
   catalytic promiscuity, 63
   directed evolution, 54, 59
   distinction from lipase/protease, 93
   mammalian, 191
   microbial, 203
   structure, 78
   transesterification / principle, 27
   zymogram, 93

F

FACS, 56
Ferric NHases, 264
Ferrocenes
   resolution of, 140
Flap *see lid*
Fluorimetric assays, 88
Focused directed evolution, 61

G

GCL
   availability, 73
   classification, 74
   effect of vinyl acetate, 85
   resolution of
      amino acids, 183
Gene-shuffling *see DNA-shuffling*
Geotrichum candidum lipase *see GCL*
Glycidol, 160
Glycosidase, 279
   asymmetrization of *meso*-diols, 282
   resolution of racemic alcohols, 282
Glycosynthases, 281

H

Haloalcohol dehalogenase, 283
Head group exchange, 244
High-throughput screening *see HTS*
HLE

resolution with, 203
HLL
   availability, 73
   classification, 74
Horse liver esterase *see HLE*
HTS, 88
Humicola lanuginosa lipase *see HLL*
Hydantoinase
   application, 236
   commercial reactions, 240
   directed evolution, 59
α/β-Hydrolase fold, 78
Hydrolases *see lipase, esterase, protease, nitrilase, nitrile hydratase, epoxide hydrolase*
   classification, 74
Hydrolysis
   principle, 27

I

Ibuprofen
   resolution of
      in reverse micelles, 38
      in supercritical carbon dioxide, 40
      with CAL-B, 145
Immobilization, 43
   adsorption and entrapment, 43
   CLEA's, 46
   CLEC's, 45
   covalent, 44
   protein-coated micro-crystals, 44
   sol gel-entrapment, 44
In situ racemization *see dynamic kinetic resolution*
In situ recycling, 276
in vitro compartmentalization *see IVC*
Inducers, 263
Inositols, 119
Interfacial activation, 79, 94
Ionic liquids, 35
Isopropenyl acetate, 84
IVC, 54

K

Kinetic control
   in peptide synthesis, 215, 216
Kinetic resolution
   principle, 5
   recycling, 6
   sequential, 6

L

Lactones
   macrolactonization, 158
   resolution of
      with esterase, 199, 202
      with lipase, 155
   rule, 157
Lid, 79
Lipase
   acyl donors, 84
   assay, 87
   asymmetric synthesis, 25
   availability, 69, 73
   catalytic promiscuity, 63
   chemoselective reactions, 163
   classification, 74
   commercial reactions, 160, 161
   commercially available, 70
   directed evolution, 57
   distinction from esterase/protease, 93
   hydrolysis
      principle, 27
   in lipid modification, 188
   in oligomerization, 188
   in polymerization, 188
   in reverse micelles, 37
   mechanism, 79
   primary hydroxyls
      in sugars, 165
   protection/deprotection of
      amines, 184
      carboxyl groups, 185
   regioselective reactions, 163
   regioselectivity
      solvent effects, 28
   resolution of
      alcohols
         axially-disymmetric, 135
         non-carbon stereocenters, 139
         primary, 121
         remote stereocenter, 137
         secondary, 94
         spiro, 135
         tertiary, 130
      amines, 141
      anhydrides, 154
      carboxylic acids, 145
         quaternary stereocenter, 152
         remote stereocenter, 153
         sulfur stereocenter, 152
         α-stereocenter, 145
         β-stereocenter, 150, 151
      lactones, 155
      organometallics, 139

peroxides, 144
thiols, 144
triglycerides, 130
secondary hydroxyls in sugars, 175
soluble in organic solvents, 46
structure, 81
substrate binding site, 82, 83
supercritical carbon dioxide, 39, 42
supercritical fluids, 39
synthesis of
alkyl glycosides, 167, 174
diltiazem, 160
peroxycarboxylic acids, 191
pharmaceutical precursors, 163
surfactants/detergents, 166, 169
transesterification / principle, 27
zymogram, 93
Lipid modification, 188
logP, 30

M

Mammalian epoxide hydrolase, 252
resolution with, 253, 254
Mammalian esterase, 202 *see also PLE, AChE*
Mandelic acid, 276
Mass spectroscopy assay, 92
mEH's *see microsomal epoxide hydrolase*
metagenome approach, 49
Metal carbonyl complexes
resolution of, 140
Mevalonolactone, 256
Michael addition, 16, 64
Microbial epoxide hydrolase
bacterial, 254
fungal, 257
yeast, 260
Microbial esterase, 203
others, 210
Microsomal epoxide hydrolase, 250
Mitsonubo inversion, 23
Molecular evolution *see directed evolution*
Mutator strain, 52

N

Naproxen
synthesis of
with Carboxylesterase NP, 147, 208
with nitrilase, 273
NHase *see nitrile hydratase*
Nitrilase, 263
from metagenome, 49, 276
mechanism, 264

properties, 264
Nitrile hydratase, 263
enantioselectivity, 273
mild conditions, 269
properties, 264, 268
regioselectivity, 272
Nitriles
enzymatic hydrolysis, 263
enantioselectivity, 273
induction of enzymes, 263
mild conditions, 269
regioselectivity, 272
Nocardia sp.
epoxide hydrolase, 256
Non-steroidal anti-inflammatory *drugs see also ibuprofen, naproxen*
synthesis of
with carboxylesterase NP, 208
with nitrile hydrolyzing enzymes, 274
NSAID *see non-steroidal anti-inflammatory drugs*

O

Occurrence
amidases, 213
esterase, 77, 193
lipase, 70
nitrilase, 263
nitrile hydratase, 263
proteases, 213
Oligomerization, 188
Organometallics
resolution of, 139
Organophosphorus compounds, 284
Organophosphorus hydrolase, 284
Oxime esters, 85
Oxyanion, 78

P

Pantolactone, 210
Papain
availability, 213
Paroxonases, 284
PcamL
availability, 73
classification, 74
PCL
availability, 73
classification, 75
dynamic kinetic resolution, 11
polymerization, 189
properties, 77
reaction with

phenolic hydroxyls, 180
resolution of
  amines, 142
  amino acids, 183
  anhydrides, 154
  carboxylic acids
    α-stereocenter, 148
    β-stereocenter, 150, 151
  ferrocenes, 140
  peroxides, 144
  primary alcohols, 122, 123
    rule, 121
  secondary alcohols, 107, 108
    rule, 95
  thiols, 144
secondary hydroxyls in sugars, 176
structure, 81
substrate binding site, 82, 83
supercritical carbon dioxide, 39
PCL-AH *see also PCL*
  availability, 73
PEG
  for lipase modification, 46
Penicillin amidase
  availability, 213
  general features, 220
  in cephalosporin synthesis, 217
  resolution of
    alcohols, 228
    amines, 228
    carboxylic acids, 240
  structure, 222
Penicillin G acylase *see penicillin amidase*
Penicillium camembertii lipase *see PcamL*
Penicillium roquefortii lipase *see ProqL*
Peroxides
  resolution with lipase, 144
Peroxycarboxylic acids, 191
PFL *see also PCL*
  availability, 73
PfragiL
  availability, 73
PGA *see penicillin amidase*
phage display, 56
Phenolic hydroxyls, 180
Phospholipase, 243 *see also PLA$_1$, PLA$_2$, PLC, PLD*
  regioselectivity, 243
Phospholipase A$_1$ *see PLA$_1$*
Phospholipase A$_2$ *see PLA$_2$*
Phospholipase C *see PLC*
Phospholipase D *see PLD*
Phospholipids, 243
Phosphotriesterases, 284
pH-stat assay, 88

Pig liver esterase, 191 *see PLE*
Pipecolic acid, 270
PLA$_1$, 243
PLA$_2$, 243
PLC, 244
PLD, 244
  in head group exchange, 246
PLE, 191
  active-site model, 194
  biochemical properties, 192
  preparation, 192
  resolution of
    alcohols/lactones, 199
    carboxylic acids, 200
      α-stereocenter, 197
      β-stereocenter, 198
    miscellaneous substrates, 200
    primary or secondary *meso*-diols, 198
  substrate model, 196
  substrate spectra, 194
Polyethylene glycol *see PEG*
Polymerization, 188
Porcine pancreatic lipase *see PPL*
PPL
  availability, 73
  classification, 74
  in dynamic kinetic resolution, 11
  in polymerization, 188
  primary hydroxyls
    in diols, 181
    in sugars, 165
  properties, 75
  protection/deprotection of
    carboxyl groups, 187
  reaction with
    phenolic hydroxyls, 180
  resolution of
    carboxylic acids
      α-stereocenter, 149
      β-stereocenter, 150, 151
    lactones, 156
    peroxides, 144
    primary alcohols, 126, 127
    secondary alcohols, 105
      rule, 95
    thiols, 144
  secondary hydroxyls in sugars, 175
  sugar deprotection, 164
Prolidase, 286
Propranolol, 120
ProqL
  availability, 73
Protease
  activity assay, 220
  availability, 213

commercial reactions, 240
distinction from lipase/esterase, 93
enantioselective reactions, 224
occurrence, 213
resolution of
   alcohols, 224
   amides, 224
   amines, 224
   carboxylic acids, 230
structure, 222
substrate binding nomenclature, 214
synthesis of amides, 215
Proteinase K
   availability, 213
   structure, 222
Pseudomonas aeruginosa
   esterase, 210
   lipase
      resolution of
         amines, 142
Pseudomonas cepacia lipase *see PCL*
Pseudomonas fluorescens
   esterase
      resolution with, 210
   lipase *see PCL*
Pseudomonas fragi lipase *see PfragiL*
Pseudomonas glumae lipase *see CVL*
Pseudomonas lipase *see also PCL, P. aeruginosa lipase*
   dynamic kinetic resolution, 11
   resolution of
      primary alcohols, 122
         rule, 121
      secondary alcohols, 107, 108
Pseudomonas marginata esterase, 210
Pseudomonas putida esterase, 210
Pseudomonas sp. KWI-56 lipase *see also PCL*
   directed evolution, 60

Q

Quaternary stereocenters
   resolution with lipase, 130, 133
Quick E, 89

R

Rabbit liver esterase *see RLE*
Racemase
   lactate, 14
   mandelate, 14, 23
Racemases, 13
Random mutagenesis, 50
Recycling
   in kinetic resolution, 6
Regioselective reactions, 163
Regioselectivity
   in lipase-catalyzed reactions, 28
Resolution of
   alcohols
      with amidase, 224
      with esterase, 198, 202, 210
      with lipase
         survey, 94, 121
      with protease, 224
   amino acids, 229
   anhydrides, 154
   carboxylic acids
      with amidase, 230
      with esterase, 197, 202, 210
      with lipase, 145
      with protease, 230
   epoxides, 253
   lactones
      with esterase, 202
      with lipase, 156
   nitriles, 273
Resorufin, 90
Retaining glycosidase, 279
Reverse micelles, 37
Rhizopus delemar lipase *see ROL*
Rhizopus javanicus lipase *see RJL, ROL*
Rhizopus niveus lipase *see ROL*
Rhodococcus rhodochrous
   synthesis of
      acrylamide, 269
      nicotinamide, 269
      others, 270
   SP409
      epoxide hydrolase activity, 254
      nitrile hydrolysis, 271
   species
      NCIMB11216
         epoxide hydrolase, 255
      nitrile hydrolyzing enzymes, 263
Rhodococcus butanica see R. rhodochrous
RJL
   availability, 73
RLE
   resolution with, 203
RML
   availability, 73
   classification, 74
   in dynamic kinetic resolution, 11
   polymerization, 189
   properties, 76
   resolution of
      ferrocenes, 140
      primary alcohols, 128

secondary alcohols, 116
substrate binding site, 83
supercritical carbon dioxide, 39
synthesis of
   alkyl glycosides, 170
ROL
   availability, 73
   classification, 74
   protection/deprotection of
      carboxyl groups, 185
   resolution of
      primary alcohols, 128

S

SAM-II
   lipase *see PCL*
$SCCO_2$ *see supercritical carbon dioxide*
Screening, 56
sEH *see soluble epoxide hydrolase*
Selection, 54
Selenosubtilisin, 66
sequence-based discovery, 49
Sequential kinetic resolution, 6
Shvo's catalyst, 18
Sol gel entrapment *see immobilization*
Soluble epoxide hydrolase, 250
Solvent
   effect on enantioselectivity, 31
   effect on regioselectivity, 28
   selection criteria, 30
Spectrophotometric assays, 88, 220
Structures
   AChE, 201
   amidases, 222
   esterase, 78
   proteases, 222
Subtilisin
   availability, 213
   chemical modification, 60
   directed evolution, 60
   general features, 218
   in peptide synthesis, 217
   occurrence, 213
   resolution of
      alcohols, 225, 226
      amines, 227
      carboxylic acids, 232, 233, 238
   site-directed mutagenesis, 60
   structure, 222, 223
Sugars
   alkyl glycosides, 167
   primary hydroxyls, 165
   secondary hydroxyls, 175
   synthesis of

surfactants/detergents, 166
Supercritical carbon dioxide, 39, 42
Supercritical fluids, 39
Synthesis activity assay, 92

T

Tertiary alcohols
   resolution with lipase, 130
TG *see triglyceride*
Thermitase
   availability, 213
   in deprotection of carboxyl groups, 185
   structure, 222
Thermodynamic control
   in peptide synthesis, 215, 216
Thermolysin
   availability, 213
   general features, 219
   in aspartame synthesis, 216
   structure, 222
Thioglycoligase, 281
Thioglycosynthase, 282
Thiols
   resolution of
      with lipase, 144
Transesterifications
   principles, 27
Transglycosylation, 280
Transphosphatidylation, 244
Triglycerides, 188
   stereochemical numbering, 130

U

Umbelliferone, 56, 90

V

Vinyl acetate, 84, 85

W

Water activity, 33
Water content, 33
Wheat germ lipase
   deprotection of sugars, 171

X

XAD-8 *see Amberlite*

Z

Zymogram, 9